Parasitology: An Integrated Approach

Edited by Henry Evans

SYRAWOOD
PUBLISHING HOUSE

New York

Published by Syrawood Publishing House,
750 Third Avenue, 9th Floor,
New York, NY 10017, USA
www.syrawoodpublishinghouse.com

Parasitology: An Integrated Approach
Edited by Henry Evans

International Standard Book Number: 978-1-68286-405-0 (Hardback)

The publisher's policy is to use permanent paper from mills that operate a sustainable forestry policy. Furthermore, the publisher ensures that the text paper and cover boards used have met acceptable environmental accreditation standards.

Trademark Notice: Registered trademark of products or corporate names are used only for explanation and identification without intent to infringe.

Cataloging-in-publication Data

Parasitology : an integrated approach / edited by Henry Evans.
   p. cm.
Includes bibliographical references and index.
ISBN 978-1-68286-405-0
1. Parasitology. 2. Medical parasitology. 3. Molecular parasitology. I. Evans, Henry.
QL757 .P37 2017
577.857--dc23

Printed in the United States of America.

TABLE OF CONTENTS

Permissions

List of Contributors

Index

PREFACE

Study of parasites, their hosts and the relationship between the host and parasites is called parasitology. As a field of study it draws techniques from interdisciplinary disciplines like cell biology, bioinformatics, molecular biology, immunology, etc. This book is a compilation of chapters that discuss the most vital concepts and emerging trends in the field of parasitology. The text covers not only the vital components of parasitology but it also presents an integrated approach to study the varied aspects of this discipline. It aims to present to its readers multiple fields that are integrative in nature for example medical parasitology, veterinary parasitology, parasite ecology, etc. As this field is emerging at a rapid pace, the contents of this book will help the readers understand the modern concepts and applications of the subject.

Every book is a source of knowledge and this one is no exception. The idea that led to the conceptualization of this book was the fact that the world is advancing rapidly; which makes it crucial to document the progress in every field. I am aware that a lot of data is already available, yet, there is a lot more to learn. Hence, I accepted the responsibility of editing this book and contributing my knowledge to the community.

While editing this book, I had multiple visions for it. Then I finally narrowed down to make every chapter a sole standing text explaining a particular topic, so that they can be used independently. However, the umbrella subject sinews them into a common theme. This makes the book a unique platform of knowledge.

I would like to give the major credit of this book to the experts from every corner of the world, who took the time to share their expertise with us. Also, I owe the completion of this book to the never-ending support of my family, who supported me throughout the project.

Editor

An Experimental *Toxoplasma gondii* Dose Response Challenge Model to Study Therapeutic or Vaccine Efficacy in Cats

Jan B. W. J. Cornelissen[1]*, Joke W. B. van der Giessen[2,3], Katsuhisa Takumi[3], Peter F. M. Teunis[4], Henk J. Wisselink[1]

1 Central Veterinary Institute of Wageningen UR, Department of Infection Biology, Lelystad, The Netherlands, 2 Central Veterinary Institute of Wageningen UR, Department of Bacteriology and TSEs, Lelystad, The Netherlands, 3 National Institute of Public Health and the Environment (RIVM), Centre for Zoonoses and Environmental Microbiology, Bilthoven, The Netherlands, 4 National Institute for Public Health and the Enviroment (RIVM), Centre for Epidemiology, Bilthoven, The Netherlands

Abstract

High numbers of *Toxoplasma gondii* oocysts in the environment are a risk factor to humans. The environmental contamination might be reduced by vaccinating the definitive host, cats. An experimental challenge model is necessary to quantitatively assess the efficacy of a vaccine or drug treatment. Previous studies have indicated that bradyzoites are highly infectious for cats. To infect cats, tissue cysts were isolated from the brains of mice infected with oocysts of *T. gondii* M4 strain, and bradyzoites were released by pepsin digestion. Free bradyzoites were counted and graded doses (1000, 100, 50, 10), and 250 intact tissue cysts were inoculated orally into three cats each. Oocysts shed by these five groups of cats were collected from faeces by flotation techniques, counted microscopically and estimated by real time PCR. Additionally, the number of *T. gondii* in heart, tongue and brains were estimated, and serology for anti *T. gondii* antibodies was performed. A Beta-Poisson dose-response model was used to estimate the infectivity of single bradyzoites and linear regression was used to determine the relation between inoculated dose and numbers of oocyst shed. We found that real time PCR was more sensitive than microscopic detection of oocysts, and oocysts were detected by PCR in faeces of cats fed 10 bradyzoites but by microscopic examination. Real time PCR may only detect fragments of *T. gondii* DNA without the presence of oocysts in low doses. Prevalence of tissue cysts of *T. gondii* in tongue, heart and brains, and anti *T. gondii* antibody concentrations were all found to depend on the inoculated bradyzoite dose. The combination of the experimental challenge model and the dose response analysis provides a suitable reference for quantifying the potential reduction in human health risk due to a treatment of domestic cats by vaccination or by therapeutic drug application.

Editor: Laura J. Knoll, University of Wisconsin Medical School, United States of America

Funding: These authors have no support or funding to report.

Competing Interests: The authors have declared that no competing interests exist.

* Email: jan.cornelissen@wur.nl

Introduction

Toxoplasmosis, caused by the protozoan parasite *Toxoplasma gondii* has a worldwide distribution, one-third of the global human population has been exposed to this parasite [1]. The integrated public health impact defined as disease burden expressed in Disability Adjusted Life Years (DALYs) is globally considered to be very high [2]. In the USA, *T. gondii* ranked third out of 14 foodborne pathogens [3] and in the Netherlands, the total burden of toxoplasmosis was estimated 3620 DALYs, ranking *T. gondii* as the first among 14 enteral pathogens examined [4]. Because of this high public health impact of toxoplasmosis intervention measures need to be implemented.

Cats and other Felidae are the primary source of a *T. gondii* infection [5,6]. Cats become infected by feeding on infected meat (wild rodents and birds) or, less effectively, by ingestion of sporulated oocysts [7]. This results in an, for cats only

enteroepithelial sexual cycle that leads to the shedding of millions of oocysts into the environment within a period of two to three weeks [8,9,10,11]. The oocysts may contaminate the environment and can resist extreme environmental conditions, remaining infectious for periods up to 18 months [12] or longer [13,14].

Toxoplasma may be transmitted to humans through the consumption of raw or undercooked meat from livestock (e.g. pigs, cows and sheep) containing tissue cysts [15] or by ingesting of food or water contaminated with oocysts from infected cat faeces [16,17]. *T gondii* can also be transmitted vertically by a primary infection with *T. gondii* during pregnancy and entering of the parasite into the foetal circulation by infection of the placenta [18].

Since *T. gondii* can be considered a major foodborne pathogen, the development of strategies to prevent humans to become infected is of increasing importance. The ultimate control strategy would be to prevent infected cats from shedding oocysts in the environment. Although vaccines nor other drug treatments in cats

Table 1. The dynamics of oocyst shedding by cats after challenge infection with mouse bradyzoites or tissue cysts of *T. gondii* strain M4.

Experimental group	Dose	Stage	No of cats	No. of cats positive		Days after infection						Total of oocysts shed	
						Mean pre-patent period		Mean patent period		Mean peak oocyst shedding			
				ME**	Real Time PCR	ME	Real Time PCR	ME	Real Time PCR	ME	Real Time PCR	ME	Real-Time PCR**
1	10	Bradyzoite	3	0	3	NA*	5.3	NA*	2.7	NA*	6.7	0	1.3E-07
2	50	Bradyzoite	3	1	3	7	6	9	4.3	9	7	1.9E+07	4.3E-05
3	100	Bradyzoite	3	1	3	7	5.7	9	4	8	7	6.9E+07	9.0E-05
4	1000	Bradyzoite	3	3	3	5.3	5	9.3	10	7.3	7.3	2.5E+08	4.0E-04
5	250	Tissue cysts	3	3	3	5	4	10.3	10.7	6	7	9.1E+07	1.0E-04

*Not Applicable.
**Microscopical examination.

are yet commercially available, such control strategies may become available in the future [19,20,21].

To evaluate the efficacy of vaccines or drugs, a standardised animal model is needed. Cats can be experimentally infected with tissue cysts [22] whereby only a few tissue cysts are necessary to infect cats [23]. Isolated bradyzoites from tissue cysts are also very infectious for cats [25,26,27], after ingesting a few bradyzoites cats can shed millions of oocysts [25].

Several studies have been published studying *T. gondii* vaccine development in cats [28,29,30,31,32,33,34]. In these studies, vaccines were evaluated using 200 to1000 brain tissue cysts produced in mice to challenge orally cats. However, a tissue cyst may contain 2 to 1,000 bradyzoites [24] indicating that the dose is not well defined when tissue cysts are used to infect cats.

It was our aim to develop a standardised challenge model in cats. Cats were experimentally infected with various doses of bradyzoites harvested from tissue cysts of experimentally infected mice. The results were used to estimate the infectivity, by means of a dose response model, appropriate for challenge studies in cats. We found that shedding of oocysts by cats after experimental infection is dose- and time-dependent.

Materials and Methods

2.1. Toxoplasma strain

Oocysts from *T. gondii* genotype II strain M4 were obtained from Prof. Dr. E.A. Innes of the Moredun Research Institute (Edinburgh, Scotland).

2.2 *T. gondii* infection in mice: preparing of inoculum for infection of cats

To prepare tissue cysts and bradyzoites for experimentally infecting cats, Swiss Webster mice at the age of six weeks were orally infected with 100 oocysts of *T. gondii* M4 strain in a volume of 0.25 ml PBS. At seven weeks p.i. three mice were sacrificed, brains were collected and brain tissue cysts were harvested by a discontinuous 30–90% Percoll gradient [35] according to the modified protocol of Fritz et al. [27]. Briefly, three-quarters of each brain were passed through a 100 μm cell strainer into a 50 ml conical tube. The plunger of a six ml syringe was used to press the brain tissue through the strainer, thereby retaining the fatty tissue in the strainer. Brain tissue was washed with PBS and resuspended in a volume of 4 ml. A density gradient was prepared for each sample in a 50 ml conical tube by carefully layering (from bottom to top) 9 ml 90% Percoll, followed by 9 ml 30% Percoll and finally followed by 10 ml brain suspension. Percoll dilutions were made using one×PBS. Gradient preparations were centrifuged at 1200× g for 15 min at 4°C. Tissue cysts were harvested from the 30% and 30%/90% interface, suspended in 45 ml PBS and centrifuged at 3000× g for 15 min at 4°C. The supernatant was removed and the pellet containing the tissue cysts was transferred to a 1.5 ml micro centrifuge tube. The volume was brought up to one ml with Hank's Balanced Salt Solution (HBSS; Life Technologies Europe BV Bleiswijk, the Netherlands). A 25 μl aliquot was used for tissue cyst enumeration under light microscopy at a 40× magnification. To release the bradyzoites from the tissue cysts an equal volume of pepsin digestive fluid (0.26 g pepsin; 5 g NaCl; 7 g HCl; distilled water to 500 ml) was added to the tissue cysts (final concentration 0.026%) [36]. Subsequently, the suspension was incubated at 37°C for 10 min, neutralized with Na_2CO_3 and suspended in Dulbecco's modified Eagle's medium (DMEM) plus 3% foetal calf serum (FCS). Bradyzoites were counted in a haemocytometer and adjusted to a concentration of 2.5×10^3/ml.

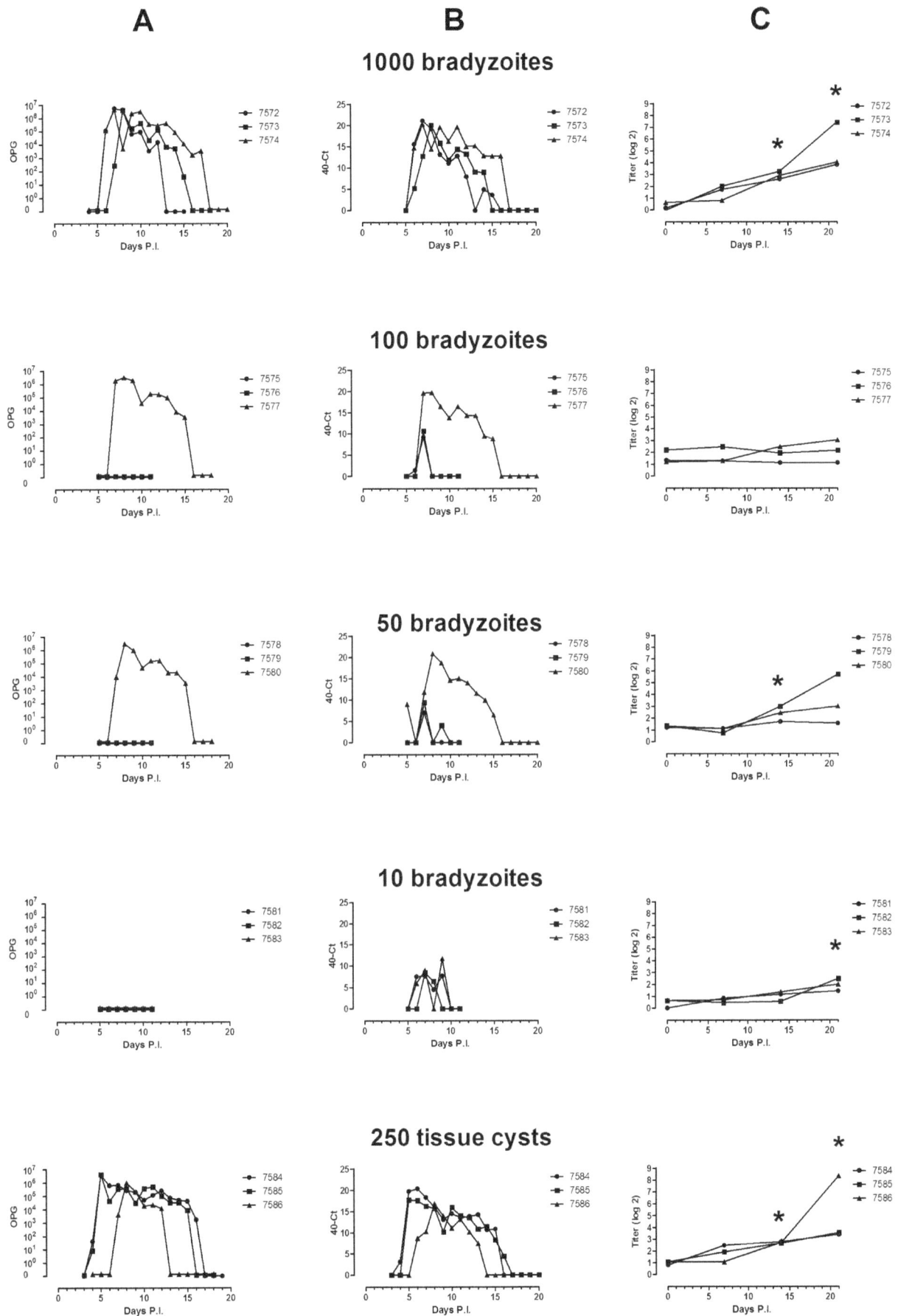

A   **B**   **C**

1000 bradyzoites

100 bradyzoites

50 bradyzoites

10 bradyzoites

250 tissue cysts

Figure 1. Oocysts shedding (A), real time PCR values of the oocysts shedding (B) and antibody response (log2 titres) (C) from the kittens infected with 10, 50, 100 and 1000 bradyzoites or 250 tissue cysts of T. *gondii* Strain M4. The groups contained three cats each. *P<0.05 significant to the anti T. *gondii* IgG response at week 0.

To prepare bradyzoites and tissue cysts for infection of cats, ten mice were orally inoculated with 100 oocysts of T. *gondii* strain M4. At seven weeks p.i. three mice were sacrificed. Per mouse brain 80–1200 tissue cysts were collected and per tissue cyst 166–275 bradyzoites were collected. To infect cats with bradyzoites, a stock solution of $2.5 * 10^3$ bradyzoites per ml was prepared and further diluted preparing inoculation doses of 1000, 100, 50, and 10 bradyzoites in a volume of 2.5 ml. To infect cats with tissue cysts a solution of 250 tissue cysts in a volume of 2.5 ml was prepared.

2.3 T. *gondii* infection in cats

Fifteen cats (10 male and 5 female) at the age of 11 weeks were obtained from Isoquimen, Sant Feliu de Codines, Spain. In a health monitoring report, the supplier of the cats declared that the cat colony is free for T. *gondii* as determined by blood serum testing for anti T. *gondii* antibodies and examination of faeces for the presence of oocysts. For verification of absence of T. *gondii* infection, blood serum of the mothers (n = 8) of the kittens was obtained in advance of delivery of the kittens and tested for anti T. *gondii* antibodies by an indirect ELISA with tachyzoite antigen as described below. After transport to the animal facilities of CVI (Lelystad, The Netherlands), the kittens received commercially available dry food and tap water ad libitum. Kittens were housed individually in an accommodation according to EU regulation 2007/526/EC. General health observations were conducted daily during cleaning and feeding activities and during social interaction with biotechnicians. Kittens were acclimatized to the animal experimental facility for seven days. During the acclimatization period, oocysts counts were performed in faeces as described below to demonstrate the absence of a T. *gondii* infection. At the age of 12 weeks, five groups of three kittens were orally challenged with 1000, 100, 50 and 10 bradyzoites of the T. *gondii* M4 strain in 2.5 ml DMEM-3% FCS using a curved blunt cannula. One group of three SPF kittens was orally challenged with 250 brain tissue cysts from mice infected with T. *gondii* M4 strain in 2.5 ml PBS. Between 3 and 21 days after inoculation, all faeces were collected daily from a litter box filled with grit. From each kitten, all faeces was sieved out, collected and sent to the laboratory. The litter box was cleaned thereafter.

From each kitten, faeces was weighed and oocysts were purified from 2 gram out of 25 gram mixed faeces by 100 µM sieving, and saturated NaCl (d = 1.18 g/ml) flotation according to Wainwright et al. [37]. Oocysts were floated by centrifugation at 2500 g for 10 minutes (non-braked). A sample of 200 µl with floated oocysts was taken by carefully pipetting from the uppermost part of the contents of the centrifuge tube. In this sample the number of floated oocysts per gram (OPG) was counted using a haemocytometer and the quantity of T. *gondii* DNA was determined by real time PCR for 529 T. *gondii* gen as described below. Determination of oocyst shedding was finished when no oocysts had been detected for three consecutive days. Serum samples were collected from a jugular vein of the kittens at the day of challenge and at one, two, and three weeks post-challenge (p.c.). At three weeks post challenge, when no more shedding occurred the experiment was terminated. Kittens were brought under full narcosis with 40 mg/kg Ketamine (Alfasan, Woerden, The Netherlands) plus 2 mg/kg Xylazine (Eurovet, Putten, The Netherlands) and 0.05 mg/kg Atropine (Eurovet, Putten, The

Netherlands). Subsequently, kittens were bled and from each kitten, samples were obtained from brains, tongue and heart for detection of T. *gondii*. A quarter of the brains of each kitten was passed through a 100 µm cell strainer into a 50 ml conical tube using the plunger of a six ml syringe to press the tissue through the strainer (retaining the fatty tissue in the strainer) and washed with PBS to a total volume of four ml. Three-quarters of the tongue and heart of each kitten were homogenized in four ml PBS with a disposable Omni Tissue homogenizer during one minute. A 25 µl aliquot was taken for cyst determination under light microscopy at a magnification of 40× and an aliquot of 200 µl was taken for DNA extraction.

2.4 DNA extraction and real time PCR conditions

For detection of T. *gondii* by real time PCR, 200 µl floated oocysts preparations from faeces of cats, heart, brain and/or tongue tissue samples of kittens were further homogenized by vortexing the samples with 50–100 glass beads (0.4 mm) for one to two minutes. Microscopically it was confirmed that after this treatment most oocysts and tissue cysts had been broken. DNA was extracted with the Qiagen DNA easy Blood & Tissue Kit (Qiagen GMBH, Hilden, Germany) according to the manufacturer's manual. One µl DNA was tested by the PCR with on-line detection using Syber Green PCR Master Mix (Applied Biosystems, Foster City, CA, USA) in an ABI 7500 Real-Time PCR system (PE Applied Biosystems, Foster City, CA, USA). A standard reaction mixture contained 12.5 µl of two× QuantiTect SYBR Green PCR Master Mix, one µl (10 µM) of the primers, one µl of DNA template and 10.5 µl PCR grade water. The Tox-oligonucleotides are complementary to the 529-bp repeat element (GenBank AF146527) [38] and the forward and reverse Toxoplasma primers were described previously as Tox-9, and Tox-11 [39]. The cycling profile involved an initial PCR activation step at 95°C for 10 min, followed by 40 cycles of denaturation at 94°C for 15 s, primer annealing at 59°C for 30 s and extension at 72°C for 30 s. The fluorescence was measured at the end of each cycle. Following amplification, a melt curve analysis was performed to verify the specificity of the amplified products by their specific melting temperatures (Tm). Post amplification melting curve with SYBR Green dye of Toxoplasma positive DNA samples revealed one peak. For quantification of the amount of Toxoplasma DNA in the samples, a standard curve of a plasmid containing the 529 repeat gen in pGEM-T easy (Promega Benelux b.v. Leiden, The Netherlands) was used. Amounts of DNA were based on the log 10 dilutions of the standard. All samples analysed had a blank PCR control and samples negative for Toxoplasma DNA are depicted in Fig. 1B as 40-Ct = 0, whereas samples with 40-Ct>0 were considered as positive. Data acquisition and analysis of the results were performed using the 7500 System SDS Software Version 2.0.1 (Applied Biosystems).

2.5 Serology for T. *gondii*

Serum samples of the infected cats were processed with an in-house developed ELISA test for determining anti T. *gondii* antibody titers. ELISA plates (Nunc Polysorp 475094; Sanbio, Uden, The Netherlands) were coated with a T. *gondii* tachyzoites lysate (5 µg/ml in Na_2CO_3 buffer pH 9.5) overnight at 37°C. Wells were incubated for one h at 37°C with sera in two-fold dilutions from 1/25 to 1/3200 in PBS containing 0.05% Tween-

Table 2. Prevalence of tissue cysts as revealed by real time PCR of the heart, tongue and brains of cats infected with bradyzoites or tissue cysts of *T. gondii* strain M4.

Kitten no.	Dose	Infectious Material	40-Ct value		
			Heart	Tongue	Brains
7581	10	Bradyzoites	0	0	0
7582	10	Bradyzoites	0	0	0
7583	10	Bradyzoites	0	0	0
7578	50	Bradyzoites	0	0	0
7579	50	Bradyzoites	9.0	13.4	0
7580	50	Bradyzoites	0.2	0.0	0
7575	100	Bradyzoites	0	0.0	0
7576	100	Bradyzoites	0	0.0	4.1
7577	100	Bradyzoites	5	11.7	0
7572	1000	Bradyzoites	0	12.8	0
7573	1000	Bradyzoites	11.6	14.4	0
7574	1000	Bradyzoites	12.2	15.5	0
7584	250	Tissue cysts	3.2	0	0
7585	250	Tissue cysts	0	0	0
7586	250	Tissue cysts	11.7	0	0

80. The next incubation was for one h at 37°C with a peroxidase-conjugated Goat anti Cat-h&l (clone 10MG1-2, Bio connect, Huisen, The Netherlands) HRPO conjugate in a dilution of 1:10.000 in ELISA buffer. Serum from cats either infected or free from a *T. gondii* infection were used as positive and negative control. As substrate, 3,3',5,5'-tetramethylbenzidine (TMB) with H_2O_2 was used. After incubation for 5 min at room temperature, the reaction was stopped by the addition of sulphuric acid. The plate was read by the use of a Bio-Kinetics ELISA reader (Bio-Tek Instruments Inc., Winooski, VT) at 450 nm. Antibody titres were expressed as the 2-log of the regression coefficient of the optical density vs. serum concentration. Per group of three cats, ELISA results were taken to calculate the geometric means and standard deviations. Antibody titres (2-log) were compared to antibody titre in serum samples at day 0, and were analysed for statistical significance by the Mann-Whitney U test in the GraphPad Prism version 5.0 software, and where considered significant if the P-value<0.05.

2.6 Dose-Response Model

The infection status of cats inoculated with *T. gondii* M4, as judged by OPG count >0 was fitted by a Beta-Poisson dose-response model [40]. This model is based on the assumption that any single viable bradyzoite may cause infection, and accounts for variability in the probability of infection resulting from variation in host susceptibility or pathogen infectivity, due to genetic or other factors. Sampling effects in the applied inoculum were accounted for by assuming that the ingested number of bradyzoites is a Poisson sample from a suspension with mean equal to the target dose. Given a certain probability of infection at each dose, the numbers of infected cats at that dose are binomially distributed. Using the binomial likelihood the two parameters (α, β) can be estimated. A Monte Carlo sample of dose response parameters was obtained based on the oocysts output using JAGS (3.2.0), a program for analysis of Bayesian hierarchical models using

Markov chain Monte Carlo (MCMC) analysis (http://mcmc-jags.sourceforge.net).

2.7 Animal Ethics Committee

The animal study has been conducted with permission of the Animal Ethics Committee of the Animal Sciences Group (Lelystad, the Netherlands) dated 3 October 2012, registered under number 2012073.c.

3. Results

3.1 *T. gondii* infection in kittens

None of the cats showed any clinical symptom (depression, central nervous behaviour, gastro intestinal signs, diarrhoea) throughout the study period.

Oocysts in faeces. The dynamics of oocyst shedding by cats after challenge infection with mouse bradyzoites or tissue cysts of *T. gondii* strain M4 is shown in Table 1. All three kittens from the group infected with 250 tissue cysts shed oocysts in the faeces from day 4 till day 16 (Fig. 1A). The three kittens infected with 1000 bradyzoites shed oocysts from day 6 till day 17. At peak shedding on day 6–7 p.i, the kittens in this group produced 5* 10^6 oocysts per gram faeces (OPG). From the kittens infected with 100 or 50 bradyzoites only one out of three shed oocysts at day 7–16 p.i., and from the kittens infected with 10 bradyzoites none shed oocysts, as determined microscopically.

Real time PCR of faeces. All kittens which tested positive by microscopic examination were also positive by real time PCR. Moreover, all OPG-negative animals: the two negative kittens infected with 50 or 100 bradyzoites, and all kittens challenged with 10 bradyzoites became also positive for *T. gondii* by real rime PCR (Fig. 1B). The Ct values in these lower dose groups were higher showing that the concentration of available DNA was lower.

Serology. Compared to kittens in the other groups, the kittens that were infected with 250 tissue cysts or 1000 bradyzoites

M4 OPG

M4 OPG

M4 RT-PCR

M4-RT-PCR

VEG OPG

VEG OPG

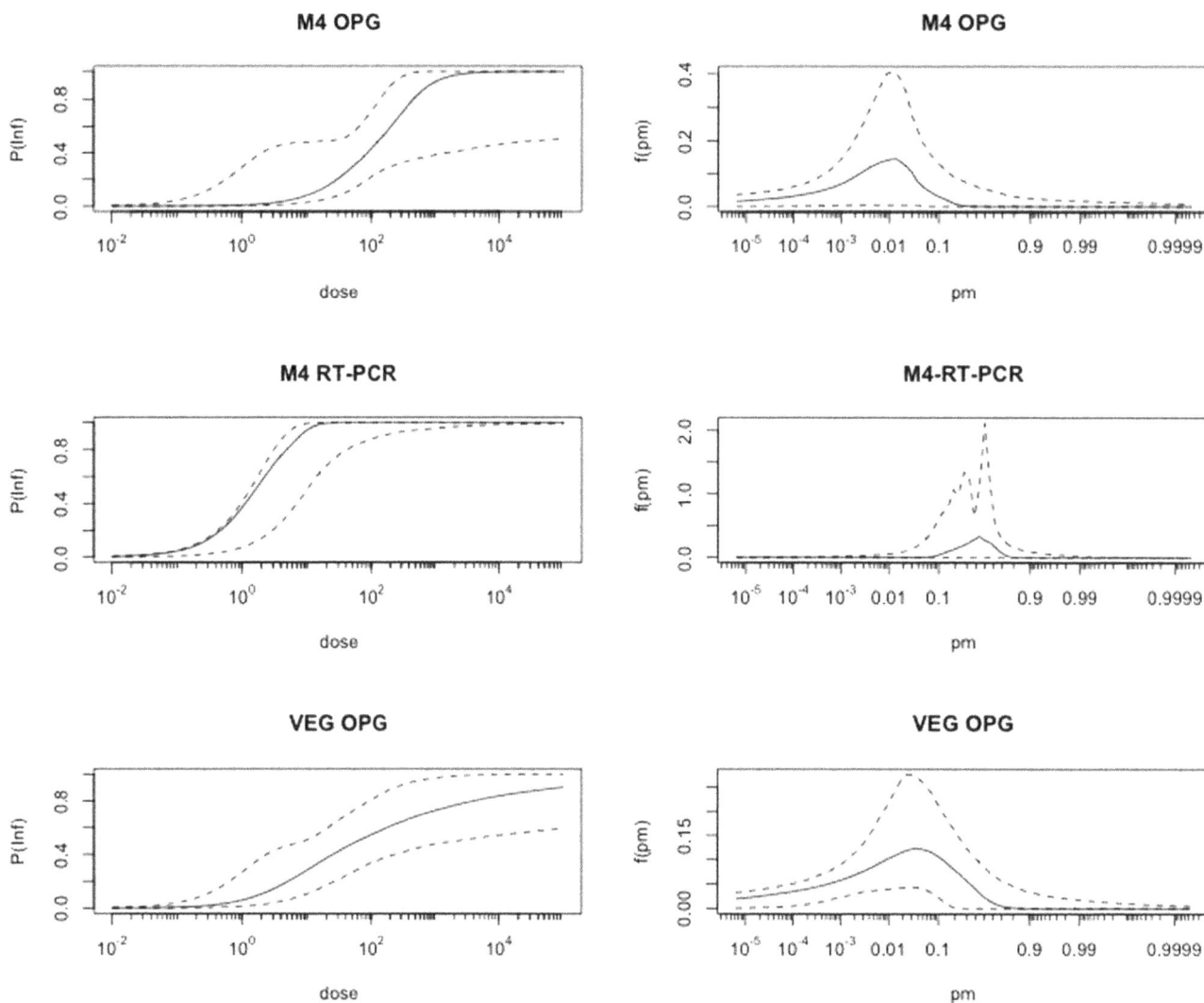

Figure 2. Panels in column A illustrate the dose-response for *Toxoplasma gondii* **in cats: Horizontal axes show mean number of bradyzoites; vertical axes show probability that a cat is infected following ingestion.** Each panel refers to: OPG of M4 strain (top), real time-values of M4 strain (middle) and OPG of VEG strain (bottom). Panels in column B illustrate posterior density for the dose response parameter (probability of infection by a single bradyzoites): Horizontal axes show possible estimate for the parameter; vertical axes show probability density. Solid line indicates the mean and dashed lines are 2.5% and 97.5% percentiles.

had the highest titres (Fig. 1C). In both groups, antibody titres at 14 and 21 days p.c. appeared to be significantly higher ($P<0.05$) compared to day 0. In the other dose groups the increase was much smaller and in the kittens infected with 10 or 50 bradyzoites we found a significant increase in IgG titre ($P<0.05$) respectively at week three and two compared to antibody titre in serum samples obtained at day 0. One kitten (nr 7579) of the group infected with 50 bradyzoites showed an IgG titre of 5.7 (2log) at day 21 p.c., shedding of oocysts in faeces of this cat was only confirmed by real time PCR and not microscopically.

Real Time PCR and microscopy of brain, tongue and heart. The real time PCR results detected *T. gondii* DNA in five tongue and in seven heart samples of the infected kittens. Of all the brain samples tested, only one kitten (100 bradyzoites) had very low Ct values in the real time PCR on brain samples, the other 14 kittens were negative (Table 2). Ct values after PCR testing of heart and tongue of the cats infected with 1000 bradyzoites were comparable, indicating that there was no

difference in distribution of tissue cysts over the tongue, and heart. (Table 2). Microscopically, we did not find any tissue cysts in brain, heart and tongue samples.

3.2 *T. gondii* dose response model in cats

Infection was defined as: 1. counting one or more oocysts in faecal samples (oocysts per gram, OPG), and 2. positive results of the real time PCR of faecal samples. The estimated mean ID50 based on the OPG counts is 181 bradyzoites (36–806 bradyzoites, 95% credible interval (ci), Fig. 2A). The estimated probability of infection per single bradyzoite is 0.02 (95% ci 0.003–0.13 per single bradyzoites, Fig. 2B). Based on the infection as determined by real time PCR, the estimated mean ID50 is two bradyzoites with a 95% credible interval of 0.044–11 bradyzoites. Equivalently, the probability of infection per single bradyzoite is 0.38 (95% ci 0.08–0.52 per single bradyzoite). To investigate whether infectivity is specific to the M4-strain, we also applied this dose response model to a similar experimental challenge model in

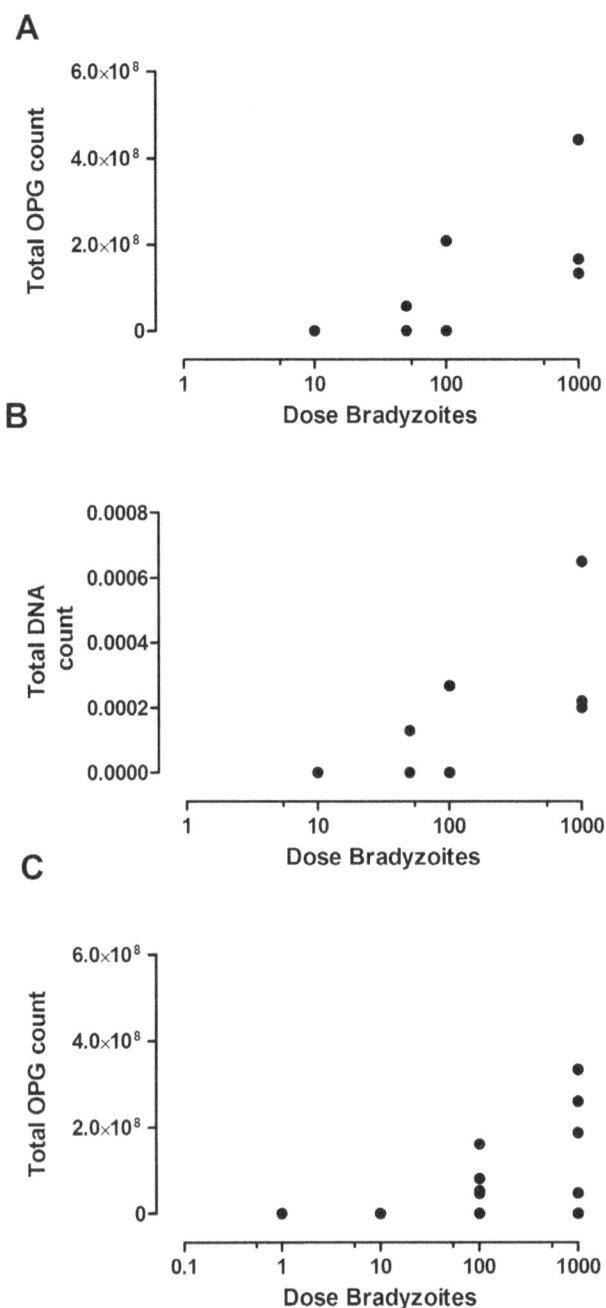

Figure 3. Total amount of *T. gondii* shedding per individual infected cat: M4 strain based on the OPG faecal output (A), the real time PCR DNA values in the faeces (B), the VEG strain (C) based on the OPG faecal output.

which the VEG strain was used (Fig. 2C; the datasets of experiment I and II from Dubey [25]). Using OPG>0 to characterize infection, the estimated mean ID50 was 214 bradyzoites (95% ci 7–1800 bradyzoites). To investigate whether the estimated infectivities of the M4 and VEG strains (Fig. 2A; M4 and VEG OPG) are different, a likelihood ratio test was done, showing that a model assuming equal infectivities of M4 and VEG did not provide a significantly worse fit than the two separate infectivity estimates (*P*-value = 0.47: likelihood ratio test, 2 d.f.). The total OPG and DNA count shed by cats in the experimental

groups using 10, 100 and 1000 bradyzoiets are shown in Fig. 3A and B. In addition, the total OPG counts shed by cats in the experimental infections as described by Dubey [25] are shown in Fig. 3C indicating that total OPG shedding was indeed in the same order of magnitude for the two studies.

Discussion

The aim of our study was to establish a standardised challenge model in cats as a reference for studying the effectiveness of vaccines or therapeutic drugs against T. *gondii* infection in cats.

We selected *T. gondii* strain M4. This strain belongs to genotype II, the most predominant *T. gondii* genotype in Europe [41]. Strain M4 is able to produce brain tissue cysts in mouse and oocysts in the gut of cats [27]. Strain M4 is therefore a suitable candidate for a standardised challenge model in cats.

To prepare the challenge inoculum to experimentally infect cats, a sufficiently concentrated suspension of fresh bradyzoites and tissue cysts was produced in mice, seven weeks after oral infection with oocysts. This finding is in agreement with other studies [22] in which harvesting tissue cysts after 6–8 weeks is sufficient time for infectious stages to develop.

We inoculated cats at an age of 12 weeks orally with four different doses (from low to high) of bradyzoites and determined a dose-response model for *T. gondii*. None of the challenged cats had clinical symptoms. This is in agreement with previous findings that cats older than three months rarely develop clinical signs after oral infection with tissue cysts [22].

We found that cats can shed millions of oocysts in their faeces after ingesting as few as 50 bradyzoites. These results are in agreement with results obtained by Dubey [25] who found that 100 bradyzoites of *T. gondii* strain VEG orally inoculated could lead to shedding of oocysts in the faeces. Our findings confirm the high infectivity of the bradyzoite stage of the parasite for cats [25,26,27]; therefore this parasite stage is an important candidate for use as challenge inoculum in vaccination or treatment-challenge studies in cats. Other authors [26,29,32,34] used mouse tissue cysts to infect cats, however as found by others and confirmed in this study, tissue cysts may contain five to several hundreds of bradyzoites, "depending on the age of the cysts" [24]. Suspended bradyzoites in contrast to tissue cysts provide a reproducible dose. Using the different *T. gondii* strain M4 bradyzoites doses in cats, we found a dose response relation. In addition, there was no significant difference in the dose response between the *T. gondii* M4 strain used in this study and the *T. gondii* VEG strain as reported by Dubey [25] using OPG as a response. Concerning M4 strain only, a dose response relation shifted toward higher infectivity when real time PCR positivity instead of OPG is chosen as an endpoint. Real time PCR and microscopy both detect oocyst shedding, but the 529 repeat element detected by real time PCR is present in 200–300 copies per parasite and is therefore more sensitive [38]. At low doses (100 – 50 bradyzoites), real time PCR detects more positives, and even in the lowest dose group (10 bradyzoites) the real time PCR detects low levels of shedding of *T. gondii* genomes. Interestingly, these low dose shedding responses appear different from those in OPG positive animals. The Ct values are 10–15 cycles lower than those in OPG positive, and the duration of shedding is 5–10 days shorter. The question is whether this low-level shedding represents true infection, producing infectious oocysts. The real time PCR assay may detect the DNA extracted from infectious oocysts in all dose groups (as confirmed with OPG shedding) or it may only detect fragments of *T. gondii* DNA without the presence of oocysts in low dose groups. This might have consequences in case of challenge studies to evaluate the vaccine effectiveness for example, since only infectious oocysts shed by cats are a risk for humans and are thus of epidemiological relevance.

In previous studies, it was described that pre-patent periods in cats differs from 3 to10 days after the ingestion of tissue cysts (bradyzoites) and 18 days or more after the ingestion of oocysts or tachyzoites [23,42,43,44]. Our study confirmed that the prepatent period was 5–7 days among all bradyzoite challenge doses. The patent period increased with the increasing dose of bradyzoites. Hence, it is an important feature to determine the optimal dose for vaccination challenge or therapy experiments.

It has been described that IgG antibodies are initially absent during primary *T. gondii* infection in cats, but then start to increase after a few weeks to reach protective levels and then remain detectable for years. IgM antibodies rise within days, and usually decrease over the following few weeks. Dubey et al. [45] reports that cats seroconverted 10 days p.i and high titres persisted even after 6 years, however, in some cats with a chronic infection IgM persists. In the present study we detected significantly higher IgG titres as early as three weeks after infection in cats that received the highest dose of *T. gondii* bradyzoites and the tissue cysts challenge group.

We detected *T. gondii* DNA by real time PCR in tongue, heart and in one brain tissue of infected cats. Dubey described earlier the detection of *T. gondii* tissue cysts in heart, skeletal muscles, diaphragm, brain, spleen, kidneys; pancreas, stomach, adrenals, lungs, thymus, cervical and mesenteric lymph nodes salivary glands and eyes and tongue of bradyzoites infected cats [26], of which tongue, heart and brain were important target organs [46].The cats with the highest infection dose revealed a high *T. gondii* DNA load in tongue and heart.

In summary, we conclude that the oocysts of *T. gondii* strain M4 can be used to produce high numbers of bradyzoites in the brain of the mouse to be used as a challenge model in cats. Moreover, the bradyzoite stage of *T. gondii* strains can be used to establish a dose response model for oral challenge with *T. gondii* in the cat. The dose response in cats is *T. gondii* strain independent at least for the M4 and VEG strains. However, the results using microscopy or real time PCR will have influence on the outcomes of the dose response, most likely caused by differences in detection limit of each assay. Both the probability of infection and the numbers of oocysts shedded by infected cats appeared to depend on the challenge dose. Therefore, we can use this M4 challenge model quantitatively to determine the effects of any intervention, vaccine or drug induced, expressed in a reduction of the total oocyst output of *T. gondii* clonal types II and III infected cats. For the future, our *T. gondii* dose response model in cats can be exploited to evaluate different strategies aiming at cats to reduce oocysts shedding.

Acknowledgments

We thank prof. dr. E.A. Innes of the Moredun Research Institute in Edinburgh (Scotland) for providing oocysts from *T. gondii* strain M4.

Author Contributions

Conceived and designed the experiments: JC HW JG. Performed the experiments: JC HJW. Analyzed the data: JC HW JG PT KT. Contributed reagents/materials/analysis tools: JC PT KT. Contributed to the writing of the manuscript: JC HW JG PT KT.

References

1. Verma R, Khanna P (2012) Development of *Toxoplasma gondii* vaccine: A global challenge. Hum Vaccin Immunother 30: 291–293.
2. Torgerson PR, Mastroiacovo P (2013) The global burden of congenital toxoplasmosis: systematic review. Bull World Health Organ 91:501–508.
3. Batz MB, Hoffmann S, Morris JG Jr (2012) Ranking the disease burden of 14 pathogens in food sources in the United States using attribution data from outbreak investigations and expert elicitation. J Food Prot 75:1278–1291.
4. Havelaar AH, van Rosse F, Bucura C, Toetenel MA, Haagsma JA, et al. (2010) Prioritizing emerging zoonoses in the Netherlands. PLoS One 15: e13965.
5. Hill DE, Chirukandoth S, Dubey JP (2005) Biology and epidemiology of *Toxoplasma gondii* in man and animals. Anim Health Res Rev 6: 41–61.
6. Tenter AM, Heckeroth AR, Weis LM (2000) *Toxoplasma gondii*: from animals to humans. Int J Parasitol 30: 1217–1258.
7. Dubey JP (2006) Comparative infectivity of oocysts and bradyzoites of *Toxoplasma gondii* for intermediate (mice) and definitive (cats) hosts. Vet Parasitol 40: 69–75.
8. Dubey JP, Beattie CP (1988). Toxoplasmosis of Animals and Man. CRS Press, Boca Raton, FL 220 p.
9. Dubey JP (1994) Toxoplasmosis. J Am Vet Med Assoc 205: 1593–1598.
10. Dubey JP (1995). Duration of immunity to shedding of *Toxoplasma gondii* oocysts by cats. J Parasitol 81: 410–415.
11. Dabritz HA, Conrad PA (2010) Cats and Toxoplasma: implications for public health. Zoonoses Public Health 57: 34–52.
12. Dumètre A1, Le Bras C, Baffet M, Meneceur P, Dubey JP, et al. (2008) Effects of ozone and ultraviolet radiation treatments on the infectivity of *Toxoplasma gondii* oocysts. Vet Parasitol 153: 209–213.
13. Dubey JP (1998) *Toxoplasma gondii* oocyst survival under defined temperatures. J Parasitol 84: 862–865.
14. Lindsay DS, Dubey JP (2009) Long-term survival of *Toxoplasma gondii* sporulated oocysts in seawater. J Parasitol 95: 1019–1020.
15. Dubey JP, Jones JL (2008) *Toxoplasma gondii* infection in humans and animals in the United States. Int J Parasitol 38: 1257–1278.
16. Bowie WR, King AS, Werker DH, Isaac-Renton JL, Bell A, et al. (1997) Outbreak of toxoplasmosis associated with municipal drinking water. The BC Toxoplasma Investigation Team. Lancet 350: 173–177.
17. Bahia-Oliveira LM1, Jones JL, Azevedo-Silva J, Alves CC, Oréfice F, et al. (2003) Highly endemic, waterborne toxoplasmosis in north Rio de Janeiro state, Brazil. Emerg Infect Dis 9: 55–62.
18. Pfaff AW, Candolfi E (2008) New insights in toxoplasmosis immunology during pregnancy. Perspective for vaccine prevention. Parassitologia 50: 55–58.
19. Dubey JP (1996) Strategies to reduce transmission of *Toxoplasma gondii* to animals and humans. Vet Parasitol 64: 65–70.
20. Mateus-Pinilla NE, Dubey JP, Choromanski L, Weigel RM (1999) A field trial of the effectiveness of a feline *Toxoplasma gondii* vaccine in reducing T. gondii exposure for swine. J Parasitol 85: 855–860.
21. Innes EA, Bartley PM, Maley S, Katzer F, Buxton D (2009) Veterinary vaccines against *Toxoplasma gondii*. Mem Inst Oswaldo Cruz 104: 246–251.
22. Dubey JP (2010) Toxoplasmosis of animals and humans. By Taylor and Francis Group, LLC CRC Press, Boca Raton 38.
23. Dubey JP, Frenke JK (1972) Cyst-induced toxoplasmosis in cats. J Protozool 19: 155–177.
24. Dubey JP, Lindsay DS, Speer CA (1998) Structures of *Toxoplasma gondii* Tachyzoites, Bradyzoites, and Sporozoites and Biology and Development of Tissue Cysts Clin Microbiol Rev 11: 267–299.
25. Dubey JP (2001) Oocyst shedding by cats fed isolated bradyzoites and comparison of infectivity of bradyzoites of the VEG strain *Toxoplasma gondii* to cats and mice. J Parasitol 87: 215–219.
26. Dubey JP (2006) Comparative infectivity of oocysts and bradyzoites of *Toxoplasma gondii* for intermediate (mice) and definitive (cats) hosts. Vet Parasitol 140: 69–75.
27. Fritz H, Bar B, Packham A, Melli APA, Conrad PA (2012) Methods to produce and safely work with large numbers of *Toxoplasma gondii* oocysts and bradyzoite cysts. J Microbiol Methods 88: 47–52.
28. Frenkel JK, Pfefferkorn ER, Smith DD, Fishback JL (1991) Prospective vaccine prepared from a new mutant of *Toxoplasma gondii* for use in cats. Am J Vet Res 52:759–764.
29. Frenkel JK, Pfefferkorn ER, Smith DD, Fishback JL (1993) Immunisation of cats with tissue cysts, bradyzoites and tachyzoites of the T-263 strain of *Toxoplasma gondii*. J Parasitol 79: 716–719.
30. Mateus-Pinilla NE, Dubey JP, Choromanski L, Weigel RM (1999) A field trial of the effectiveness of a feline *Toxoplasma gondii* vaccine in reducing T. gondii exposure for swine. J Parasitol 85: 855–860.
31. Mishima M, Xuan X, Yokoyama N, Igarashi I, Fujisaki K, et al. (2002) Recombinant feline herpesvirus type I expressing *Toxoplasma gondii* ROP2 antigen inducible protective immunity in cats. Parasitol Res 88:144–149.
32. Omata Y, Aihara Y, Kanda M, Saito A, Igarashi I, et al. (1996) *Toxoplasma gondii* experimental infection in cats vaccinated with 60 Co-irradiated tachyzoites. Vet Parasitol 65: 173–183.
33. Garcia JL, Navarro IT, Biazzono L, Freire RL, da Silva Guimarães J Jr, et al. (2007) Protective activity against oocyst shedding in cats vaccinated with crude

rhoptry proteins of the *Toxoplasma gondii* by the intranasal route. Vet Parasitol 145: 197–206.

34. Zulpo DL, Headley SA, Biazzono L, da Cunha IA, Igarashi M, et al. (2012) Oocysts shedding in cats vaccinated by the nasal and rectal routes with crude rhoptry proteins of *Toxoplasma gondii*. Exp Parasitol 131: 223–230.

35. Cornelissen AW, Overdulve JP, Hoenderboom JM (1981) Separation of Isospora (Toxoplasma) gondii cysts and cystozoites from mouse brain tissue by continuous density-gradient centrifugation. Parasitology 83: 103–108.

36. Popiel I, Gold MC, Booth KS (1982) Quantification of *Toxoplasma gondii* bradyzoites. J Parasitol 82: 330–332.

37. Wainwright KE, Miller MA, Barr BC, Gardner IA, Melli AC, et al. (2007) Chemical inactivation of *Toxoplasma gondii* oocysts in water. J Parasitol 93: 925–931.

38. Homan WL, Vercammen M, De Braekeleer J, Verschueren H (2000) Identification of a 200- to 300-fold repetitive 529 bp DNA fragment in *Toxoplasma gondii*, and its use for diagnostic and quantitative PCR. Int J Parasitol 30: 69–75.

39. Reischl U1, Bretagne S, Krüger D, Ernault P, Costa JM (2003) Comparison of two DNA targets for the diagnosis of Toxoplasmosis by real-time PCR using fluorescence resonance energy transfer hybridization probes. BMC. Infect Dis 2: 3–7.

40. Teunis P, Takumi K, Shinagawa K (2004) Dose response for infection by *Escherichia coli* O157:H7 from outbreak data. Risk analysis: an official publication of the Society for Risk Analysis 24: 401–407.

41. Ajzenberg D, Yera H, Marty P, Paris L, Dalle F, et al. (2009) Genotype of 88 *Toxoplasma gondii* isolates associated with toxoplasmosis in immunocompromised patients and correlation with clinical findings. J Infect Dis 15:1155–1167.

42. Freyre A, Dubey JP, Smith DD, Frenkel JK (1989) Oocyst-induced *Toxoplasma gondii* infections in cats. J Parasitol 75:750–755.

43. Dubey JP (1996) Infectivity and pathogenicity of *Toxoplasma gondii* oocysts for cats. J Parasitol 82: 957–961.

44. Dubey JP (2005) Unexpected oocyst shedding by cats fed *Toxoplasma gondii* tachyzoites: in vivo stage conversion and strain variation. Vet Parasitol 133: 289–298.

45. Dubey JP, Lappin MR, Thulliez P (1995) Long-term antibody responses of cats fed *Toxoplasma gondii* tissue cysts. J Parasitol 81: 887–893.

46. Dubey JP (1997) Tissue cyst tropism in *Toxoplasma gondii*: a comparison of tissue cyst formation in organs of cats, and rodents fed oocysts. Parasitology 115: 15–20.

2

Gametocytocidal Screen Identifies Novel Chemical Classes with *Plasmodium falciparum* Transmission Blocking Activity

Natalie G. Sanders, David J. Sullivan, Godfree Mlambo, George Dimopoulos, Abhai K. Tripathi*

W. Harry Feinstone Department of Molecular Microbiology and Immunology, Johns Hopkins Bloomberg School of Public Health, Johns Hopkins University, Baltimore, Maryland, United States of America

Abstract

Discovery of transmission blocking compounds is an important intervention strategy necessary to eliminate and eradicate malaria. To date only a small number of drugs that inhibit gametocyte development and thereby transmission from the mosquito to the human host exist. This limitation is largely due to a lack of screening assays easily adaptable to high throughput because of multiple incubation steps or the requirement for high gametocytemia. Here we report the discovery of new compounds with gametocytocidal activity using a simple and robust SYBR Green I- based DNA assay. Our assay utilizes the exflagellation step in male gametocytes and a background suppressor, which masks the staining of dead cells to achieve healthy signal to noise ratio by increasing signal of viable parasites and subtracting signal from dead parasites. By determining the contribution of exflagellation to fluorescent signal and using appropriate cutoff values, we were able to screen for gametocytocidal compounds. After assay validation and optimization, we screened an FDA approved drug library of approximately 1500 compounds, as well as the 400 compound MMV malaria box and identified 44 gametocytocidal compounds with sub to low micromolar IC_{50}s. Major classes of compounds with gametocytocidal activity included quaternary ammonium compounds with structural similarity to choline, acridine-like compounds similar to quinacrine and pyronaridine, as well as antidepressant, antineoplastic, and anthelminthic compounds. Top drug candidates showed near complete transmission blocking in membrane feeding assays. This assay is simple, reproducible and demonstrated robust Z-factor values at low gametocytemia levels, making it amenable to HTS for identification of novel and potent gametocytocidal compounds.

Editor: Sanjai Kumar, Food and Drug Administration, United States of America

Funding: This work was supported by a grant from the national Institutes of Health R01AI061576, the Johns Hopkins Malaria Research Institute and The Bloomberg Family Foundation. Publication of this article was funded in part by the Open Access Promotion Fund of the Johns Hopkins University Libraries. The funders had no role in study design, data collection and analysis, decision to publish, or preparation of the manuscript.

* Email: atripat2@jhu.edu

Introduction

Malaria is a historically relentless public health problem and continues in the present day to contribute to severe morbidity and mortality worldwide, impeding development in many of the world's poorest countries. *Plasmodium falciparum* malaria is associated with the highest fatality rates, resulting in an estimated 200 million cases and more than one million deaths in 2012 [1]. Efforts to control, eliminate, and ultimately eradicate this disease have only been partially successful, with failure due in large part to the development of drug resistance in both the *Anopheles* mosquito vector, as well as the parasite [2,3]. Sustainable interventions and control measures have also posed a challenge, and a multi-faceted strategy targeting both transmission and disease is necessary if there is any hope of controlling this devastating disease [2–4].

Of particular interest is the discovery of new chemical entities and classes targeting the sexual stage of the parasite, gametocytes, which are responsible for transmission back to the mosquito

vector. To this end, a variety of assays have been developed, each utilizing different measures of parasite viability including alamar blue to detect metabolic activity, detection of parasite proteins such as lactate dehydrogenase, or bioluminescence of viable transgenic parasites [5–11]. While the reported assays are more high-throughput than the gold standard of counting Giemsa-stained blood films, they still have limitations including the requirement for transgenic parasites or multiple incubation and transfer steps.

Here we describe a simple assay using the SYBR-green I DNA probe along with a background suppressor to assay for live gametocytes. To achieve robust signal to noise ratio we use a combination of exflagellation, to increase DNA content from viable male gametocytes, and background suppressor to mask the signals from drug killed gametocytes. Incubation time after drug treatment is minimal with no transfer or centrifugation steps and can be easily adapted to higher throughput formats such as 384 or 1536-well plates. In addition, this assay does not require transgenic parasites and thus could be used to screen field isolates. After validating the assay, we screened an FDA-approved library of 1584 compounds as well as the MMV malaria box of 400 confirmed antimalarials that are active against asexual blood stages in *P. falciparum*. We report the results of both drug screens, with particular emphasis on the novel classes of active compounds identified by the assay: quaternary ammonium compounds and acridine-like compounds.

Materials and Methods

Ethics statement

This study was carried out in strict accordance with the recommendations in the Guide for the Care and Use of Laboratory Animals of the National Institutes of Health. Mice were only used for mosquito rearing as a blood source according to approved protocol. The protocol was approved by the Animal Care and Use Committee of the Johns Hopkins University (ACUC MO12H76).

P. falciparum gametocyte cultivation

The *P. falciparum* NF54 strain was cultured according to the method described by Trager and Jenson with minor modifications. Briefly parasites were cultured using O^+ human erythrocytes at 4% hematocrit in parasite culture medium (RPMI 1640 supplemented with 25 mM HEPES, 10 mM Glutamine, 0.074 mM hypoxanthine and 10% O^+ human serum. Cultures were maintained under standard conditions of 37°C in a candle jar made of glass desiccators. Gametocyte cultures were initiated at 0.5% mixed stage parasitemia from low passage stock and cultures were maintained up to day 15 with daily media changes. To achieve greater level of asexual parasitemia before gametocytogenesis, hematocrit was reduced to 2% between days 3 to 6. After day 6 hematocrit was brought back to approximately 4%. To block reinvasion of remaining asexual parasites and obtain pure and near synchronous gametocytes, cultures were treated with 50 mM *N*-acetyl-D-glucosamine (NAG) for 72 hours between days 8 to 11.

Development and optimization of gametocytocidal assay

To determine the effect of drugs on mature gametocytes we developed a SYBR Green I based DNA quantification assay. Because gametocytes do not multiply, we utilized the decrease in live gametocytes after killing combined with an increase in DNA content of male gametes after exflagellation to quantitate the effect of drugs. For determination of live gametocytes, our assay uses a background suppressor, a live-cell impermeable dye (CyQUANT direct cell proliferation assay kit, Life Technologies, Grand Island, NY, USA) which can enter dead cells and specifically blocks green fluorescence, resulting in the subtraction of SYBR Green I signal from drug killed parasites. For assay optimization mature gametocytes were enriched using Percoll density gradient centrifugation. Enriched gametocytes were plated in 96 well plates and serially diluted with uninfected 2% hematocrit erythrocytes to obtain serial gametocytemia values. Triplicate wells of each parasite dilution were either treated with 10 µM pyrvinium pamoate or 0.1% DMSO (vehicle control) for 48 hrs at 37°C in candle jar as described above. After drug exposure, 11 µl of 10x exflagellation medium (RPMI 1640 with 200 mM HEPES, 40 mM sodium bicarbonate, 100 mM glucose pH 8.0) was added and plates were incubated at room temperature for 30 min. Next 11 µl of 10x CyQUANT direct background suppressor and SYBR Green I in PBS was added and plate was incubated at room temperature for 2 hrs. After addition of detection reagents plates

were protected from light. Fluorescence was then measured at excitation and emission wavelengths of 485 and 535 in a plate reader (HTS7000 Perkin Elmer). To achieve consistent reads, special care was taken not to disturb the settled layer of gametocyte infected erythrocytes for consistent readings, during addition of reagents, incubations and detection. To determine the respective contribution of background suppressor (gametocyte killing) and exflagellation (increase in DNA content), each assay was performed in parallel plates with and without exflagellation. To prevent spontaneous exflagellation, the plate without exflagellation was maintained at 37°C until fluorescence values were determined.

Screening of JHU FDA approved compound library

The Johns Hopkins University Clinical Compound Library version 1.3 is comprised of more than 1500 drugs, which were approved by the FDA for treatments of different diseases or medical conditions. The JHU drug library is stocked in 96 well plates at 10 mM in 100% DMSO. In order to achieve dispensable concentration we diluted compounds in incomplete RPMI to new master plates at 400 µM. We dispensed 5 µl of Compound library to the 96-well plates, to a final compound concentration of 20 µM and DMSO concentration of 0.1%. Columns 1 and 12 of each plate were used as in-plate controls and contained 0.1% DMSO (negative control, 0% inhibition) and 20 µM clotrimazole (positive control, ~70% inhibition), respectively. Ninety-five µl of day 15 pure gametocyte cultures at approximately 3–5% gametocytemia were then added to the compound containing plates at 1% hematocrit. Plates were incubated for up to 48 hrs at 37°C in microaerophilic conditions of a candle jar. After 48 hrs of drug treatment, gametogenesis was induced by adding 11 µl of 10x exflagellation media and incubation for 30 min at room temperature. Next 11 µl of 10x CyQUANT direct background suppressor and SYBR Green I (Life Technologies, Grand Island, NY, USA) in PBS was added to the plates and further incubated at room temperature for 2 hrs in the dark. Plates were then read at excitation and emission wavelengths of 485 and 535 nm respectively and raw data was transferred to Microsoft Excel. Fluorescence signals from both negative (0.1% DMSO) and positive (clotrimazole) wells were used for quality control of the assay and to determine percent inhibition by each compound. Compounds which showed values between positive and negative controls were considered primary hits. All the hits from the primary screen were retested at 10 µM for microscopic examination and at multiple concentrations for IC_{50} determinations.

Screening of the MMV Malaria Box

The Medicines for Malaria Venture (MMV) kindly provided the Malaria Box which was comprised of 200 drug like and 200 probe like inhibitors of *P. falciparum* asexual stage. The MMV Box was supplied in 96-well plates at 10 mM stocks in 100% DMSO. We diluted the compound library by 50 fold to make master plates at 200 µM and 5 µl of each compound was dispensed into the duplicate assay plates, to achieve final concentration of 10 µM. The first and last columns on each plate were used for the negative (0.1% DMSO) and positive (10 µM pyrvinium pamoate) controls. Plate set up and detection of fluorescence was performed as described above in the method section for the FDA approved compound library. Compounds showing >50% inhibition in both replicates were considered primary hits. All the primary hits were then retested at multiple doses and IC_{50} values were determined.

A

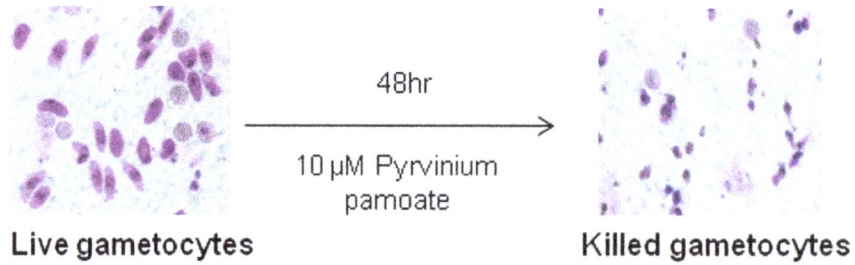

Live gametocytes

48hr

10 µM Pyrvinium pamoate

Killed gametocytes

B

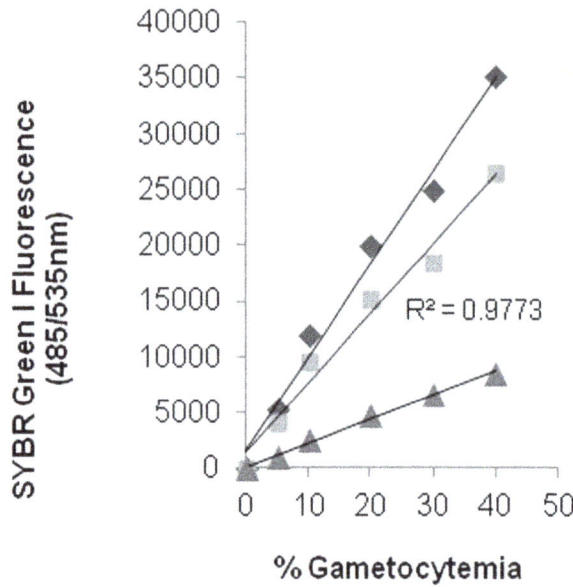

SYBR Green I Fluorescence (485/535nm)

R² = 0.9773

% Gametocytemia

C

Z-factors for SYBR Green I Assay plus background suppressor	
%Gametocytes (2%hct)	Z-factor
40	0.86
30	0.81
20	0.68
10	0.57
5	0.37

$$Z = 1 - \frac{3(\sigma_z + 3\sigma_c)}{|\mu_z - \mu_c|}$$

D

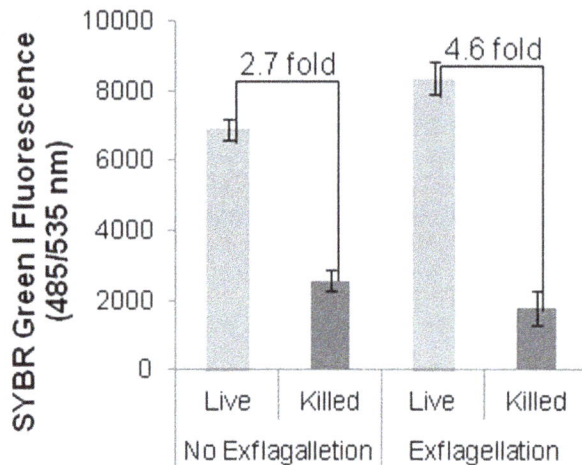

SYBR Green I Fluorescence (485/535 nm)

2.7 fold   4.6 fold

Live   Killed   Live   Killed

No Exflagalletion   Exflagellation

E

Sample wells		s (no drug)	c (10 µM PP)
12051	6459	13088	6034
10100	11971	13613	6175
11136	11561	13205	6412
12074	12142	12877	6295
	µ	13196	6229
	σ	310	162

$$100 - \left(\frac{sample - \mu_c}{\mu_s - \mu_c}\right) \times 100 = \% \ inhibition$$

F

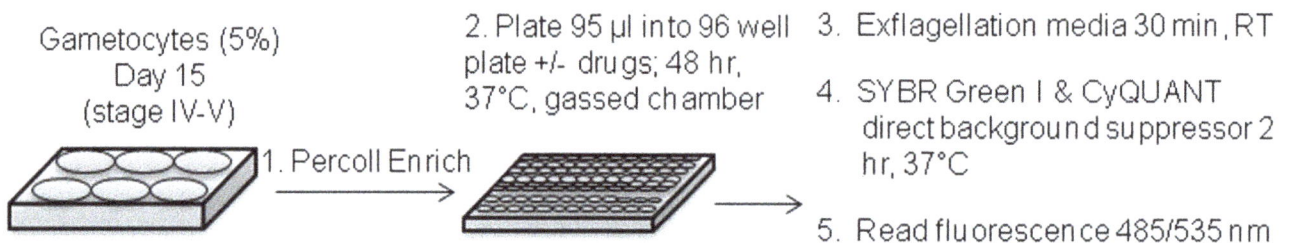

Gametocytes (5%) Day 15 (stage IV-V)

1. Percoll Enrich

2. Plate 95 µl into 96 well plate +/- drugs; 48 hr, 37°C, gassed chamber

3. Exflagellation media 30 min, RT

4. SYBR Green I & CyQUANT direct background suppressor 2 hr, 37°C

5. Read fluorescence 485/535 nm

Figure 1. SYBR Green I-Background suppressor gametocytocidal assay. (A) Giemsa stained culture before and after treatment with 10 μM pyrvinium pamoate. (B) SYBR Green I fluorescence of gametocytes, total (diamond), killed (triangle) and live gametocytes after drug treatment (total minus killed, square) with decreasing number of gametocytes per uninfected cell, diluted with 2% hematocrit RBCs in media in presence of background suppressor. (C) Z-factors calculated for each gametocyte dilution. Z factors were calculated using the equation shown, described previously for validating high throughput assays (σ = standard deviation, μ = mean, s = sample, c = control or in this case zero gametocytes) [10]. (D) SYBR Green I fluorescence of live or pyrvinium pamoate-killed gametocytes in the presence of CyQUANT background suppressor, with and without exflagellation with background well fluorescence (no parasites) subtracted out as a blank. (E) Example of assay plate SYBR Green I fluorescence in the presence of background suppressor and calculations for % inhibition. Green indicates a positive hit with high inhibition attributable to gametocyte killing and red indicates an intermediate hit, with inhibition attributable to potentially exflagellation inhibition and/or moderate gametocyte killing. (F) Overall assay setup with five steps: 1. Culture and enrich gametocytes, 2. Incubate with drug for 48 hr, 3. Add exflagellation media and incubate 30 min, 4. Add SYBR Green I and background suppressor and incubate 2 hr, 5. Read SYBR Green I fluorescence at excitation 485 nm and emission 535 nm.

Z factor determinations

Z factors were calculated using an equation described previously for validating high throughput assays (Figure 1C) [12]. In all assays, means and standard deviations were calculated from a minimum of four replicates for each sample well (positive pyrvinium pamoate control) and four to eight replicates for control well (negative no drug control). There were two biological replicates completed for every drug/plate (one in the second screen of the FDA drug library) and the Z-factors reported were mean Z-factors calculated from the sum of all Z-factors calculated for each assay in each screen. For example, the Z-factor for the FDA approved drug library screen was calculated as a mean of the

Z-factors calculated for each assay plate (with four positive control wells and four to eight negative control wells per plate) in that screen.

Mosquito rearing and membrane feeding assay

Anopheles gambiae Keele strain mosquitoes were maintained on a 10% sugar solution at 27°C and 80% humidity with a 12-h light/dark cycle according to standard rearing methods. Day 15 gametocytes were treated with gametocytocidal compounds or 0.1% DMSO for 48 hr and then were centrifuged and diluted to 0.3% final gametocytemia in a mixture of RBCs supplemented with human serum for mosquito membrane feeding assays. Unfed

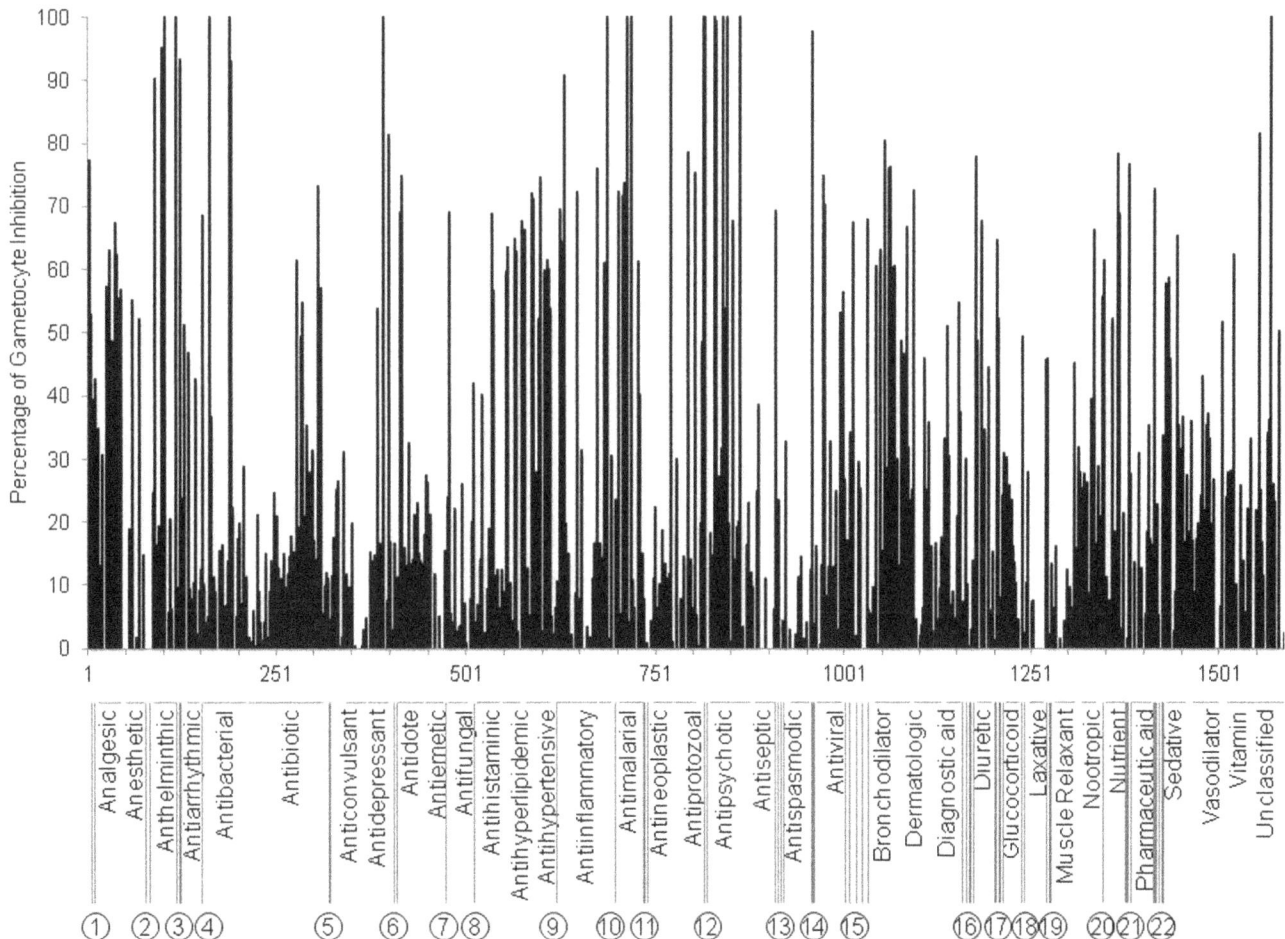

Figure 2. Inhibition by FDA drug library. SYBR Green I assay results for the Johns Hopkins Clinical Compound Library version 1.3 of FDA approved drugs screened at 20 μM. Plot of percentage of gametocytocidal activity of 1,584 compounds compared to clotrimazole control.

mosquitoes were removed after feeding, and midguts were dissected 7 days later and stained with 0.1% mercurochrome. The number of oocysts per midgut was determined with a light-contrast microscope, and the median infection intensity was calculated for the control and each experimental group.

Results

Assay optimization

Using SYBR Green I as a live-cell permeable fluorescent probe, we were able to detect gametocytes based on DNA content, with exflagellation as a means to increase DNA content in viable male gametes. To increase our signal to noise ratio, we used a background suppressor from the CyQUANT Direct Cell Proliferation Assay kit which works specifically by entering permeabilized or dead cells and masking green fluorescence. By using SYBR Green I in conjunction with the background suppressor, we were able to mask the signal from dead or damaged gametocytes and only read SYBR Green I fluorescence from live or intact cells.

The assay was optimized to determine sensitivity comparing drug treated and untreated parasites. SYBR Green I fluorescent signal from total and killed (10 μM pyrvinium pamoate treated, Figure 1A) gametocytes was shown to increase linearly with increasing gametocytemia (Figure 1B) and after subtracting out signal from killed gametocytes, retained fluorescent signal with a coefficient of determination of 0.97, indicating strong predictive value of gametocyte number on fluorescent signal. To determine the limit of detection and sensitivity of the assay, a Z-factor was calculated for serially diluted gametocyte culture at 2% hematocrit, which showed increase in Z-factor values with increasing gametocyetmia levels (Figure 1C). Addition of the CyQUANT background suppressor dye greatly increased the sensitivity of the assay compared to exflagellation, which marginally enhanced the signal of live gametocytes (Figure 1D). Specifically, beginning with an average ratio of 4:1 female to male mature gametocytes, exflagellation increased live gametocyte signal from 7000 to 8000 fluorescent units, suggesting a contribution of 10–20% of

Table 1. Gametocytocidal compounds identified in JHU FDA-approved drug library screen with greater than 70% inhibition and/or IC50≤20 μM.

Compound	Indication	Gametocyte		Asexual	
		20 μM % inh	Avg μM IC$_{50}$	10 μM % inh (SD)	Avg μM IC$_{50}$
Melphalan	Antineoplastic	151	4.0	22.9 (2.3)	20.0
Gentian violet	Antiseptic	148	8.5	99.2 (0.6)	0.6
Homidium (Ethidium) bromide	Anthelminthic	148	0.38	99.3 (0.0)	0.1
Ifosfamide	Antineoplastic	136	2.0	0.0 (0.0)	NA
Pentamidine	Antiprotozoal	129	0.7	97.7 (0.0)	1.0
Thonzonium bromide	Antiseptic	113	6.0	98.1 (0.0)	6.3
Cetalkonium chloride	Antiseptic	112	6.0	93.9 (0.1)	12.6
Benzethonium chloride	Antiseptic	112	6.0	98.6 (0.0)	4.0
Cetylpyridinium bromide	Antiseptic	110	9.0	86.4	6.3
Benzalkonium chloride	Antiseptic	109	7.0	98.5 (0.0)	5.0
Methylbenzethonium chloride	Antiseptic	108	10.0	98.3 (1.4)	-
Pyrvinium pamoate	Anthelminthic	103	4.0	99.6	0.6
Maprotiline	Antidepressant	102	0.9	37.4 (0.9)	20
Anastrozole	Antineoplastic	102	0.6	-	NA
Cetylpyridinium chloride	Antiseptic	99	7.0	66.4 (8.8)	-
Benzododecinium chloride	Antiseptic	98	5.0	98.0[†]	0.1[‡]
Tilorone	Antiviral	98	5.5	99.2 (0.0)	0.2[‡]
Dithiazanine iodide	Anthelminthic	95	7.0	92.9 (0.28)	3.2
Pyrithione zinc	Antiseptic	93	0.6	98.6 (0.9)	-
Antimony potassium tartrate	Anthelminthic	90	3.5	97.7 (0.0)	NA
Primaquine	Antimalarial	76	20	84.2 (10.0)	1.3
Anazolene sodium	Diagnostic aid	72	0.6	−6.3 (8.9)	20.0
Megestrol acetate	Progestogen	69	3.5	−0.7 (5.7)	NA
Acetomenaphthone	Pharmaceutic aid	66	8.5	-	12.6
1-Pentanol	Dermatologic	61	0.7	5.2 (12.5)	-
Clotrimazole	Antifungal	55	-	-	1.3

Asexual stage 10 μM inhibition data was obtained from the Collaborative Drug Discovery Database (CDDD), 10 μM drug 3D7 48 hr, ^{3}H hypoxanthine assay for parasite inhibition protocol, and asexual IC$_{50}$ data was obtained from from Eastman et al. or from the CDDD WRAIR IC 50 nM D6 protocol as noted [47,48]. Gametocytocidal IC$_{50}$ values were calculated from one experiment with three replicates for top compounds.
[†]Data only available for 96 hr assay,
[‡]WRAIR D6 data, – Unavailable, NA not active.

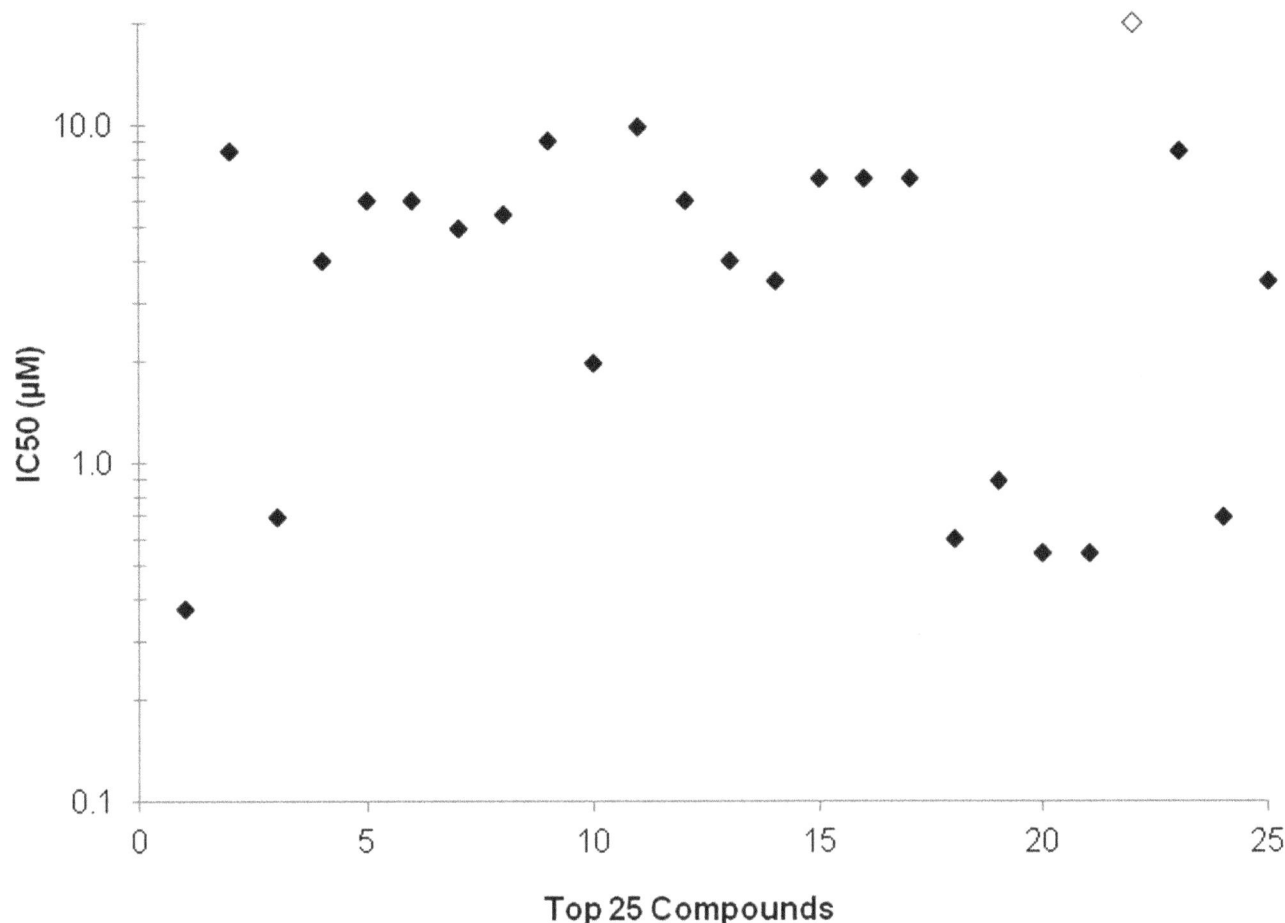

Figure 3. Plot of IC$_{50}$ results. IC$_{50}$ values less than or equal to 20 μM of 25 hits from FDA approved drug library screen. Primaquine (open diamond) demonstrated an IC$_{50}$ value equal to 20 μM.

exflagellation to overall fluorescent signal (Figure 1D). Drugs inhibiting exflagellation but not killing the parasites would result in low to intermediate inhibition in this assay (red highlighted value, Figure 1E), with anything greater than 20% inhibition indicative of some gametocyte killing (green highlighted value, Figure 1E). Making blood films of positive hits can further differentiate whether parasites are being killed or damaged or whether exflagellation inhibition is occurring. For our assay, we set a cutoff value of greater than 70% inhibition (equal to or better than clotrimazole) for the FDA drug library screen and greater than 50% inhibition (using pyrvinium pamoate as a positive control) for the MMV box screen to capture gametocytocidal compounds. The final assay setup for drug screening is briefly illustrated in Figure 1E.

Assay validation and FDA-approved drug library screen

The Johns Hopkins University Clinical Compound Library version 1.3 of FDA-approved drugs was screened using the assay described above to identify compounds that had gametocytocidal activity, confirmed with Giemsa stained smears of drug treated cultures for the top hits. During the initial screening, clotrimazole was identified as a moderately active gametocytocidal compound, showing 70% inhibition at 20 μM and was then used as a lower cutoff control for identification of screening hits, in order to screen for compounds that were gametocytocidal and did not only inhibit exflagellation. Uninfected erythrocytes were used as baseline for

the initial screening. The FDA approved drug library was screened at 20 μM and we initially selected the top 70 compounds showing more than 50% inhibition for evaluation using the gold standard of microscopic examination and IC$_{50}$ determination at multiple concentrations (Figure 2, Table S1). As expected most hits showing more than 70% inhibition during initial screening were confirmed to be gametocytocidal by microscopic examination and they showed a clear dose dependent response. Overall we identified 25 compounds with IC$_{50}$ values less than 20 μM, with most less than 10 μM (Table 1, Figure 3). Most of the compounds with intermediate activity were determined to inhibit exflagellation (data not shown) but were not gametocytocidal as indicated by Giemsa smears of drug-treated gametocytes. The mean Z-factor calculated from the FDA approved drug library screen was 0.52 (SEM = 0.07, Table S2).

Major drug classes with gametocytocidal activity

As a result of the FDA drug library screen, several drug classes were identified that showed activity against gametocytes, including a known antimalarial, primaquine, as well as other classes including antiseptics, antineoplastics, antihelminthics, antivirals, antiprotozoals, antidepressants, and pharmaceutical aids (Figure 4, complete list of indications is in Table S3). Eight of the twenty-five positive hits were identified as a single class of drugs, quaternary ammonium compounds (QACs) which were classified as antiseptics. Pyrvinium pamoate, an anthelminthic, demonstrat-

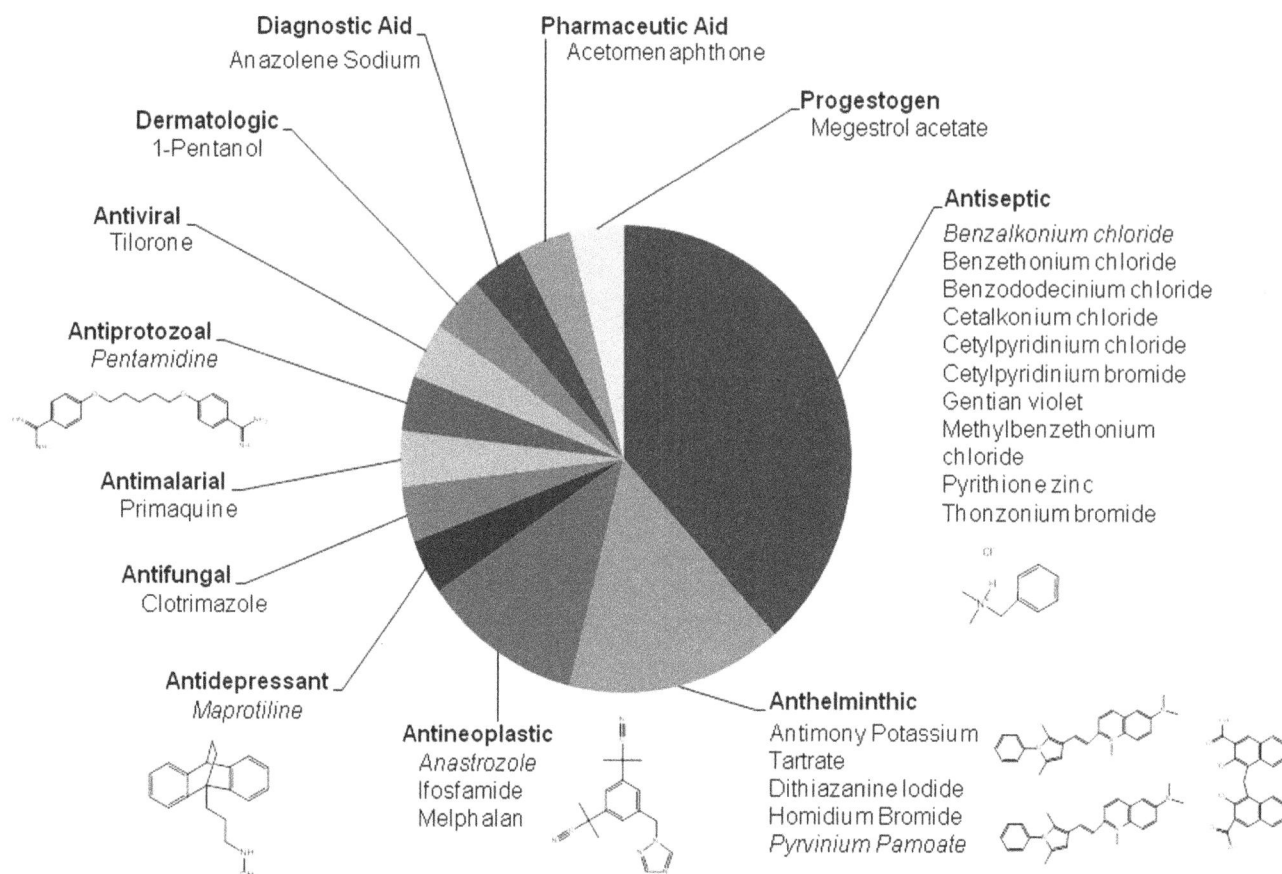

Figure 4. Drug class representation of active molecules. Classes of drug indications identified with activity against gametocytes with an $IC_{50} \leq$ 20 μM. Structures shown correspond to italicized compounds.

ed 100% inhibition at 10 μM and was used for further assays as a positive control for 100% inhibition or 'killed' parasite control in the presence of the background suppressor. By using a positive control of killed parasites in conjunction with the background suppressor rather than uninfected red blood cells, we were able to prevent artificially high inhibition values and screen for live gametocytes, not total gametocytes.

Validation of gametocytocidal compounds with membrane feeding assay

In order to validate the transmission blocking activity of compounds exhibiting the most potent gametocytocidal activity, mosquito infections through feeding on treated and untreated gametocyte cultures were performed. Gametocyte cultures were treated with methylene blue, a known gametocytocidal compound, clotrimazole, pyrvinium pamoate, and one of the quaternary ammonium compounds cetalkonium chloride for 48 hrs prior to ingestion by mosquitoes through a membrane feeder, and mosquito infections were determined 7 days later as a measure of oocyst stage parasite on the mosquito midgut tissue (Figure 5). All compounds demonstrated dose dependent transmission blocking activity, with pyrvinium pamoate showing the highest potency with 100% efficacy at 500 nM.

MMV malaria box screen

Medicines for Malaria Venture (MMV) has generously put together a 'malaria box' of four hundred compounds with proven

antimalarial activity against asexual blood stage parasites and made them freely available for use in the development of effective antimalarial compound screens, particularly those designed to identify liver stage and transmission blocking drugs. We screened these four hundred compounds using our gametocytocidal assay, this time using pyrvinium pamoate as a positive control due to increased efficacy compared to clotrimazole (Figure 6, Table S4). Our initial screen of the MMV box identified eighteen compounds with greater than 80% inhibition at 10 μM which we further screened to determine their IC_{50}s (Table 2). Seventeen of the compounds were confirmed as having greater than 50% inhibition at 10 μM and IC_{50}s less than 10 μM, with one compound MMV019918 showing a submicromolar IC_{50}. Of these seventeen compounds with gametocytocidal activity, seven were drug-like, while ten were probe-like, as described by MMV [13]. In addition, compounds with the greatest activity against gametocytes also showed nanomolar IC_{50}s against the asexual stage parasite as reported with the compound information by MMV. The mean Z-factor calculated from the MMV malaria box screen was 0.57 (SEM = 0.04).

Furthermore, we compared our MMV box hits with hits from four other assays using different reporters, including luciferase-expressing parasites, alamar blue, or confocal fluorescence microscopy [10,14–16]. We found that all of our eighteen hits overlapped between different assays (Figure 7, Table S5).

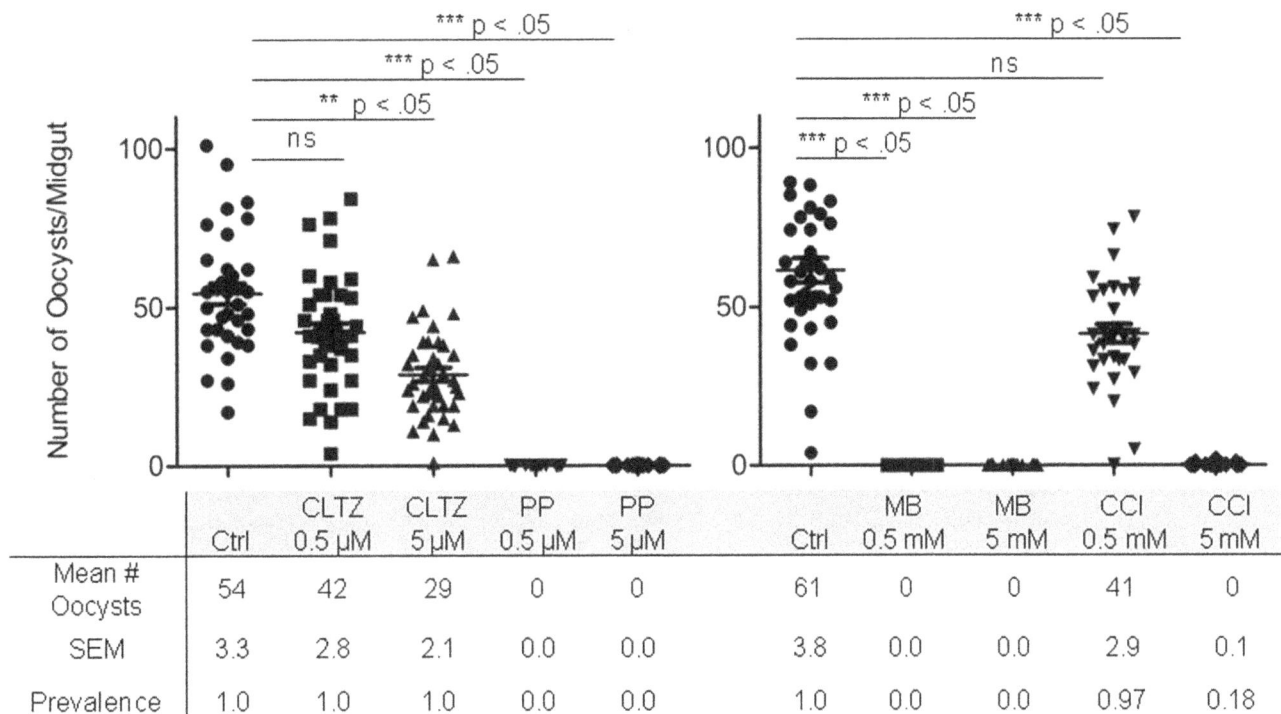

Figure 5. Inhibition of oocyst development of top compounds from JHU FDA-approved clinical compound library. Dose dependent transmission blocking activity of clotrimazole (CLTZ), pyrvinium pamoate (PP), methylene blue (MB) and cetalkonium chloride (CCl) measured by number of oocysts per mosquito midgut.

Discussion

To realize the goal of malaria elimination and eradication we need to add new and potent weapons active against multiple life stages of the parasite. Because most of the currently licensed antimalarials target only the asexual intra-erythrocytic stage, which is responsible for the pathology of disease, we urgently need to expand our antimalarial arsenal. Drugs which can effectively kill sexual gametocyte stages, responsible for transmission to the mosquito vector, will be required for malaria elimination. In order to find new tools we have established a simple and robust HTS gametocytocidal assay based on DNA content of live gametocytes. Because gametocytes do not multiply we have utilized male gametocyte exflagellation and a background suppressor to subtract the DNA fluorescence signals from dead cells to achieve a robust signal to noise ratio. As emphasized earlier in our description of assay optimization, we carefully took into consideration the contribution of exflagellation to fluorescent signal, and set cutoff values for our assay which allowed us to screen for compounds with gametocytocidal activity and not merely exflagellation inhibition. However, it should be noted that our assay does not allow us to distinguish between male and female gametocyte killing, but instead looks at overall live gametocytes, and at lower inhibition levels, male gametocyte viability. Because sex ratio tends to be biased in all *Plasmodium* species [17], we start out with two to four times as many females as males, and quantification is not skewed to a male exflagellation assay. Linearity of the assay was determined as a function of percent gametocytes at 2% HCT, which showed a linear relationship with an R2 value of ≥0.95. While many gametocytocidal assays have been developed, many of these assays have features that make them difficult to adapt to high-throughput screening such as multiple incubations steps, requirement for high gametocytemia, or transgenic parasites,

making it impossible to use field isolates without further genetic manipulation. Our assay is simple enough to be used in any laboratory with access to malaria culture and a fluorescence plate reader, while also maintaining the sensitivity and robustness required for a high-throughput screening assay. We utilized our assay to screen an FDA-approved drug library of 1500 compounds as well as the MMVs Malaria box of 400 compounds to identify new pharmacophores with gametocytocidal activity.

Screening of the FDA approved drug library at 20 μM led to the identification of several classes of compounds with gametocytocidal activity. Most of the hits from FDA approved library were antiseptic, anthelminthic, and antineoplastic compounds as well as some antimicrobials and an antidepressant drug. Clotrimazole, an antifungal, was identified as having 70% inhibition against gametocytes with an asexual IC_{50} of 1.3 μM, and was recently reported as a hit in another gametocytocidal screen [11]. Pyrithione zinc is an antiseptic which showed activity in our assay with high efficacy against both sexual and asexual stages of the parasite and was also recently reported in the screening of a different library for gametocytocidal drugs [10]. As expected our screen identified primaquine as a gametocytocidal compound, albeit at a higher than reported IC_{50} due to lack of metabolism to the highly effective phenolic metabolites of primaquine required for inhibition [18]. The antineoplastic compounds, anastrozole, ifosfamide, and melphalan, demonstrated greater than 50% inhibition at 0.55 to 4 μM concentrations against gametocytes (Table 1). The data available for melphalan showed 20% inhibition of 3D7 at 10 μM and an IC_{50} of 20 μM compared to an IC_{50} of 4 μM against *P. falciparum* gametocytes, suggesting that melphalan shows slightly less efficacy against asexual compared to sexual parasites. Of the anthelminthics, homidium bromide and pyrvinium pamoate demonstrated the highest efficacy against gametocytes, with 100% inhibition at 20 μM

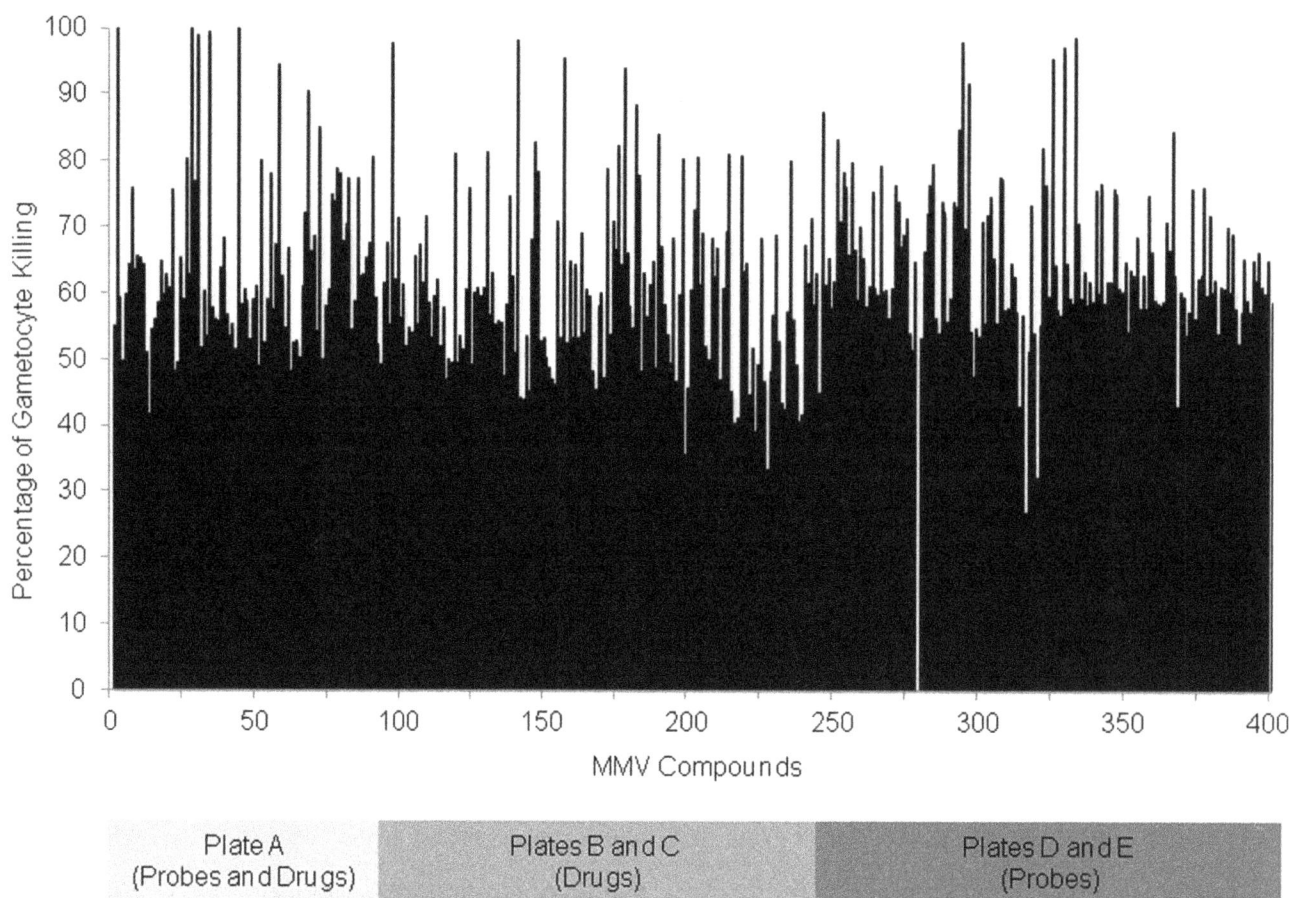

Figure 6. Inhibition by MMV Malaria Box. SYBR Green I assay results for the MMV box screened at 10 µM. Plot of percentage of gametocytocidal activity of 400 compounds compared to pyrvinium pamoate control.

and IC$_{50}$ values of 0.38 µM and 4 µM respectively, while also effectively inhibiting 70–100% of asexual stages at 10 µM. Homidium bromide (ethidium bromide) is a well-known fluorescent DNA-intercalating agent used in molecular biology and is known to be mutagenic, whereas pyrvinium pamoate is an FDA-approved anthelminthic compound used to treat pinworm, with activity against *Cryptosporidium parvum*, and thought to inhibit mitochondrial NADH-fumarate reductase [19–21]. A recent study demonstrates nanomolar inhibition of pyrvinium pamoate against both 3D7 and K1 strains of *P. falciparum* asexual blood stage parasites with further derivatization studies suggesting the quaternary amino group in the quinoline ring is not required for antimalarial activity [22]. Removing the positive charge from the molecule may allow better bioavailability of pyrvinium pamoate, and further investigation of gametocytocidal activity of uncharged derivatives is warranted. The other anthelminthics antimony potassium tartrate and dithiazanine iodide inhibited 90% of gametocytes at 20 µM and 90% of asexual stages at 10 µM. Dithiazanine iodide has some structural similarity to pyrvinium pamoate and also possesses a quaternary amine, which raises the question of whether a positive charge is critical for gametocytocidal activity. Interestingly, maprotiline, a tetracyclic antidepressant similar to the tricyclic antidepressant methylene blue, demonstrated nanomolar inhibition of both gametocyte and asexual stages of *P. falciparum*, but showed greater efficacy against gametocytes. Methylene blue has reported efficacy against gametocytes *in vitro* and also showed *in vivo* efficacy against

asexual parasites in multiple murine models of cerebral malaria, protecting 75% of mice at 10 mg/kg for five days post-infection [9,23–26]. Our observations suggest further exploration of tetracyclic and tricyclic antidepressants for gametocytocidal activity.

The antiseptic QACs were the most highly represented class of drugs in the hits from the FDA approved library screen, comprising eight out of twenty five hits. Most of the QACs identified in the screen, including cetalkoniumchloride, thonzonium bromide, and benzododecinium chloride, demonstrated almost 100% efficacy against gametocytes at 20 µM with low micromolar IC$_{50}$s. QACs with antimicrobial activities were identified as early as the 1930s and are among the most useful antiseptics and disinfectants, and have been used for a variety of clinical purposes [27–31]. These drugs can function as choline analogs and can inhibit the de novo phosphatidylcholine biosynthetic pathway of the malaria parasite. QACs have previously been shown to inhibit asexual blood stages of *P. falciparum* at nanomolar concentrations, with greater activity seen with long alkyl side chains and increased steric hindrance around the nitrogen atom [32]. Phosphatidylcholine, the predominant phospholipid produced by malaria parasites, plays essential structural and regulatory roles in parasite development and differentiation. Previous studies in *P. falciparum* have demonstrated the presence of two pathways for phosphatidylcholine biosynthesis, the cytidine diphosphate (CDP)-choline pathway, which uses host choline and fatty acids as precursors, and the

Table 2. Gametocytocidal compounds identified from MMV Malaria Box with greater than 50% inhibition at 10 μM and available corresponding data on asexual stage inhibition from MMV.

MMV #	Gametocyte			Asexual	
	% inh 10 μM	IC$_{50}$ (SD, μM)	% inh 5 μM	EC$_{50}$ (nM)	CHEMBL EC$_{50}$ (μM)
MMV665941	122	1.8 (0.2)	96	255	0.62
MMV000448	110	5.4 (1.4)	95	235	0.03, 1.04, 0.53
MMV006172	104	2.6 (0.3)	97	142	0.057, 0.64
MMV396797	100	8.8 (1.2)	-	477	NA
MMV665878	100	1.1 (0.6)	99	139	0.27
MMV667491	99	4.5 (1.5)	-	1230	NA
MMV665830	98	3.3 (0.6)	98	1005	0.25
MMV019780	98	3.8 (0.7)	98	697	0.84
MMV019555	97	3.4 (0.5)	100	376	0.20
MMV019881	96	5.5 (0.9)	98	646	1.04
MMV019918	92	0.9 (0.3)	96	801	1.51
MMV019690	90	>10	97	935	0.78
MMV000445	86	10.00	98	1135	1.97
MMV007591	85	5.4 (1.0)	85	ND	1.12
MMV000848	85	3.6 (3.2)	97	660	1.08
MMV020505	83	6.6 (1.1)	96	876	0.80
MMV006303	82	2.1 (0.6)	-	391	0.03
MMV396794	82	8.2 (0.9)	NA	1160	NA

serine decarboxylase-phosphoethanolamine methyltransferase (SDPM) pathway, which uses host serine and fatty acids as precursors. Recent studies have shown that QACs inhibit multiple steps during phospholipid biosynthesis by targeting the choline carrier as well as enzymes of both the SDPM and the CDP–choline pathways [33,34]. A recently published study demonstrates the essentiality of phosphotidylcholine synthesis for gametocyte development and transmission by knocking out or inhibiting the key enzyme in this pathway, phosphoethanolamine methyl transferase, which results in inhibition of gametocyte maturation and also blocks transmission [35]. These observations strongly suggest a critical role for phospholipid metabolism during *P. falciparum* gametocyte stages and may present a unique target for multistage drug development. While challenges with poor absorption have been associated with this group of compounds due to a net positive charge, improvements using a prodrug approach have shown promise [32]. A choline analog, Albitiazolium is already in clinical trials for complicated malaria using intra-peritoneal or intra-muscular routes, and efforts are underway to develop this compound for uncomplicated malaria using an oral route [36]. Thus we have not only identified a class of compounds with efficacy against both asexual and sexual stages but also a shared target which can be utilized to identify new pharmacophores active against both asexual and transmission stages of malaria parasites.

In regards to cytotoxicity, route of drug administration and approved drug levels for the aforementioned hits, many of the compounds identified, including the QACs, are topical agents which are not approved for oral drug use. Anthelminthic compounds such as pyrvinium pamoate and dithiazanine iodide are approved for oral administration, but are not absorbed to appreciable levels by the GI tract and thus are not available in the bloodstream. Antineoplastics such as melphalan can be given orally or intravenously, but perhaps not surprisingly have side effects including bone marrow suppression. Maprotiline, however, is an orally administered antidepressant with an LD$_{50}$ of 90 mg/kg in women, according to DrugBank, and approved prescription of 75–150 mg daily, depending on the severity of depression [37,38]. While many of these FDA approved drug hits may not be immediately available or appropriate for oral antimalarial chemotherapy, they do provide novel pharmacophores with gametocytocidal and/or asexual activity, and are suggestive of new drug targets.

The successful screening and hit identification from FDA approved library led us to request the 400 compound Malaria Box of asexual blood stage active compounds from MMV. We screened the Malaria Box at 10 μM in duplicate, this time using 10 μM pyrvinium pamoate as a positive control (100% inhibition) and 0.1% DMSO as a vehicle control. As compared to the FDA approved library, we observed a higher number of compounds showing inhibition, which was expected as all these compounds have potent activity against the asexual blood stages. In all we obtained 18 hits, 17 of which showed a dose dependent response against mature gametocytes. The majority of the active compounds were very similar in structure, with seven containing acridine-like structures, three fused benzene rings with a central nitrogen, with varied side chains, one similar to that of chloroquine (MMV665830). Quinacrine and pyronaridine are both acridine-based compounds which have been proven clinically effective against malaria [39]. Multiple mechanisms of action have been proposed and proven for the various acridine-like compounds, including inhibition of hemozoin crystallization [40–42], mitochondrial bc_1 complex [43,44], DNA Topoisomerase II [45,46], and also DNA intercalation, though the latter has not been correlated with increased antimalarial activity [39]. Of note, pyronaridine and other Topo II inhibitors have been shown to

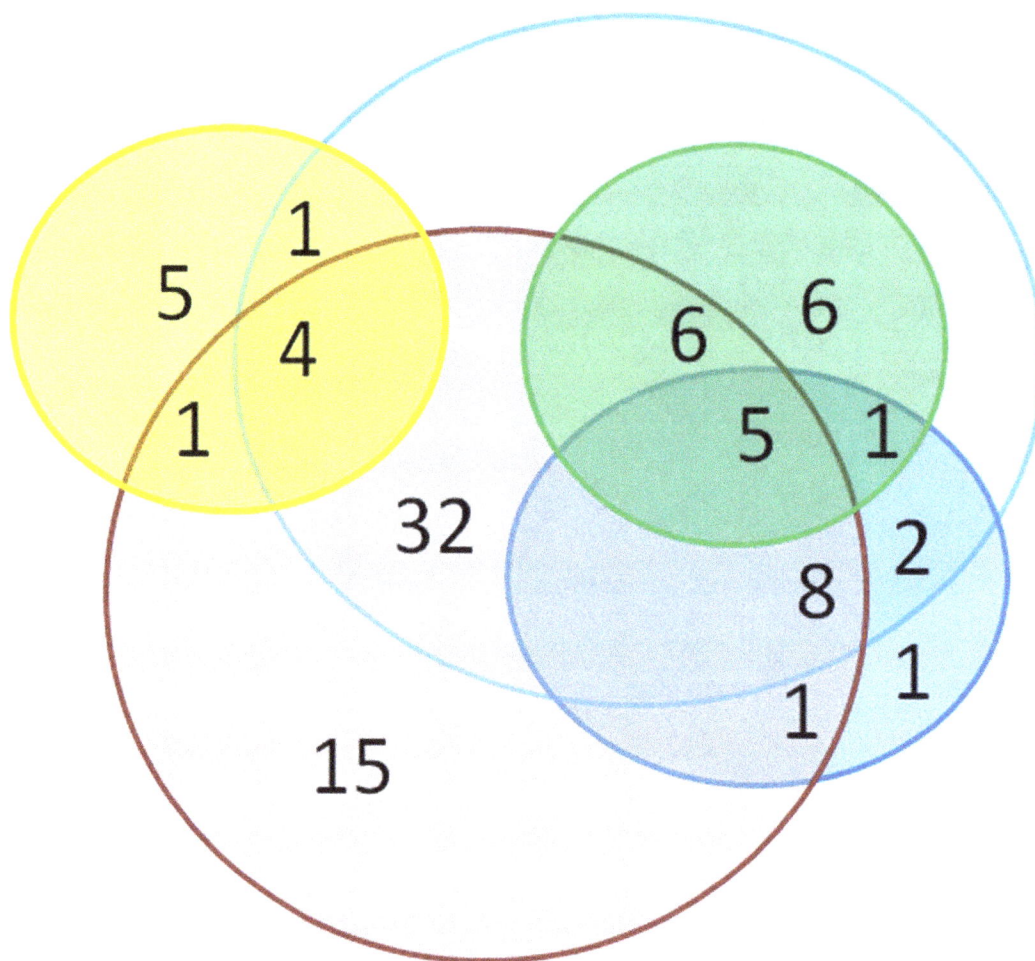

Figure 7. Overlap of recent screening assays for Malaria Box. Venn Diagram comparing our SYBR Green I assay (green) MMV box hits with hits from four other assays: Confocal fluorescence microscopy (red), Alamar blue early (dark blue) and late (light blue) and Luciferase (yellow) [10,14–16].

inhibit both asexual and sexual stages of *P. falciparum* in a previous study, suggesting that Topoisomerase II inhibitors may be utilized to target multiple parasite stages including gametocytes [45]. Towards the end of our library screening and data analysis, four manuscripts describing results of gametocytocidal screening of the MMV malaria box were published. Comparing our MMV hits with these four recent assays, we found that all of our hits overlapped with either the early or late alamar blue or confocal microscopy assays or both, but no hits were shared with the early gametocyte luciferase-based assay (Figure 7, Table S5) [10,14–16]. MMV019918 was a top hit identified by four assays, including our SYBR Green I, the alamar blue and confocal microscopy assays, with nanomolar inhibition against late and early stages (IC$_{50}$s ranging from 320–890 nM depending on the assay). Four other compounds including MMV000448, MMV006172, MMV007591 and MMV019555 were also identified by all four assays.

Conclusions

We have successfully produced and validated a gametocytocidal drug screening assay that will be easily adaptable to high-throughput format using SYBR Green I and a background suppressor to read DNA content of live gametocytes after exposure to drug. Using this assay we screened an FDA-approved drug

library and the MMV Malaria box, totaling approximately 2000 compounds and identified two highly represented classes of compounds, QACs and acridine-like compounds, which were effective against both sexual and asexual stages of the parasite. Further target validation is required to ascertain the mechanism of action of these compounds in gametocytes.

Supporting Information

Table S1   Top 70 Compounds FDA library screen IC50 data and Giemsa stained blood films.

Table S2   Controls and Z-factor analysis.

Table S3   Complete list of indications of all compounds in the Johns Hopkins Clinical Compound Library version 1.3.

Table S4   MMV Malaria Box Screen Data and Analysis.

Table S5 Compilation of top MMV malaria box hits from four gametocyte assays with three different reporters compared to the SYBR Green assay hits.

Table S6 FDA drug library SYBR Green I fluorescence data and analysis.

Table S7 Membrane feeding assay data.

Acknowledgments

We thank the Johns Hopkins Malaria Research Institute insectary and parasitology core facilities for assistance with parasite cultures and mosquito rearing. We also thank the MMV for the gift of the malaria box compounds. Publication of this article was funded in part by the Open Access Promotion Fund of the Johns Hopkins University Libraries.

Author Contributions

Conceived and designed the experiments: NGS DJS GM GD AKT. Performed the experiments: NGS GM AKT. Analyzed the data: NGS DJS GM GD AKT. Contributed to the writing of the manuscript: NGS DJS GM GD AKT.

References

1. Murray CJL, Rosenfeld LC, Lim SS, Andrews KG, Foreman KJ, et al. (2012) Global malaria mortality between 1980 and 2010: a systematic analysis. Lancet 379: 413–431.
2. License A, Malaria G, Program E (2011) A research agenda for malaria eradication: vector control. PLoS Med 8: e1000401.
3. Mendis K, Rietveld A, Warsame M, Bosman A, Greenwood B, et al. (2009) From malaria control to eradication: The WHO perspective. Trop Med Int Health 14: 802–809.
4. Birkholtz L-M, Bornman R, Focke W, Mutero C, de Jager C (2012) Sustainable malaria control: transdisciplinary approaches for translational applications. Malar J 11: 431.
5. Peatey CL, Spicer TP, Hodder PS, Trenholme KR, Gardiner DL (2011) A high-throughput assay for the identification of drugs against late-stage *Plasmodium falciparum* gametocytes. Mol Biochem Parasitol 180: 127–131.
6. D'Alessandro S, Silvestrini F, Dechering K, Corbett Y, Parapini S, et al. (2013) A *Plasmodium falciparum* screening assay for anti-gametocyte drugs based on parasite lactate dehydrogenase detection. J Antimicrob Chemother 68 (9): 2048–2058.
7. Tanaka TQ, Williamson KC (2011) A malaria gametocytocidal assay using oxidoreduction indicator, alamarBlue. Mol Biochem Parasitol 177: 160–163.
8. Lelièvre J, Almela MJ, Lozano S, Miguel C, Franco V, et al. (2012) Activity of clinically relevant antimalarial drugs on *Plasmodium falciparum* mature gametocytes in an ATP bioluminescence "transmission blocking" assay. PLoS One 7: e35019.
9. Adjalley SH, Johnston GL, Li T, Eastman RT, Ekland EH, et al. (2011) Quantitative assessment of *Plasmodium falciparum* sexual development reveals potent transmission-blocking activity by methylene blue. Proc Natl Acad Sci U S A 108: E1214–23.
10. Sun W, Tanaka TQ, Magle CT, Huang W, Southall N, et al. (2014) Chemical signatures and new drug targets for gametocytocidal drug development. Sci Rep 4: 3743.
11. Tanaka TQ, Dehdashti SJ, Nguyen D-T, McKew JC, Zheng W, et al. (2013) A quantitative high throughput assay for identifying gametocytocidal compounds. Mol Biochem Parasitol 188: 20–25.
12. Zhang J-H, Chung TDY, Oldenburg KR (1999) A Simple Statistical Parameter for Use in Evaluation and Validation of High Throughput Screening Assays. J Biomol Screen 4: 67–73.
13. Spangenberg T, Burrows JN, Kowalczyk P, McDonald S, Wells TNC, et al. (2013) The open access malaria box: a drug discovery catalyst for neglected diseases. PLoS One 8: e62906.
14. Duffy S, Avery VM (2013) Identification of inhibitors of *Plasmodium falciparum* gametocyte development. Malar J 12: 408.
15. Bowman JD, Merino EF, Brooks CF, Striepen B, Carlier PR, et al. (2013) Anti-apicoplast and gametocytocidal screening to identify the mechanisms of action of compounds within the Malaria Box. Antimicrob Agents Chemother 58: 811–819.
16. Lucantoni L, Duffy S, Adjalley SH, Fidock DA, Avery VM (2013) Identification of MMV Malaria Box Inhibitors of *Plasmodium falciparum* Early-Stage Gametocytes Using a Luciferase-Based High-Throughput Assay. Antimicrob Agents Chemother 57: 6050–6062.
17. Baker DA (2010) Malaria gametocytogenesis. Mol Biochem Parasitol 172: 57–65.
18. Pybus BS, Marcsisin SR, Jin X, Deye G, Sousa JC, et al. (2013) The metabolism of primaquine to its active metabolite is dependent on CYP 2D6. Malar J 12: 1.
19. Downey AS, Chong CR, Graczyk TK, Sullivan DJ (2008) Efficacy of pyrvinium pamoate against *Cryptosporidium parvum* infection in vitro and in a neonatal mouse model. Antimicrob Agents Chemother 52: 3106–3112.
20. Chong CR, Chen X, Shi L, Liu JO, Sullivan DJ (2006) A clinical drug library screen identifies astemizole as an antimalarial agent. Nat Chem Biol 2: 415–416.
21. Tomitsuka E, Kita K, Esumi H (2012) An anticancer agent, pyrvinium pamoate inhibits the NADH-fumarate reductase system–a unique mitochondrial energy metabolism in tumour microenvironments. J Biochem 152: 171–183.
22. Teguh SC, Klonis N, Duffy S, Lucantoni L, Avery VM, et al. (2013) Novel conjugated quinoline-indoles compromise *Plasmodium falciparum* mitochondri-

al function and show promising antimalarial activity. J Med Chem 56: 6200–6215.
23. Coulibaly B, Zoungrana A, Mockenhaupt FP, Schirmer RH, Klose C, et al. (2009) Strong gametocytocidal effect of methylene blue-based combination therapy against falciparum malaria: a randomised controlled trial. PLoS One 4: e5318.
24. Pascual A, Henry M, Briolant S, Charras S, Baret E, et al. (2011) In vitro activity of Proveblue (methylene blue) on *Plasmodium falciparum* strains resistant to standard antimalarial drugs. Antimicrob Agents Chemother 55: 2472–2474.
25. Dormoi J, Briolant S, Desgrouas C, Pradines B (2013) Impact of methylene blue and atorvastatin combination therapy on the apparition of cerebral malaria in a murine model. Malar J 12: 127.
26. Dormoi J, Briolant S, Desgrouas C, Pradines B (2013) Efficacy of proveblue (methylene blue) in an experimental cerebral malaria murine model. Antimicrob Agents Chemother 57: 3412–3414.
27. D'Arcy PF, Taylor EP (1962) Quaternary ammonium compounds in medicinal chemistry I. J Pharm Pharmacol 14: 129–146.
28. D'Arcy PF, Taylor EP (1962) Quaternary ammonium compounds in medicinal chemistry II. J Pharm Pharmacol 14: 193–216.
29. Bhattacharya BK, Sen AB (1965) Chemotherapeutic Properties of Some New Quaternary Ammonium Compounds; Their Cesticidal Action Against Hymenolepis Nana. Br J Pharmacol Chemother 24: 240–244.
30. Krawczyk J, Keane N, Swords R, O'Dwyer M, Freeman CL, et al. (2013) Perifosine–a new option in treatment of acute myeloid leukemia? Expert Opin Investig Drugs 22: 1315–1327.
31. Dorlo TPC, Balasegaram M, Beijnen JH, de Vries PJ (2012) Miltefosine: a review of its pharmacology and therapeutic efficacy in the treatment of leishmaniasis. J Antimicrob Chemother 67: 2576–2597.
32. Peyrottes S, Caldarelli S, Wein S, Périgaud C, Pellet A, et al. (2012) Choline Analogues in Malaria Chemotherapy. Curr Pharm Des 18: 3454–3466.
33. Tischer M, Pradel G, Ohlsen K, Holzgrabe U (2012) Quaternary ammonium salts and their antimicrobial potential: targets or nonspecific interactions? Chem Med Chem 7: 22–31.
34. Calas M, Cordina G, Bompart J, Ben Bari M, Jei T, et al. (1997) Antimalarial activity of molecules interfering with *Plasmodium falciparum* phospholipid metabolism. Structure-activity relationship analysis. J Med Chem 40: 3557–3566.
35. Bobenchik A (2013) *Plasmodium falciparum* phosphoethanolamine methyltransferase is essential for malaria transmission. Proc Natl Acad Sci 110: 18262–18267.
36. Wein S, Maynadier M, Bordat Y, Perez J, Maheshwari S, et al. (2012) Transport and pharmacodynamics of albitiazolium, an antimalarial drug candidate. Br J Pharmacol 166: 2263–2276.
37. Drugs.com Maprotiline Dosage (n.d.). Available: http://www.drugs.com/dosage/maprotiline.html. Accessed 2014 Jul15.
38. Drug Bank: Maprotiline (n.d.). Available: http://www.drugbank.ca/drugs/DB00934. Accessed 2014 Jul 15.
39. Valdés AF (2011) Acridine and Acridinones: Old and New Structures with Antimalarial Activity. Open Med Chem J 5: 11–20.
40. Yu X-M, Ramiandrasoa F, Guetzoyan L, Pradines B, Quintino E, et al. (2012) Synthesis and biological evaluation of acridine derivatives as antimalarial agents. Chem Med Chem 7: 587–605.
41. Fernández-Calienes A, Pellón R, Docampo M, Fascio M, D'Accorso N, et al. (2011) Antimalarial activity of new acridinone derivatives. Biomed Pharmacother 65: 210–214.
42. Guetzoyan L, Yu X-M, Ramiandrasoa F, Pethe S, Rogier C, et al. (2009) Antimalarial acridines: synthesis, in vitro activity against *P. falciparum* and interaction with hematin. Bioorg Med Chem 17: 8032–8039.
43. Biagini GA, Fisher N, Berry N, Stocks PA, Meunier B, et al. (2008) Acridinediones: Selective and Potent Inhibitors of the Malaria Parasite Mitochondrial bc 1 Complex. Molec Pharm 73: 1347–1355. doi:10.1124/mol.108.045120.1991.
44. Barton V, Fisher N, Biagini GA, Ward SA, O'Neill PM (2010) Inhibiting *Plasmodium* cytochrome bc1: a complex issue. Curr Opin Chem Biol 14: 440–446.

45. Chavalitshewinkoon-Petmitr P, Pongvilairat G (2000) Gametocytocidal activity of pyronaridine and DNA topoisomerase II inhibitors against multidrug-resistant *Plasmodium falciparum in vitro*. Parasitol Int. 48(4): 275–80.

46. Auparakkitanon S, Wilairat P (2000) Cleavage of DNA induced by 9-anilinoacridine inhibitors of topoisomerase II in the malaria parasite *Plasmodium falciparum*. Biochem Biophys Res Commun 269: 406–409.

47. Collaborative Drug Discovery (n.d.). Available: https://www.collaborativedrug.com/. Accessed 2013 Aug 5.

48. Eastman RT, Pattaradilokrat S, Raj DK, Dixit S, Deng B, et al. (2013) A Class of Tricyclic Compounds Blocking Malaria Parasite Oocyst. Antimicrob Agents Chemother 57 (1): 425–535. doi:10.1128/AAC.00920-12.

3

Expression Levels of *pvcrt-o* and *pvmdr-1* Are Associated with Chloroquine Resistance and Severe *Plasmodium vivax* Malaria in Patients of the Brazilian Amazon

Gisely C. Melo[1], Wuelton M. Monteiro[1,2], André M. Siqueira[1], Siuhelem R. Silva[3], Belisa M. L. Magalhães[1], Aline C. C. Alencar[1], Andrea Kuehn[2,4], Hernando A. del Portillo[4,5], Carmen Fernandez-Becerra[4]*, Marcus V. G. Lacerda[1,2]*

1 Universidade do Estado do Amazonas, Manaus, Amazonas, Brazil, 2 Fundação de Medicina Tropical Dr. Heitor Vieira Dourado, Manaus, Amazonas, Brazil, 3 Universidade Paulista UNIP, Manaus, Amazonas, Brazil, 4 Barcelona Centre for International Health Research (CRESIB, Hospital Clínic-Universitat de Barcelona), Barcelona, Spain, 5 Institució Catalana de Recerca i Estudis Avançats (ICREA), Barcelona, Spain

Abstract

Molecular markers associated with the increase of chloroquine resistance and disease severity in *Plasmodium vivax* are needed. The objective of this study was to evaluate the expression levels of *pvcrt-o* and *pvmdr-1* genes in a group of patients presenting CQRPv and patients who developed severe complications triggered exclusively by *P. vivax* infection. Two different sets of patients were included to this comprehensive study performed in the Brazilian Amazon: 1) patients with clinically characterized chloroquine-resistant *P. vivax* compared with patients with susceptible parasites from *in vivo* studies and 2) patients with severe vivax malaria compared with patients without severity. Quantitative real-time PCR was performed to compare the transcript levels of two main transporters genes, *P. vivax* chloroquine resistance transporter (*pvcrt-o*) and the *P. vivax* multidrug resistance transporter (*pvmdr-1*). Twelve chloroquine resistant cases and other 15 isolates from susceptible cases were included in the first set of patients. For the second set, seven patients with *P. vivax*-attributed severe and 10 mild manifestations were included. Parasites from patients with chloroquine resistance presented up to 6.1 (95% CI: 3.8–14.3) and 2.4 (95% CI: 0.53–9.1) fold increase in *pvcrt-o* and *pvmdr-1* expression levels, respectively, compared to the susceptible group. Parasites from the severe vivax group had a 2.9 (95% CI: 1.1–8.3) and 4.9 (95% CI: 2.3–18.8) fold increase in *pvcrt-o* and *pvmdr-1* expression levels as compared to the control group with mild disease. These findings suggest that chloroquine resistance and clinical severity in *P. vivax* infections are strongly associated with increased expression levels of the *pvcrt-o* and *pvmdr-1* genes likely involved in chloroquine resistance.

Editor: Takafumi Tsuboi, Ehime University, Japan

Funding: This study was supported by Conselho Nacional de Desenvolvimento Científico e Tecnológico (CNPq) (grant number 575788/2008–9). CFB received a visiting fellowship from the Strategic Program for Science, Technology & Innovation of FAPEAM (PECTI-SAÚDE). MVGL has a Level 1 Productivity Fellowship from the National Council for Scientific and Technological Development (CNPq). The funders had no role in study design, data collection and analysis, decision to publish, or preparation of the manuscript.

Competing Interests: The authors have declared that no competing interests exist.

* Email: marcuslacerda.br@gmail.com (MVGL); carmen.fernandez@cresib.cat (CFB)

Introduction

Mechanisms underlying severe malaria triggered by *Plasmodium vivax* have been poorly appreciated, mainly the likely higher and more-severe disease burden imposed by increasingly chloroquine-resistant *P. vivax* parasites (CQRPv) [1,2]. Most common manifestations of severe vivax malaria include severe anemia and respiratory distress, being particularly associated to young age [2–4]. Of interest, the majority of reports on severe vivax malaria come from regions where drug-resistant *P. vivax* parasites significantly thread the radical cure and control of this infection [5]. The first reports of CQRPv were made in 1989 [6,7] from Papua New Guinea and Indonesia. In Brazil, the first case of properly ascertained *in vivo* CQRPv was from a patient treated in Manaus, in the Brazilian Amazon [8]. In this city, a subsequent trial assessed the efficacy of standard supervised CQ monotherapy

and the proportion of failures was 10.1% [9]. In parallel to the emergence of CQRPv in the Brazilian Amazon, reports of clinical severity exclusively associated with *P. vivax* infection increased substantially [3,10–12].

Based on observational studies, an association between CQRPv and disease severity in *P. vivax* was proposed in the Indonesian Papua [2], although this has not been properly and systematically evaluated. CQR in *P. falciparum* is associated with mutations in the *pfcrt-o* and *pfmdr-1* genes [13,14] but similar studies with the orthologous genes of *P. vivax*, *pvcrt-o* and *pvmdr-1*, failed to demonstrate a similar association [15–17]. Interestingly, a pioneer case of severe disease exclusively associated with *P. vivax*, demonstrated a significant increase on expression levels, not mutations, of these genes [18]. Altogether, these studies suggest the possibility of a significant association of CQRPv and disease severity and that expression levels of *pvcrt-o* and *pvmdr-1* should be better explored as molecular markers of both phenomena to further enhance the understanding of the clinico-epidemiological. The aim of this study was to evaluate the expression levels of *pvcrt-o* and *pvmdr-1* genes in a well-characterized group of patients presenting CQRPv and in a group of patients who developed severe complications triggered by *P. vivax* infection.

Methods

Ethics statement

The study was approved by the Ethics Review Board of the *Fundação de Medicina Tropical Heitor Vieira Dourado* (FMT-HVD) (approval number 343/2009). Participants were instructed about the objectives of the study and signed an informed consent. In the case of under 18 years, the consent form was signed by the guardians. Patients diagnosed with malaria were treated according to the guidelines of the Brazilian Ministry of Health [19].

Study area

The study was carried out from June 2011 to December 2012 in the FMT-HVD, an infectious disease referral center located in Manaus, Western Brazilian Amazon.

Selection of patients for the CQRPv study

The study included patients with *P. vivax* malaria of both sexes, aged 6 months-60 years, bodyweight greater than 5 kg, presenting blood parasite density from 250 to 100,000 parasites/ml and axillary temperature $\geq 37.5°C$ or history of fever in the last 48 hours. Exclusion criteria were: use of antimalarials in the previous 30 days, refusal to be followed up for 42 days and any clinical complication [19]. Patients received supervised treatment with 25 mg/kg of chloroquine (CQ) phosphate over a 3-day period (10 mg/kg on days 0 and 7.5 mg/kg on days 1 and 2). Primaquine was prescribed at the end of follow-up or if disease recurred for 7-day period in the dosage of 0.5 mg/kg per day. Patients who vomited the first dose within 30 minutes after drug ingestion were re-treated with the same dose. Patients were evaluated on days 0, 1, 2, 3, 7, 14, 28 and 42 and, if they felt ill, at any time during the follow-up period. In all sample, blood smear, full blood counts and PCR were performed. On admission (D0) and in the day of recrudescence (DR), 20 ml of intravenous whole blood was also collected for DNA and RNA storage. CQ and desethylchloroquine (DCQ) plasmatic levels were determined only in case of parasitological failure [20]. Three aliquots of 100 µl of whole blood from DR samples were spotted onto filter paper for later analysis by high performance liquid chromatography (HPLC) to estimate the levels of CQ and DCQ as previously described [21,22]. CQRPv was defined if: 1) recurrences of the parasite with plasma concentrations of CQ higher than 100 ng/dl, or 2) recurrences with smaller concentrations but presenting DR and D0 with the same alleles. The control group consisted of patients with no parasitemia recurrence during the same follow-up period.

Selection of patients with severe *P. vivax* malaria and patients with mild symptoms

During the same period, admitted patients classified as severe if presenting any of the WHO severity criteria [23], and patients with mild manifestations from the outpatient clinics. On admission, 20 ml of intravenous whole blood was collected for complete blood count, blood biochemistry and for DNA and RNA extraction. Other tests were requested at physician's discretion. To exclude patients with infection other than *Plasmodium*, all patients also had a blood culture performed for aerobic bacteria

Table 1. Demographic and clinical characteristics of the study participants admitted to a tertiary health center, Manaus, Amazonas, Brazil.

Code	Sex/Age	DR	Haemoglobin level in D0 (g/dL)	Haemoglobin level in DR (g/dL)	Chloroquine concentrations (ng/dL)	Asexual malarial parasites/µl in D0	Asexual malarial parasites/µl in DR	Day of RNA Isolation	Genotyping
R1	M/55	27	14	15.1	249	690.0	5695.0	D0/DR	Same
R2	F/9	32	10.2	11.6	61	4347.0	6871.0	D0/DR	Same
R3	M/23	34	13.4	12.6	68	297.0	3734.6	D0/DR	Same
R4	M/42	36	14.2	13.5	256	686.0	436.8	D0/DR	Same
R5	M/13	38	12.4	12.0	103	2582.1	2613.2	D0/DR	Same
R6	M/25	25	15.2	15.6	533	1974.7	5385.6	D0/DR	Same
R7	M/37	28	14.0	14.4	263	2548.0	50.0	D0/DR	Same
R8	M/26	32	13.5	12.9	188	78.0	11096.3	D0	Same
R9	M/9	31	9.4	10.0	230	1136.8	4968.0	D0	Same
R10	M/47	31	12.9	12.0	130	5059.2	4160.0	D0	Different
R11	M/36	34	14.7	14.5	118	2105.6	2853.6	DR	Different
R12	F/44	34	13.6	13.1	111	676.8	2613.2	DR	Same

Day admission (D0). Day of recrudescence (DR).

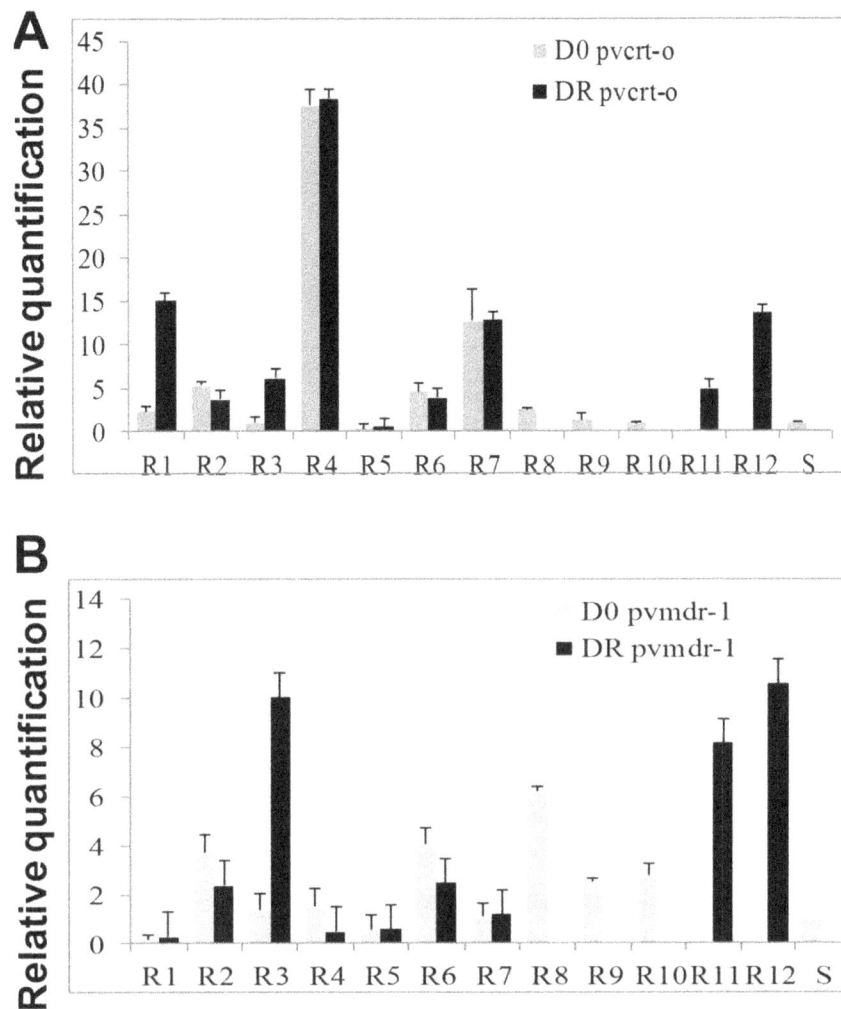

Figure 1. Expression levels of chloroquine resistance genes in patients with CQRPv parasites. Relative quantification of *pvcrt-o* (A) and *pvmdr1* (B) transcripts in total RNA obtained from parasites from patients with chloroquine-resistant *P. vivax* vs a pool of total RNA obtained from 5 patients with CQ-susceptible parasites. Chloroquine-resistant *P. vivax* parasites (R). Chloroquine-susceptible *P. vivax* parasites (S). Day of admission (D0). Day of recrudescence (DR). The error bars in Figures 1 and 3 reflect the average standard error of the Ct.

and serological tests for leptospirosis (IgM), HIV-1/HIV-2, hepatitis A (anti-HAV IgM), hepatitis B (HBsAg), hepatitis C (anti-HCV), hepatitis D (total anti-HDV) and dengue (RT-PCR). Patients presenting any sign of severity were treated with intravenous artesunate, as recommended by the WHO guidelines [19].

P. vivax malaria diagnosis

Thick blood smear was prepared as recommended by the Walker technique [24] and evaluated by an experienced microscopist [25]. Parasite densities (parasites/µl) were calculated by counting the number of parasites per 500 leukocytes in high magnification fields, and the number of leukocytes/µl per patient. In addition, differential counting of asexual forms (ring-stage parasites, mature trophozoites and schizonts) was made to ensure that there was no difference between groups of cases and controls.

Afterwards, real-time PCR was performed to confirm *P. vivax* mono-infection. The extraction of total DNA from whole blood was performed using the QIAamp DNA Blood Mini Kit (Qiagen, USA), according to the manufacturer's protocol. The DNA was amplified in an Applied Biosystems 7500 Fast System using

primers and TaqMan fluorescence labeled probes for real time PCR [26].

RNA extraction and expression of *pvcrt-o* and *pvmdr-1* genes

For RNA extraction, 5 ml of whole blood were collected and processed using saponin. Total RNA was purified using Trizol reagent (Invitrogen) according to the manufacturer's instructions. Furthermore, 2% agarose gel electrophoresis was made to guarantee RNA quality. Amplification reactions were performed using Power SYBR Green PCR Master Mix (Applied Biosystems) and 45 ng of template cDNA prepared from each sample. PCR products were amplified and detected on a 7500 FAST (Applied Biosystems). The primers were

*pvmdr-1*F (AAGGATCAAAGGCAACCCA),
*pvmdr-1*R (TCAGGTTGTTACTGCTGTTGCTATT),
*pvcrt-o*F (ATGTCCAAGATGTGCGACGAT),
*pvcrt-o*R (CTGGTCCCTGTATGCAACTGAC),
*pvtubulin*F (CCAAGAATATGATGTGTGCAAGTG),
*pvtubulin*R (GGCGCAGGCGGTTAGG) [18]. Cycling parameters for PCR were an initial denaturation step at 95°C for

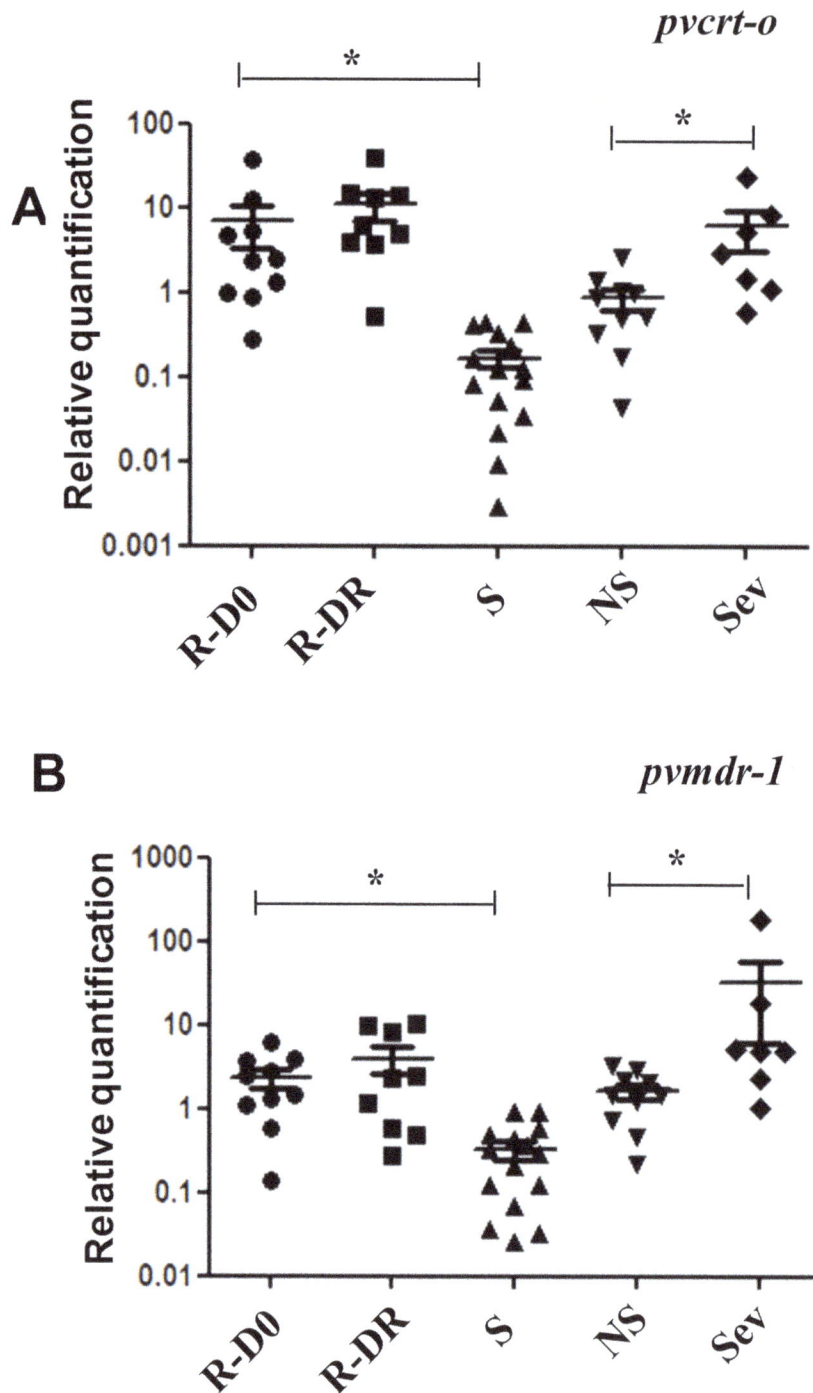

Figure 2. Expression gene levels of *pvcrt-o* and *pvmdr-1* in all the groups. Relative quantification of *pvcrt-o* (A) and *pvmdr1* (B) transcript levels in total RNA obtained from parasites from severe patients with chloroquine-resistant *P. vivax*, patients susceptible to CQ, severe patients and patients without severity symptoms. Chloroquine-resistant *P. vivax* parasites (R). Chloroquine-susceptible *P. vivax* parasites at D0 (S). Severe cases (Sev). Non-severe cases (NS). *p<0.05.

10 minutes, followed by 40 cycles of 95°C for 15 seconds, and 60°C for 1 minute. Samples were run in triplicate. To analyze the relative transcript levels, the threshold cycle value (Ct) of each sample was used to calculate and compare the ΔCt of each sample to that of the *P. vivax* housekeeping gene Sal I β-tubulin; the ΔΔCt was also calculated as $N\text{-}fold = 2^{-\Delta\Delta CT}$ [27] to compare the transcript levels of *pvcrt-o* and *pvmdr-1* in the different groups of patients included in this study versus CQ-susceptible patients. After the PCR reaction in real time, an additional step of decoupling the amplicon to generate a melting curve was performed. The presence of only one peak in the dissociation curves reveals the absence of unspecific bands, confirming the specificity of the amplification primers in real time PCR.

Table 2. Frequency of polymorphisms in the *pvmdr1 and pvcrt-o* genes for the different set of patients.

Genotype	Number of samples (%)		P	Number of samples (%)		P
	Chloroquine-resistant	Chloroquine-susceptible		Severe malaria	Non-severe malaria	
pvmdr-1						
Mutant L1076 codon	0/12 (0.0)	0/15 (0.0)	NS	1/7 (14.3)	0/10 (0.0)	NS
Wild-type Y976 codon	12/12 (100.0)	15/15 (100.0)	NS	7/7 (100.0)	10/10 (100.0)	NS
pvcrt-o						
Wild-type without K10 insert	12/12 (100.0)	15/15 (100.0)	NS	7/7 (100.0)	10/10 (100.0)	NS

Not significant (NS).

Reproducibility of the method (accuracy inter-runs) was assessed through measures of Cts obtained by cDNA quantification of one isolate analyzed 10 times in two runs that were performed on different days, using the same equipment. The results were expressed as mean, standard deviation (SD) and coefficient of variation (CV%). The repeatability of the method (accuracy intra-run) was evaluated by measures of Cts obtained by cDNA quantification of one isolate measured 10 times in the same run. The results were expressed by mean, standard deviation (SD) and coefficient of variation (CV%).

Sequencing analysis of *pvcrt-o* and *pvmdr-1* genes

All samples from patients presenting CQRPv or patients with severe symptoms and the respective control groups, obtained prior to treatment (on D0), were sequenced for searching mutations suspected to be associated to CQR. Extraction of whole DNA was carried out using a QIAamp DNA Mini Kit (QIAGEN, Germany) according to the manufacturer's protocol. PCR primers and different reaction conditions used to amplify *pvcrt-o* and *pvmdr-1* gene sequences were made as previously described (Table S1). The amplification products were quantified by NanoDrop 2000 (Thermo Scientific) and sent for sequencing to Macrogen (South Korea). Amino acid sequences were compared with sequences of Sal I (Salvador I): GenBank accession nos. ADE74979.1 for *pvcrt-o* [17] and EU333979.1 for *pvmdr-1* [28]. The deduced amino acid sequences were aligned and analyzed using Mutation Surveyor. The *pvmdr-1* and *pvcrt-o* gene sequences were deposited in GenBank under the accession numbers KM016489 to KM016517 and KM016518 to KM016532, respectively.

Polymorphism analysis of *msp1*F3 and MS2 molecular markers for patients with CQRPv

PCRs were performed in 20 μl reactions containing 10 μM of each primer, 2 μl of 10×buffer B, 2.5 mM each dNTP, 25 mM MgCl2, 1.5 U Taq DNA polymerase and 5 μl genomic DNA (Table S2). 3 μl of diluted primary PCR product was used as template for the nested PCR (MS2 and *msp1*F3). These polymorphism markers were chosen based on preliminary data from other patients from Manaus (unpublished data). PCRs were performed in a thermocycler with conditions as follows: initial denaturation at 95°C for 1 min, followed by 29 cycles (primary PCR) or 24 cycles (nested PCR) of denaturation at 95°C for 30 seconds, annealing at 59°C for 45 sec, elongation at 72°C for 1 min, followed by a final elongation at 72°C for 5 min. PCR products were stored in the dark (plates wrapped in foil) at 4°C until be sent to Macrogen for capillary electrophoresis-based sequencing and analyzed with GeneMarker version 2.6.0.

Statistical Analyses

Data were analyzed using SPSS version 16.0 for Windows (SPSS Inc Chicago, IL, USA). Normal distribution of data of hemoglobin and gene expression was evaluated with the Kolmogorov-Smirnov test. Linear regression was performed to correlate this variable with hemoglobin and gene expression among hospitalized patients. Spearman's correlation test was performed to correlate CQ concentration with gene expression in the DR and gene expression at D0 and DR with concentration of RNA. The Wilcoxon rank pairs test was performed to compare CQR patients had paired expression levels at D0 and DR. Chi-square or Fisher's test was used to test differences in proportions of polymorphism and distribution of asexual blood stages. The genetic variation for each microsatellite locus was measured by calculating the expected heterozygosity (H_E). H_E was calculated using D0 and DR for each locus as $= [n/(n-1)] [1-\sum p_i^2]$ where n is the number of isolates sampled and p_i is the frequency of allele i.

Results

Reproducibility, repeatability of measurements of Cts and specificity

For reproducibility, amplification of *pvcrt-o* and *pvmdr-1* genes showed average Ct values ranging from 35.14 to 37.93 (mean Ct value 36.82) and from 34.99 to 36.19 (mean Ct value 35.44), respectively. Among the twenty measured values, doing in to different runs, the maximum coefficient of variation was 2.34% and 1.12%, for *pvcrt-o* and *pvmdr-1* respectively. For the repeatability, the coefficient of variation was 1.35% for the *pvcrt-o* and 0.75% for *pvmdr-1* in the first run, and 2.77% for the *pvcrt-o* and 1% for *pvmdr-1* in the second run (Table S3).

After the relative quantification, only one peak was observed in the dissociation curves, confirming a good specificity of the primers in real time PCR.

Gene expression levels from CQR parasites

In this period, in a parallel study to evaluate efficacy of CQ with the 42-day follow-up, with supervised drug administration, 135 patients were enrolled, and 16 were CQR (11.8%). Of these, 12 with adequate RNA storage were included in the analysis. Clinical and laboratorial characteristics of the 12 patients with CQRPv included in this study are presented in Table 1. Out of this total, 83.3% patients were male, with a mean age of 29.0 years. The mean hemoglobin was 13.1 g/dl at D0 and 12.7 g/dl in DR. The geometric mean parasitemia was 1781.7 parasites/μl at D0 and 4206.5 parasites/μl at DR. It was found that mean blood levels of CQ plus DCQ were greater than 100 ng/ml on the DR in 10 of

Table 3. Description of 7 patients presenting severe vivax malaria admitted to a tertiary health center, Manaus, Amazonas, Brazil.

Code	Sex/age	PCR	Microscopy	WHO severe malaria criterion	Duration of fever (days)	Hemoglobin (g/dl)	Bilirubin total (mg/dl)	Serum creatinine (mg/dl)	Asexual malarial parasites/mm³
S1	M/43	Positive	Positive	Severe anemia	6	4.9	3.1	1.0	11160
S2	F/34	Positive	Positive	Severe anemia	1	4.8	8.1	0.9	34673
S3	F/41	Positive	Positive	ARDS	8	8.9	2.3	0.8	24814
S4	F/22	Positive	Positive	Severe anemia	8	6.4	0.4	0.6	264
S5	M/12	Positive	Negative	Prostation, acute renal failure	6	5.4	0.9	7.2	0
S6	M/74	Positive	Positive	Severe anemia, ARDS, acute renal failure	5	7.0	1.7	4.4	325
S7	M/44	Positive	Positive	Severe anemia	10	6.7	1.7	1.5	7000

Acute respiratory distress syndrome (ARDS).

12 (83.3%) patients with recurring infection. Five patients with CQ-susceptible parasites and mild malaria were selected to generate the relative expression of 12 CQRPv and 15 CQ-susceptible isolates.

In patients R1 to R7 gene expression was analysed at D0 and DR, while in patients R8 to R10 and R11 to R12, due to insufficient RNA amount, gene expression was measured only at D0 or DR, respectively. Patients with CQ-resistant *P. vivax* parasites presented a higher gene expression of *pvcrt-o* and *pvmdr-1* at D0 and DR when compared to the susceptible group (Figure 1A and 1B). For the CQR patients, median gene expression values at D0 and DR, presented 2.4 fold (95% CI: 0.96–7.1) and 6.1 fold (95% CI: 3.8–14.3) increase in *pvcrt-o* levels (Figure 1A and 2A) compared to the susceptible patients at D0 with 0.12 fold (95% CI: 0.034–0.324) (Figure S1A and 2A). Median gene expression for *pvmdr-1* presented 2.0 fold (95% CI: 0.95–3.8) and 2.4 fold (95% CI: 0.53–9.1) increase levels at D0 and DR, for the CQR patients (Figure 1B and 2B) versus 0.288 fold (95% CI: 0.068–0.497) for the susceptible patients at D0 (Figure S1B and 2B). Patients R2, R4, R6 and R7 showed a higher expression level of both genes in relation to the control group at D0 and DR (Figure 1). It was not observed significant differences in expression levels between D0 and DR for resistant parasites (p = 0.375 for *pvcrt-o* and p = 0.844 for *pvmdr-1*). However, when we compared expression levels between resistant and sensible parasites at D0 for both genes, values were significantly greater for treatment failures (p < 0.05 for both genes). No correlation was found between CQ concentration and gene expression of DR for *pvcrt-o* (p = 0.286) and *pvmdr-1* (p = 0.433) genes and between gene expression of *pvcrt-o* and *pvmdr-1* and concentration of RNA in D0 and DR (p > 0.05).

To discard that difference in gene expression levels was not due to difference in the distribution of asexual blood stages, parasites distribution was counted in both, CQRPv and CQ-susceptible samples (Tables S4 and S5). Of notice, no significant differences were observed between immature and mature asexual forms in both groups (p > 0.05). Moreover, it was found that the *pvcrt-o* and *pvmdr-1* gene mutations studied were not related to CQ-resistance (p > 0.05) (Table 2).

In order to characterize the genetic variation for each individual population, two neutral microsatellite loci (*msp1*F3 and MS2) were analyzed in D0 and DR samples. The analysis of microsatellites showed at least one concordant allele (83.3%) in primary infection and recrudescence (Table 1). Patients which presented CQ concentration less than 100 ng/ml, showed the same microsatellite allele in D0 and DR. The expected heterozygosity (H_E) was different in parasites from primary and recrudescence episodes (for *msp1*F3, primary: $H_E = 0.915$ and recrudescence: $H_E = 0.681$; and for MS2, primary: $H_E = 0.531$ and recrudescence $H_E = 0.771$).

Gene expression levels in severe vivax malaria parasites

In the period of study, 70 patients were hospitalized with *P. vivax*. However, due to many comorbidities, and absence of any criterion of severity, only 16 could be characterized as severe vivax malaria. Of these, 7 patients presenting mono-infection by *P. vivax* with any criterion of severity were included. Of these patients, 4 had severe anemia, 1 had severe anemia paralleled with acute respiratory distress syndrome (ARDS) and acute renal failure, 1 had acute renal failure and prostration and 1 had ARDS alone. Demographic and clinical characteristics of the patients enrolled are presented in Table 3. Of these patients, 51.0% were male, with a mean age of 38.6 years. The mean disease duration was 6.3 days, median hemoglobin was 6.3 g/dl and mean parasitemia was 11,180 parasites/µl.

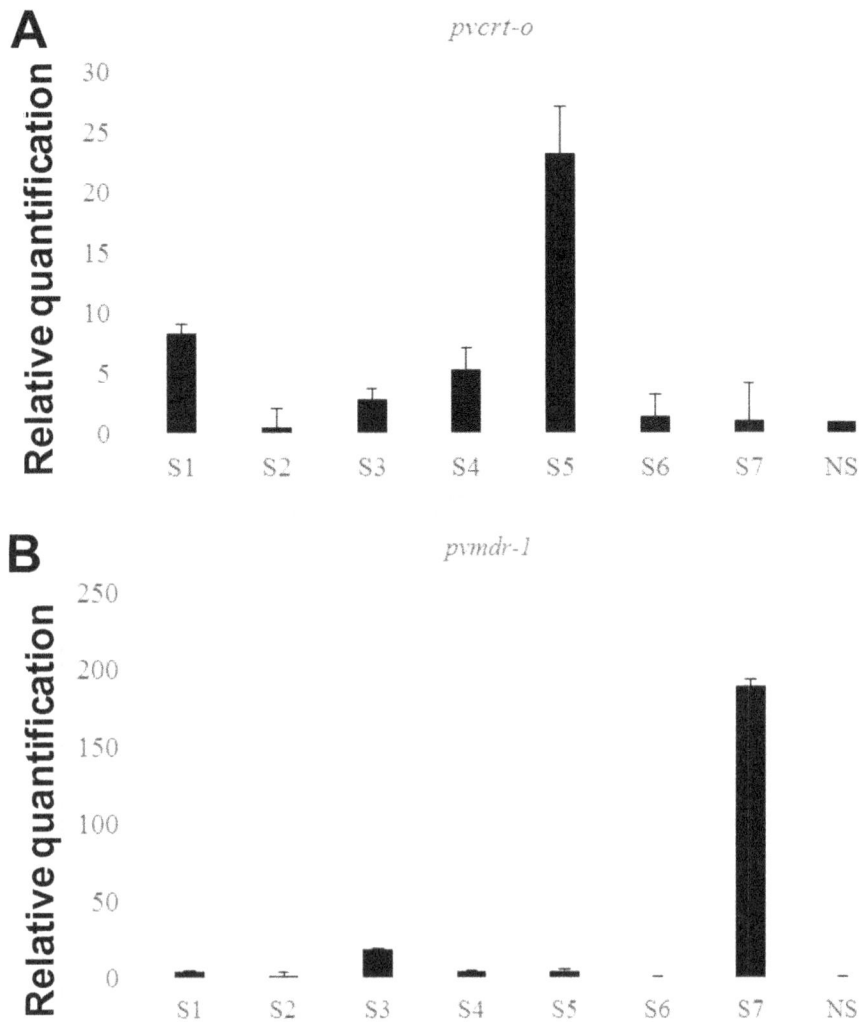

Figure 3. Expression level of chloroquine resistance genes in severe patients. Relative quantification of *pvcrt-o* (A) and *pvmdr1* (B) transcript levels in total RNA obtained from parasites from severe patients vs a pool of total RNA obtained from parasites susceptible to CQ. Severe cases (S). Non-severe cases (NS). The error bars in Figures 1 and 3 reflect the average standard error of the Ct.

Ten patients diagnosed with non-severe malaria were selected and included as controls. The same 5 CQ-susceptible isolates from patients with mild malaria above cited were used to generate the relative expression of seven severe malaria patients and 10 non-severe malaria patients. The group of patients with severe vivax malaria had a 2.9 (95% CI: 1.1–8.3) and 4.9 (95% CI: 2.3–18.8) fold increase in *pvcrt-o* and *pvmdr-1* expression level versus 0.7 (95% CI: 0.3–1.1) and 1.5 (95% CI: 0.7–2.4) for the control group with mild disease (p = 0.007 and p = 0.01) (Figure 2, Figure 3 and Figure S2). To discard that difference in gene expression levels was not due to difference in the distribution of asexual blood stages, parasites distribution was counted in both, severe vivax malaria (Table S6) and non-severe. No significant difference was found between immature and mature asexual forms among patients with severe and non-severe malaria (p>0.05) (Table S6). The *pvcrt-o* and *pvmdr1* gene mutations studied were not related to severity (p>0.05) (Table 2).

In a linear regression analysis, it was found a negative correlation between hemoglobin values (of severe and non-severe patients) and the log of the expression of the *pvcrt-o* (p<0.001) and *pvmdr-1* (p = 0.007) genes among hospitalized patients (Figure 4).

Discussion

There have been recent reports on CQR from different regions of the world [8,9,15,28–31], including Brazil [3,10,11]. Although the relevance of this problem, the molecular mechanisms of CQRPv are poorly understood. In this work, expression levels of *pvcrt-o* and *pvmdr-1* genes were determined in patients with CQRPv from a well-characterized cohort under supervised treatment, pointing to a link between *in vivo* CQR and overexpression of both genes. Furthermore, these patients showed higher concentrations of CQ and microsatellite revealed the presence of the same clonal nature at D0 and DR, bringing evidence that the re-emergent infection is the same as the primary infection, therefore indicating recrudescence.

It is likely that CQR contributes to the burden of severe vivax malaria in regions where resistance is emerging [4,32], but there is paucity of available data regarding this issue. In this work, the group of patients with severe vivax malaria presented an increase

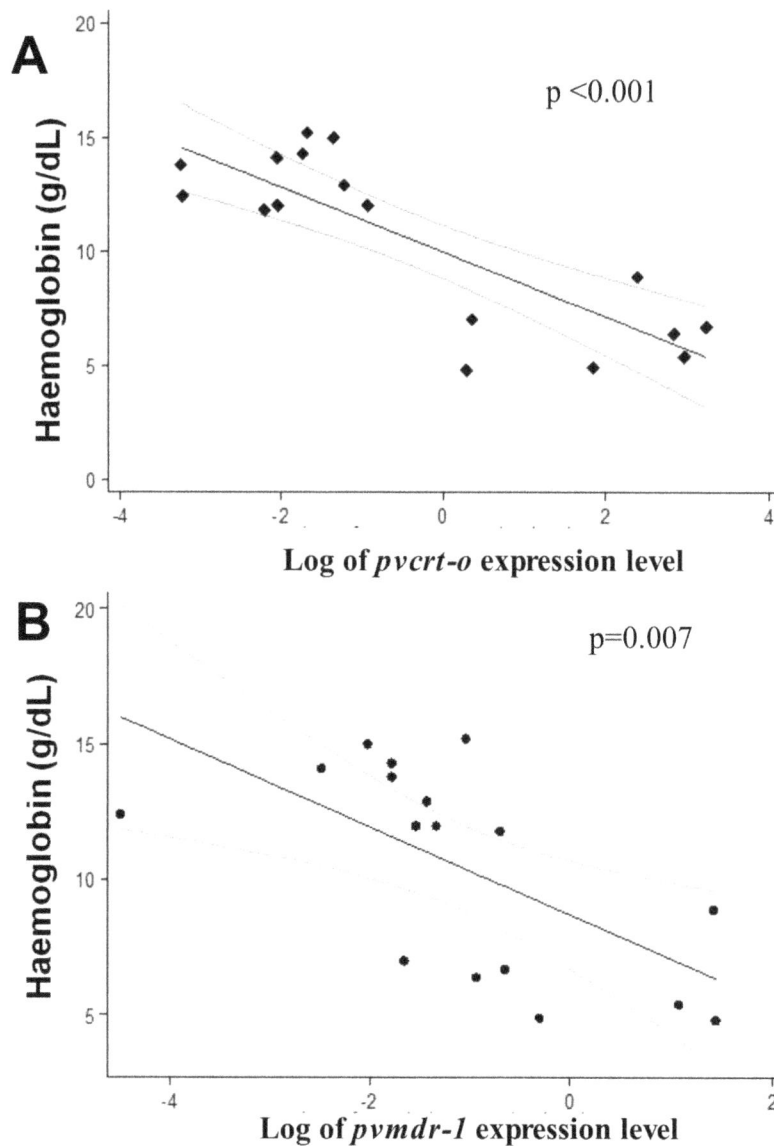

Figure 4. Correlation between expression gene levels *pvcrt-o* **and** *pvmdr-1* **genes and hemoglobin concentration.** Correlation between expression levels of chloroquine resistance genes, *pvcrt-o* (A) and *pvmdr-1* (B) and hemoglobin in hospitalized patient.

in *pvcrt-o* and *pvmdr-1* expression levels, as compared to the control group with mild disease. These two genes have been suggested as genetic markers of CQR in *P. vivax* [33,34].

Our results are in concordance with those reported by Fernandez-Becerra et al. [18], where high levels of *pvcrt-o* and *pvmdr-1* gene expression were observed in a patient with severe vivax malaria compared to three non-severe. A prospective study conducted in the Papua New Guinea highlighted *P. vivax* as a major cause of severe malaria, particularly in settings with established or emerging chloroquine resistance, supporting indirectly our findings [2]. One speculates that in areas in which CQR *P. vivax* circulates, acute febrile illness potentially becomes chronic, recurrent and severe, due to recurrent parasitemia [5]. Moreover, chronic infections could lead to diserythropoiesis, as is seen in *P. falciparum*-induced anemia [35].

In this investigation, CQR and severe *P. vivax* malaria were observed in parallel with a high expression of *pvcrt-o* and/or

pvmdr-1 genes in patients seen at a reference center in the Brazilian Amazon. Even though it is tempting to conclude about a causal association, a clear limitation of this study is that CQR has not been evaluated in the same patients with severe vivax malaria, especially because severe patients were in use of intravenous artesunate (ACT), making it impossible to evaluate CQ-susceptibility *in vivo* phenotype. Actually, patients with CQRPv were identified from following WHO criteria, excluding patients with any clinical complication [19]. Another clear limitation is the small number of samples used in this study. Moreover, so far it is not possible to rule out a differential virulence concurring with dissimilar drug resistance phenotypes. Studies on *P. falciparum* laboratory strains and field isolates of *P. vivax* suggest that there is marked variability in growth rates of isolates *ex vivo* with CQR isolates growing faster than CQ-susceptible isolates [36,37]. Since parasites from patients with severe disease have greater *ex vivo* multiplication rates compared to those from patients with non-severe disease [38], the possibility arises that the highly CQR isolates may be more virulent.

Mefloquine (MQ) was used in the Brazilian Amazon until the beginning of the 2000s. Indeed, CQ and MQ resistance parallel in this region [8]. Thus, even after more than one decade of non-exposure to MQ, we could not discard this effect in the present. Actually, increased *pvmdr-1* copy number was found in two isolates in the Western Brazilian Amazon [39]. Unfortunately, *pvmdr-1* copy numbers were not measured in our samples.

In conclusion, our study suggests that CQR and severe *P. vivax* malaria are associated with increased expression levels of the *pvcrt-o* and *pvmdr-1* genes. CQRPv could be rising in the Brazilian Amazon and probably this fact could be contributing to the simultaneous spread of vivax malaria and clinical severity related to this parasite. Further studies are needed with larger number of patients and using other molecular markers to confirm the underlying pathogenesis of severe disease, and the degree to which this is related to the emergence of multidrug resistant strains of *P. vivax*.

Supporting Information

Figure S1 Expression levels of chloroquine resistance genes in patients susceptible to chloroquine treatment. Relative quantification of *pvcrt-o* (A) and *pvmdr1* (B) transcripts in total RNA obtained from parasites from patients with chloroquine-resistant *P. vivax* vs a pool of total RNA obtained from parasites susceptible to CQ. Chloroquine-susceptible *P. vivax* parasites (S). Day admission (D0). Day of recrudescence (DR). The error bars reflect propagated error calculated with the average standard error of the Ct.

Figure S2 Expression level of chloroquine resistance genes in non-severe *P. vivax* malaria. Relative quantification of *pvcrt-o* (A) and *pvmdr1* (B) transcript levels in total RNA obtained from parasites from severe patients vs a pool of total RNA obtained from parasites susceptible to CQ. Non-severe cases (NS). The error bars reflect propagated error calculated with the average standard error of the Ct.

Table S1 Oligonucleotide primers used for DNA sequencing of *P. vivax* orthologs genes.

Table S2 Oligonucleotides used for genotyping *P. vivax* parasites.

Table S3 Reproducibility and repeatability of measurements of Cts amplification of targets *pvcrt-o* and *pvmdr-1*.

Table S4 Gene expression and different intra-erythrocytic stages of chloroquine resistance *P. vivax* parasites admitted to a tertiary health center, Manaus, Amazonas, Brazil.

Table S5 Different intra-erythrocytic stages of chloroquine-susceptible *P. vivax* parasites admitted to a tertiary health center, Manaus, Amazon, Brazil.

Table S6 Gene expression and different intra-erythrocytic stages of patients presenting severe vivax malaria admitted to a tertiary health center, Manaus, Amazonas, Brazil.

Acknowledgments

We acknowledge Omaira Veras for helpful discussion on microsatelite analysis, Nélida Teresa Sánchez Rúa, Maria Raimunda Costa, Wellington Silva and Kim Machado for samples collection and José Luiz Vieira and Marly Marques de Melo for drugs measurement.

Author Contributions

Conceived and designed the experiments: GCM CFB MVGL. Performed the experiments: GCM. Analyzed the data: GCM WMM AMS AK HAP CFB. Contributed reagents/materials/analysis tools: GCM AMS SRS BMLM ACCA. Contributed to the writing of the manuscript: GCM WMM HAP CFB MVGL. Critical revision of the manuscript: GCM WMM AMS HAP CFB MVGL.

References

1. Baird JK (2011) Resistance to chloroquine unhinges vivax malaria therapeutics. Antimicrob Agents Chemother 55: 1827–1830.
2. Tjitra E, Anstey NM, Sugiarto P, Warikar N, Kenangalem E, et al. (2008) Multidrug-resistant *Plasmodium vivax* associated with severe and fatal malaria: a prospective study in Papua, Indonesia. PLoS Med 5: e128.
3. Lanca EF, Magalhaes BM, Vitor-Silva S, Siqueira AM, Benzecry SG, et al. (2012) Risk factors and characterization of *Plasmodium vivax*-associated admissions to pediatric intensive care units in the Brazilian Amazon. PLoS One 7: e35406.
4. Genton B, D'Acremont V, Rare L, Baea K, Reeder JC, et al. (2008) *Plasmodium vivax* and mixed infections are associated with severe malaria in children: a prospective cohort study from Papua New Guinea. PLoS Med 5: e127.
5. Price RN, Douglas NM, Anstey NM (2009) New developments in *Plasmodium vivax* malaria: severe disease and the rise of chloroquine resistance. Curr Opin Infect Dis 22: 430–435.
6. Rieckmann KH, Davis DR, Hutton DC (1989) *Plasmodium vivax* resistance to chloroquine? Lancet 2: 1183–1184.
7. Baird JK, Basri H, Purnomo, Bangs MJ, Subianto B, et al. (1991) Resistance to chloroquine by *Plasmodium vivax* in Irian Jaya, Indonesia. Am J Trop Med Hyg 44: 547–552.
8. Alecrim MdG, Alecrim W, Macedo V (1999) *Plasmodium vivax* resistance to chloroquine (R2) and mefloquine (R3) in Brazilian Amazon region. Rev Soc Bras Med Trop 32: 67–68.
9. de Santana Filho FS, Arcanjo AR, Chehuan YM, Costa MR, Martinez-Espinosa FE, et al. (2007) Chloroquine-resistant *Plasmodium vivax*, Brazilian Amazon. Emerg Infect Dis 13: 1125–1126.
10. Alexandre MA, Ferreira CO, Siqueira AM, Magalhaes BL, Mourao MP, et al. (2010) Severe *Plasmodium vivax* malaria, Brazilian Amazon. Emerg Infect Dis 16: 1611–1614.
11. Lacerda MV, Fragoso SC, Alecrim MG, Alexandre MA, Magalhaes BM, et al. (2012) Postmortem Characterization of Patients With Clinical Diagnosis of *Plasmodium vivax* Malaria: To What Extent Does This Parasite Kill? Clin Infect Dis 55: e67–74.
12. Lacerda MV, Mourao MP, Alexandre MA, Siqueira AM, Magalhaes BM, et al. (2012) Understanding the clinical spectrum of complicated *Plasmodium vivax* malaria: a systematic review on the contributions of the Brazilian literature. Malar J 11: 12.
13. Das S, Chakraborty SP, Hati AK, Roy S (2013) Association between prevalence of chloroquine resistance and unusual mutation in *pfmdr-1* and *pfcrt* genes in India. Am J Trop Med Hyg 88: 828–834.
14. Ngo T, Duraisingh M, Reed M, Hipgrave D, Biggs B, et al. (2003) Analysis of *pfcrt*, *pfmdr1*, *dhfr*, and *dhps* mutations and drug sensitivities in *Plasmodium falciparum* isolates from patients in Vietnam before and after treatment with artemisinin. Am J Trop Med Hyg 68: 350–356.
15. Suwanarusk R, Russell B, Chavchich M, Chalfein F, Kenangalem E, et al. (2007) Chloroquine resistant *Plasmodium vivax*: in vitro characterisation and association with molecular polymorphisms. PLoS One 2: e1089.
16. Barnadas C, Ratsimbasoa A, Tichit M, Bouchier C, Jahevitra M, et al. (2008) *Plasmodium vivax* resistance to chloroquine in Madagascar: clinical efficacy and polymorphisms in *pvmdr1* and *pvcrt-o* genes. Antimicrob Agents Chemother 52: 4233–4240.
17. Nomura T, Carlton JM, Baird JK, del Portillo HA, Fryauff DJ, et al. (2001) Evidence for different mechanisms of chloroquine resistance in 2 Plasmodium species that cause human malaria. J Infect Dis 183: 1653–1661.

18. Fernandez-Becerra C, Pinazo MJ, Gonzalez A, Alonso PL, del Portillo HA, et al. (2009) Increased expression levels of the *pvcrt-o* and *pvmdr1* genes in a patient with severe Plasmodium vivax malaria. Malar J 8: 55.

19. WHO (2006) Guidelines for the treatment of malaria. Geneva.

20. Soto J, Toledo J, Gutierrez P, Luzz M, Llinas N, et al. (2001) *Plasmodium vivax* clinically resistant to chloroquine in Colombia. Am J Trop Med Hyg 65: 90–93.

21. Ruebush TK, 2nd, Zegarra J, Cairo J, Andersen EM, Green M, et al. (2003) Chloroquine-resistant *Plasmodium vivax* malaria in Peru. Am J Trop Med Hyg 69: 548–552.

22. Naing C, Aung K, Win DK, Wah MJ (2010) Efficacy and safety of chloroquine for treatment in patients with uncomplicated *Plasmodium vivax* infections in endemic countries. Trans R Soc Trop Med Hyg 104: 695–705.

23. WHO (2012) Management of severe malaria: a practical handbook Geneva.

24. WHO (2010) Malaria Report: 2010.

25. MS (2009) Manual de Diagnóstico Laboratorial da Malária. In: Epidemiológica DdV, editor. Brasilia.

26. Snounou G, Viriyakosol S, Zhu XP, Jarra W, Pinheiro L, et al. (1993) High sensitivity of detection of human malaria parasites by the use of nested polymerase chain reaction. Mol Biochem Parasitol 61: 315–320.

27. Livak KJ, Schmittgen TD (2001) Analysis of relative gene expression data using real-time quantitative PCR and the 2 (-Delta Delta C(T)) Method. Methods 25: 402–408.

28. Orjuela-Sanchez P, de Santana Filho FS, Machado-Lima A, Chehuan YF, Costa MR, et al. (2009) Analysis of single-nucleotide polymorphisms in the crt-o and mdr1 genes of *Plasmodium vivax* among chloroquine-resistant isolates from the Brazilian Amazon region. Antimicrob Agents Chemother 53: 3561–3564.

29. Gama BE, Oliveira NK, Souza JM, Daniel-Ribeiro CT, Ferreira-da-Cruz Mde F (2009) Characterisation of *pvmdr1* and *pvdhfr* genes associated with chemoresistance in Brazilian *Plasmodium vivax* isolates. Mem Inst Oswaldo Cruz 104: 1009–1011.

30. Gama BE, Lacerda MV, Daniel-Ribeiro CT, Ferreira-da-Cruz Mde F (2011) Chemoresistance of Plasmodium falciparum and *Plasmodium vivax* parasites in

Brazil: consequences on disease morbidity and control. Mem Inst Oswaldo Cruz 106 Suppl 1: 159–166.

31. Baird JK (2004) Chloroquine resistance in *Plasmodium vivax*. Antimicrob Agents Chemother 48: 4075–4083.

32. Marfurt J, Mueller I, Sie A, Maku P, Goroti M, et al. (2007) Low efficacy of amodiaquine or chloroquine plus sulfadoxine-pyrimethamine against *Plasmodium falciparum* and *P. vivax* malaria in Papua New Guinea. Am J Trop Med Hyg 77: 947–954.

33. Sa JM, Yamamoto MM, Fernandez-Becerra C, de Azevedo MF, Papakrivos J, et al. (2006) Expression and function of *pvcrt-o*, a *Plasmodium vivax* ortholog of *pfcrt*, in *Plasmodium falciparum* and *Dictyostelium discoideum*. Mol Biochem Parasitol 150: 219–228.

34. Brega S, Meslin B, de Monbrison F, Severini C, Gradoni L, et al. (2005) Identification of the *Plasmodium vivax* mdr-like gene (*pvmdr1*) and analysis of single-nucleotide polymorphisms among isolates from different areas of endemicity. J Infect Dis 191: 272–277.

35. Menendez C, Fleming AF, Alonso PL (2000) Malaria-related anaemia. Parasitol Today 16: 469–476.

36. Reilly HB, Wang H, Steuter JA, Marx AM, Ferdig MT (2007) Quantitative dissection of clone-specific growth rates in cultured malaria parasites. Int J Parasitol 37: 1599–1607.

37. Russell B, Chalfein F, Prasetyorini B, Kenangalem E, Piera K, et al. (2008) Determinants of in vitro drug susceptibility testing of *Plasmodium vivax*. Antimicrob Agents Chemother 52: 1040–1045.

38. Chotivanich K, Udomsangpetch R, Simpson JA, Newton P, Pukrittayakamee S, et al. (2000) Parasite multiplication potential and the severity of Falciparum malaria. J Infect Dis 181: 1206–1209.

39. Vargas-Rodriguez RD, da Silva Bastos M, Menezes MJ, Orjuela-Sanchez P, Ferreira MU (2012) Single-Nucleotide Polymorphism and Copy Number Variation of the Multidrug Resistance-1 Locus of *Plasmodium vivax*: Local and Global Patterns. Am J Trop Med Hyg.

4

Inhibition of the SR Protein-Phosphorylating CLK Kinases of *Plasmodium falciparum* Impairs Blood Stage Replication and Malaria Transmission

Selina Kern[1,2], Shruti Agarwal[1], Kilian Huber[3], André P. Gehring[3], Benjamin Strödke[3], Christine C. Wirth[2], Thomas Brügl[1], Liliane Onambele Abodo[2], Thomas Dandekar[4], Christian Doerig[5,6], Rainer Fischer[2], Andrew B. Tobin[7], Mahmood M. Alam[7], Franz Bracher[3], Gabriele Pradel[2]*

1 Research Center for Infectious Diseases, University of Würzburg, Würzburg, Germany, 2 Institute of Molecular Biotechnology, RWTH Aachen University, Aachen, Germany, 3 Department of Pharmacy – Center for Drug Research, Ludwig-Maximillians University, Munich, Germany, 4 Bioinformatics, Biocenter, University of Würzburg, Würzburg, Germany, 5 INSERM U609, Global Health Institute, Ecole Polytechnique Fédérale de Lausanne (EPFL), Lausanne, Switzerland, 6 Department of Microbiology, Monash University, Clayton, Victoria, Australia, 7 Department of Cell Physiology and Pharmacology, MRC Toxicology Unit, University of Leicester, Leicester, United Kingdom

Abstract

Cyclin-dependent kinase-like kinases (CLKs) are dual specificity protein kinases that phosphorylate Serine/Arginine-rich (SR) proteins involved in pre-mRNA processing. Four CLKs, termed PfCLK-1-4, can be identified in the human malaria parasite *Plasmodium falciparum*, which show homology with the yeast SR protein kinase Sky1p. The four PfCLKs are present in the nucleus and cytoplasm of the asexual blood stages and of gametocytes, sexual precursor cells crucial for malaria parasite transmission from humans to mosquitoes. We identified three plasmodial SR proteins, PfSRSF12, PfSFRS4 and PfSF-1, which are predominantly present in the nucleus of blood stage trophozoites, PfSRSF12 and PfSF-1 are further detectable in the nucleus of gametocytes. We found that recombinantly expressed SR proteins comprising the Arginine/Serine (RS)-rich domains were phosphorylated by the four PfCLKs in *in vitro* kinase assays, while a recombinant PfSF-1 peptide lacking the RS-rich domain was not phosphorylated. Since it was hitherto not possible to knock-out the *pfclk* genes by conventional gene disruption, we aimed at chemical knock-outs for phenotype analysis. We identified five human CLK inhibitors, belonging to the oxo-β-carbolines and aminopyrimidines, as well as the antiseptic chlorhexidine as PfCLK-targeting compounds. The six inhibitors block *P. falciparum* blood stage replication in the low micromolar to nanomolar range by preventing the trophozoite-to-schizont transformation. In addition, the inhibitors impair gametocyte maturation and gametogenesis in *in vitro* assays. The combined data show that the four PfCLKs are involved in phosphorylation of SR proteins with essential functions for the blood and sexual stages of the malaria parasite, thus pointing to the kinases as promising targets for antimalarial and transmission blocking drugs.

Editor: Luzia H. Carvalho, Centro de Pesquisa Rene Rachou/Fundação Oswaldo Cruz (Fiocruz-Minas), Brazil

Funding: The project was funded by the EU 7th framework programme (MALSIG (GP, CD), ANTIMAL (CD) and EVIMALAR (CD)) and by the Emmy Noether programme PR905/1-2(GP) and the SFB630/C6 (TD) of the Deutsche Forschungsgemeinschaft. The project was additionally supported by Inserm (CD) and the MRC-Toxicology Unit programme leader grant (ABT and MMA). SK received a fellowship from the program for female scientists of the University of Würzburg. The funders had no role in study design, data collection and analysis, decision to publish, or preparation of the manuscript.

* Email: gabriele.pradel@molbiotech.rwth-aachen.de

Introduction

The protozoan parasite *Plasmodium falciparum* is responsible for more than 600,000 fatal cases caused by the tropical disease malaria per annum [1]. During life cycle progression from humans to mosquitoes, *P. falciparum* switches between stages with high replication rates and ones arrested in their cell cycle and also passes through a phase of sexual reproduction. These rapid transformations require fine-tuned mechanisms of gene expression, and the importance of post-transcriptional regulation of gene expression in *Plasmodium* parasites has previously been highlighted [2]. These include the alternative splicing (AS) of pre-mRNA, enabling the parasite to express functionally different protein isoforms. Two genome-wide studies implied that more than 200 AS events occur during blood stage replication of *P. falciparum* [3,4].

AS involves multiple auxiliary factors that control splice site selection, spliceosome assembly as well as the splicing reaction [5,6]. One group of proteins regulating the selection of alternatively spliced exonic or intronic pre-mRNA sequences is that of Serine/Arginine-rich (SR) proteins [5]. SR proteins usually contain a RNA recognition motif and an Arginine/Serine-rich (RS) domain required for protein-protein interactions during splicing. SR proteins are phosphorylated by several protein kinase families, including the cyclin-dependent kinase-like kinases (CLKs) [7,8].

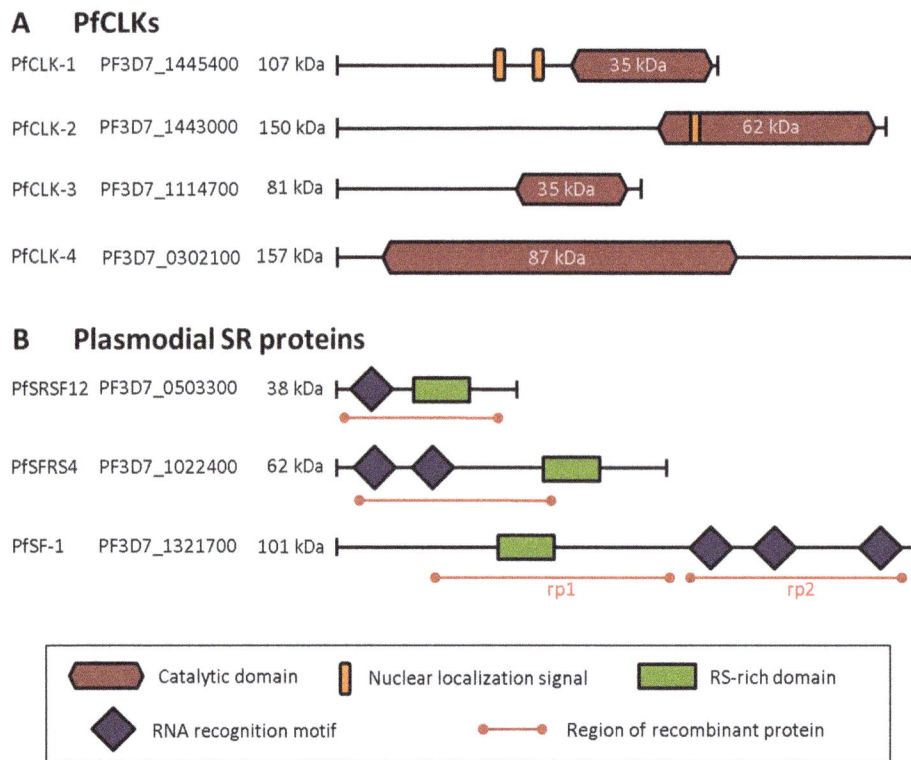

Figure 1. Schematic of the plasmodial PfCLKs and SR proteins. A. Domain structures of the PfCLKs. B. Domain structures of the plasmodial SR proteins investigated in this study.

The genome of *P. falciparum* encodes four members of the CLK family, which were previously termed PfCLK-1-4 [9–11]. For PfCLK-1 (originally described as LAMMER kinase) [12] and PfCLK-2 homologies with the yeast SR protein kinase Sky1p were shown [11]. Both kinases are expressed in the *P. falciparum* blood stages and phosphorylate a number of substrates *in vitro*, including the Sky1p substrate, SR protein Npl3p, and the plasmodial alternative splicing factor PfASF-1 [11]. Similarly, PfCLK-4 (also known as SRPK1) is expressed in the blood stages of *P. falciparum*, where it phosphorylates the plasmodial protein SR1 [13], an AS factor required for parasite proliferation [14]. Previous reverse genetics studies indicated an important role of the four PfCLKs for completion of the asexual replication cycle, since disruption of the respective genes was not achievable, while the loci were amenable to a modification that does not cause loss-of-function [11,15].

Here we aimed to investigate the involvement of the four PfCLKs in the phosphorylation of plasmodial SR proteins and to determine the crucial role of the kinases for the blood and transmission stages of *P. falciparum* via chemical knock-outs using a variety of newly identified CLK inhibitors.

Materials and Methods

Gene IDs and data analysis

The following PlasmoDB gene identifiers (plasmodb.org; previous IDs set in brackets) [16,17] are assigned to the CLKs and SR proteins investigated in this study (shown in Fig. 1): PfCLK-1, PF3D7_1445400 (PF14_0431); PfCLK-2, PF3D7_1443000 (PF14_0408); PfCLK-3, PF3D7_1114700 (PF11_0156); PfCLK-4, PF3D7_0302100 (PFC0105w); PfPKRP, PF3D7_0311400 (PFC0485w); PfSFRS4, PF3D7_1022400

(PF10_0217); PfSRSF12, PF3D7_0503300 (PFE0160c); PfSF-1, PF3D7_1321700 (MAL13P1.120).

Bioinformatics

The following computer programs and databases were used for the *in silico* studies: For gene sequence annotation, PlasmoDB (www.plasmodb.org) [16,17], the SMART program (www.smart.embl-heidelberg.de) [18,19] and NCBI sequence analysis software and databanks [20] were used. Multiple sequence alignment involved programs ClustalW (www.ebi.ac.uk/clustalw) [21] and Clone Manager 9, and formatting of multiple sequence alignments was pursued according to standard methods (espript.ibcp.fr).

CLK inhibitors

Chlorhexidine (CHX) was purchased from Sigma-Aldrich. The spiropiperidino-β-carbolines KH-CARB-10, KH-CARB-11, and KH-CARB-13xHCl were prepared as described previously (Fig. 2A) [22]. The aminopyrimidyl β-carboline C-117 and the aminopyrimidyl carbazole gea-27 were prepared starting from known methyl ketones as precursors (Fig. 2B). In short, treatment of 1-acetyl-β-carboline (1; see Fig. 2B) [23] with tert-butoxy-bis(dimethylamino)methane (Bredereck's reagent) in refluxing dimethylformamide, followed by addition of 4-methylpiperazine-1-carboxamidinium sulfate and potassium carbonate gave the target compound C-117 in good yield in one single operation [24]. For the synthesis of gea-27 the acetylcarbazole (2) [25] was protected at the pyrrole nitrogen with the SEM (2-(trimethylsilyl)-ethoxymethyl) group to give (3), then heated with Bredereck's reagent and subsequently with guanidinium carbonate and potassium carbonate. The resulting aminopyrimidine intermediate was deprotected with HF to give the target compound. Syntheses of C-117 and gea-27 are described in detail in (Methods S1). All

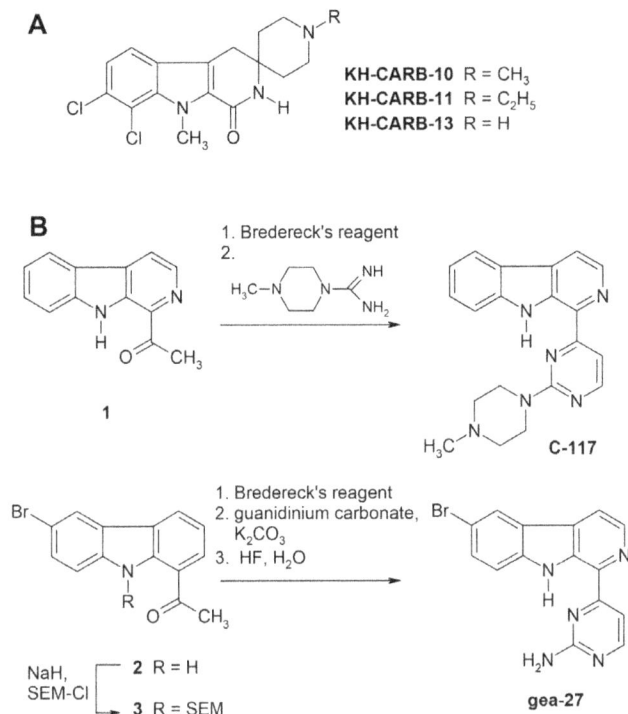

Figure 2. Chemical structures of CLK inhibitors. A. Structures of the spiropiperidino-β-carbolines KH-CARB-10, KH-CARB-11, and KH-CARB-13xHCl. B. Synthesis of the aminopyrimidyl β-carboline C-117 and the aminopyrimidyl carbazole gea-27.

inhibitors were prepared as 100 mM stock solutions in dimethyl sulfoxide (DMSO).

Parasite culture

Asexual blood stage parasites and gametocytes of the *P. falciparum* NF54 [26] isolate and asexual blood stage parasites of the *P. falciparum* strains 3D7 [27] and F12 [28] were cultivated in human erythrocytes *in vitro* as described [29–31]. The following parasite lines were obtained through the MR4 as part of the BEI Resources Repository, NIAID, NIH: *Plasmodium falciparum* NF54, MRA-1000, deposited by M Dowler, Walter Reed Army Institute of Research and *Plasmodium falciparum* 3D7, MRA-102, deposited by DJ Carucci. Parasite line F12 was kindly provided by Pietro Alano, Istituto Superiore di Sanità, Rome. Human A$^+$ erythrocyte sediment and serum were purchased from the University Hospital Aachen, Germany (PO no. DKG-NT 9748). The erythrocyte and sera samples were pooled and the donors remained anonymous; the work on human blood was approved by the ethics commission of RWTH Aachen University. RPMI medium 1640 (Gibco) was supplemented with either A$^+$ human serum (for NF54 and F12) or 0.5% Albumax II (for 3D7; Invitrogen), hypoxanthine (Sigma-Aldrich) and genta-micin (Invitrogen) and cultures were maintained at 37°C in an atmosphere of 5% O$_2$, 5% CO$_2$, 90% N$_2$. Gametogenesis was induced by incubating mature gametocyte cultures in 100 µM xanthurenic acid for 15 min at room temperature (RT) [32,33]. For synchronization, parasite cultures with 3–4% ring stages were centrifuged to obtain the pellet, which was resuspended in five times pellet's volume of 5% prewarmed sorbitol (AppliChem) in RPMI medium (Invitrogen) and incubated at RT for 10 min [34]. The cells were washed once with RPMI medium to remove the

sorbitol, diluted to 5% vol. hematocrit and cultured as described above.

Recombinant protein expression

Recombinant proteins were expressed as fusion proteins either with a glutathione S-transferase (GST)-tag using the pGEX 4T1 vector (Amersham Bioscience) or with a maltose binding protein (MaBP)-tag using the pIH-vector (kindly provided by Kim Williamson, Loyola University Chicago). Cloning into the pGEX4T1 vector was mediated by *Eco*RI/*Not*I restriction sites added at the ends of PCR-amplified gene fragments and cloning into the pIH-vector was mediated by *Eco*RI/*Sal*I restriction sites. Primers used for cloning are listed in (Table S1). Recombinant proteins were expressed in *E. coli* BL21 (DE3) RIL according to the manufacturer's protocol (Stratagene). GST-fusion proteins were purified from bacterial extracts using glutathione-sepharose according to the manufacturer's protocol (GE Healthcare). MaBP-tagged recombinant proteins were purified using amylose resin (New England Biolabs) as described previously [35] with following modifications of the procedure: pelleted bacteria were directly resuspended in lysis buffer containing complete, EDTA-free protease inhibitor cocktail (Roche), incubated on ice for 20 min and homogenized by 4 min of sonication (50 cycles/50% intensity). DNAse treatment was not deployed. Amylose-bound fusion protein was eluted during batch purification according to the manufacturer's protocol.

Generation of antisera

GST-tagged recombinant PfPKRP protein was prepared from bacterial inclusion bodies as described previously [36], while GST-tagged recombinant PfCLK-4 and MaBP-tagged recombinant SR proteins were purified via chromatography as described above. Specific immune sera against the recombinant fragments of the respective kinases and the SR-proteins were generated by the initial immunization of 6 weeks-old female NMRI mice (Charles River Laboratories) with 100 µg recombinant protein emulsified in Freund's incomplete adjuvant (Sigma-Aldrich) followed by a boost with 30 µg recombinant protein 4 weeks after immunization. Mice were anesthetized by intraperitoneal injection of a mixture of ketamine and xylazine according to the manufacturer's protocol (Sigma-Aldrich), and immune sera were collected 10 days after the second immunization (boost) via heart puncture. Following sera collection the anesthetized mice were sacrificed via severing the cervical spine. The immune sera of three mice immunized with the same antigen were pooled; sera of three non-immunized mice were used as negative control. The antisera recognized the cognate recombinant protein (data not shown). Experiments for the generation of antisera in mice were approved by the animal welfare committees of the government of Lower Franconia, Germany (ref. no. 55.2-2531.01-58/09), and of the District Council of Cologne, Germany (ref. no. 84-02.05.30.12.097 TVA). The generation of mouse anti-PfCLK-1 and anti-PfCLK-2 antisera was described previously [11]. As a second source of PfCLK1-specific antibody, sera directed against the peptide sequence NRTKTSDTEDKKER (AA508-521) upstream of the catalytic domain were produced by immunization of two rabbits (Biogenes, Berlin). Specific PfCLK-3 rat antibody was raised by immunizing rats with peptide YKSKHEENSPDGDSY (AA30-44) and purified by protein G as described [15].

Indirect immunofluorescence assay

Parasite preparations for indirect immunofluorescence assays (IFAs) included mixed asexual blood stages of *P. falciparum* F12 strain or mature gametocytes of NF54 strain. Preparations were air-dried on slides and fixed for 10 min either in -80°C methanol or, to label the SR proteins, with 4% paraformaldehyde (pH 7.4). For membrane permeabilization and blocking of non-specific binding, methanol-fixed cells were incubated in 0.01% saponin, 0.5% bovine serum albumin faction V (BSA) and 1% neutral goat serum (Sigma-Aldrich) in PBS for 30 min. Paraformaldehyde-fixed samples were permeabilized with 0.1% vol. Triton X-100 and 125 mM glycine (Carl Roth) in PBS for 30 min, followed by blocking with 3% BSA in PBS for 1 h. Preparations were then incubated for 2 h at 37°C with rat antisera against PfCLK-3 or mouse antisera against PfCLK-4 and the SR proteins. Antisera dilutions of 1:50 to 1:100 were used. Binding of primary antibody was visualized using fluorophore-conjugated goat anti-rat or anti-mouse antibodies (Alexa Fluor 488; Molecular Probes). Asexual blood stage parasites were highlighted with rabbit immune sera specific for the merozoite surface protein PfMSP-1 (ATCC), while gametocytes were highlighted with rabbit anti-Pfs230 antisera, followed by incubation with fluorophore-conjugated goat anti-rabbit antibodies (Alexa Fluor 594; Molecular Probes). In the co-localization experiments, PfCLK-1 was immunolabelled in the young schizont stages using the respective rabbit antisera (dilution of 1:100) in combination with Alexa Fluor 594 goat anti-rabbit antibody. Nuclei were highlighted by incubating the specimens with Hoechst nuclear stain 33342 (Molecular Probes) for 1 min. Labelled specimens were examined by confocal laser scanning microscopy using a LEICA TCS SP5 or an Olympus BX41 fluorescence microscope in combination with a ProgRes Speed XT5 camera. Digital images were processed using Adobe Photoshop CS software. Quantifications of schizonts (highlighted by anti-MSP-1 antisera or Hoechst nuclear stain) or gametocytes (highlighted by anti-Pfs230 antisera) positive for the PfCLKs or the SR proteins were performed in triplicate (40–60 parasites counted per well).

Western blot analysis

Mixed blood stage parasites of *P. falciparum* strain NF54 were harvested and treated with 0.15% saponin for erythrocyte lysis. Parasite nuclear pellet and cytoplasmic fractions were prepared as described previously [11]. Parasite pellets, the nuclear pellet and cytoplasmic fractions as well as pellets of non-infected erythrocytes or immunoprecipitated bead-bound proteins (see below) were washed with PBS, resuspended and sonicated in lysis buffer (20 mM Tris-HCl pH 8.0, 10 mM EDTA pH 8.0, 400 mM NaCl, 1 mM PSMF, 10 mM β-glycerophosphate, 10 mM NaF, 0.25% Triton X-100) supplemented with a protease inhibitor cocktail (Roche Diagnostics). Parasite proteins were separated by SDS-PAGE electrophoresis and transferred to Hybond ECL nitrocellulose membrane (Amersham Biosciences) according to the manufacturer's protocol. Membranes were blocked for non-specific binding by incubation in Tris-buffered saline containing 5% skim milk and 1% BSA, followed by immune recognition for 2 h at RT with rabbit anti-PfCLK-1, rat anti-PfCLK-3 or mouse anti-PfCLK-4, anti-Pf39 [36] or anti-Pfalpha-5 antisera [37]. For control sera from non-immunized animals were used. After washing, membranes were incubated for 1 h at RT with an alkaline phosphatase-conjugated secondary antibody (Sigma-Aldrich) and developed in a solution of nitroblue tetrazolium chloride (NBT) and 5-bromo-4-chloro-3-indoxyl phosphate (BCIP; Sigma-Aldrich) for 5–30 min. Scanned blots were processed using Adobe Photoshop CS software.

Immunoprecipitation assay

Mixed *P. falciparum* NF54 asexual blood stage parasites were harvested and treated with 0.15% saponin for erythrocyte lysis.

Parasite pellets were resuspended in 150 µl lysis buffer and sonicated with 50% amplitude and 50 cycles followed by centrifugation at 13,000×g at 4°C for 10 min. Pre-clearing of lysates was carried out by consecutive incubation with 5% vol. serum of non-immunized mouse, rabbit or rat, and 20 µl of protein G-beads (Santa Cruz Biotechnology) for 30 min each at 4°C. Lysates were then incubated for 2 h with the respective kinase-specific antisera (see above) or anti-Pf39 antisera under constant agitation at 4°C. Subsequently the antibody-antigen complexes were incubated over night with 20 µl of protein G-beads at 4°C. The beads were centrifuged, washed four times with 1×PBS and were subsequently deployed in Western blotting (see above) or the kinase activity assay (see below). Gels were stained after resolving with Coomassie Blue Stain and scanned for digital processing using Adobe Photoshop CS software.

Kinase activity assay

Kinase reactions of 30 µl were carried out in a standard kinase buffer (20 mM Tris-HCl, pH 7.5, 20 mM MgCl$_2$, 2 mM MnCl$_2$, 10 mM NaF, 10 mM β-glycerophosphate, 10 µM ATP and 0.1 MBq [γ-^{32}P]-ATP), using immunoprecipitated endogenous kinases as well as 5–10 µg of the respective substrate. When the PfPKRP-specific immunoprecipitates were used in the assays, 50 mM of bovine calmodulin (CaM; dissolved in water; CalBiochem) were added to the reaction. Recombinantly expressed His$_6$-tagged protein kinase 6 (rPK6), purified as previously reported [38], was used as a positive control for the kinase activity assay utilizing exogenous substrates. For testing of the CLK inhibitors, sorbitol-synchronized ring stages were incubated with the inhibitors at IC$_{50}$ concentrations for 12 h before immunoprecipitation. In the assays, exogenous substrates (histone H1, myelin basic protein (MBP), and α- and β-casein; Sigma-Aldrich), and the recombinant SR proteins used as phosphoacceptor substrates were added to the reaction. The recombinant SR proteins were purified via chromatography as described above. For negative control, purified MaBP-tag alone was used for substrate in the kinase activity assays. An additional negative control, in which the parasite lysate was replaced by the same volume of PBS (PBS control), was used to exclude unspecific phosphorylation of reaction components. Reactions were incubated at 37°C for 1 h under constant agitation and terminated by addition of reducing Laemmli buffer for 10 min at 95°C. Samples were separated on 12% SDS-PAGE, dried by means of vacuum gel drying and exposed to X-ray films for 48–90 h at -20°C. For quantification of inhibition of phosphorylation, the mean grey values (MGV) of the phosphorylation bands were measured using the ImageJ program. Background values were subtracted and the MGVs for the DMSO control were set to 100% to calculate the relative MGV (rMGV).

Malstat assay

The CLK inhibitors were screened for antiplasmodial activity against *P. falciparum* strain 3D7 at concentrations of 6.4 nM–500 µM, using the Malstat assay as described [39–41]. Sorbitol-synchronized ring stages were plated in triplicate in 96-well plates (200 µl/well) at a parasitemia of 1% in the presence of the compounds. Chloroquine (diphosphate salt, Sigma-Aldrich; dissolved in double-distilled water) served as positive control in all experiments. Incubation of parasites with DMSO alone at a concentration of 0.5% vol. was used as negative control. Treated parasites were cultivated *in vitro* for 72 h, resuspended and aliquots of 20 µl were transferred to a new plate. 100 µl of the Malstat reagent was added to initiate the conversion of lactate to pyruvate by parasite lactate dehydrogenase (pLDH) in the

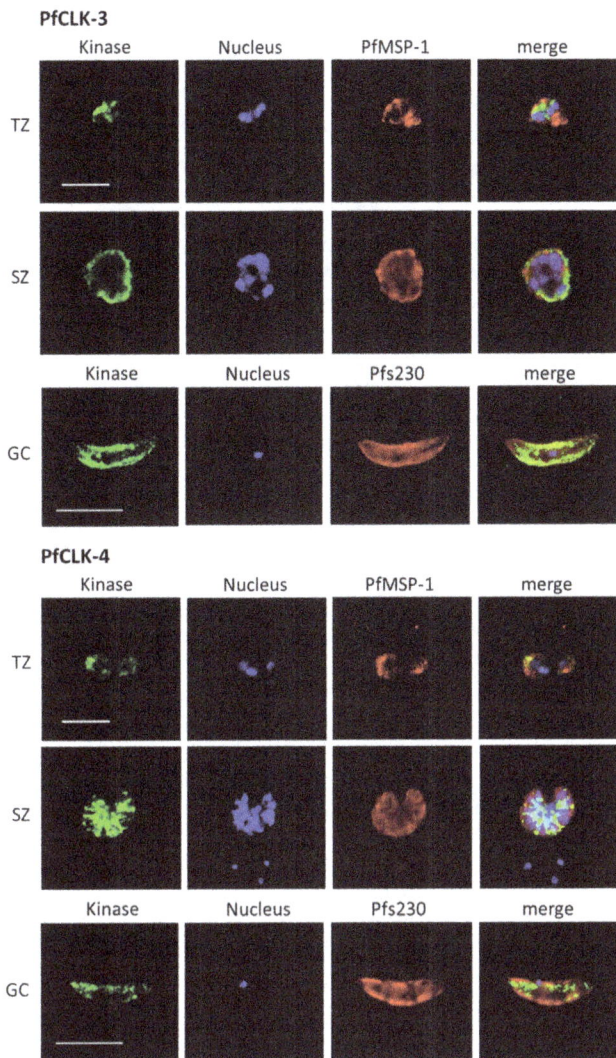

Figure 3. Subcellular localization of PfCLK-3 and PfCLK-4 in the blood and gametocyte stages. Mixed asexual blood stage cultures containing trophozoites (TZ) and schizonts (SZ) and mature gametocyte (GC) cultures were fixed with methanol and prepared for IFA, using rat antisera against PfCLK-3 and mouse antisera against PfCLK-4 (green). The parasite nuclei were highlighted by Hoechst staining (blue). The asexual blood stages were labelled with rabbit antisera against PfMSP-1 and gametocytes with rabbit antisera against Pfs230 (red). Bar, 5 μm.

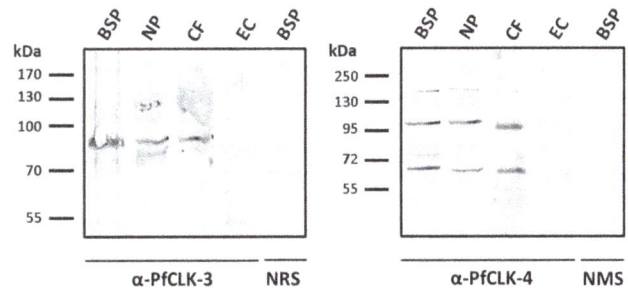

Figure 4. Expression of PfCLK-3 and PfCLK-4 in the cytoplasmic and nuclear fractions of blood stage parasites. Western-blot analyses on lysates of mixed blood stage parasites (BSP) as well as of nuclear pellet (NP) or cytoplasmic fraction (CF) of BSP using rat antisera against PfCLK-3 or mouse antisera against PfCLK-4 detected the full length protein bands for PfCLK-3 and PfCLK-4 of approximately 80 and 150 kDa, respectively. For PfCLK-4, two additional protein bands of approximately 100 and 70 kDa are present. No proteins were detected by the anti-PfCLK antisera in lysates of non-infected erythrocytes (EC). Immunoblotting with sera from non-immunized rat (NRS) or mouse (NMS) did not result in the labelling of any protein bands.

presence of the co-factor APAD. The reduced APAD (APADH) formed during the reaction is used for the assessment of pLDH activity by adding a 20 μl of a mixture of NBT (nitro blue tetrazolium)/Diaphorase (1:1; 1 mg/ml stock each) to the Malstat reaction. The APADH and the NBT form a purple coloration and the absorbance was measured at OD (630 nm). Each compound was tested 2 to 4 times, and the IC_{50} values were calculated from variable-slope sigmoidal dose-response curves using the GraphPad Prism program version 5.

Stage of inhibition assay

The compounds CHX, KH-CARB-10, KH-CARB-11, KH-CARB-13xHCl, and gea-27 were added to synchronized ring stage parasites (T0) in IC_{50} (0.8, 7.5, 6.1, and 4.0 μM, respectively) and IC_{80} (4.0, 37.0, 30.0, and 20.0 μM respectively) concentrations, and blood smear samples were taken at 24, 36, 48, and 60 h of incubation with CHX and at 24 h of incubation in case of the other inhibitors. The numbers of ring stages, trophozoites and schizonts as well as of dead parasites were counted for a total number of 100 infected erythrocytes for each setting.

Gametocyte toxicity assay

P. falciparum NF54 parasites were grown at high parasitemia to favour gametocyte formation. Upon appearance of stage II gametocytes, 1 ml of culture was aliquoted in triplicate in a 24-well plate in the presence of compounds at the respective IC_{50} concentrations. The gametocytes were cultivated for 7 d and the medium was replaced daily. For the first 48 h of cultivation, the gametocytes were treated with CLK inhibitors; subsequently the medium was compound-free. At day 7, Giemsa-stained blood smears were prepared and the gametocytemia was evaluated by counting the numbers of gametocyte stages IV and V in a total number of 1000 erythrocytes. Negative controls were performed with 0.5% vol. of DMSO. Chloroquine was used as additional negative control, while epoxomicin was used as positive control [41]. Two independent experiments in triplicates were conducted and the mean gametocytemia was calculated for each compound. Data from the experimental cultures was normalized to the DMSO control, which was set to 100%, and the student's t-test was performed for statistical analysis using the Microsoft Excel 2010 program.

Exflagellation inhibition assay

A volume of 100 μl of mature NF54 gametocyte cultures was pre-incubated with CLK inhibitors in concentrations ranging between 0.1 μM-1 mM for 15 min at 37°C. Each sample was then transferred to RT and 100 μM of xanthurenic acid was added for activation. After another 15 min, the numbers of exflagellation centers were counted in 30 optical fields using a Leica DMLS microscope by 400-fold magnification. Two independent experiments were performed in duplicates and the inhibition of exflagellation was calculated as a percentage of the number of exflagellation centers in compound-treated cultures in relation to the number of exflagellation centers in untreated controls with 0.5% vol. of DMSO. The IC_{50} values were calculated from variable-slope sigmoidal dose-response curves using the GraphPad Prism program version 4.

PfSRSF12

PfSFRS4

PfSF-1

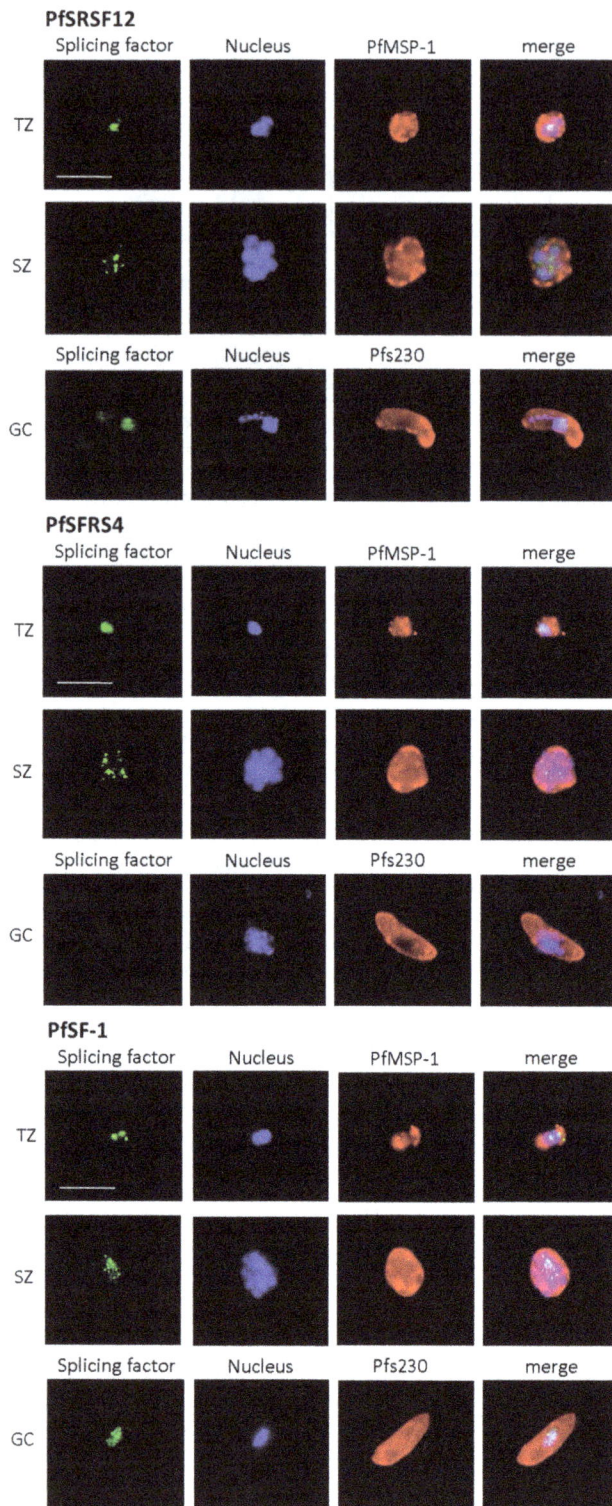

Figure 5. Subcellular localization of the SR proteins in the blood and gametocyte stages. Mixed asexual blood stage cultures containing trophozoites (TZ) and schizonts (SZ) and mature gametocyte (GC) cultures were fixed with paraformaldehyde and prepared for IFA, using mouse antisera against PfSRSF12, PfSFRS4 and PfSF-1 (green). The parasite nuclei were highlighted by Hoechst staining (blue). The asexual blood stages were labelled with rabbit antisera against PfMSP-1 and gametocytes with rabbit antisera against Pfs230 (red). Bar, 5 μm.

Results

The PfCLKs show homologies with yeast SR protein kinase Sky1p

The genome of *P. falciparum* encodes four proteins clustering within the CLK family, termed PfCLK-1-4 [9]. In PfCLK-1-3, the predicted kinase catalytic domains are located at the C-terminus; the catalytic domain of PfCLK-4, on the other hand, starts at the N-terminus and spans more than half of the protein (see Fig. 1A). While *in silico* analysis of gene sequences previously revealed two nuclear localization signals for PfCLK-1 and one for PfCLK-2 upstream of the C-terminal catalytic domains [11], PfCLK-3 and PfCLK-4 do not comprise such signals.

The catalytic domain sequences of the four PfCLKs showed homologies with the *Saccharomyces cerevisae* SR protein kinase Sky1p [42], as was shown by sequence alignment (Fig. S1) [11]. The sequence DLKPxN with the conserved Aspartate 126 is present and is considered to be the catalytic base. The loop at positions 169–193 signifies the activation segment, starting with Aspartate 169 and ending with sequence APE in PfCLK-1, PfCLK-3, PfCLK-4 and SPE in PfCLK-2 and Sky1p. The ATP-binding domain GXGXXG is present at positions 8–13 for PfCLK-1 and GXGXXS for PfCLK-3, PfCLK-4 and Sky1p, but is missing in PfCLK-2. Furthermore, sequence alignment with Sky1p revealed matches between substrate binding residues of the kinases with the substrate binding site of Sky1p, including Arginine 187, Tyrosine 189, Arginine 190 and Glutamate 215 (Fig. S1).

The PfCLKs and the SR proteins are expressed in the blood stages of *P. falciparum*

Previous studies showed that PfCLK-1 and PfCLK-2 are present in the asexual blood stages and in gametocytes of *P. falciparum*. Here, PfCLK-1 can be found in the nucleus of trophozoites, while in schizonts and gametocytes the kinase is present in nucleus and cytoplasm. PfCLK-2 is detected in the nucleus and the cytoplasm of the asexual blood and gametocyte stages [11]. Via IFA we now show that PfCLK-3 and PfCLK-4 are mainly present in the nucleus of trophozoites, while in schizonts and gametocytes both kinases are primarily located in the cytoplasm (Fig. 3). Particularly for PfCLK-3 a rim-associated labelling pattern was observed in the latter stages. In the IFAs, the asexual blood stage parasites and the gametocytes were highlighted by labelling of plasmalemma-associated proteins, i.e. PfMSP-1 and Pfs230, respectively.

The presence of PfCLK-3 and PfCLK-4 in the nucleus and cytoplasm of blood stage parasites was subsequently confirmed by Western blot analysis. Immunoblotting of blood stage parasite lysate with rat anti-PfCLK-3 antisera labelled the full length kinase of approximately 80 kDa (Fig. 4). Full length PfCLK-3 was further detected, when nuclear pellet and cytoplasmic fractions were immunoblotted with the respective antibody. Immunoblotting of the blood stage lysate as well as nuclear pellet and cytoplasmic fractions with mouse anti-PfCLK-4 antisera, on the other hand, resulted in the labelling of three bands, the full-length kinase band of approximately 160 kDa and two additional bands with molecular weights of approximately 100 and 70 kDa, which might represent processing products (Fig. 4). Lysate of non-infected erythrocytes were used for control, and no protein bands were detected after immunoblotting with the respective anti-PfCLK antisera. Similarly, no protein bands were detected, when the blood stage parasite lysates were immunoblotted with sera of non-immunized animals (Fig. 4).

We then investigated the blood stage-specific localization of three previously identified SR proteins, i.e. PfSRSF12, PfSFRS4

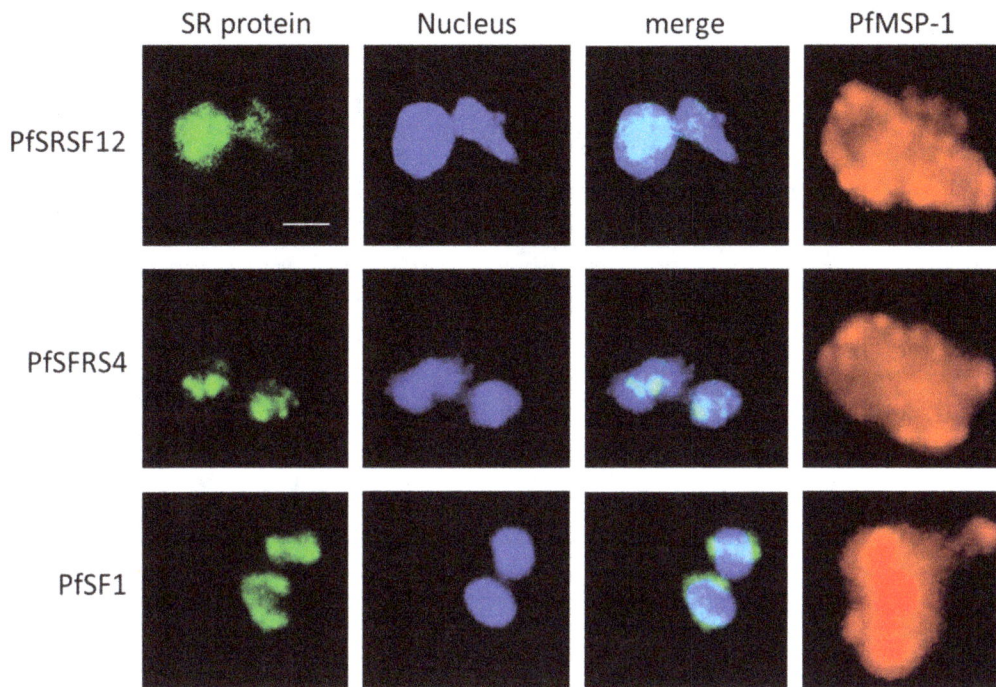

Figure 6. Localization of the SR proteins in the parasite nucleus. Transforming trophozoites (2-nuclei stages) were prepared for IFA as described in Fig. 5 and the localization of the SR proteins was investigated in the Hoechst-positive parasite nuclei. Bar, 5 μm.

and PfSF-1 (see Fig. 1B) [14]. Transcriptome data available at PlasmoDB point to a predominant transcript expression in the trophozoite stage for all three SR proteins [17]. In accord with these data, IFAs, using respective antisera raised in mice, detected the three SR proteins in the trophozoite nucleus (Fig. 5, top rows). An additional minor labelling was observed in the nuclei of the schizont stages (Fig. 5, center rows). Furthermore, PfSRSF12 and PfSF-1 were present in the nucleus of gametocytes, while PfSFRS4 was not detected in these stages (Fig. 5, bottom rows). In depth analysis of the localization of the SR proteins in transforming trophozoites (2-nuclei stage) confirmed that the splicing factors are present in distinct areas of the parasite nuclei, while they cannot be detected in the cytoplasm of the parasites (Fig. 6).

We subsequently performed co-localization experiments between the three SR proteins, using the respective mouse antisera, and PfCLK-1, using antisera raised in rabbit. The co-localization experiments confirmed that the SR proteins are solely present in the nuclei, while in the early schizont stage, PfCLK-1 is present both in the parasite nuclei and the cytoplasm. Co-localization of PfCLK-1 with the three SR proteins can be detected in distinct nuclear regions (Fig. S2A).

In a subsequent step, the numbers of parasites positive for the PfCLKs and the SR proteins were determined. When blood stage schizonts were highlighted by immunolabelling with anti-MSP-1 antibody or by Hoechst nuclear staining, $99\pm1.0\%$ of schizonts labelled for PfCLK-1-3 or PfSF-1, $96\pm2\%$ of schizonts labelled for PfCLK-4, $92\pm2.8\%$ of schizonts labelled for PfSFRS4, and $94\pm0.6\%$ of schizonts labelled for PfSRSF12. Further $100\pm0.9\%$ of Pfs230-positive gametocytes labelled for PfCLK-1, $98\pm1.1\%$ of gametocytes labelled for PfCLK-2, $84\pm2.3\%$ of gametocytes labelled for PfCLK-3, $94\pm1.1\%$ of gametocytes labelled for PfCLK-4, $96\pm1.6\%$ of gametocytes labelled for PfSF-1, and $96\pm0.6\%$ of gametocytes labelled for PfSRSF12.

No labelling was detected, when serum of non-immunized mice was used in the IFAs. Further, IFAs using mouse antisera directed against the GST- and MaBP-tags did not result in any labelling of the blood stage parasites (Fig. S2B).

The PfCLKs mediate phosphorylation of SR proteins

We aimed to determine the phosphorylation activity of precipitated PfCLKs obtained from asexual blood stage cultures. As described previously, immunoblotting of blood stage parasite lysate with mouse or rabbit anti-PfCLK-1 antisera resulted in a processed kinase band of approximately 60 kDa (Fig. S3A), while mouse anti-PfCLK-2 antisera detected a full-length protein band at 150 kDa [11]. Furthermore, immunoblotting of blood stage parasite lysate with rat anti-PfCLK-3 or mouse anti PfCLK-4 antisera detected full-length kinase bands of approximately 80 and 160 kDa, respectively (see above, Fig. 4). When the PfCLKs were immunoprecipitated from the blood stage parasite lysate using the respective antisera, the full-length kinase bands for PfCLK-2-4 were detected, when the same antisera were used for immuno-blotting (Fig. S3B). It was not possible, however, to detect the processed 60 kDa band typical for PfCLK-1, since protein bands of this molecular weight interfere with the protein band of the heavy chain of the precipitating antibody, which is running as a smear at a molecular weight of approximately 55 kDa (Fig. S3B). For control, sera of non-immunized animals were used in immunoblotting and no precipitating protein bands (with the exception of the heavy chain antibody band) were detected (Fig. S3C). Similarly, no precipitated protein bands were detected, when mouse antisera directed against the endoplasmic reticulum-associated protein Pf39 [43] or the proteasome subunit alpha 5 (Pfalpha-5) [37] were used for immunoblotting (data not shown).

The immunoprecipitated PfCLK proteins were used in the kinase activity assays as described previously [11], adding as

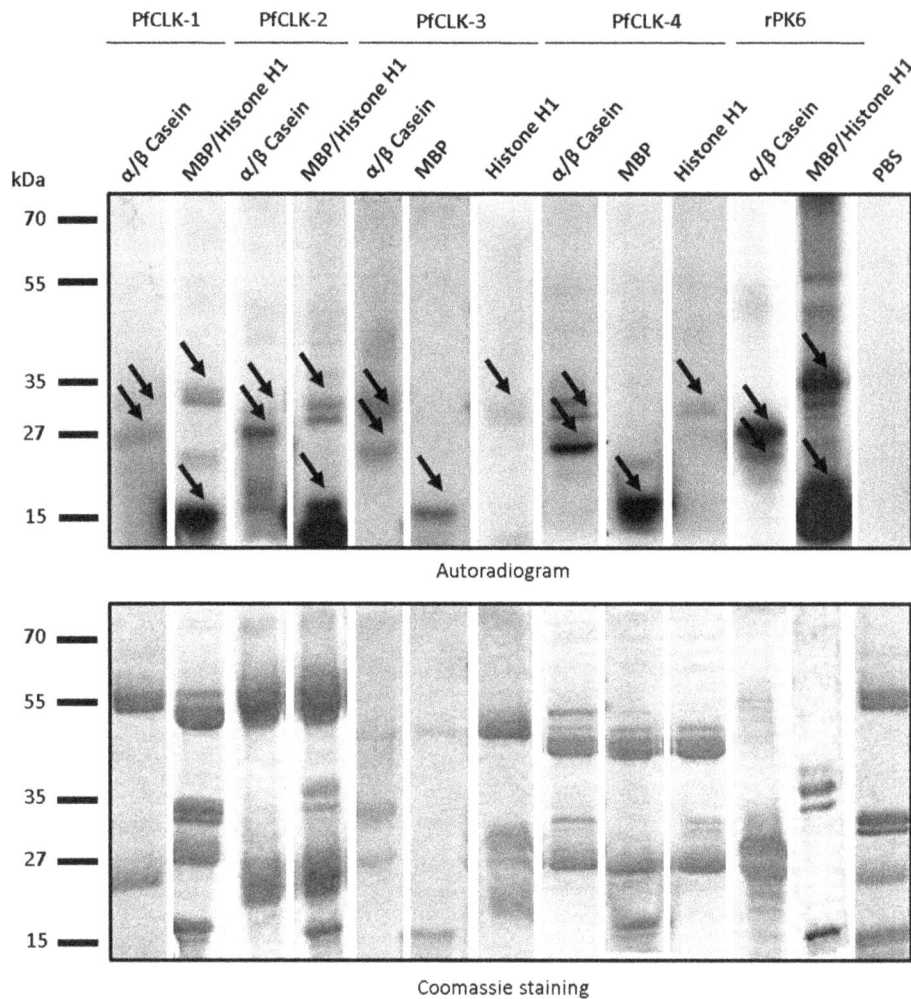

Figure 7. Phosphorylation of exogenous substrates by immunoprecipitated PfCLKs. Kinase activity assays were deployed to detect phosphorylation of the substrates histone H1, MBP and α/β casein (~33, 18, and 28/34 kDa, respectively; indicated by arrows) by the four immunoprecipitated PfCLKs (autoradiogram, upper panel), using the PfCLK-specific respective mouse, rabbit or rat antisera. Assays without precipitated proteins (PBS control) were used for negative controls. Recombinant protein kinase 6 (rPK6) was used for positive control. Coomassie blue staining (lower panel) of radiolabelled SDS-gels was used as a loading control.

exogenous substrates histone H1, MBP, or α/β casein, as well as radiolabelled [γ-^{32}P] ATP. All PfCLKs were associated with phosphorylation activity of all three substrates (Fig. 7, upper panel). Recombinant PfPK6 was used as a positive control [38] and also exhibited strong phosphorylation activity of the three exogenous substrates. As negative control, the assay was performed without precipitated kinases (PBS control). Coomassie blue staining of radiolabelled gels were used for loading control (Fig. 7, lower panel).

Subsequently we wanted to determine if the plasmodial SR proteins PfSRSF12, PfSFRS4, and PfSF-1 function as *in vitro* substrates for the immunoprecipitated PfCLKs. Peptides comprising the RS-rich domains were bacterially expressed as tagged fusion proteins. For PfSF-1, two recombinant proteins were expressed, comprising either the N-terminal part with the RS-rich domain or the C-terminal part, which includes the RNA recognition motif (see Fig. 1B). The SR proteins were purified via affinity chromatography before deploying these as substrates in the above described kinase activity assays. The assays revealed that recombinant PfSRSF12 was phosphorylated by anti-PfCLK-2 and anti-PfCLK-3 precipitate (Fig. 8A, upper panel), whilst recombi-

nant PfSFRS4 was phosphorylated by all PfCLK-specific precipitates. When the immunoprecipitates were incubated with the N-terminal fraction of PfSF-1, phosphorylation signals were detected with the exception of the PfCLK-3 precipitate. In contrast, no phosphorylation signals were detected, when the immunoprecipitates were incubated with the C-terminal part of PfSF-1, which did not comprise the RS-rich motif (Fig. 8B, upper panel). Purified MaBP-tag was used as a substrate instead of the recombinant SR proteins and served as a negative control in the assays, and no phosphorylation of MaBP was observed (Fig. 8C, upper panel). Coomassie blue staining of radiolabelled gels was used as a loading control (Fig. 8, lower panels).

CLK inhibitors impair the phosphorylation activity of the PfCLKs

Previous attempts to knock-out the *pfclk* genes via single cross-over gene disruption in order to study the function of the PfCLKs was unsuccessful [11,15]. However, locus modification (tagging each sequence with a Myc epitope to the 3′-end) was achieved [11,15] (S. Kern, G. Pradel, unpublished observations), demonstrating that the genomic loci are accessible for recombination.

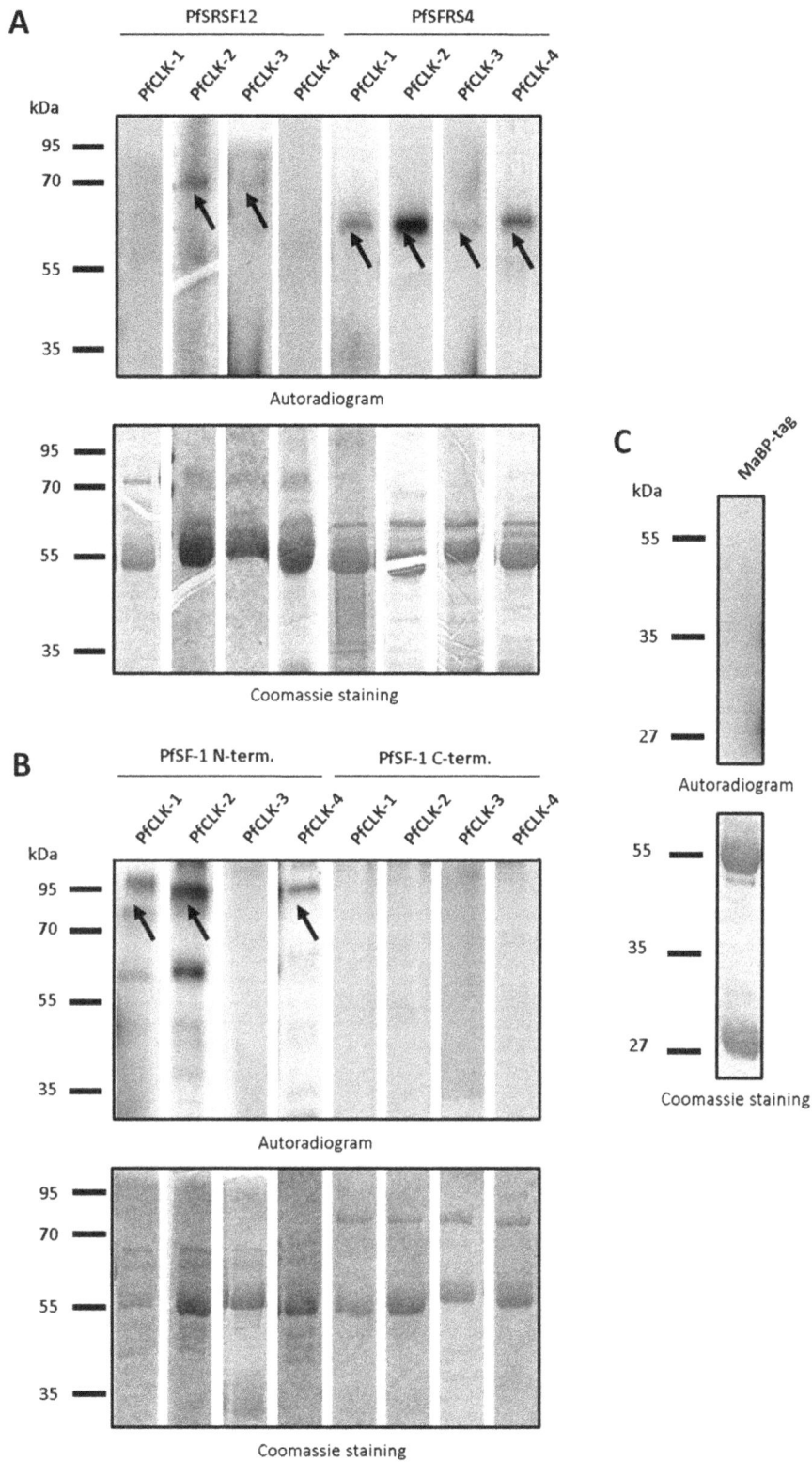

Figure 8. Phosphorylation of plasmodial SR proteins by immunoprecipitated PfCLKs. A. Kinase activity assays were deployed to detect phosphorylation of recombinant PfSRSF12 and PfSFRS4 (~73 and 65 kDa, respectively; indicated by arrows) by two or more of the PfCLKs (autoradiogram; upper panel). B. The N-terminal part (~95 kDa; indicated by arrows), but not the C-terminal part (86 kDa) of recombinant PfSF-1 was phosphorylated by immunoprecipitated PfCLKs. An additional phosphorylation signal of truncated N-terminal PfSF-1 was visible at approximately 60 kDa. C. MaBP-tag alone (43 kDa) as substrate was used as negative control. Shown here is an assay using PfCLK-3-specific immunoprecipitate, similar results were obtained with immunoprecipitates of other PfCLKs (not shown). Coomassie blue staining (lower panels) of radiolabelled SDS gels was used as a loading control.

Table 1. Antiplasmodial and transmission blocking activities of the CLK inhibitors.

Compound	Class	IC$_{50}$ (Malstat) [μM]	IC$_{50}$ (EIA) [μM]
KH-CARB10	oxo-β-carboline	7.2±2.40	14.0±3.72
KH-CARB11	oxo-β-carboline	6.1±1.09	15.8±0.84
KH-CARB13xHCl	oxo-β-carboline	4.4±1.84	13.8±6.22
C-117	aminopyrimidine	7.5±2.76	67.5±13.03
gea-27	aminopyrimidine	5.2±0.35	154.2±56.43
CHX	antiseptic	0.6±0.20	19.8±0.93
Chloroquine	4-aminoquinoline	0.04±0.013	N/A

These data lead to the conclusion that the PfCLKs have essential roles for the parasite blood stages, prohibiting the survival of PfCLK knock-out mutants.

In order to functionally analyse the role of the PfCLKs for the *P. falciparum* blood and sexual stages, we aimed at a chemical inhibition of the kinases. We screened a small compound library of 63 human CLK inhibitors [22,44] for antiplasmodial activity, using the Malstat assay (Table S2). We identified five compounds, i.e. the oxo-β-carbolines KH-CARB-10, KH-CARB-11, KH-CARB-13xHCl, and the aminopyrimidines C-117 and gea-27, which exhibited half-maximal growth inhibitory concentrations (IC$_{50}$) in the low micromolar range (Table 1). We further screened the antiseptic chlorhexidine (CHX) for antiplasmodial activities, because it has previously been reported to act on *P. falciparum* [45] as well as on CLKs [46,47]. In the Malstat assay, CHX exhibited an IC$_{50}$ value in the nanomolar range (Table 1).

In order to confirm that the inhibitors act on the PfCLKs, we tested these for their inhibitory effect on PfCLK-mediated phosphorylation. Blood stage parasites were incubated with the three inhibitors gea-27, KH-CARB-13xHCl, and CHX at IC$_{50}$ concentrations for 12 h. DMSO was used for negative control. The kinase activity assays were conducted as described above, using MBP for substrate. The signals for phosphorylated MBP were detected by autoradiography and the rMGV of the phosphorylation signals was measured. The assays revealed a reduction in the rMGV by 24.1–76.4%, when the parasites were incubated with the inhibitors prior to the assay (Fig. 9A). For gea-27, no effect on the phosphorylation activity of PfCLK-4 was detected.

As control, the assays were carried out on immunoprecipitates using antisera against the CaM-dependent protein kinase-related protein PfPKRP [15]. The 295 kDa PfPKRP has an N-terminal catalytic domain (Fig. S4A) and is a homolog of the *P. berghei* PKRP, which plays a role for parasite transmission to the mosquito [48]. Because protein expression of PfPKRP has not yet been described in the *P. falciparum* blood stages, we tested the antisera in IFA prior to use in the kinase activity assays. PfPKRP was detected in the asexual blood stage schizonts (Fig. S4B). In gametocytes, the kinase is present throughout maturation from stage II to stage V, and here is localized in the cytoplasm (Fig. S4B, C). Because PfPKRP was annotated as a CaM-dependent kinase, the experiments were conducted with and without addition of 50 mM CaM. In the presence of CaM, the PfPKRP-specific immunoprecipitate was able to phosphorylate MBP (Fig. 9B). In reactions lacking CaM, no specific phosphorylation of MBP was detected, indicating that the activity of PfPKRP is CaM-dependent. Noteworthy, no differences in the MBP phosphorylation signals were detected between precipitates of parasites treated with the CLK inhibitors and DMSO-treated control parasites (Fig. 9B). The PfPKRP control experiment on the one hand demonstrates that the CLK inhibitors do not inhibit PfPKRP, and on the other hand proves that the decrease in the phosphorylation signal of the PfCLK-specific precipitates seen after treatment with the CLK inhibitors is not caused by a general reduced viability of the inhibitor-treated parasites. As a second negative control, antisera against Pf39 [43] was used for immunoprecipitation. Pf39-specific precipitate did not phosphorylate MBP, showing that in the absence of immunobound kinases, the parasite precipitate has no phosphorylation activity (Fig. 9C).

PfCLK inhibitors block schizogony and sexual stage development

After confirming the PfCLK-specific inhibitory activity, we aimed to determine at which developmental stage the parasites become arrested by the inhibitors. This was investigated using the stage-of-inhibition assay as previously described [41,49]. The most active inhibitor, CHX, was given to the synchronized ring stage parasites (T0) at approximate IC$_{50}$ and IC$_{80}$ concentrations and Giemsa smears were taken at 24–60 h of CHX incubation. The Giemsa smears showed that the CHX-treated parasites developed to trophozoites but that the majority of the parasites died before entering the schizont stage. While a few parasites escaped the killing during the first replication cycle at the given concentrations, these died during trophozoites-to-schizont transformation of the second replication cycle (Fig. 10A). To investigate, if a similar killing mechanism applies to the other CLK inhibitors, we determined the stage of inhibition for KH-CARB-10, KH-CARB-11, KH-CARB-13xHCl, and gea-27 after 24 h of incubation of parasites with the respective compound at the approximate IC$_{50}$ and IC$_{80}$ concentrations. In all cases, the blood stage parasites died, once they entered schizogony (Fig. 10B). The lowest killing effect at 24 h of compound incubation was observed for gea-27, here schizonts were observed in parasite samples treated with IC$_{50}$ and IC$_{80}$ concentrations of the compound (Fig. 10B).

Once the antiplasmodial activity of the CLK inhibitors was demonstrated, we investigated their effect on the sexual transmission stages of *P. falciparum*. Firstly, the gametocytocidal activity of the compounds was tested, using the gametocyte toxicity test as previously described [41]. The CLK inhibitors significantly reduced the maturation of gametocytes by 40–60% compared to DMSO-treated controls. KH-CARB-11 showed the highest gametocytocidal activity with a reduction of mature gametocytes by 62.5% (Fig. 10C). Chloroquine was used as a negative control, while the proteasome inhibitor epoxomicin was used as a positive control. The CLK inhibitors further impaired exflagellation of *P.*

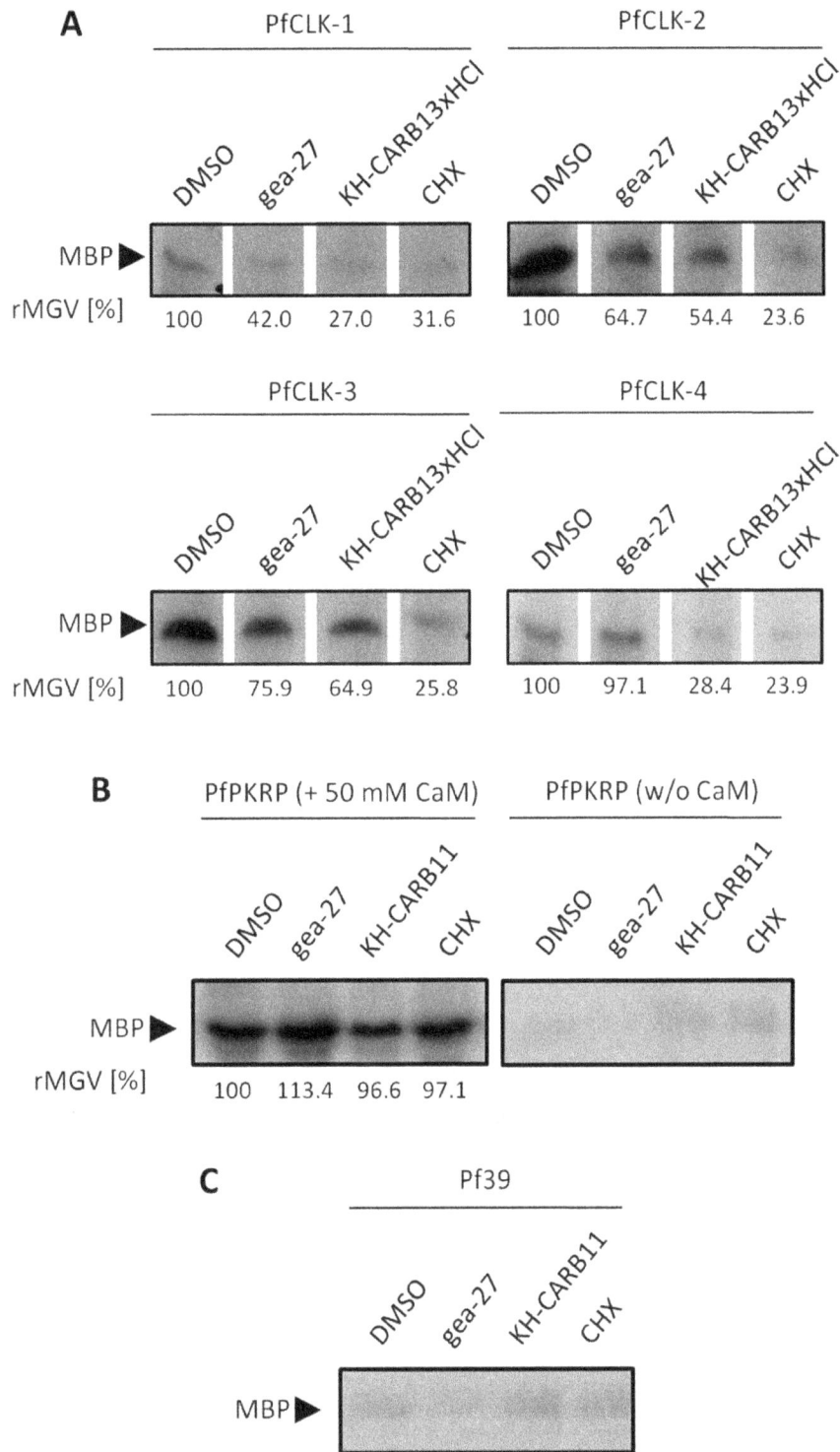

Figure 9. Effect of CLK inhibitors on MBP phosphorylation. A. Kinase activity assays were deployed to detect MBP phosphorylation by inhibitor-treated immunoprecipitated PfCLKs. The parasites were incubated with the CLK inhibitors at IC_{50} concentrations or 0.5% vol. DMSO for 12 h prior to the assays. The phosphorylation signals were measured as rMGV (MGV of DMSO set to 100%). B. PfPKRP-specific immunoprecipitate phosphorylates MBP in the presence but not absence of 50 mM CaM, independent from prior incubation of the parasites with the CLK inhibitors. C. Pf39-specific immunoprecipitate was used as a negative control in the assays.

falciparum, as demonstrated by exflagellation inhibition assay. The strongest inhibitory activities on male gametogenesis were shown for KH-CARB-10, KH-CARB-11, and KH-CARB-13xHCl with IC_{50} values between 10–20 µM (Table 1). The combined data indicate that the PfCLKs play essential roles for erythrocytic schizogony, gametocyte differentiation and gametogenesis.

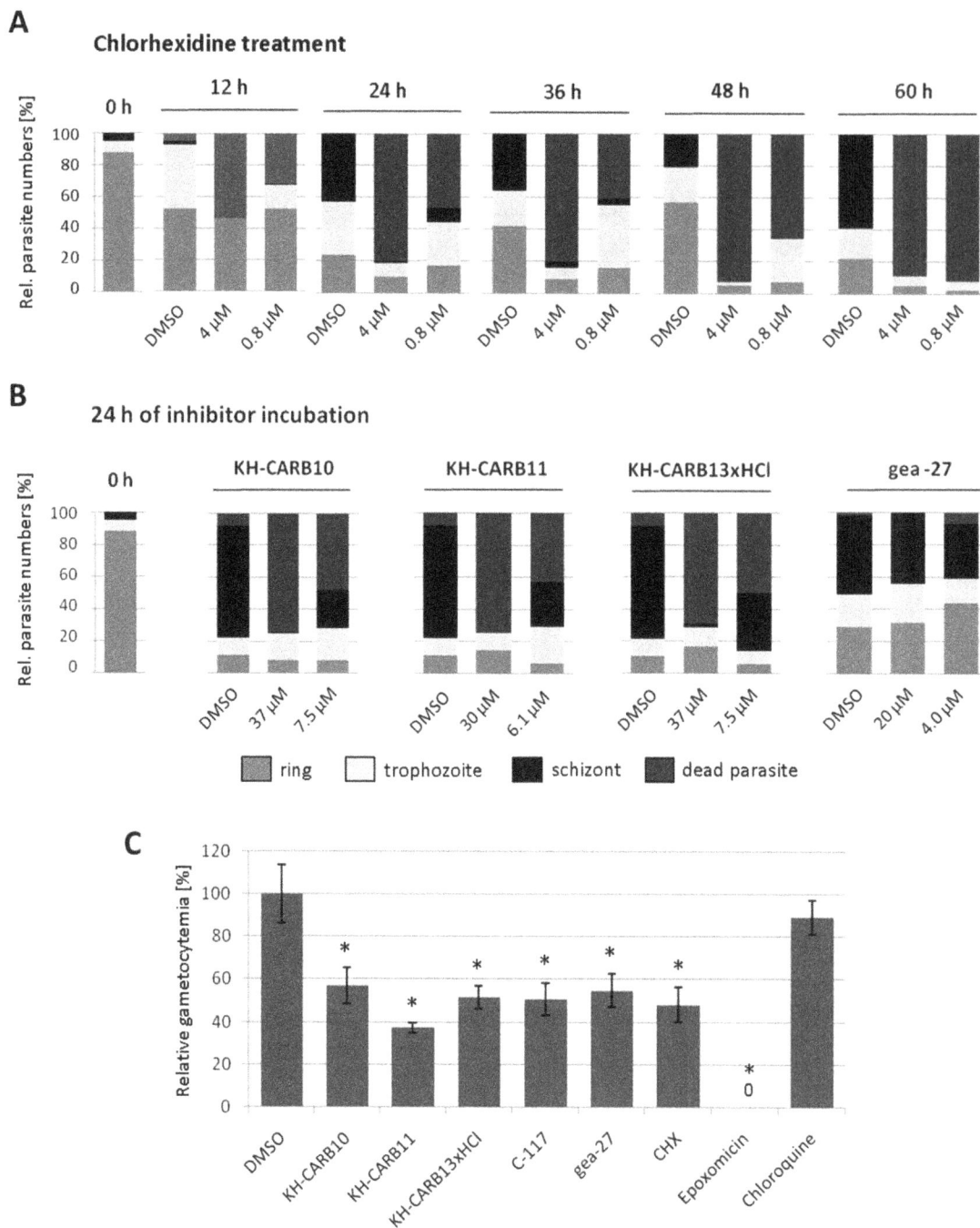

Figure 10. Effect of CLK inhibitors on blood stage parasites. A. Stage of growth inhibition of asexual blood stages between 12–60 h of CHX treatment. CHX at approximate IC_{50} and IC_{80} concentrations or 0.5% vol. of DMSO was added to synchronized parasites at the ring stage. Giemsa stained blood smears were prepared at six time points between 0–60 h of incubation with CHX and the numbers of ring stages, trophozoites, schizonts and dead parasites were counted. Histograms indicate the percentages of developmental stages present in the respective blood smears. B. Parasites were treated with CLK inhibitors as described above and the stage of growth inhibition was determined at 24 h of compound incubation. A total number of 100 parasites were counted in A and B for each condition. C. Compounds at IC_{50} concentrations or 0.5% vol. of DMSO were added to stage II gametocyte cultures for two days. After seven days, the numbers of stage IV and V gametocytes were counted in a total of 1000 erythrocytes and correlated to the gametocyte numbers of DMSO control (set to 100%). *, significant reduction of gametocyte numbers (p<0.001, student's t-test). Epoxomicin was used as positive and chloroquine was used as a negative control.

Discussion

CLKs are dual specificity protein kinases that phosphorylate SR proteins involved in pre-mRNA processing. These, when not in action, reside in the nuclear speckles, but shuttle between the nucleus and the cytoplasm during splicing. The activity of SR proteins is controlled by their phosphorylation status, and a change in phosphorylation changes the ability of SR proteins to interact with other proteins. Because the choice of splice sites during pre-mRNA processing is regulated by the concentration and the phosphorylation status of the SR proteins, CLKs play an indirect role in governing splice site selection emphasizing their importance for AS (reviewed in [50]).

The four PfCLKs of *P. falciparum* show homologies with yeast Sky1p, a kinase involved in mRNA splicing and mRNA transport [42,51]. The kinases are abundantly expressed in the asexual blood stages of *P. falciparum*. In the trophozoites, the PfCLKs are present in the nucleus, where they co-localize with the plasmodial SR proteins PfSRSF12, PfSFRS4 and PfSF-1. During schizogony and gametocyte development, the kinases relocate from the nucleus to the parasite cytoplasm. Noteworthy, PfCLK-1 and PfCLK-2 possess nuclear localization signal sites and are predicted to be localized in the parasite nuclear speckles. A nucleus-cytoplasm passage was demonstrated for both kinases on the ultrastructural level [11]. The combined immunohistological data let us conclude that the PfCLKs accompany splicing factors during the shuttle between nucleus and cytoplasm or phosphorylate substrate in the cytoplasm in dependence on the developmental stage of the blood stage parasite.

In previous studies we were unable to generate *P. falciparum* parasites with disrupted *pfclk* loci, while the insertion of a tagging sequence was possible for all *pfclk* genes [11,15] (S. Kern, G. Pradel, unpublished observation). The inability to knock out a kinase gene locus, together with the ability to modify the allele in a way that does not cause loss-of-function of the gene product, are strongly indicative of an essential role during the asexual blood cycle. The inability to disrupt a single *pfclk* gene further demonstrates that none of the PfCLKs can compensate for the loss of another PfCLK, indicating that the kinases have non-redundant functions. This is in accordance with the findings that multiple kinases are involved in SR protein phosphorylation in different cellular compartments. For example human ASF becomes phosphorylated by SRPK1 and SRPK2 in the cytosol, thus facilitating its nuclear import and accumulation in the nuclear speckles. There, a family of CLKs hyperphosphorylates the SR protein, mediating its release from the nuclear speckles and relocation to the region, in which splicing events occur [52–54]. Noteworthy, in the rodent malaria model *P. berghei*, the *pbclk4* gene (termed *srpk1* in this study) can be knocked out, resulting in impaired exflagellation [55]. In accordance with these findings we recently showed that PfCLK-4 is abundantly expressed in gametocytes, but down-regulated once gametogenesis has ceased [56], indicating that PfCLK-4 has a particular function during parasite transmission from the mammalian host to the mosquito vector.

We identified three plasmodial SR proteins, PfSRSF12, PfSFRS4, and PfSF-1, which function as substrates for the PfCLKs *in vitro*, and the four kinases appear to have different phosphorylation preferences for the three SR proteins. The three SR proteins are predominantly present in the nucleus of blood stage parasites, where they co-localize with the PfCLKs, and PfSRSF12 and PfSF1 can further be detected in gametocytes. Additionally, PfCLK-4 (SRPK1) was previously reported to phosphorylate AS factor SR1 [13]. For all of the plasmodial SR proteins, phosphorylation sites have been identified (Table S3) and the proteins are phosphorylated *in vivo*, as determined in the *P. falciparum* schizont stages [15,57–59]. Noteworthy, a peptide of PfSF-1, which comprises the RNA recognition motifs, but not the RS-rich motifs, cannot be phosphorylated by any of the four PfCLKs, demonstrating the importance of the RS-rich motif for the interaction between SR protein and kinase.

Due to the inability to knock-out the *pfclk* genes by reverse genetics, we aimed to chemically knock-out the PfCLKs for functional studies. We identified five human CLK inhibitors as well as the antiseptic CHX as antiplasmodial agents and demonstrated their inhibitory effect on the PfCLKs via kinase activity assay. The inhibitors act on the *Plasmodium* parasites in the low micromolar range. While we demonstrate that the CLK inhibitors have no activity against the calmodulin-dependent kinase PfPKRP, other off-target effects of the inhibitors can currently not be excluded.

Morphological analyses on drug-treated parasites showed that the inhibitors arrest the parasites during the trophozoites-to-schizont transition. Furthermore, the inhibitors impaired gametocyte development and exflagellation. Particularly the KH-CARB inhibitors acted on blood stage replication and on exflagellation in similar concentrations. These data indicate that the PfCLKs have an important role during schizogony and are further crucial during parasite transmission from the human to the mosquito. These findings make them potential candidates as targets for antimalarials with transmission blocking properties.

Supporting Information

Figure S1   Alignment of kinase domains of the four PfCLKs with SRPK Sky1p of *Saccharomyces cerevisae*.

Figure S2   Co-localization and control IFAs.

Figure S3   Controls of immunoprecipitation assays.

Figure S4   The CaM-dependent kinase PfPKRP.

Table S1   Primers used for cloning.

Table S2   Chemical structures and antimalarial activities of the CLK inhibitors tested in this study.

Table S3   Phosphorylation sites identified in the SR proteins of *P. falciparum* blood stage schizonts.

Methods S1   Synthesis of CLK inhibitors (experimental and analytical details).

Acknowledgments

We thank Ludmilla Sologub (University of Würzburg) for technical assistance and Makoah Nigel Aminake (University of Cape Town) for critical reading of the manuscript. The project was funded by the EU 7th framework programs MALSIG (GP, CD), ANTIMAL (CD) and EVIMALAR (CD) as well as by the Emmy Noether program (GP) and the SFB630 (TD) of the Deutsche Forschungsgemeinschaft. The project was additionally supported by INSERM (CD) and the MRC-Toxicology Unit program leader grant (ABT and MMA). SK received a fellowship from the program for female scientists of the University of Würzburg. SA was supported by the BioMedTec International Graduate School of Science "Lead structures of cell function" of the Elite Network Bavaria, Germany.

Author Contributions

Conceived and designed the experiments: CD RF ABT FB GP. Performed the experiments: SK SA TB CCW LOA MMA. Analyzed the data: TD MMA GP. Contributed reagents/materials/analysis tools: KH APG BS FB. Wrote the paper: SK GP.

References

1. WHO World Malaria Report (2013) http://www.who.int/malaria/world_malaria_report_2013.

2. Deitsch K, Duraisingh M, Dzikowski R, Gunasekera A, Khan S, et al. (2007) Mechanisms of gene regulation in *Plasmodium*. Am J Trop Hyg 77: 201–208.

3. Otto TD, Wilinski D, Assefa S, Keane TM, Sarry LR, et al. (2010) New insights into the blood-stage transcriptome of *Plasmodium falciparum* using RNA-Seq. Mol Microbiol 76: 12–24.

4. Sorber K, Dimon MT, DeRisi JL (2011) RNA-Seq analysis of splicing in *Plasmodium falciparum* uncovers new splice junctions, alternative splicing and splicing of antisense transcripts. Nucleic Acids Res 39: 3820–3835.

5. Bourgeois CF, Lejeune F, Stévenin J (2004) Broad specificity of SR (serine/arginine) proteins in the regulation of alternative splicing of pre-messenger RNA. Prog Nucleic Acid Res Mol Biol 78: 37–88.

6. Wahl MC, Will CL, Lührmann R (2009) The spliceosome: design principles of a dynamic RNP machine. Cell 136: 701–718.

7. Gui JF, Lane WS, Fu XD (1994) A serine kinase regulates intracellular localization of splicing factors in the cell cycle. Nature 369: 678–682.

8. Colwill K, Pawson T, Andrews B, Prasad J, Manley JL, et al. (1996) The Clk/Sty protein kinase phosphorylates SR splicing factors and regulates their intranuclear distribution. EMBO J 15: 265–275.

9. Ward P, Equinet L, Packer J, Doerig C (2004) Protein kinases of the human malaria parasite *Plasmodium falciparum*: the kinome of a divergent eukaryot. BMC Genomics 5: 79.

10. Anamika, Srinivasan N, Krupa A (2005) A genomic perspective of protein kinases in *Plasmodium falciparum*. Proteins 58: 180–189.

11. Agarwal S, Kern S, Halbert J, Przyborski JM, Baumeister S, et al. (2011) Two nucleus-localized CDK-like kinases with crucial roles for malaria parasite erythrocytic replication are involved in phosphorylation of splicing factor. J Cell Biochem 112: 1295–1310.

12. Li JL, Targett GAT, Baker DA (2001) Primary structure and sexual stage-specific expression of a LAMMER protein kinase of *Plasmodium falciparum*. Int J Parasitol 31: 387–392.

13. Dixit A, Singh PK, Sharma GP, Malhotra P, Sharma P (2010) PfSRPK1, a novel splicing-related kinase from *Plasmodium falciparum*. J Biol Chem 285: 38315–38323.

14. Eshar S, Allemand E, Sebag A, Glaser F, Muchardt C, et al. (2012) A novel *Plasmodium falciparum* SR protein is an alternative splicing factor required for the parasites' proliferation in human erythrocytes. Nucleic Acids Res 40: 9903–9916.

15. Solyakov L, Halbert J, Alam MM, Semblat JP, Dorin-Semblat D, et al. (2011) Global kinomic and phospho-proteomic analyses of the human malaria parasite *Plasmodium falciparum*. Nat Commun 2: 565.

16. Fraunholz MJ, Roos DS (2003) PlasmoDB: exploring genomics and post-genomics data of the malaria parasite, *Plasmodium falciparum*. Redox Rep 8: 317–320.

17. Aurrecoechea C, Brestelli J, Brunk BP, Dommer J, Fischer S, et al. (2009) PlasmoDB: a functional genomic database for malaria parasites. Nucleic Acids Res 37: D539–D543.

18. Schultz J, Milpetz F, Bork P, Ponting C (1998) SMART, a simple modular architecture research tool: Identification of signaling domains. Proc Natl Acad Sci USA 95: 5857–5864.

19. Letunic I, Doerks T, Bork P (2012) SMART 7: recent updates to the protein domain annotation resource. Nucleic Acids Res 40: D302–D305.

20. Sayers EW, Barrett T, Benson DA, Bolton E, Bryant SH, et al. (2010) Database resources of the National Center for Biotechnology Information. Nucleic Acids Res 38: D5–D16.

21. Higgins DG, Thompson JD, Gibson TJ (1996) Using CLUSTAL for multiple sequence alignments. Methods Enzymol 266: 383–402.

22. Huber K, Brault L, Fedorov O, Gasser C, Filippakopoulos P, et al. (2012) 7,8-Dichloro-1-oxo-β-carbolines as a versatile scaffold for the development of potent and selective kinase inhibitors with unusual binding modes. J Med Chem 55: 403–413.

23. Bracher F, Hildebrand D (1993) β-Carboline alkaloids, II. Tributyl(1-ethoxyvinyl)stannan als C₂-Baustein für die Synthese von β-Carbolin-Alkaloiden. Liebigs Ann Chem 1993: 837–839.

24. Puzik A, Bracher F (2009) A convenient approach to the canthin-4-one ring system: Total synthesis of the alkaloids tuboflavine and norisotuboflavine. J Heterocyclic Chem 46: 770–773.

25. Gehring AP, Bracher F (2012) A convenient conversion of substituted cyclohexenones into aryl methyl ketones. Synthesis 44: 2441–2447.

26. Ponnudurai T, Leeuwenberg AD, Meuwissen JH (1981) Chloroquine sensitivity of isolates of *Plasmodium falciparum* adapted to *in vitro* culture. Trop Geogr Med 33: 50–54.

27. Walliker D, Quakyi IA, Wellems TE, McCutchan TF, Szarfman A, et al. (1987) Genetic analysis of the human malaria parasite *Plasmodium falciparum*. Science 236: 1661–1666.

28. Alano P, Roca L, Smith D, Read D, Carter R, et al. (1995) *Plasmodium falciparum*: parasites defective in early stages of gametocytogenesis. Exp Parasitol 81: 227–235.

29. Trager W, Jensen JB (1976) Human malaria parasites in continuous culture. Science 193: 673–675.

30. Ifediba T, Vanderberg JP (1981) Complete *in vitro* maturation of *P. falciparum* gametocytes. Nature 294: 364–366.

31. Cranmer SL, Magowan C, Liang J, Coppel RL, Cooke BM (1997) An alternative to serum for cultivation of *Plasmodium falciparum in vitro*. Trans R Soc Trop Med Hyg 91: 363–365.

32. Billker O, Lindo V, Panico M, Etienne AE, Paxton T, et al. (1998) Identification of xanthurenic acid as the putative inducer of malaria development in the mosquito. Nature 392: 289–292.

33. Garcia GE, Wirtz RA, Barr JR, Woolfitt A, Rosenberg R (1998) Xanthurenic acid induces gametogenesis in *Plasmodium*, the malaria parasite. J Biol Chem 273: 12003–12005.

34. Lambros C, Vanderberg JP (1979) Synchronization of *Plasmodium falciparum* erythrocytic stages in culture. J Parasitol 65: 418–420.

35. Williamson KC, Keister DB, Muratova O, Kaslow DC (1995) Recombinant Pfs230, a *Plasmodium falciparum* gametocyte protein, induces antisera that reduce the infectivity of *Plasmodium falciparum* to mosquitoes. Mol Biochem Parasitol 75: 33–42.

36. Scholz SM, Simon N, Lavazec C, Dude MA, Templeton TJ, et al. (2008) PfCCp proteins of *Plasmodium falciparum*: gametocyte-specific expression and role in complement-mediated inhibition of exflagellation. Int J Parasitol 38: 327–340.

37. Aminake MN, Arndt HD, Pradel G (2012) The proteasome of malaria parasites: A multi-stage drug target for chemotherapeutic intervention? Int J Parasitol Drugs Drug Resist 2: 1–10.

38. Bracchi-Ricard V, Barik S, Delvecchio C, Doerig C, Chakrabarti R, et al. (2000) PfPK6, a novel cyclin-dependent kinase/mitogen-activated protein kinase-related protein kinase from *Plasmodium falciparum*. Biochem J 347: 255–263.

39. Makler MT, Hinrichs DJ (1993) Measurement of the lactate dehydrogenase activity of *Plasmodium falciparum* as an assessment of parasitemia. Am J Trop Med Hyg 48: 205–210.

40. Makler MT, Ries JM, Williams JA, Bancroft JE, Piper RC, et al. (1993) Parasite lactate dehydrogenase as an assay for *Plasmodium falciparum* drug sensitivity. Am J Trop Med Hyg 48: 739–741.

41. Aminake MN, Schoof S, Sologub L, Leubner M, Kirschner M, et al. (2011) Thiostrepton and derivatives exhibit antimalarial and gametocytocidal activity by dually targeting parasite proteasome and apicoplast. Antimicrob Agents Chemother 55: 1338–1348.

42. Nolen B, Yun CY, Wong CF, McCammon JA, Fu XD, et al. (2001) The structure of Sky1p reveals a novel mechanism for constitutive activity. Nat Struct Biol 8: 176–183.

43. Templeton TJ, Fujioka H, Aikawa M, Parker KC, Kaslow DC (1997) *Plasmodium falciparum* Pfs40, renamed Pf39, is localized to an intracellular membrane-bound compartment and is not sexual stage-specific. Mol Biochem Parasitol 90: 359–365.

44. Fedorov O, Huber K, Eisenreich A, Filippakopoulos P, King O, et al. (2011) Specific CLK inhibitors from a novel chemotype for regulation of alternative splicing. Chem Biol 18: 67–76.

45. Geary TG, Jensen JB (1983) Effects of antibiotics on *Plasmodium falciparum in vitro*. Am J Trop Hyg 32: 221–225.

46. Younis I, Berg M, Kaida D, Dittmar K, Wang C, et al. (2010) Rapid-response splicing reporter screens identify differential regulator of constitutive and alternative splicing. Mol Cell Biol 30: 1718–1728.

47. Wong R, Balachandran A, Mao AYQ, Dobson W, Gray-Owen S, et al. (2011) Differential effect of CLK SR kinases on HIV-1 gene expression: potential novel targets for therapy. Retrovirology 8: 47.

48. Purcell LA, Leitao R, Ono T, Yanow SK, Pradel G, et al. (2010) A putative kinase-related protein (PKRP) from *Plasmodium berghei* mediates infection in the midgut and salivary glands of the mosquito. Int J Parasitol 40: 979–988.

49. Barthel D, Schlitzer M, Pradel G (2008) Telithromycin and quinupristin-dalfopristin induce delayed death in *Plasmodium falciparum*. Antimicrob Agents Chemother 52: 774–777.

50. Stamm S (2008) Regulation of alternative splicing by reversible protein phosphorylation. J Biol Chem 283: 1223–1227.

51. Siebel CW, Feng L, Guthrie C, Fu XD (1999) Conservation in budding yeast of a kinase specific for SR splicing factors. Proc Natl Acad Sci USA 96: 5440–5445.

52. Misteli T, Cáceres JF, Clement JQ, Krainer AR, Wilkinson MF, et al. (1998) Serine phosphorylation of SR proteins is required for their recruitment to sites of transcription *in vivo*. J Cell Biol 143: 297–307.

53. Lai MC, Lin RI, Huang SY, Tsai CW, Tarn WY (2000) A human importin-beta family protein, transportin-SR2, interacts with the phosphorylated RS domain of SR proteins. J Biol Chem 275: 7950–7957.

54. Aubol BE, Chakrabarti S, Ngo J, Shaffer J, Nolen B, et al. (2003) Processive phosphorylation of alternative splicing factor/splicing factor 2. Proc Natl Acad Sci USA 100: 12601–12606.

55. Tewari R, Straschil U, Bateman A, Böhme U, Cherevach I, et al. (2010) The systematic functional analysis of *Plasmodium* protein kinases identifies essential regulators of mosquito transmission. Cell Host Microbe 8: 377–387.

56. Ngwa CJ, Scheuermayer M, Mair GR, Kern S, Brügl T, et al. (2013) Changes in the transcriptome of the malaria parasite *Plasmodium falciparum* during the initial phase of transmission from the human to the mosquito. BMC Genomics 14: 256.

57. Treeck M, Sanders JL, Elias JE, Boothroyd JC (2011) The phosphoproteomes of *Plasmodium falciparum* and *Toxoplasma gondii* reveal unusual adaptations within and beyond the parasites' boundaries. Cell Host Microbe 10: 410–419.

58. Lasonder E, Green JL, Camarda G, Talabani H, Holder AA, et al. (2012) The *Plasmodium falciparum* schizont phosphoproteome reveals extensive phospha-
tidylinositol and cAMP-protein kinase A signaling. J Proteome Res 11: 5323–5337.

59. Pease BN, Huttlin EL, Jedrychowski MP, Talevich E, Harmon J, et al. (2013) Global analysis of protein expression and phosphorylation of three stages of *Plasmodium falciparum* intraerythrocytic development. J Proteome Res 12: 4028–4045.

5

Paleomicrobiology: Revealing Fecal Microbiomes of Ancient Indigenous Cultures

Raul J. Cano[1]*, Jessica Rivera-Perez[2], Gary A. Toranzos[2], Tasha M. Santiago-Rodriguez[3], Yvonne M. Narganes-Storde[4], Luis Chanlatte-Baik[4], Erileen García-Roldán[2], Lucy Bunkley-Williams[5], Steven E. Massey[2]

1 Center for Applications in Biotechnology, California Polytechnic State University, San Luis Obispo, California, United States of America, 2 Department of Biology, University of Puerto Rico, San Juan, Puerto Rico, 3 Department of Pathology, University of California San Diego, San Diego, California, United States of America, 4 Center for Archaeological Research, University of Puerto Rico, Rio Piedras Campus, San Juan, Puerto Rico, 5 Department of Biology, University of Puerto Rico, Mayaguez Campus, San Juan, Puerto Rico

Abstract

Coprolites are fossilized feces that can be used to provide information on the composition of the intestinal microbiota and, as we show, possibly on diet. We analyzed human coprolites from the Huecoid and Saladoid cultures from a settlement on Vieques Island, Puerto Rico. While more is known about the Saladoid culture, it is believed that both societies co-existed on this island approximately from 5 to 1170 AD. By extracting DNA from the coprolites, followed by metagenomic characterization, we show that both cultures can be distinguished from each other on the basis of their bacterial and fungal gut microbiomes. In addition, we show that parasite loads were heavy and also culturally distinct. Huecoid coprolites were characterized by maize and Basidiomycetes sequences, suggesting that these were important components of their diet. Saladoid coprolite samples harbored sequences associated with fish parasites, suggesting that raw fish was a substantial component of their diet. The present study shows that ancient DNA is not entirely degraded in humid, tropical environments, and that dietary and/or host genetic differences in ancient populations may be reflected in the composition of their gut microbiome. This further supports the hypothesis that the two ancient cultures studied were distinct, and that they retained distinct technological/cultural differences during an extended period of close proximity and peaceful co-existence. The two populations seemed to form the later-day Taínos, the Amerindians present at the point of Columbian contact. Importantly, our data suggest that paleomicrobiomics can be a powerful tool to assess cultural differences between ancient populations.

Editor: Bryan A. White, University of Illinois, United States of America

Funding: This study was partially funded by NIH RISE Program (NIH Grant No. 5R25GM061151-12). The authors also would like to thank Mark Brolaski of Mo Bio Laboratories for providing the necessary supplies and support for the extraction of fossil DNA. The funders had no role in study design, data collection and analysis, decision to publish, or preparation of the manuscript.

Competing Interests: The authors have declared that no competing interests exist.

* Email: rcano@calpoly.edu

Introduction

Coprolites are fossilized fecal specimens that give us an opportunity to infer on an extinct organism's diet and intestinal microbiota if the DNA is well preserved. Taphonomic conditions such as a highly biomineralized environment or a rapid decline in the sample's water activity (Aw), induce the fossilization process [1]. It has been widely believed that feces are not well preserved in tropical environments due to the high humid conditions. This may likely be one of the reasons why coprolite studies of indigenous Caribbean and other tropical/subtropical cultures are scarce. However, we obtained human coprolites from Vieques, an island approximately 8 Km off the southeastern coast of Puerto Rico. Located in the Caribbean Sea, 2,020 km from the Equator, the climate in Vieques is generally humid with yearly temperatures ranging from approximately 24 to 28°C.

Puerto Rico is considered an important area for archaeological studies in the Caribbean due to the variety of ancient deposits that have been found on the island. Over 3,000 ancient settlements have been discovered to date, of which, 250 are located in the Island of Vieques and correspond to at least four different ancient cultures that inhabited the island. Among these cultures were the Saladoids and Huecoids, two horticulturalist cultures that coexisted in Vieques (an Island off the coast of Puerto Rico) for over 1,000 years (from 5 AD to 1170 AD) after migrating from South America. Originally from present-day Venezuela, the Saladoids migrated to the island of Vieques by 160 BC and to the main Island of Puerto Rico by 430 BC [2,3,4]. While living in this region they maintained their ancestral heritage, as shown by their signature use of white and red painted pottery. However, they also incorporated different traits that they gradually learned/observed from other cultures present on the island. In contrast, little is known about the origins of the Huecoid culture, but they are believed to be originally from the eastern Andes in present-day Bolivia and Peru and are known to have settled in Puerto Rico by

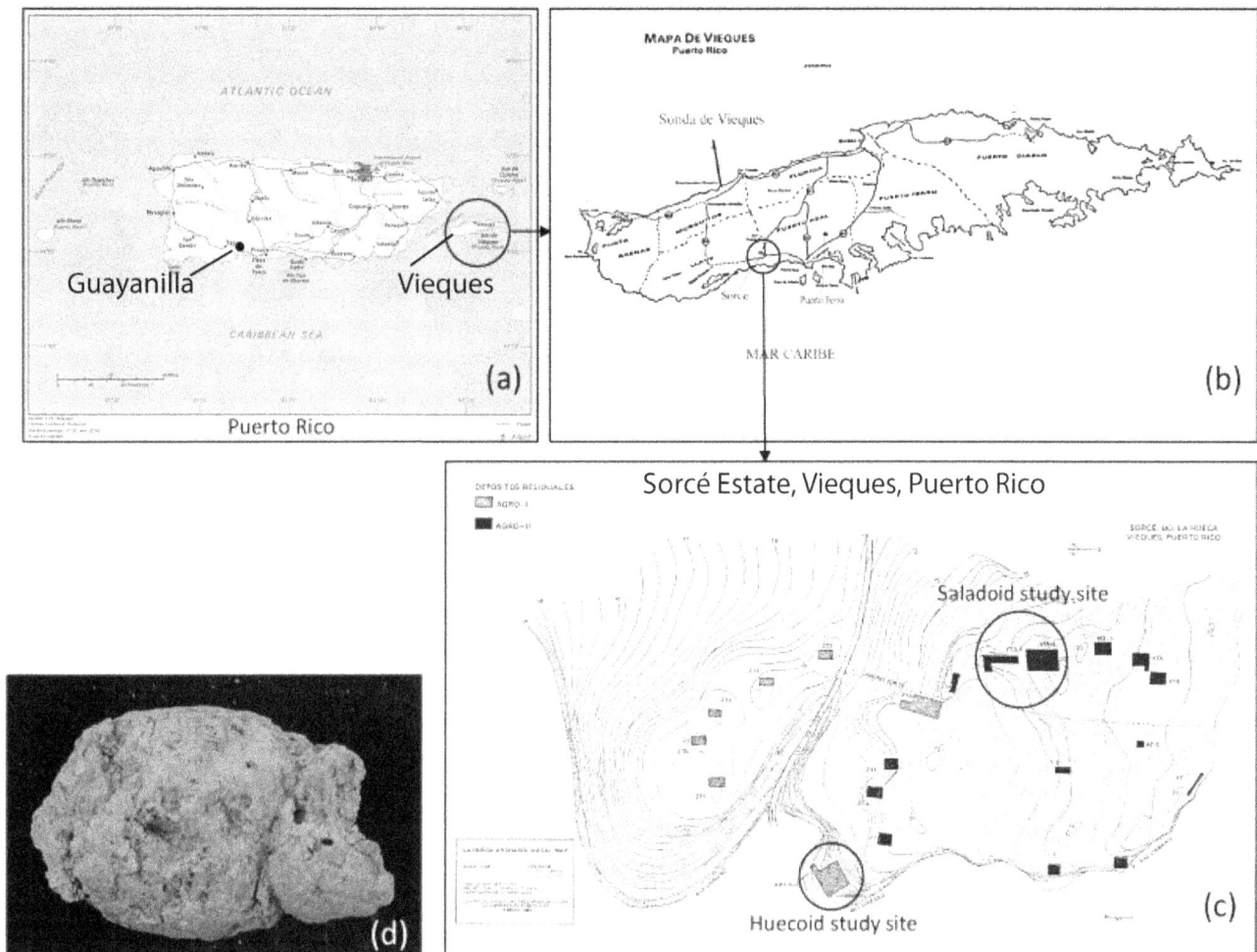

Figure 1. Location and obtainment of coprolites used in this study. Panels (a) and (b) show the sampling sites, located in Sorcé, Vieques, an island off the eastern coast of Puerto Rico. Panel (c) shows the Huecoid and Saladoid archaeological study sites (namely AGRO-I and AGRO-II, respectively). Panel (d) shows a coprolite extracted from these archeological sites.

at least 5 AD. The Huecoids are characterized by their delicate carvings of semiprecious stones and by their resilience to incorporate material or cultural traits from other cultures [5]. The apparent representation of a pair of Andean condors among their amulets suggests to archaeologists their ancestral residence in the Andes. This is also supported by the possible practice of cranial deformation. Both of these cultures greatly impacted other indigenous cultures present on the island, and are thought to have played a part in the development of the predominant culture colloquially referred to as the "Taínos" [6].

Paleomicrobiological studies have shown marked differences between the microbial communities present in Huecoid and Saladoid coprolites [7]. These studies were performed using Terminal Restriction Fragment analysis (TRFL-P), a technique that, although extremely useful for community profiling, has some intrinsic limitations. For example, its total scope extends to the relatively limited database used for downstream analyses. This intrinsic bias implies that microorganisms not in the database will remain hidden to this type of analysis. Also these analyses are often biased towards the most predominant species, leaving out important information on the rare microbiota present in the sample [8]. In light of these limitations, next-generation sequencing (NGS) represents a more appropriate technique for analyzing

these type of samples, mainly due to the high resolving power that characterizes NGS platforms conjointly with bioinformatics tools. Although microbial community profiling when using NGS also has some bias towards the most predominant DNA in the sample, this happens to a lesser extent. Also, NGS is capable of detecting and amplifying previously unidentified/uncultured microorganisms that could be relevant to analysis of the sample's microbial community [9].

The principal aim of this study was to compare these two ancient populations and corroborate whether they present marked differences in their fecal microbiomes due to diet and/or cultural factors. In order to achieve this, we compared the microbiota found in the cores and cortices of coprolites from both cultures using NGS.

Materials and Methods

Sample Description

Coprolites were found in two excavations directed by the archaeologists Dr. Luis A. Chanlatte-Baik and Yvonne M. Narganes-Storde. One excavation site was located at the Sorcé Estate in La Hueca, Vieques, Puerto Rico, the second in Tecla, Guayanilla, PR (**Figure 1**). A total of thirty-four coprolites were used in this study. All thirty-four samples were used for parasitological studies (Huecoid n = 12; Saladoid n = 22). Five of the Saladoid samples used were from Sorce, Vieques and the remaining from Tecla, Guayanilla. In addition, fifteen of the coprolites screened for parasites were also used for microbiome analyses (**Table 1**). Coprolites were provided for this study by Luis Chanlatte-Baik and Yvonne Narganes at the Center for Archaeological Research at the University of Puerto Rico, Rio Piedras Campus. All necessary permits for the collection of samples used in this study were obtained from the Center for Archaeological Research at the University of Puerto Rico, Rio Piedras Campus., complying with all relevant regulations. Repository information, including the nomenclature for precise identification of each specimen containing geographical location, excavation site and archaeological depth for coprolites used in both the DNA and parasitological analyses are described in **Table 1**. Coprolites dated from 180 A.D. to 600 A.D., as indicated by previous ^{14}C dating of material obtained from the same or an equivalent archaeological excavation quadrant and depth of each coprolite (e.g. charcoal and mollusk shells) [10] (**Table 1**). [14]C dating was conducted by Teledyne Isotopes (Westwood, NJ) and BETA Analytic, Inc. (Miami, FL) using standard methods.

Sample Handling and Processing

Coprolites selected for DNA analyses were cut in half; one portion was used for DNA extraction and the other for parasitological studies. All sample processing and DNA extractions for the microbiome analyses were performed in an Ancient DNA laboratory where DNA extraction is conducted in class II hoods assigned exclusively for ancient DNA use. Hoods were exposed to UV light for at least 20 minutes before and after every use. Lab coats designated exclusively for ancient DNA use were routinely decontaminated overnight with commercial chlorine (Clorox). Other aseptic measures include the routine decontamination of the working space with chlorine, the use of sterile, baked and autoclaved DNA-free instruments to extract the DNA, as well as gloves. Controls were done *ad-libitum* for the absence of extraneous DNA.

DNA extraction

To minimize the presence of environmental DNA, approximately 3 mm of the outermost exterior shell of the coprolites was first removed using sterilized brushes as described previously [7]. Approximately 0.25 g of cortex and core samples were separated from each coprolite in the above hoods and DNA was extracted using the PowerSoil DNA Isolation Kit following the manufacturer's instructions (Mo Bio Laboratories, Carlsbad, CA).

Table 1. Description of coprolite samples employed in study.

Sample ID	Specimen Number[1]	Culture	Sampling Area	Radiocarbon date
1a	SV_YTA-1_I-5_6– 60 cm	Saladoid	Cortex	335–395 A.D.
1b			Core	
2a	SV_YTA-2_J-22_80 cm	Saladoid	Cortex	270–385 A.D.
2b			Core	
3a	SV_YTA-2_M-25_40 cm	Saladoid	Cortex	230–385 A.D
3b			Core	
4a	SV_Z-W_2.0 m	Huecoid	Cortex	1300–220 A.D.
4b			Core	
5a	SV_YTA-2_I-24_1-1.2 m	Saladoid	Cortex	Circa 385 A.D.
5b			Core	
6a	SV_Z-M_1.2 m	Huecoid	Cortex	Circa 450 A.D.
6b			Core	
7a	SV_YTA-2_I-15_1 m	Saladoid	Cortex	285–375 A.D
7b			Core	
8a	SV_Z-W_1.8 m	Huecoid	Cortex	Circa 245 A.D.
8b			Core	
9a	SV_Z-W_ 1.6 m	Huecoid	Cortex	Circa 385 A.D.
9b			Core	
10a	SV_Z-C_1.8 cm	Huecoid	Cortex	Circa 245 A.D.
10b			Core	
11b	SV_Z-20_2.0 m	Huecoid	Core	Circa 245 A.D.
12b	SV_Z-X_60 cm	Huecoid	Core	470–600 A.D.
13b	SV_Z-L_70 cm	Huecoid	Core	Circa 385 A.D.
14b	SV_ Z-T_S-1_0.9 m	Saladoid	Core	Circa 385 A.D.
15b	SV_YTA-2_H-21_1.2 m	Saladoid	Core	230–385 A.D

[1]Prefix SV indicates the sampling site was in the Sorcé Estate (S) the Island of Vieques (V) Puerto Rico. The remaining characters refer to the specific excavation site from which the specimens were obtained (e.g., YTA-1_I-5-6_) and archaeological depth (e.g., 60 cm).

DNA amplification and sequencing

Fifteen coprolites encompassing paired samples of cortex and core (for a total of 30 samples) were sequenced using the Ion Torrent PGM System for sequencing (Life Technologies Corp.) of 16S rRNA gene reads and the Roche 454 FLX Titanium instrument for detection of 18S rRNA genes.

The 16S rRNA gene V4 variable region was amplified using the PCR primers 515f (GTGCCAGCMGCCGCGGTAA)/806r (GGACTACHVGGGTWTCTAAT) [11]. This particular region was selected in order to target both bacteria and archaea present in the samples. PCR amplifications were conducted using a single-step 30 cycle PCR using the HotStarTaq Plus Master Mix Kit (Qiagen, USA) under the following conditions: 94°C for 3 minutes, followed by 28 cycles of 94°C for 30 seconds, 53°C for 40 seconds and 72°C for 1 minute, after which a final elongation step at 72°C for 5 minutes was performed. Sequencing was performed at Molecular Research Laboratory, (www.mrdnalab.com), (Shallowater, TX, USA) on an Ion Torrent PGM following the manufacturer's guidelines. Similarly, the fungal 18S rRNA gene was amplified using SSUfungiF (TGGAGGGCAAGTCTGGTG) / SSUFungiR (TCGGCATAGTTTATGGTTAAG) (Hume et al., 2012). A single-step 30 cycle PCR using HotStarTaq Plus Master Mix Kit (Qiagen, Valencia, CA) was used under the following conditions: 94°C for 3 minutes, followed by 28 cycles of 94°C for 30 seconds; 53°C for 40 seconds and 72°C for 1 minute; after which a final elongation step at 72°C for 5 minutes was performed. All amplicon products from each samples were mixed in equal concentrations and purified using Agencourt AMPure beads (Agencourt Bioscience Corporation, MA, USA). Samples were sequenced utilizing the Roche 454 FLX Titanium instrument and reagents according to manufacturer's guidelines

Ancient and Extant Sequence analyses

Sequences of extant Amazonian indigenous cultures were obtained from the Short Read Archive (SRA) database (Accession numbers: ERX115092, ERX115316, ERX115218, ERX115130, and ERX115095). The sequences were microbiomes obtained from amplification of the V4 region of the 16S rRNA gene. These sequences were downloaded in FASTA format and used for all comparative studies as described below.

Raw sequence data were prepared for microbiome analysis using QIIME [12]. A total of 3.4 million multiplexed reads from both the 454 and PGM runs were assigned to samples based on their corresponding barcode using *split_libraries.py* using default filtering parameters. 16S rDNA sequences from coprolites were analyzed, individually or merged with modern stool microbiomes. Coprolite 16S rDNA demultiplexed sequences were sorted based on sample ID using the QIIME script *extract_seqs_by_sample_id.py* and grouped into core and cortex subsets for further analysis. (**Table 2**). *De novo* Operational Bacterial and fungal operational taxonomic units (OTUs) were selected using *pick_de_novo_otus.py* workflow, obtaining a total of n = OTUs. For *18S* data set from QIIME-formatted Silva 111 reference database for Quast et al 2013; (http://www.arb-silva.de/) genetic reference database for eukaryotes for OTU picking and taxonomy assignments (*assign_taxonomy.py*) was used. *16S* and *18S* taxonomy was defined by ≥97% similarity to reference sequences. The phylogenetic composition of the micro-communities present in the samples was characterized using *summarize_taxa_through_plots.py* up to the genus (L7) level.

Alpha and beta diversity

Alpha diversities and rarefaction curves of communities found in coprolite cores and cortices were computed using the *alpha_rarefaction.py* workflow with a custom parameters file that included Shannon statistical analysis. Alpha diversity metrics such as Chao1, which estimates the species richness, Observed_Species, which counts the unique OTUs in a sample and PD_Whole_Tree, which is based on phylogeny, were used for this analysis. Beta diversity distance matrices, UPGMA trees and PCoA plots were computed using *jackknifed_beta_diversity.py*, with default parameters. Distance matrices between sample types were also computed using Primer E v6 software. For comparative purposes, these data were analyzed in parallel with extant fecal microbiomes

Statistical Analysis

Comparisons between coprolite core and cortex microbiomes were made using Procrustes and Adonis analysis (per mutational multivariate analysis of variance using distance matrices). For Adonis, an unweighted UniFrac distance matrix was used using the QIIME script *compare-categories.py* For Procrustes analysis, the beta diversity of the coprolites' cortices and respective cores were compared using QIIME 1.8. The principal coordinate matrices from unweighted UniFrac PCoA plots from core and cortex samples were transformed using the *transform_coordinate_matrices.py* script and the resulting matrices compared using the script *compare_3d_plots.py*.

Comparison of the microbiomes from coprolites of Saladoid and Huecoid origin were done using the *compare_categories.py* script of QIIME 1.8. For this comparison both the Adonis and Permanova methods were conducted using the unweighted UniFrac distance matrix generated by *beta_diversity_through_plots.py* with 999 permutations.

Microscopic analysis for Parasite Eggs

A total of 34 coprolites were used to search for parasites in both cultures. One gram of each coprolite was rehydrated in 14 mL of an aqueous solution of trisodium phosphate 0.5% for 72 h [13]. Samples were shaken vigorously, screened through a 1,500 μm mesh separating all macroscopic material from the sample. To the resultant filtrate, 1 ml of 10% acetic acidic-formalin solution per 10 g of filtrate was added (10:1) to avoid bacterial and fungal growth [14]. The filtered sample was allowed to settle for 72 h after which ten microscope slides were prepared. 50 μL of sediment from each sample was deposited on a slide and mixed with a drop of glycerin. A cover slip was placed on top and the slide was scanned in a serpentine manner covering the whole slide[15,16]. Each parasite egg and larvae found was photographed and measured at 40× and 60× with a calibrated ocular micrometer. For the lack of a taxonomic key to identify parasite eggs, morphological characters such as projections, shape, and presence of larva inside were used for identification. Other non-parasitic organisms were also photographed. All parasite studies were done at the University of Puerto Rico, Mayagüez, Campus.

Results

16S rRNA gene sequences for the coprolites studied ranged from 11,834–281,055 with a median of 26,308 and 151,135 for core and cortex samples respectively (**Table 2**). Fungal 18S rRNA gene sequences ranged from 1,510 to 10,485 with a median of 3,037 for core samples and 4,595 for cortical samples (**Table 2**). Alpha rarefaction plots showed a sampling depth of 5,000, representing approximately 75% of the sample with lowest species count (**Figure 2**). Alpha diversity metrics were consistently higher for *16S* than *18S* in all samples. Diversity indices are depicted in **Table 3**.

Table 2. Coprolite sequence statistics.

16S SSU

Coprolite Core			Coprolite Cortex		
SampleID	Seq Count	Mean Length	SampleID	Seq Count	Mean Length
1b	112,710	207±87	1a	281,055	210±87
2b	235,943	204±85	2a	112,178	219±83
3b	97,550	227±80	3a	153,768	212±85
4b	302,010	206±86	4a	66,082	218±79
5b	15,788	159±83	5a	311,595	213±86
6b	30,782	227±81	6a	222,080	220±85
7b	226,285	222±82	7a	148,502	206±84
8b	248,994	212±86	8a	254,785	213±84
9b	23,862	207±88	9a	114,920	222±83
10b	18,049	189±96	10a	104,132	219±82
11b	26,308	225±87	11a	ND	ND
12b	15,002	213±81	12a	ND	ND
13b	11,834	221±80	13a	ND	ND
14b	23,487	208±87	14a	ND	ND
15b	18,073	201±85	15a	ND	ND
MEDIAN	26,308		MEDIAN	151,135	

18S Fungi SSU

Coprolite Core			Coprolite Cortex		
SampleID	Seq Count	Mean Length	SampleID	Seq Count	Mean Length
1b.ssu	1,930	457±106	1a.ssu	10,485	463±107
2b.ssu	3,187	458±106	2a.ssu	3,242	434±101
3b.ssu	2,295	450±111	3a.ssu	2,021	449±104
4b.ssu	5,191	462±107	4a.ssu	1,895	410±102
5b.ssu	5,088	458±106	5a.ssu	5,587	421±98
6b.ssu	1,510	452±105	6a.ssu	4,761	446±103
7b.ssu	2,577	469±105	7a.ssu	5,273	445±103
8b.ssu	8,119	468±109	8a.ssu	9,753	449±112
9b.ssu	2,886	471±108	9a.ssu	4,429	453±99
10b.ssu	3,411	460±107	10a.ssu	1,991	440±112
MEDIAN	3,037		MEDIAN	4,595	

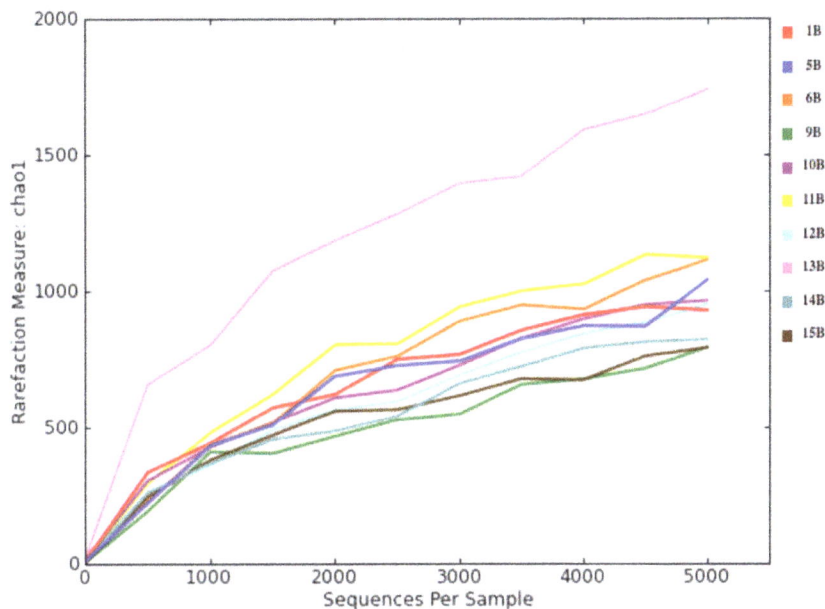

Figure 2. Rarefaction plots the 16S rRNA gene microbiome of the coprolite samples. Rarefaction plots for Huecoid (4B, 6B, 8B, 10B, 11B, 12B, 13B) and Saladoid (1B, 2B, 3B, 5B, 7B, 9B, 14B, 15B) coprolites are shown. Plots were generated using the chao1metic of QIIME 1.8 *alpha_rarefaction.py* with a sampling depth of 5,000. All 15 samples were obtained from the core region (B) of each coprolite.

Statistical Analysis

Procrustes analyses of core and cortex samples were conducted as a control study to assess the fecal microbiome in the coprolites and to see any differences from obvious soil contaminants. Procrustes results showed differing beta diversities when comparing the cortices of the samples to their corresponding cores (**Figure 3**). Cortices showed higher proportions of soil-associated microbes (e.g. 65% Actinobacteria and 11% Firmicutes) compared to the coprolite cores (49% Actinobacteria and 6% Firmicutes). Based on these results, all further studies were conducted using core samples exclusively

Similarly, Adonis analysis showed a significant difference between core and cortex microbiomes with an R^2 of 0.191 (p = 0.001). Adonis and Permanova analyses showed significant differences between the microbiomes of Huecoid and Saladoid coprolites. The Adonis test yielded an R^2 value of 0.287 and a p value of 0.001. Similarly, the Permanova analysis resulted in a Pseudo-F value of 1.98 (p = 0.001).

Fecal microbiome in each culture

16S rDNA analysis. Figure 4 illustrates the differences in taxonomic composition of the Huecoid and Saladoid fecal microbiota. Extant Amazonian fecal microbiome was included for comparative purposes. The proportions of key microorganisms showed major variations between cultures, suggesting possible differences in their diets. While the Burkholderiales, Sphingomonadales and Lactobacillales were more represented in both the Huecoid and Saladoid cultures, the Neisseriales and Bacteroidales were more represented in the Amazonian gut.

Table 4 compares the percent similarities (A) and differences (B) in the fecal microbiomes of the Huecoid and Saladoid. For instance, Bacteroidetes were found to be 13% of the Saladoid fecal microbiota, in comparison Bacteroidetes in the Huecoid comprised approximately only 3% of the microbiota. As there are limitations in targeted *16S* and *18S* sequencing and the limited data available in gene databases, some of the microbiota could

only be confidently identified at a phylum or class level, while others were identified at the Order and Family level.

Figure 5 illustrates the Principal Coordinate Analysis (PCoA) of the bacterial communities found in Huecoid and Saladoid coprolite samples. PCoA of 16S rDNA sequences generated two distinct clusters, where samples originating from the same culture grouped together. Clustering of the samples may have been mainly due to the microbes described in **Table 4**.

18S rDNA Analysis. Figure 6 illustrates the PCoA results of the18s rDNA in Huecoid vs. Saladoid fecal microbiota. PCoA of 18SrDNA sequences showed two distinct clusters for the Huecoid and Saladoid samples. Figure 7 summarizes the relative abundance of fungi in coprolite samples. In general, proportions of Ascomycota were similar between both cultures. However, greater proportions of Basidiomycetes were detected in Huecoid coprolites (Figure 7a). *Saccharomyces spp.* and *Debaryomyces spp.*, were found to be more common in Huecoid feces, whereas *Candida spp.* and *Malasezia spp.* abundances were much higher in Saladoid feces (**Figure 7b**).

Extant vs. extinct fecal microbiotas

Figure 8 compares the Saladoid and Huecoid microbiota to that of fecal microbiota from extant indigenous cultures. Panels (a) and (b) show PCoA of the three groups using PC1 with PC2 and PC3 with PC2. PC1 and PC2 (panel a) show all three groups separating on PC1, representing 51.29% of the variation. The PC3 vs. PC2 plot (panel b) illustrates the differences in microbiome composition that exist among individual samples.

Bacteroidetes were found in higher proportions in modern stool (9.03%) compared to the coprolite samples (0.49%) (**Figure 8**). Similarly, Firmicutes, Clostridiales (specifically Ruminococcaceae, Peptostreptococcaceae, Lachnospiraceae and Eubacteriaceae, among others) made up 81.00% of the modern stool microbial community, while only 18.00% of the coprolite microbiota. In contrast, Actinobacteria were more numerous (48.50% of the microbial community) in coprolites when compared to modern

Table 3. Alpha Diversity Metrics for Coprolite Microbiomes.

Sample ID	Shannon		PD_Whole_Tree		Chao1		Observed_Species	
	16S–V4	Fungi SSU	16S–V4	Fungi SSU	16S–V4	Fungi SSU	16S–V4	Fungi SSU
1a	9.943	4.372	557.945	7.745	14000.189	329.923	10867	234
1b	9.964	3.517	434.365	2.621	11793.849	101.111	8286	56
2a	9.625	2.483	397.655	3.527	11256.698	104.250	7577	68
2b	8.839	2.513	357.141	1.935	9232.645	74.375	6993	53
3a	8.852	2.662	265.672	2.069	6953.633	75.111	5060	54
3b	8.578	3.518	214.855	1.969	5759.892	121.667	4067	70
4a	9.027	ND	224.176	ND	6032.968	ND	4066	ND
4b	9.926	3.189	589.744	2.794	15360.207	128.000	11915	97
5a	10.213	4.623	631.177	6.248	15850.235	243.410	12764	185
5b	8.807	2.360	157.685	1.698	4624.760	91.625	2698	60
6a	9.022	2.436	368.085	2.561	9811.314	87.769	7084	70
6b	6.171	1.995	75.281	0.377	1760.210	21.000	1111	11
7a	9.340	3.525	374.334	5.605	9543.409	167.625	7210	130
7b	8.912	1.046	342.978	1.316	8869.651	66.000	6756	31
8a	10.245	1.951	608.415	3.815	15947.904	248.882	12098	134
8b	9.805	1.033	500.801	3.125	12884.610	180.615	9918	101
9a	8.851	3.358	405.891	4.074	11308.664	252.154	7663	125
9b	8.756	2.257	445.816	1.397	11669.569	72.429	8986	53
10a	8.534	1.701	313.578	1.963	8843.658	105.500	5817	62
10b	8.486	2.394	318.719	2.226	8173.095	80.714	5947	61

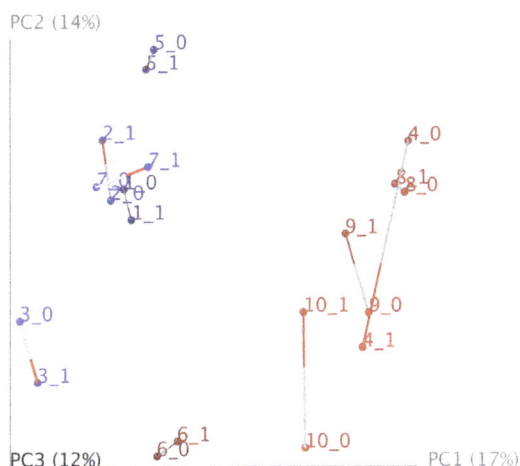

Figure 3. Procrustes analysis compares the 16S microbiome found in the cores and respective cortices of coprolite samples. Samples 6,10, 4, 9 and 8 are of Huecoid origin. Samples 1, 2, 3, 5 ND 7 are of Saladoid origin. Samples identified with number "0" were obtained from coprolite cores and those with the number "1" are from cortical surfaces of the coprolite.

stool samples (2.35%). Specifically, Actinomycetales and Rubrobacterales were found to make up 28.00–33.00% and 10.00–12.00%, respectively, of the detected microbiome.

Eukaryotic Parasites in Coprolites

Eukaryotic enteric parasites were detected in both Huecoid and Saladoid cultures (**Table 5**). There were differences, however, in parasite loads and parasite species between the two cultures. There were twice as many infected Saladoid coprolites as there were Huecoid. *Ascaris lumbricoides* and *Trichuris trichura*, were found in both cultures but with greater number among the Saladoids. *Enterobius vermicularis* and Cestodes were found also in both cultures. Hookworms were found only in Saladoid coprolites. Also, a *Paragonimus westermani* infection may have been detected in a Huecoid coprolite. The overall % positives were much higher in the Saladoid coprolites. *Dipylidium caninum* was observed in 13% of the Saladoid samples, and interestingly not in the Huecoid. Similarly, the genera *Trichostrongylus*, *Diphyllobotrium* as well as hookworms were found associated only with Saladoid coprolites. As part of the DNA analyses, we found fish parasites (*Goussia spp.*), which were detected exclusively in Saladoid coprolites.

Discussion

Though it has been assumed (correctly in some cases), that excreta is rapidly degraded in humid tropical environments, the

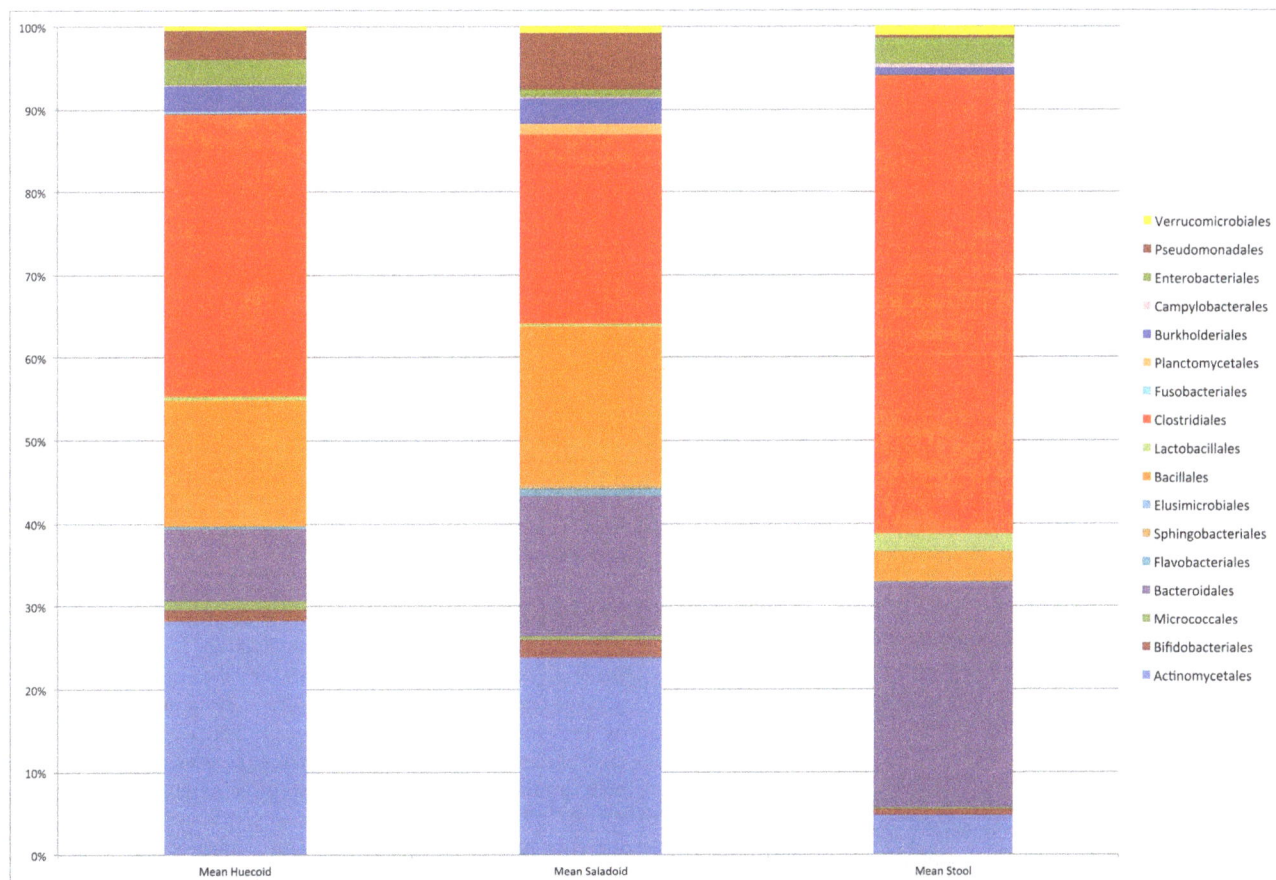

Figure 4. Taxonomic comparison of Huecoid and Saladoid Microbiomes. Figure was generated using summarize_taxa_through_plots.py workflow of QIIME 1.8. Results are illustrated at the Order level. Extant Amazonian stools microbiome was included for comparisons. Mean values for each culture represent taxa obtained from 8 Huecoid coprolite cores, 7 Saladoid samples, and 5 adult, extant Amazonian stools.

Table 4. Similarities (A) and differences (B) found in the microbial fecal communities of the Huecoid and Saladoid cultures.

Taxonomy			Huecoid (%)	Saladoid (%)
Phylum	*Class*	*Order*		
(A) % Taxa Similarities				
Actinobacteria	Nitriliruptoria	-	0.1	0.1
Actinobacteria	Thermoleophilia	-	48.6	53
Bacteroidetes	Flavobacteriia	-	0.1	0.4
Proteobacteria	Gammaproteobacteria	Xanthomonadales	0.2	0.1
Proteobacteria	Deltaproteobacteria	Entotheonellales	0.1	0.3
Proteobacteria	Deltaproteobacteria	Syntrophobacterales	0.3	0.1
Proteobacteria	Deltaproteobacteria	Myxococcales	0.1	0.1
Proteobacteria	Betaproteobacteria	MND1	0.4	0.4
Proteobacteria	Betaproteobacteria	Burkholderiales	0.2	0.5
Proteobacteria	Betaproteobacteria	Burkholderiales	0.3	0.3
Proteobacteria	Alphaproteobacteria	Sphingomonadales	0.1	0.1
Proteobacteria	Alphaproteobacteria	Rhizobiales	0.2	0.4
Proteobacteria	Alphaproteobacteria	Rhodospirillales	0.7	0.1
Planctomycetes	Planctomycetia	Gemmatales	0.5	0.4
Chloroflexi	SAR202	-	0.2	0.5
Chloroflexi	-	-	0.9	0.5
Planctomycetes	Planctomycetia	Pirellulales	0.4	0.1
Nitrospirae	Nitrospira	Nitrospirales	3.5	4.7
Firmicutes	Bacilli	Bacillales	4.7	4.7
Crenarchaeota	Thaumarchaeota	Nitrososphaerales	0.7	0.6
(B) % Taxa Differences				
Chloroflexi	Ellin6529	-	8.7	14
Bacteroidetes	-	-	13.3	2.9
Planctomycetes	-	-	4.1	2.7
Actinobacteria	Actinobacteria	-	1.2	0.4
Actinobacteria	Acidimicrobiia	-	1.2	3.3
Proteobacteria	Gammaproteobacteria	Enterobacteriales	1.2	0
Proteobacteria	Gammaproteobacteria	Pseudomonadales	1.4	3

finding of coprolites in archaeological excavation sites located in Puerto Rico clearly contradicts this assumption. Additionally, it is very uncommon for archaeologists to focus on finding coprolites during archaeological excavations in the tropics as many are not familiar with the morphologies of the typical human or animal coprolite. If we take into consideration that feces are excreted (about 500–1,500 g/person/day), excreta should be one of the most abundant organic archeological findings at human and animal dig sites, if care is taken to search for them [17].

Microbiologically, excreta have its own biota, which, exempting periods of enteritis or other gastrointestinal diseases, and should be relatively constant in diversity and composition. Fecal microbiota are a subset of the microorganisms present in the gastrointestinal tract that are shed during defecation, and as such give much information about an individual's core gut microbiome as well as allochthonous bacteria associated with ingested food, water and very likely, air. Thus, the analysis of fossilized fecal material using NGS can be an important tool in archaeological studies to determine the prevalence of certain microorganisms, pathogenic (such as parasites, for example) and non-pathogenic alike. In terms of the overall composition of the fecal microbiome, however, differences may exist both in taxon distribution and relative abundance as a result of cultural or dietary habits. In fact, it has been clearly determined in a previous study that the fecal microbiota of these particular Antillean cultures is highly dependent on ethnicity [7].

Analysis of the core vs. cortex of coprolites

Procrustes and Adonis analyses showed marked differences between the cores and cortices of the coprolites. Larger proportions of soil microbes (e.g. Actinobacteria) were observed in the cortices of the coprolites, which were likely due to environmental contamination, whereas smaller proportions were seen in the corresponding cores. In addition, cortices mostly shared soil microbes, possibly due to proximal burial sites, while the core microbiomes of each culture differed greatly from one another. This suggests that, although the outer parts of the coprolites were contaminated with the soil that surrounded the samples, the inner core of the coprolites remained largely intact. This is the principal reason for our downstream analyses to be conducted using only the core of the samples.

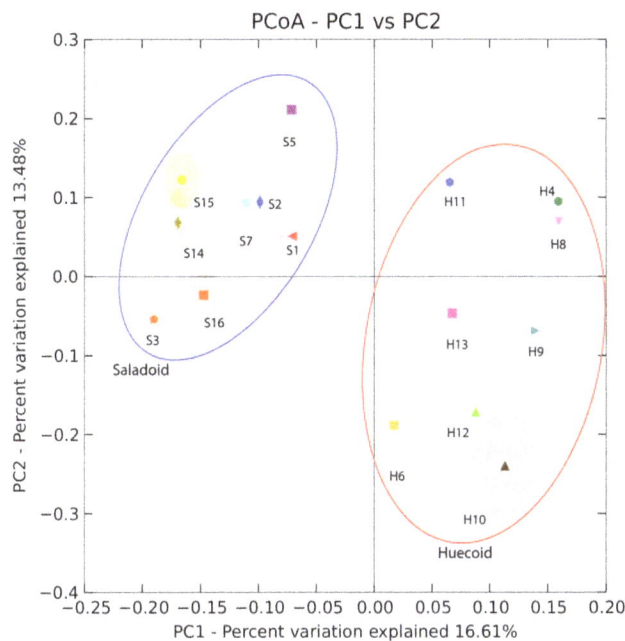

Figure 5. Principal Coordinate Analysis (PCoA) of the Bacterial Communities present in Huecoid and Saladoid Coprolites. Unweighted UniFrac and weighted UniFrac principal coordinates were generated and plotted using QIIME 1.8. Samples with the prefix S are from Saladoid coprolite cores and those with the prefix H were from Huecoid coprolite cores.

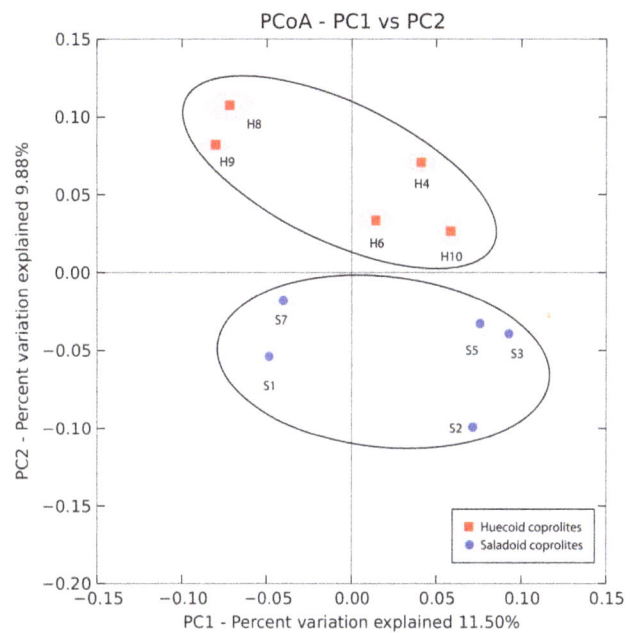

Figure 6. Principal Coordinate Analysis (PCoA) of the Fungal Communities present in Huecoid and Saladoid Coprolites. Unweighted UniFrac and weighted UniFrac principal coordinates were generated and plotted using QIIME 1.8. Samples with the prefix S are from Saladoid coprolite cores and those with the prefix H were from Huecoid coprolite cores.

Analysis of Huecoid and Saladoid coprolites

Both the Adonis and Permanova assays showed significant differences in the microbiome composition of coprolites from Huecoid and Saladoid sites. These analyses were performed only on samples obtained from the cores of the coprolites, thus eliminating possible differences due to environmental contamination of cortical material. These results support the hypothesis that Huecoid and Saladoid cultures, even though coexisted in the Island of Vieques, they had different cultural characteristics, most likely as a result of dietary practices.

Key observed differences in each culture's core microbiomes: Inferences on diet

Variations observed in common gut-associated bacteria, such as Proteobacteria and Enterobacteriaceae, which were more abundant in Saladoid and Huecoid samples, respectively, suggest variations in diet or host genetics. Variations in the abundance of intestinal Proteobacteria have been associated to differences in host diet previously [20]. Differences in Bacteroides abundances were also evident between both cultures, and have been associated to a high protein diet [21]. According to the large quantities of fish bones, bivalve and crab shells found in these archaeological deposits, both cultures seemed to ingest a great amount of seafood. However, although fish bones are associated with both cultures [22], we detected freshwater fish-associated amoebic parasites (*Goussia spp.*) exclusively in Saladoid coprolites, suggesting that this particular culture may have consumed raw fish regularly. In addition, previous studies detected *Vibrio sp.* and *Debaryomyces spp* in Saladoid coprolites, further supporting this hypothesis [7]. The presence of *Paragonimus westermani* in a Huecoid coprolite implies the consumption of fresh water invertebrates, or some type of aquatic plants, the secondary hosts for this species. It is known

that humans become infected by the consumption of raw food contaminated with these parasites. However, it is important to point out that this parasite is commonly confused morphologically with *Diphyllobothrium latum*, a fish-infecting parasite [23]. Interestingly, *Zea mays* (maize) was detected in coprolites from Huecoid origins, consistent with archaeological work showing its presence at the La Hueca site [24] and confirming its early presence in the Caribbean. Our results suggest that this culture may have helped introduce some of these maize strains to the Antilles during their migrations. Since they were detected in large proportions, Ascomycetes and Basidiomycetes also appear to have been important dietary elements of these cultures. However, it seems that the Huecoids had a preference for Basidiomycota fungi. Although these conclusions are highly hypothetical and perhaps speculative, we believe this is a good starting point if we are to compare future studies such as the one carried out here.

Enteric Parasite infections

The greater parasite load observed in Saladoid coprolites suggests a difference in living arrangements, whereby the population was more likely to be exposed to fecal material and thus parasite transmission. Parasite analyses were done microscopically in a separate laboratory, however the results also show major differences between the taxonomic compositions present in both cultures. A wider variety of parasitic species was detected in the Saladoids, which could also be associated to the way this culture handled their food (as previously mentioned). It is also known that both cultures had dogs as pets, particularly the Saladoids. We detected the canine parasite *Dipylidium caninum* in Saladoid samples, also suggesting pet ownership and perhaps very close contact. However, little if anything is known about the interactions of the Huecoids with dogs. Dogs have a tendency to eat human feces, and this may be one manner in which the parasites are passed from person to person, with the dogs being the

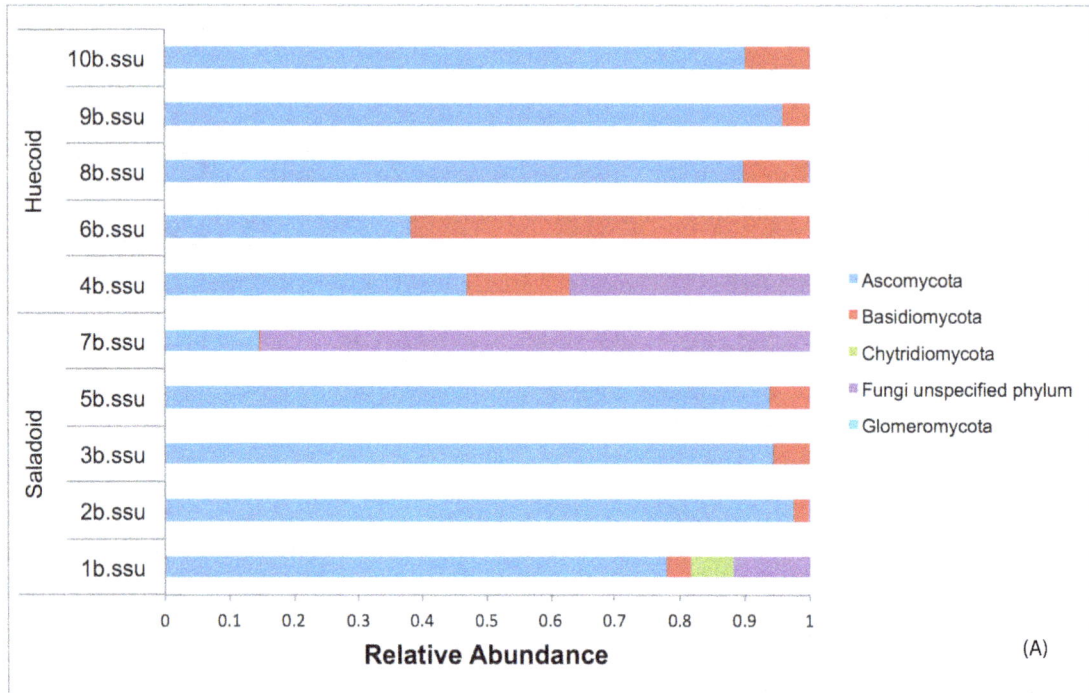

(A)

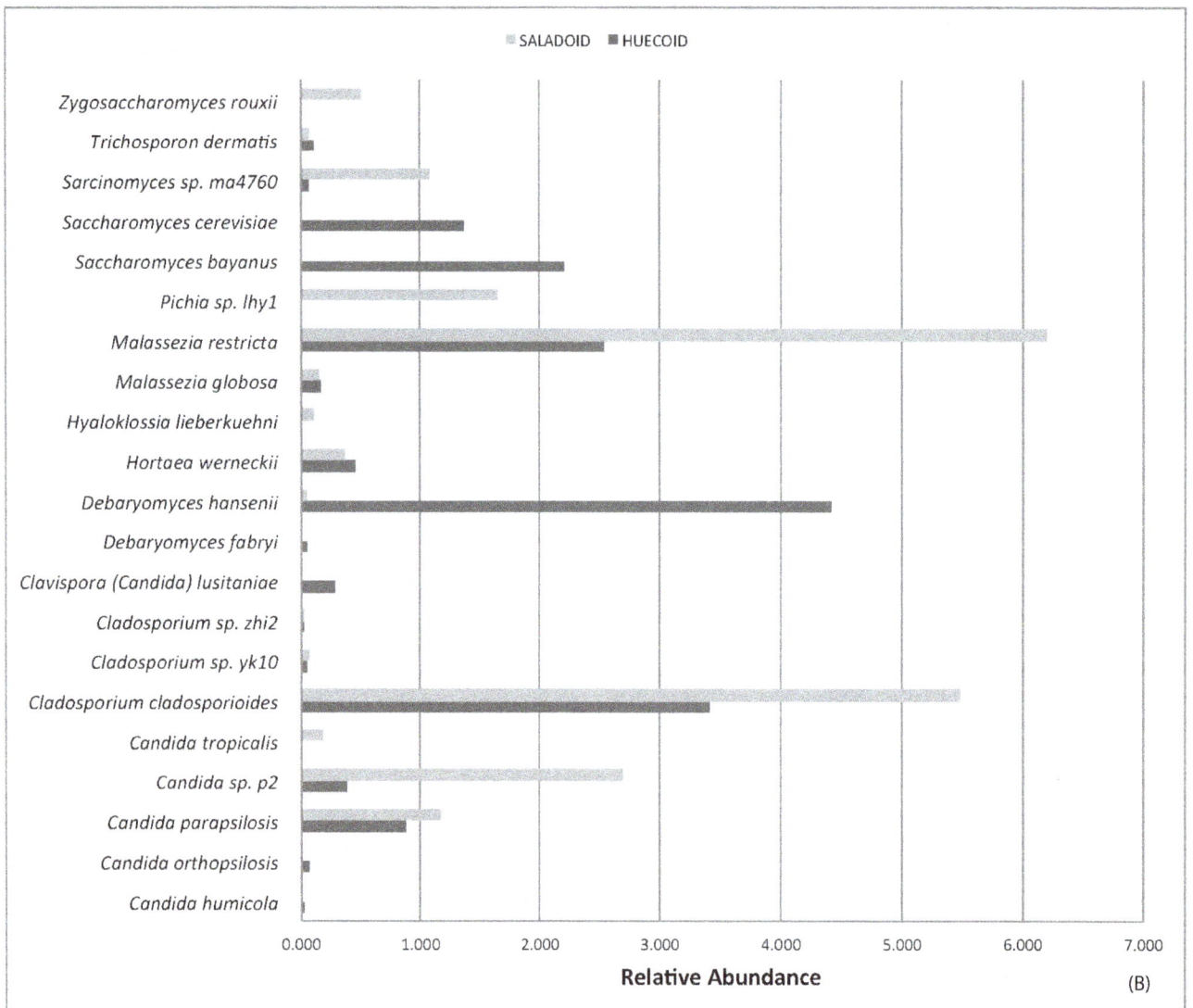

(B)

Figure 7. Relative abundance of fungi in coprolite samples. Panel (a) shows the proportions of fungi detected in Huecoid and Saladoid samples. Panel (b) shows the comparison of the proportions of yeasts detected in both cultures.

vectors. The presence of *Dipylidium caninum* supports this hypothesis, as does the high prevalence of most intestinal parasites detected in the Saladoid coprolites. In addition, present-day ethnic groups in the Amazon basin maintain close relations with dogs and even share their living space with these animals, and although highly speculative, the presence of zoonotic parasites in the Saladoid coprolites is intriguing, and requires further analysis. It is also intriguing that well-formed coprolites had such a high prevalence of enteric parasites, since many of these pathogens result in enteritis in present-day populations, which would not be amenable to the formation of coprolites. It is thus possible that multiple infections with parasites were common, and yet these infections failed to cause enteritis.

The observed differences, both in the core fecal microbiota and the detected remnant food, suggest major variations in the diets of these extinct cultures. This is further supported by the diversity and relative abundances of the parasites detected in each culture. The Saladoid and Huecoid deposits in Vieques are separated by a distance of 15–150 m [7], which suggests that the observed variations in fecal microbiota are not of geographical origins but more likely due to cultural and dietary differences.

Modern Amazonian stool vs. ancient Antillean stool: Considering the effects of taphonomic conditions

The microbiota detected in the coprolites was associated with those found in modern stool, but mainly differed in the proportions observed for each phylum. We observed two main phyla of Gram-positive bacteria in our samples (Actinobacteria and Firmicutes) and two main phyla of Gram-negative bacteria (Bacteroides and Proteobacteria).

As suggested by other studies, Gram-positive bacteria tend to have a high resistance towards dry conditions, so the presence of such a diverse array of these microorganisms preserved in these coprolites was expected [25,26]. Actinobacteria are notorious for their resistance to arid conditions; this resilience towards

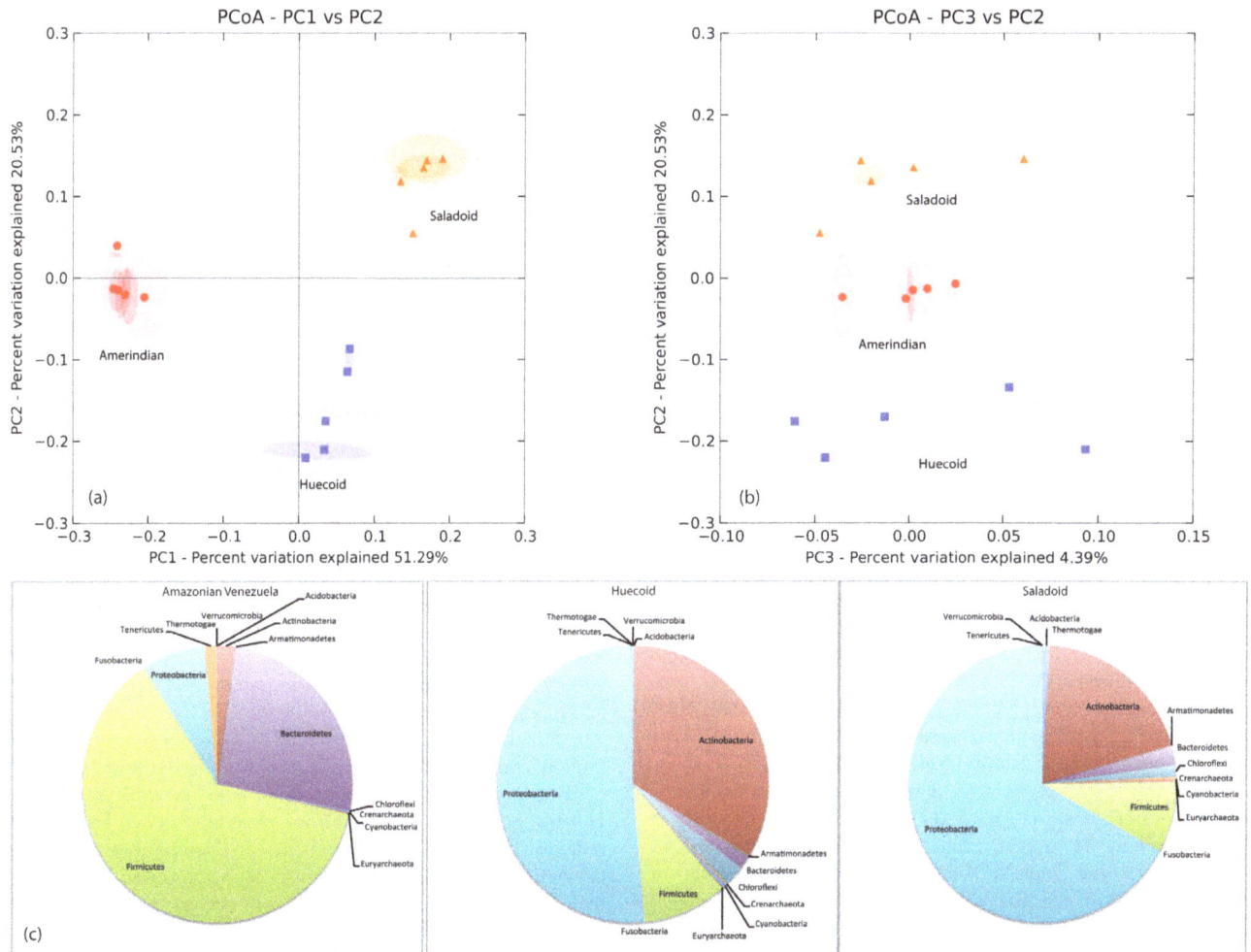

Figure 8. Comparison of Saladoid and Huecoid microbiota to the fecal microbiota of representative, extant indigenous cultures. Panel (a) shows the PCoA of coprolites and the fecal microbiota from extant indigenous cultures plotting PC1 vs. PC2. Panel (b) shows the PCoA of coprolites and the fecal microbiota from extant indigenous cultures plotting PC3 vs. PC2. Panel (c) shows the pie charts of taxa represented in coprolites and extant indigenous cultures.

Table 5. Enteric parasites as detected by microscopy.

Parasite Species	Huecoid (n = 8)		Saladoid (n = 7)	
	% Positive	Average Number Present	% Positive	Average Number Present
Ascaris lumbricoides (unfertilized)	25	56.5	43	27.0
Ascaris lumbricoides (fertilized)	25	151.5	43	44.7
Trichuris trichiura	25	57.7	57	79.3
Enterobius vermicularis	37.5	57.7	43	37.3
Trichostrongylus sp.	ND	ND	43	28.3
Hookworms	ND	ND	29	28.0
Diphyllobothrium sp.	ND	ND	14	26.0
Dipylidium caninum	ND	ND	14	30.0
Unknown Cestode	25	55.5	43	49.3
Paragonimus westermani	13	23.0	ND	ND
Unknown Trematode	ND	ND	14	26.0

taphonomic conditions makes them more likely to be detected in high abundances in archaeological samples [27]. In terms of what can or cannot be detected, we can only assume that over centuries or millennia, most, if not all of the microbiota in the coprolites has been inactivated and, unless there is rapid dehydration, the cells will be lysed and the free DNA will be rapidly degraded. It has been observed that naked DNA will remain relatively undegraded for very short periods of time in aquatic environments [18,19], however, DNA conserved intracellularly in (for example), dormant microorganisms may most likely be better preserved against taphonomic conditions than naked DNA. Again, the mere presence of coprolites indicate that there was rapid dehydration, and thus protection of the nucleic acid material from any extracellular nucleases.

Compared to modern stools, the percentage of Firmicutes detected in the coprolites was much lower. Firmicutes are known to have a low G/C content in their genomes, possibly allowing for faster DNA degradation throughout the fossilization process when compared to bacteria with high [G/C], as in the case of the Actinobacteria, (Taxonomy Browser, NCBI). Interestingly, in spite of Clostridium being spore-formers, the proportion of Firmicutes in coprolites was much lower than detected in the modern stool. However, this may be linked to previous observations suggesting a higher presence of vegetative cells when Clostridium is located in the human gut as compared to soil environments, where the conditions are stressful [28]. In addition, these cells are highly sensitive to oxygen and rapidly die when exposed to oxic conditions. Also, some bacterial spores tend to germinate when they come in contact with the gastrointestinal environment [29] thus leading to more vegetative cells rather than spores.

Low levels of Bacteroides were detected in coprolites compared to modern stool, however, this could be due to the strict anaerobes high sensitivity to oxygen. Interestingly, when compared to modern feces the coprolites showed a higher proportion of Proteobacteria in the microbial community. Though initially counter-intuitive, it is now known that dormancy is a common strategy for long term survival when Gram-negative microorganisms are faced with nutrient and water starvation, exposure to radiation and drastic changes in temperature. Nutritional stress, for example, has been shown to induce the transcription of stress proteins, which ultimately convey the cell a higher resistance towards variations in these abiotic factors [28,30]. This starvation-induced multi-stress resistance, and other bacterial dormancy mechanisms have been characterized, including a spontaneously initiated dormancy as well [31]. This dormancy state appears to be reversible in some cases [32,33,34] (though the mechanisms for resuscitation remain largely unknown), and could have played a factor in the recent isolation of viable microorganisms from coprolite samples [35].

Although these latent bacteria could account for the diversity of Gram-negative bacteria observed in the coprolites, another possibility could be that the relative half-life of these microorganisms is much higher than that of those no longer detectable in the sample. Depending on their half-life, low abundance taxa (or rare species) could have been eliminated from the sample whilst taxa initially present in higher concentrations could still be detectable after preservation for a thousand years.

Differences in coprolite microbiota and the 'Huecoid problem'

The discovery of the Huecoid culture in the 1970s by Chanlatte and Narganes led to formulation of the 'Huecoid problem', the question whether the Huecoids were ethnically distinct from the Saladoids or were simply a Saladoid subgroup. Classical archeology has provided much evidence for cultural distinctness between the two groups, but has not led to a resolution of the problem. There are many material ways in which the Huecoid and Saladoid cultures may be separated [36]. Technological differences include marked differences in pottery and lapidary carvings and more subtle differences in stone tools such as Adzes and Celts. The Saladoids apparently preferred to make their adornments in mollusk conch, in contrast the Huecoids preferred stone. Evidence of dietary differences includes the absence of turtle remains in Huecoid settlements implying they were taboo [22]. Evidence of religious differences are abundant. Firstly, burial practices are distinct, with Saladoid interments occurring within their settlements, while Huecoid burials have yet to be encountered. Iconography is distinct, unique Huecoid symbolism is seen with birds interpreted as condors, carrying human heads, not represented in Saladoid lapidary or ceramics [24]. The range of materials from which Huecoid lapidary was fashioned is much more diverse than Saladoid as are the decorative themes. Huecoid settlements have been associated with increased ritual activity, indicated by an increased incidence of implements associated with what appears to be an early version of the cohoba ritual [37,38]. Thus, the Huecoids have been described as 'religious specialists', with a spiritual role servicing the majority Saladoid community [24] (see Pagan-Jimenez 2007 for a summary). However, while physical evidence of cultural distinctness is abundant, the true role of the Huecoids remains elusive. Ancient DNA studies might hold the key to answering the 'Huecoid problem'. In our study, we have addressed the question from a unique angle, that is, of using paleomicrobiomics. Our results clearly show that the gut microbiota, prokaryotic and fungal, were distinct between the two cultures. While there exists a possibility that these differences might reflect differences in mammalian host genetics [39,40], which would indicate ethnic uniqueness, we interpret the differences to be mainly due to diet, and, perhaps to interaction with pets, as in the case of enteric parasites.

Given the apparent difference in diet indicated by differences in the intestinal microbiota, this allows us traction into the Huecoid problem. There are very few examples of groups that are differentiated by diet and live in conjoined settlements, but are still part of the same ethnic identity. Thus, it is hard to find a modern parallel of the scenario proposed by Rouse that the Huecoids were a Saladoid subgroup [41]. This does not rule out some unique arrangement within Saladoid society, and so while we would propose that both archaeological and paleomicrobiomic data strongly suggest a distinct ethnic identity to the Huecoids, this awaits confirmation.

Conclusions

We successfully extracted and sequenced DNA from archaeological fecal samples in order to assess possible differences in the fecal communities of individuals from Saladoid and Huecoid indigenous cultures. Our data show that, contrary to common belief, the formation and preservation of coprolites and DNA contained in these coprolites under humid, tropical environments for thousands of years is possible. Not only is the DNA still present, it was also detected by PCR amplification and sequenced successfully. We also demonstrate a clear difference between the fecal microbiota of these two cultures, and therefore variations in

terms of their diet and/or genetic heritage. Similar to previous results, our data also supports the hypothesis stating that the Huecoids and Saladoids originated and migrated independently from their respective origins, as opposed to having a common ancestry.

This study is one of the first in its kind and we hope will point to the importance of coprolites as important cultural markers and thus any archaeological dig should include the search and preservation of any coprolites found at the sites. This study underlines the importance of such samples for future paleomicrobiological studies. The results have several implications. First, it confirms that coprolites are not completely degraded in humid, tropical environments and thus can be formed under suitable taphonomic conditions. Second, it implies that dietary and/or host genetic differences in ancient populations may be reflected in differences in gut microbiome composition and it confirms that the two indigenous cultures were indeed distinct. Third, it demonstrates that paleomicrobiomics could be a powerful tool to assess dietary, health, genetic and cultural differences between ancient populations. Finally, it implies that these two cultures retained distinct technological/cultural differences during a period of close proximity and peaceful co-existence and suggests that the two populations, at least at this location, may have contributed to form the latter day Taínos, the Amerindians present at the point of Columbian contact.

Acknowledgments

This study was partially funded by NIH RISE Program (NIH Grant No. 5R25GM061151-12). We also would like to thank Mark Brolaski of Mo Bio Laboratories for providing the necessary supplies and support for the extraction of fossil DNA.

Author Contributions

Conceived and designed the experiments: RJC SEM. Performed the experiments: RJC JRP EGR YNS LCB. Analyzed the data: RJC SEM GAT YNS TSR LBW. Contributed reagents/materials/analysis tools: RJC SEM GAT. Contributed to the writing of the manuscript: RJC SEM TSR JRP. Obtained permissions for collecting coprolites: YNS LCB.

References

1. Sharma N, Kar RK, Agarwal A, Kar R (2005) Fungi in dinosaurian (Isisaurus) coprolites from the Lameta Formation (Maastrichtian) and its reflection on food habit and environment. Micropaleontology 51: 73–82.
2. Narganes Y (1991) Secuencia cronológica de dos sitios arqueológicos de Puerto Rico (Sorcé, Vieques y Tecla, Guayanilla); Proceedings of the Thirteenth International Congress fo the International Association for Caribbean Archaeology. Curaçao. Pp. 628–646.
3. Rodriguez-Ramos R (2010) Rethinking Puerto Rican Precolonial History. Tuscaloosa: University of Alabama Press.
4. Narganes Y (2007) Nueva cronología de varios sitios de Puerto Rico y Vieques; Proceedings of the Twenty-First Congress of the International Association for Caribbean Archaeology; Trinidad.pp. 275–294.
5. Chanlatte LA, Narganes YM (2002) La CulturaSaladoide en Puerto Rico y su rostro multicolor. Museo de Historia, Antropología y Arte: Universidad de Puerto Rico, Recinto de Río Piedras. pp. 55.
6. Crespo E (2010) Ancient Bones Tell Stories: Osteobiology of Human Remains from Tribes. In: Curet A, Stringer LM, editors.Tribes, People, Power and Rituals at the Center of the Cosmos. Tuscaloosa, AL: University of Alabama Press.
7. Santiago-Rodriguez TM, Narganes-Storde YM, Chanlatte L, Crespo-Torres E, Toranzos GA, et al. (2013) Microbial communities in pre-columbian coprolites. PLoS One 8: e65191.
8. Dunbar J, Ticknor LO, Kuske CR (2001) Phylogenetic specificity and reproducibility and new method for analysis of terminal restriction fragment profiles of 16S rRNA genes from bacterial communities. Applied and Environmental Microbiology 67: 190–197.
9. Hert DG, Fredlake CP, Barron AE (2008) Advantages and limitations of next-generation sequencing technologies: A comparison of electrophoresis and non-electrophoresis methods. Electrophoresis 29: 4618–4626.
10. Chanlatte LA, Narganes YM (2005) Cultura La Hueca. Museo de Historia, Antropologia y Arte: Universidad de Puerto Rico, Recinto de Rio Piedras. pp. 101.
11. Caporaso JG, Lauber CL, Walters WA, Berg-Lyons D, Lozupone CA, et al. (2011) Global patterns of 16S rRNA diversity at a depth of millions of sequences per sample. Proceedings of the National Academy of Sciences of the United States of America 108: 4516–4522.
12. Caporaso JG, Kuczynski J, Stombaugh J, Bittinger K, Bushman FD, et al. (2010) QIIME allows analysis of high-throughput community sequencing data. Nature Methods 7: 335–336.
13. Callen EO, Cameron TWM (1960) A prehistoric diet as revealed in coprolites. New Science 8: 35–40.
14. Goncalves MLC, Araujo A, Ferreira LF (2003) Human intestinal parasites in the past: New findings and a review. Memorias Do Instituto Oswaldo Cruz 98: 103–118.
15. Han ET, Guk SM, Kim JL, Jeong HJ, Kim SN, et al. (2003) Detection of parasite eggs from archaeological excavations in the Republic of Korea. Memorias Do Instituto Oswaldo Cruz 98: 123–126.
16. Fugassa MH, Denegri GM, Sardella NH, Araujo A, Guichon RA, et al. (2006) Paleoparasitological records in a canid coprolite from Patagonia, Argentina. Journal of Parasitology 92: 1110–1113.
17. Saito T, Hayakawa T, Nakamura K, Takita T, Suzuki K, et al. (1991) Fecal Output, Gastrointestinal Transit-Time, Frequency of Evacuation and Apparent Excretion Rate of Dietary Fiber in Young Men Given Diets Containing Different Levels of Dietary Fiber. Journal of Nutritional Science and Vitaminology 37: 493–508.
18. Alvarez AJ, Yumet GM, Santiago CL, Toranzos GA (1996) Stability of manipulated plasmid DNA in aquatic environments. Environmental Toxicology and Water Quality 11: 129–135.

19. Alvarez AJ, Yumet GM, Santiago CL, Hazen TC, Chaudhry R, et al. (1996) In situ survival of genetically engineered microorganisms in a tropical aquatic environment. Environmental Toxicology and Water Quality 11: 21–25.

20. Schwartz K, Chang H, Olson LK (2012) Metabolic Profiles of Ketolysis and Glycolysis in Malignant Gliomas: Possible Predictors of Response to Ketogenic Diet Therapy. Neuro-Oncology 14: 55–55.

21. Wu GD, Chen J, Hoffmann C, Bittinger K, Chen YY, et al. (2011) Linking Long-Term Dietary Patterns with Gut Microbial Enterotypes. Science 334: 105–108.

22. Narganes-Storde Y (1982) Vertebrate faunal remains from Sorce, Vieques, Puerto Rico. MA. Athens, GA: University of Georgia.

23. Garcia SL, Bruckner DA (1993) Liver and Lung Trematodes. Diagnostic Medical Parasitology: ASM Press. pp. 309–321.

24. Pagan-Jimenez JR (2007) De antiguos pueblos y culturas botanicas en el Puerto Rico indigena. American Archeology: Paris Monographs.

25. Pal A, Pehkonen SO, Yu LE, Ray MB (2007) Photocatalytic inactivation of Gram-positive and Gram-negative bacteria using fluorescent light. Journal of Photochemistry and Photobiology a-Chemistry 186: 335–341.

26. Chastanet A, Fert J, Msadek T (2003) Comparative genomics reveal novel heat shock regulatory mechanisms in Staphylococcus aureus and other Gram-positive bacteria. Molecular Microbiology 47: 1061–1073.

27. Zaremba-Niedzwiedzka K, Andersson SGE (2013) No Ancient DNA Damage in Actinobacteria from the Neanderthal Bone. PLoS One 8.

28. Lennon JT, Jones SE (2011) Microbial seed banks: the ecological and evolutionary implications of dormancy. Nature Reviews Microbiology 9: 119–130.

29. Casula G, Cutting SM (2002) Bacillus probiotics: Spore germination in the gastrointestinal tract. Applied and Environmental Microbiology 68: 2344–2352.

30. Kaprelyants AS, Gottschal JC, Kell DB (1993) Dormancy in Non-Sporulating Bacteria. Fems Microbiology Letters 104: 271–286.

31. Carneiro S, Ferreira EC, Rocha I (2011) A Systematic Modeling Approach to Elucidate the Triggering of the Stringent Response in Recombinant E. coli Systems. 5th International Conference on Practical Applications of Computational Biology & Bioinformatics (Pacbb 2011) 93: 313–320.

32. Koltunov V, Greenblatt CL, Goncharenko AV, Demina GR, Klein BY, et al. (2010) Structural Changes and Cellular Localization of Resuscitation-Promoting Factor in Environmental Isolates of Micrococcus luteus. Microbial Ecology 59: 296–310.

33. Cano RJ, Borucki MK (1995) Revival and Identification of Bacterial-Spores in 25-Million-Year-Old to 40-Million-Year-Old Dominican Amber (Vol 268, Pg 1060, 1995). Science 268: 1265–1265.

34. Cano RJ, Borucki MK (1995) Revival and Identification of Bacterial-Spores in 25-Million-Year-Old to 40-Million-Year-Old Dominican Amber. Science 268: 1060–1064.

35. Appelt S, Armougom F, Le Bailly M, Robert C, Drancourt M (2014) Polyphasic Analysis of a Middle Ages Coprolite Microbiota, Belgium. PLoS One 9.

36. Chanlatte-Baik LA (2013) Huecoid Culture and the Antillean Agroalfarero (Farmer-Potter) Period. In: Keegan WF, Hofman CL, Ramos RR, editors. Handbook of Carribbean Archaeology. Oxford: Oxford University Press.

37. Rodriguez M (1997) Religous beliefs of the Saladoid people. In: Wilson SW, editor. The Indigenous People of the Caribbean Gainesville University Press of Florida.

38. Wilson SM (2007). he Archeology of the Caribbean: Cambridge University Press.

39. Kashyap PC, Marcobal A, Ursell LK, Smits SA, Sonnenburg ED, et al. (2013) Genetically dictated change in host mucus carbohydrate landscape exerts a diet-dependent effect on the gut microbiota. Proceedings of the National Academy of Sciences of the United States of America 110: 17059–17064.

40. Lu K, Mahbub R, Cable PH, Ru HY, Parry NMA, et al. (2014) Gut Microbiome Phenotypes Driven by Host Genetics Affect Arsenic Metabolism. Chemical Research in Toxicology 27: 172–174.

41. Rouse I (1992) The Tainos: rise and decline of the people who greeted Columbus. New Haven: Yale University Press.

6

Continuous Variation Rather than Specialization in the Egg Phenotypes of Cuckoos (*Cuculus canorus*) Parasitizing Two Sympatric Reed Warbler Species

Szymon M. Drobniak[1], Andrzej Dyrcz[2]*, Joanna Sudyka[1], Mariusz Cichoń[1]

1 Institute of Environmental Sciences, Jagiellonian University, Kraków, Poland, **2** Department of Behavioural Ecology, Wroclaw University, Wroclaw, Poland

Abstract

The evolution of brood parasitism has long attracted considerable attention among behavioural ecologists, especially in the common cuckoo system. Common cuckoos (*Cuculus canorus*) are obligatory brood parasites, laying eggs in nests of passerines and specializing on specific host species. Specialized races of cuckoos are genetically distinct. Often in a given area, cuckoos encounter multiple hosts showing substantial variation in egg morphology. Exploiting different hosts should lead to egg-phenotype specialization in cuckoos to match egg phenotypes of the hosts. Here we test this assumption using a wild population of two sympatrically occurring host species: the great reed warbler (*Acrocephalus arundinaceus*) and reed warbler (*A. scirpaceus*). Using colour spectrophotometry, egg shell dynamometry and egg size measurements, we studied egg morphologies of cuckoos parasitizing these two hosts. In spite of observing clear differences between host egg phenotypes, we found no clear differences in cuckoo egg morphologies. Interestingly, although chromatically cuckoo eggs were more similar to reed warbler eggs, after taking into account achromatic differences, cuckoo eggs seemed to be equally similar to both host species. We hypothesize that such pattern may represent an initial stage of an averaging strategy of cuckoos, that – instead of specializing for specific hosts or exploiting only one host – adapt to multiple hosts.

Editor: Csaba Moskát, Hungarian Academy of Sciences, Hungary

Funding: This study was supported by the grant of the National Science Center no. N N304 030739 (to AD) and by funding from the Jagiellonian University within the SET project (to SMD; the project is co-financed by the European Union). The funders had no role in study design, data collection and analysis, decision to publish, or preparation of the manuscript.

Competing Interests: The authors have declared that no competing interests exist.

* Email: dyrcz@biol.uni.wroc.pl

Introduction

The common cuckoo (*Cuculus canorus*) is an obligate brood parasite utilizing a wide range of hosts that differ substantially in body size, egg morphology and reproductive behaviour. Brood parasites usually have strong impacts on host fitness, which induces strong selection pressures on the host to avoid parasites, leading to a co-evolutionary arms race [1]. The common cuckoo has particularly strong effects on its host as the parasite does not allow the host to rear any offspring. Thus, the naive hosts might quickly go extinct if they do not develop antiparasite adaptations. Egg recognition is one of the most prominent antiparasite adaptations. However, the parasite usually develops contradaptation in the form of egg mimicry. Cuckoo's potential hosts, however, have eggs that differ substantially in size, shape and colouration. In response to this diversity, cuckoos have evolved specialization to specific hosts. Within the cuckoo species in Europe, there are about 16 variants (called gentes) of egg types that match the most common host egg phenotypes [2]. Egg-phenotype matching between specific cuckoo gentes and their hosts is remarkable, and in some cases, almost perfect [2–5]. Gentes are genetically distinct and are inherited mainly through maternal genetic line [6,7].

Adaptation to parasitizing specific host species seems unquestionable in cuckoos and is apparent both on genetic and phenotypic levels. However, cuckoo females belonging to one gens may exploit a number of sympatric host species differing substantially in egg phenotypes [8–10]. In such a case, cuckoos should either 1) lay eggs that mimic the eggs of only one host species, thus laying dissimilar eggs in nests of other host species – which would be particularly expected in the early stages of a host-parasite arms race [11], or 2) show a generalist strategy by producing eggs dissimilar to eggs of any of the available host species [10]. Either strong specialization to the host species or a generalist strategy of averaging between potential hosts may be favoured by selection in such circumstances. Strong specialization should lead to the evolution of specific phenotypes matching specific host species even within a single gens, maximizing the breeding success of individual cuckoos. The evolution of such discrete egg morphs is supported by field data [12] and numerical simulations, which show that the equilibrium endpoint of an evolving brood parasite – host system is the formation of discrete egg morphs produced by the parasite [11,13]. Mathematical simulations prevent a continuous distribution of cuckoo egg-morphs from becoming a stable endpoint – discrete egg classes arise even if the initial distribution of egg phenotypes is continuous [11]. However, models allow for a continuous distribution of parasitic egg phenotypes in the initial stages of the evolutionary arms race [11,13]. Following the introduction of a new host species, producing continuously distributed egg morph, centred around an average phenotype, might be more advantageous in terms of the breeding success than immediate production of specialized eggs – particularly if such specialization requires slow genetic adaptation. Such averaging strategy should result in the disappearance of discrete mimicry to specific hosts, resulting in substantial variation in egg phenotypes observed locally (i.e. regardless of larger-scale patterns resulting from overlapping geographical variation in host egg phenotypes). In both cases, one possible scenario is the evolution of substantial variation in egg phenotype even within a single gens, but specialization should result in discontinuity of the distribution of cuckoo egg phenotypes with regard to the parasitized hosts. One has to also be aware that in the wild population, substantial imbalance in the numbers of available hosts of different species might constraint variation in cuckoo phenotypes: in the case of one prevalent host species and one or more other host species present at substantially lower numbers, cuckoos might evolve egg phenotypes that match the egg phenotype of the most common host [12,14]. Thus, the nature of observed intra- and inter-gentes variation in egg phenotypes, together with host species frequencies, are important when considering different evolutionary scenarios shaping cuckoo egg mimicry.

Here our goal was to test whether cuckoo eggs found in sympatrically occurring (Eurasian) reed warbler (*Acrocephalus scirpaceus*) and great reed warbler (*A. arundinaceus*) nests differ in size, colouration and the strength of the eggshell. Cuckoos parasitizing these two species are assigned to either the *Sylvia* egg morph or the *Acrocephalus* morph [8,15,16]. In the studied population, based on visual identification, both host species are parasitized with the same *Sylvia* morph [17,18]. Although exploited by one cuckoo egg morph, the eggs of these two host species are clearly distinct in terms of size and coloration. Reed warblers are also less vigilant towards cuckoos than great reed warbles and more readily accept cuckoo eggs [17]. In an egg discrimination experiment carried out in the study population, great reed warblers rejected 13.6% (n = 44 nests) of alien mimetic eggs (an egg of the host species taken from another nest) and reed warbler 3.6% (n = 28) of alien mimetic eggs. Similar statistics calculated for alien non-mimetic eggs (a conspecific egg painted plain brown or blue) showed 92.9% (n = 70) rejections in great reed warbler and 61.8% (n = 204) in reed warbler. Thus, clear differences in egg morphology and anti-parasite behaviour between these two hosts species provide a suitable case to study parasite egg mimicry strategies. In this paper, we try to disentangle whether the cuckoo (i) employs the averaging strategy to match the continuum of egg appearance between these two host species, (ii) shows a better match to one of the species (possibly more frequently utilized) or (iii) exhibits dichotomous specialization by laying eggs matching both host species. In our analysis, we employ morphological measurements, eggshell colour analysis [4,19], and eggshell strength measurements [20,21].

Materials and Methods

Field methods

The study was conducted from May to August in each of three consecutive years (2011–2013) in the population of two reed warbler species, located at the Stawy Milickie fishponds complex in southern Poland [17,22]. In this area reed warbler densities reach 57 breeding pairs/ha. We attempted to locate all reed warbler and great reed warbler nests in the area. Their locations were obtained through systematic search in reed-beds. Cuckoo eggs were identified during egg-laying and subsequently collected before hatching and frozen until further analysed. Shells of hatched eggs were also collected if possible. Additionally, in 2012 we collected single eggs from different nests of each of the two host species (15 of great reed warbler and 11 of reed warbler).

Egg measurements

All collected eggs were thawed and opened after the field season to collect cuckoos' genetic material. Prior to this, we photographed and colour-measured all collected eggs. Identical measurements were also done with respect to host-species eggs. Photographing was done using the Canon EOS 450D camera with a 17–70 mm lens. Photographs were taken at the focal length of approximately 25 mm with automatic shutter and iris settings. Photographs were used solely for measuring physical dimension of eggs. Colour measurements were performed using the Ocean optics JAZ portable spectrophotometer with xenon pulsed-light source (100 pulses per second) and a bifurcating optic fibre. With the probe held perpendicular to the shell surface, we lit the shell surface from a distance of 3 mm (which resulted in a lit area of approx. 3 mm^2). We measured each egg 12 times at equally-spaced spots on their surfaces. Both photographing and spectrophotometry were performed in different light conditions and at different locations each year, which might contribute to the naturally observed year differences – however all eggs were treated identically within each study year. Since all eggs were measured after the respective field seasons, their colour might change slightly due to chemical degradation of egg-shell compounds [19,23]. Unfortunately, we cannot assign precise collection rates to the eggs. We assume that the possible effects of prolonged storage (if any) would not be confounded with host-species identities, and thus would not affect subsequent comparisons. Importantly, all eggs were kept frozen in the darkness until assayed, which should further reduce any colour-related deterioration of eggshells.

Reflectance spectra obtained with JAZ were further analysed in the *pavo* package ver. 0.3–1 [24] in R ver. 2.15.0 [25]. After averaging the spectra within each measured egg and smoothing the resulting curves, we analysed resulting spectra using the average bird visual system, employing functions implemented in the *pavo* package. Briefly, individual spectra were transformed into cone receptor quantum catches (Q_i) defined as:

$$Q_i = \int_\lambda R_i(\lambda)S(\lambda)I(\lambda)d\lambda,$$

taking into account i-th receptor sensitivity, colour reflectance spectrum S and illuminant spectrum I. Quantum catches of each receptor where then transformed into Cartesian coordinates defining the position of each data point in a tetrahedral colour space spanned on the vertices of a tetrahedron (vertices represent full relative stimulations of red, blue, green and UV receptors). In this representation, the central point of the tetrahedron (equal to [0, 0] of the Cartesian coordinates system) is the achromatic point of black/white colour. Hue in this system is defined using two angles: theta θ defined on the RGB plane describes the human-visible part of the hue, whereas phi φ represent the deviation in the z-axis direction, measuring the UV component [26]. Chroma of the colour can be defined as the Euclidean distance r from this achromatic centre. In subsequent analyses, we used both θ and φ as hue measures, and r and achieved r_a (percent of maximum possible chroma for a given hue) as measures of chroma. Additionally, as tetrahedral-space colour model does not provide a measure of brilliance or brightness, we extracted mean brightness for each spectrum, defined as:

$$Br = \frac{\sum_{\lambda_{min}}^{\lambda_{max}} R_\lambda}{n},$$

which is the mean reflectance R over all n recorded wavelengths.

We also used generated quantum catches to analyse differences between eggs according to the receptor noise model, which returns colour distances between measured samples accounting for receptor-related noise in signal perception [27]. Colour contrasts (chromatic and achromatic) were generated for pairs of reed warbler vs. great reed warbler eggs, cuckoo eggs from reed warbler nests vs. cuckoo eggs from great reed warbler nests, and finally for reed warbler/great reed warbler vs. cuckoo eggs. Contrasts are expressed in the units of just noticeable differences (JND, see [24,27] and were generated assuming four colour receptors (long-, medium-, short- and very-short-wavelengths) with relative receptor densities characteristic for a blue tit retina [5,24].

Egg photographs were used to obtain their physical measurements (length and width in the widest place). Dimensions were measured to the nearest 0.1 mm in the ImageJ software [28]. These variables were then used in calculations of egg volumes (we used a commonly applied formula: $0.498 \times$ length $\times$ width^2 [29]; volume was then expressed in cm^3).

Additionally, all cuckoo eggs collected in 2011 were subject to shell-strength measurements. Briefly, eggshells were analysed using a pressure-sensitive specimeter that measures force and work required to break the shell. Shell resistance to breaking was analysed at a number of spots (sharp end, blunt end, side of the egg); both inward and outward breaking force and work were considered. Respective parameters were measured to the nearest 0.001 N (force) and 0.00001 J (work).

Statistical analyses

All measured variables for cuckoo eggs were analysed using a simple linear model with year and host species as fixed effects (*lm* procedure in R). Egg-thickness measures (available only for one year) were analysed using a two-tailed Student t test (*t.test* procedure in R). In all tests $\alpha = 0.05$ was used as the acceptable threshold type-I error. Prior to analyses, all variables were centered and standardized to have zero mean and unity variance. We ensured that all fitted models returned homoscedastic and approximately normal residuals.

All measured colour variables together with egg width, length, volume and shape, transformed for normality if necessary (see above) and standardized, were aggregated into principal components using the *prcomp* procedure in R. The first two components were used to represent each egg in a multivariate colour space to depict the presence or absence of clustering of eggs within the groups of two host species. This procedure was applied both to the cuckoo eggs and host species eggs for comparison. The resulting graph depicts the joint variation of reed warbler and cuckoo eggs in all measured variables. Additionally we provide a projection plot depicting positions of all eggs in the tetrahedral colour space projected on the circumscribed sphere of the tetrahedron. To visualize variation in hues/chromas and areas of overlapping colours, we generated in *pavo* convex hulls encapsulating respective groups of points in the tetrahedral colour space. We provide both numerical outputs from these analyses (proportions of overlapping hulls) and graphical depictions.

In total we collected and measured 15 cuckoo eggs from great reed warbler nests and 47 cuckoo eggs from reed warbler nests. Twelve eggs (great read warbler – 4, reed warbler – 8) were measured for shell-thickness.

Ethics statement

This study was carried out under the license granted by the division of the Polish Bioethical Committee at Wroclaw University (to AD). The license covered collection of the eggs, egg measurements and subsequent procedures. All procedures were performed so that they minimized the stress exhibited by birds during egg sampling and monitoring.

Results

We found no differences in shell strength in cuckoo eggs collected in nests of the two host species (Table S1). In particular, we found no significant differences in either the mean inward or outward breaking force (inward: $t_{df = 6.81} = 0.13$, $P = 0.9$; outward: $t_{df = 5.01} = 0.48$, $P = 0.65$). Eggs laid by cuckoos into nests of the two host species did not differ in their shape index (one means perfect sphere): $t_{df = 59} = 0.01$, $P = 0.98$. Overall, size variables and colour measures did not indicate any differentiation (Table 1, Figure 1a).

Host species significantly differed with respect to several egg-morphology variables. We found significant differences in shell brightness ($t_{df = 20.76} = 2.67$, $P = 0.01$), RGB hue θ ($t_{df = 12.48} = 3.77$, $P = 0.002$), realized chroma r ($t_{df = 17.68} = 5.65$, $P = 0.005$), and achieved chroma r_a ($t_{df = 15.57} = 2.47$, $P = 0.03$); no significant differences in UV hue φ were found ($t_{df = 15.44} = 0.34$, $P = 0.73$). Great reed warbler eggs are, on average, brighter, more saturated in the blue range, and generally lack the red hue (Figure 1b). Moreover, compared to reed warbler eggs, they are longer ($t = 9.97$, $df = 19.60$, $P < 0.0001$) and wider ($t = 11.38$, $df = 21.49$, $P < 0.0001$), which also translates into larger volume ($t = 14.29$, $df = 21.99$, $P < 0.0001$). Colour differences are also apparent in the tetrahedral colour space. The two host species clearly occupy separate clusters (Figure 2, Table 2), and the convex hulls surrounding colour measurements do not overlap.

Cuckoos' eggs occupy the space between the two host species, although they seem to group closer to the reed warbler eggs. This indicates that cuckoo eggs are closer to reed warbler eggs in terms of hue and chroma. This is further confirmed by a substantial overlap of convex hulls of cuckoo and reed warbler eggs (Table 2, Figure 2). In contrast, regions occupied by cuckoos' and great reed warblers' eggs do not overlap (Figure 2), however they are very close to each other. For clarity we did not present separate analogous analyses of overlap for cuckoos split by the host species – they occupy virtually the same region in the colour space and overlap extensively (over 90%).

Tetrahedral colour space representation does not account for differences in brightness. We have thus represented all measured eggs using principal components derived using all colour (i.e. hue, chroma and brightness) and size variables. Differences between host species are apparent after representing all measured host species eggs on the plane defined by the first two principal components (Figure 3). The two host species form two clearly separated clusters of points indicating they are clearly differentiated. An analogous plot representing all measured cuckoo eggs (colour and size variables) indicates that they form a uniform, non-differentiated population of points, with the variation among cuckoo eggs being substantially larger than within host species and clearly overlapping with variation in both host species (Figure 3). The first principal components represent mainly physical dimensions, chroma and RGB hue, whereas second components represent mainly brightness, RGB and UV hues (see Table S2 for loadings of all variables on the first two components). The first principal component explained 51% of total variance in eggs' morphology and the second component explained 34% of

Table 1. Comparison of differences in colour-related and size-related variables for cuckoo eggs laid in nests of reed warblers (rw) and great reed warblers (grw).

Variable name	Trait value ± SE (rw)	Trait value ± SE (grw)	t	P
Br	31.00±2.54	26.31±3.91	0.99	0.32
θ	2.80±0.04	2.81±0.06	0.20	0.84
φ	−0.48±0.06	−0.51±0.09	0.25	0.80
r	0.12±0.01	0.11±0.01	0.55	0.59
r_a	0.35±0.02	0.35±0.04	0.07	0.94
Length	22.10±0.16	22.20±0.16	0.28	0.78
Width	16.54±0.12	16.66±0.12	0.49	0.62
Volume	3.02±0.05	3.08±0.05	0.48	0.63

Estimates were obtained from linear models and are corrected for the effect of study years. All tests are performed with $df = 59$.

variance. Correlations between original variables included in the PCA are provided in Table S3.

A receptor-noise model (Vorobyev & Osorio) further confirmed the above observations (Table 3). The average chromatic contrast between cuckoo eggs and great reed warbler was similar to the chromatic contrast between great reed warbler and reed warbler. On the other hand, the achromatic contrast between cuckoo eggs and either of the two host species appeared to be similar in value and slightly higher than the contrast between the host species. Both chromatic and achromatic contrasts between cuckoo eggs found in nests of reed warbler and great reed warbler were not significantly higher than contrasts calculated within the host-species groups (all P>0.05), supporting a lack of any differentiation among cuckoo eggs.

Discussion

Our study confirms clear differentiation in egg appearance between reed warbler and great reed warbler. The eggs differ in size and colour spectra forming two distinct clusters in the principal-components analysis. Despite these clear differences between hosts, we failed to show clear clustering among cuckoos parasitizing these two species with respect to their egg morphologies. A similar conclusion comes from a study performed on four sympatrically occurring warbler species in which no differences in egg colouration were found between cuckoo eggs from nests of those host species [8]. More importantly, our data show that cuckoo eggs exhibit substantial variability that – in specific morphology components – largely overlaps the variance observed in both species. Interestingly, this overlap is less apparent when analysing only hue and chroma. In terms of chromatic components of egg colouration, all sampled cuckoo eggs were more

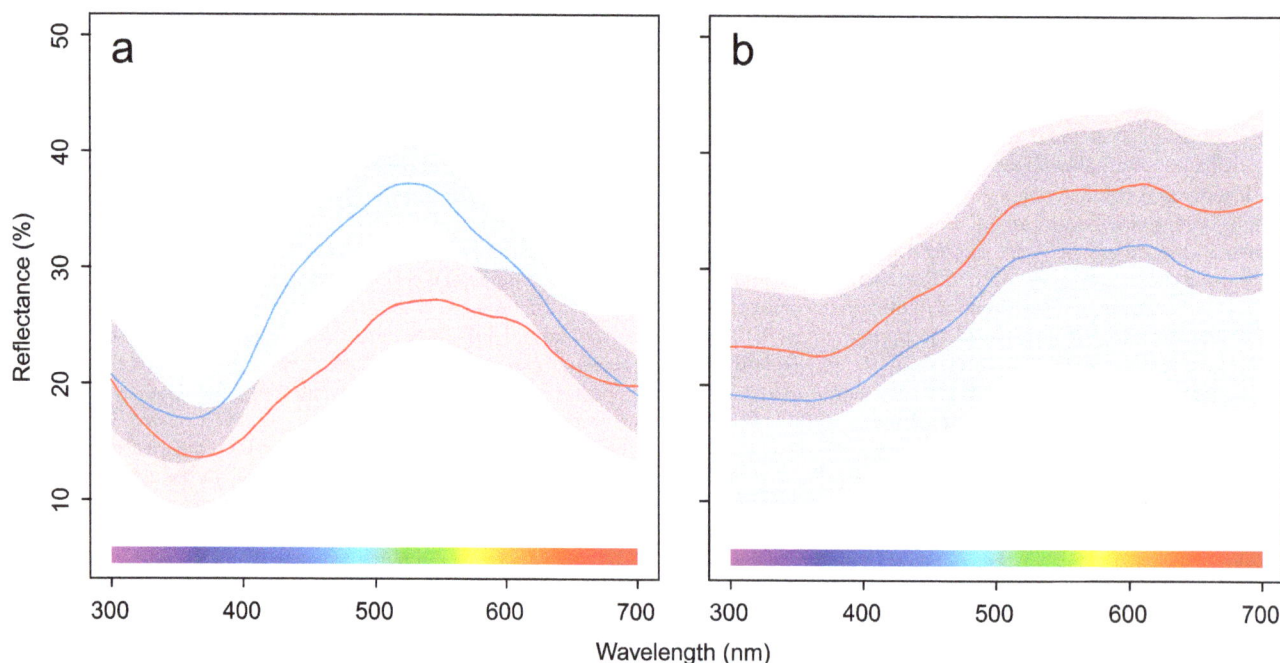

Figure 1. Average smoothed reflectance spectra of warbler eggs (a) and cuckoo eggs (b). Blue lines: eggs from great reed warbler nests; red lines: eggs from reed warbler nests. Bands represent 95% confidence regions.

Table 2. Summary of colour variables measured in the tetrahedral colour space for three analysed species.

Species	h_m	mean r_a	Convex hull overlap [%] with		
			A.arundinaceus	*A.scirpaceus*	*C.canorus*
A.arundinaceus	0.033	0.390	–	–	–
A.scirpaceus	0.047	0.325	0	–	–
Cuckoo	0.071	0.351	0	47	–

For each species we present: mean hue span (h_m), average saturation (mean r_a), and overlap of convex hulls calculated as: *volume_of_overlap/volume_of_smaller_hull* (thus the value of 1 would indicate perfect inclusion of the smaller hull in the bigger one; see [24] for more details).

similar to reed warbler host eggs than to great reed warbler eggs. However, achromatic colour components of cuckoo eggs not only exhibited larger variation, but they also overlapped both host species egg phenotypes. To our knowledge this is the first observation of such variation overlapping multiple hosts in the cuckoo brood parasitic system.

These results clearly indicate that sampled cuckoo eggs comprised phenotypes closely matching both host species as well as a range of intermediate phenotypes. This is very interesting since the proportion of parasitized nests clearly differs between these two host species, so one may expect cuckoo eggs to show a better match to the more frequently utilized host species. During the study, only 5.8% of nests of great reed warbler were parasitized

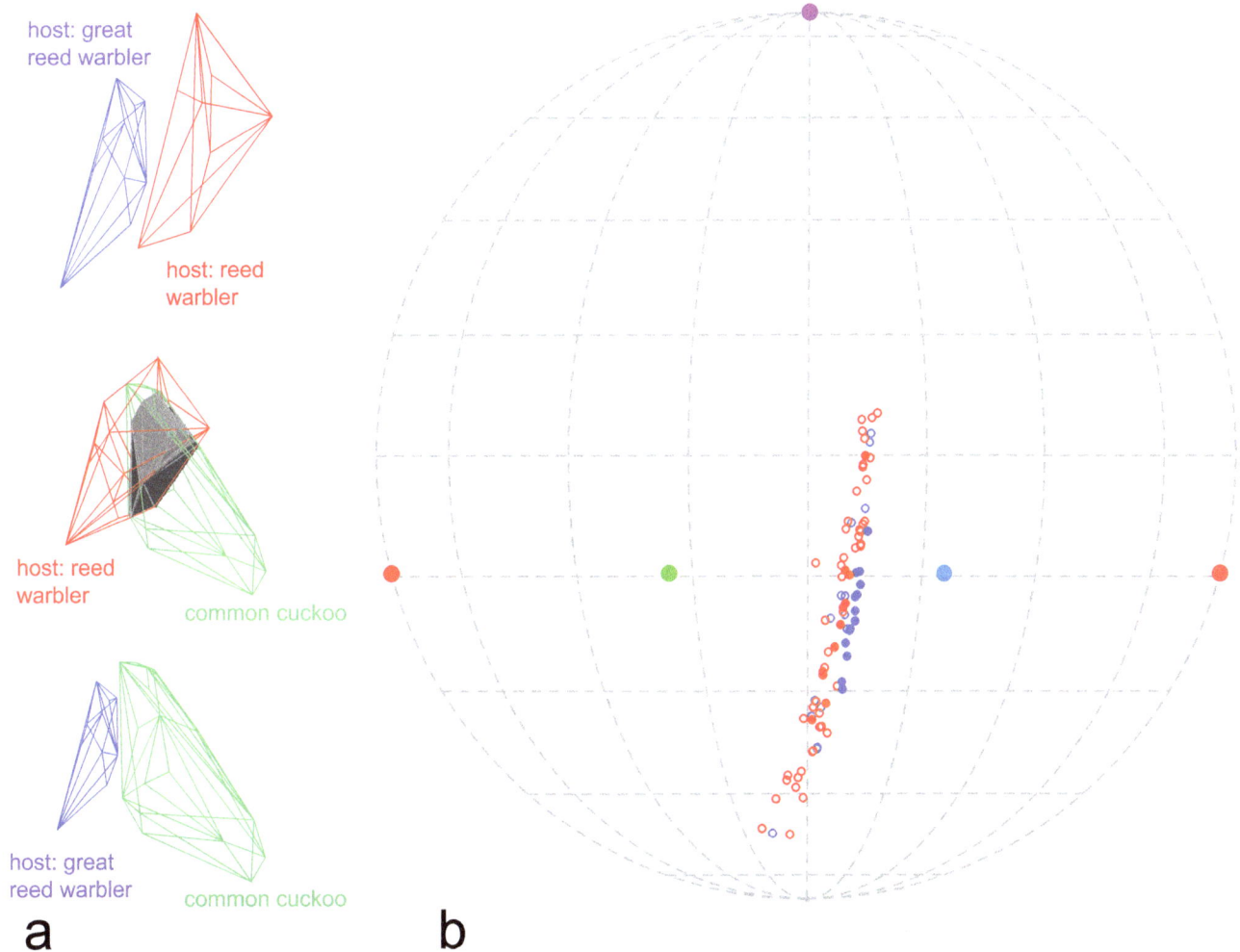

Figure 2. Cuckoo and host eggs in tetrahedral bird average visual colour space. (a) Overlap of convex hulls (grey polyhedron) encapsulating point measurements of host (red and blue) and cuckoo (green) eggs. Second and third plot are presented in identical orientations for ease of comparison and rotated to maximize the visibility of differences on the third plot (i.e. where there is no overlap). (b) Projection of colour measurement on a circumscribed sphere of the colour space tetrahedron; open symbols – cuckoos, filled symbols – hosts, red – reed warbler nests, blue – great reed warbler nests.

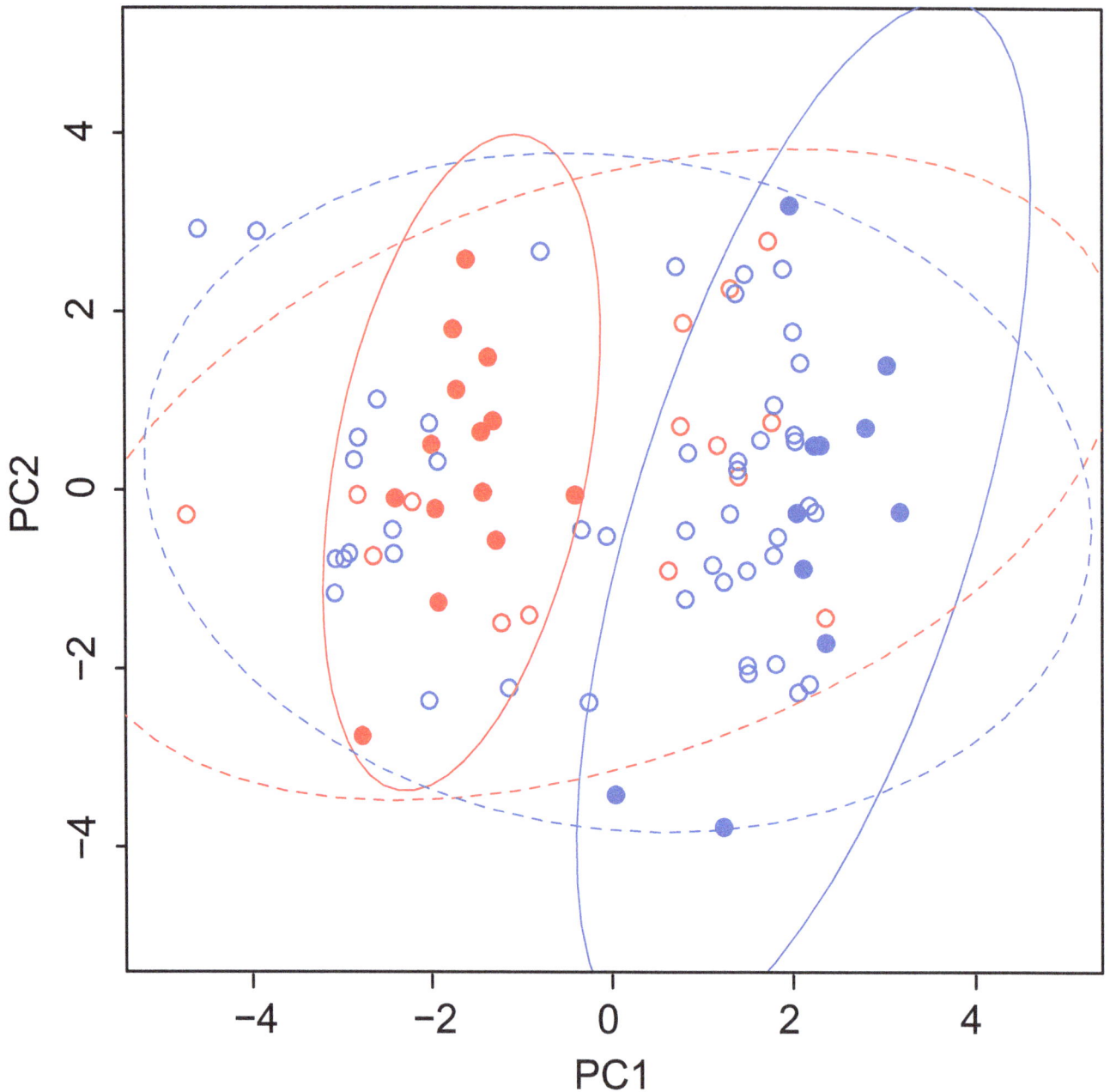

Figure 3. The first two principal components describing colour and size of cuckoo and warbler eggs. Blue symbols - great reed warbler eggs (filled) or cuckoo eggs found in great warbler nests (open); red symbols - reed warbler eggs (filled) or cuckoo eggs from reed warbler nests (open). Lines represent bivariate 95% confidence ellipses. Line colours correspond to circle colours. Solid lines - host species; dashed lines - cuckoos.

while 11.7% of reed warbler nests were parasitized. Moreover, our long-term observations suggest that great reed warbler has started to be utilized as a potential host only recently. Cuckoo eggs were found in only 2 out of over 700 nests of great reed warbler recorded in the years 1970–2006 [18]. In 2007 – out of 154 nests, 4 contained cuckoo eggs (2.6%); in 2009 – out of 136 nests, 6 cuckoo eggs were found (4.4%). In 2012–2013, 20 cuckoo eggs were found in 356 nests of the great reed warbler (5.6%). Although these proportions are similar to some other European populations (e.g. 5.3% in Moravia (Czech Republic); [30]), they are at the same time in contrast to figures from a number of other locations (Hungary and Bulgaria: 64% and 40% nests parasitized, respectively [7,31]). These figures indicate that host switching

may be an underestimated but important phenomenon in brood-parasites occurring in areas where several suitable hosts are present. Cases of host-switching have rarely been reported (but see Jelinek et al. 2014) – clearly, more research is required in this matter as host switching may represent an important component of continuously an evolving host-parasite system of brood-parasitic species. Our colour measurements are consistent with the process of cuckoos slowly adapting to the newly acquired host species – interestingly, this adaptation seems to occur firstly in achromatic components of colouration. The picture of some components of egg morphology specializing and acquiring mimicry faster than the others is an important indication that egg phenotypes are multivariate entities, with their individual components undergoing

Table 3. Receptor-noise model output for chromatic and achromatic contrasts between cuckoos and host species.

Comparison	Chromatic contrast	Achromatic contrast
Host eggs		
GRW – RW	1.23±0.05	6.05±0.29
GRW – GRW	0.58±0.06	3.43±0.40
RW – RW	0.90±0.08	4.95±0.49
Cuckoo eggs		
CC_RW – CC_GRW	1.67±0.05	2.61±0.07
CC_RW – CC_RW	1.27±0.02	2.09±0.04
CC_GRW – CC_GRW	1.27±0.21	2.07±0.31
Host eggs vs. cuckoo eggs[1]		
CC – GRW	1.28±0.01	6.26±0.09
CC – RW	0.97±0.01	6.32±0.06

All values are in the units of just-noticeable differences (JND) and are presented with respective standard errors. Shortcuts: RW – reed warbler, GRW – great reed warbler, CC – common cuckoo (all eggs), CC_GRW – cuckoo eggs from great reed warbler nests, CC_RW – cuckoos eggs from reed warbler nests.
[1]Cuckoo eggs exhibited no significant differences in terms of respective chromatic and achromatic contrasts and thus all cuckoo eggs were compared with respective host species eggs.

partly independent evolutionary changes. Obviously our data lack the dynamic aspect as the egg morphologies have not been tracked in the past. However, the existence of partial specialization in some morphology components and its complete absence in the others suggests that our system may represent and early stage of acquiring a new host, which would be consistent with patterns of egg-morph variability predicted by numerical models [11,13]. Whether the observed patterns represent a genuine snapshot from a dynamic trajectory of egg-colouration traits evolving in response to host-switching remains to be confirmed. It is possible, that the observed variation is a part of a larger pattern of egg morphological variation. Numerical simulations indicate that host egg morphologies can fluctuate and follow a cyclic patterns changing relatively rapidly [32]. It is possible that similar dynamics occur in co-evolving brood parasites; long term tracking of egg morphologies in brood parasitic systems should allow these exciting questions to be answered. Importantly such tracking would not only put evolving parasitic eggs in the appropriate temporal scale – it would also reveal how parasites shape the evolution of host eggs. It has been recently shown, that egg mimicry can have profound effects on the evolution of host egg phenotype [33–35]. In our system a substantial difference exists between the two close cuckoo hosts – further, long-term studies are needed to see how these differences are maintained and evolve in the absence of cuckoo specialization.

The sole presence of substantial variance in cuckoo egg phenotypes suggests that this trait has large evolutionary potential. Rapidly evolving egg mimicry has been recently shown among cuckoos parasitizing reed warblers in Zealand (Denmark) [36]. The authors observed a significant increase in similarity of egg coloration of the cuckoo and their reed warbler host over a 24-year study period. However, the substantial variance in cuckoo egg phenotypes may constitute an adaptation on its own. It is possible that benefits related to the ability to exploit other, less typical host species may overcome any potential costs of failure due to imperfect egg mimicry.

Our analyses focus on size and background colouration of cuckoo eggs, ignoring possible differences in the spotting pattern of cuckoo eggs. However, studies published to date have usually demonstrated background colouration to be the most important

trait differentiating gentes [4,37]. In contrast, studies considering spottiness of eggs failed to show any differences between gentes [36,38]. Similarly, Antonov et al. [39] found no significant differences in egg spotting between cuckoos parasitizing different hosts, in spite of clear differences in the analysed host species (great reed warbler, marsh warbler and corn bunting). We believe that ignoring egg spottiness in our study does not significantly bias our inference.

Lack of any differentiation in cuckoo egg phenotypes with respect to two parasitized host species was also apparent in the case of shell strength. It is well known that cuckoo eggs are equipped with strong egg shells; a number of hypotheses have been suggested to explain this observation [20,21,40]. One hypothesis is that stronger shells protect cuckoo eggs from being punctured and ejected from host nests. In our study, we predicted that cuckoo eggs laid in nests of great reed warblers would have stronger shells, which would be an adaptation to parasitizing hosts with larger and stronger beaks. We found no evidence of such differentiation. The sample size used for shell thickness measurements was relatively small, however given observed effect sizes, it is unlikely that lack of observed differentiation is a matter of power. Also, in this study we used dynamometry, whereas in a number of studies egg shell thickness was measured as a reliable proxy of egg shell strength [41,42]. In our opinion, the dynamometric approach is more appropriate as it more directly relates measured characteristics to the way the egg shell affects host anti-parasite responses and cuckoo hatching processes [20,21]. Lack of observed differences between cuckoos parasitizing hosts that differ greatly in their egg puncturing potential may on one hand suggest that the puncture-resistance hypothesis is not adequate in this system. On the other hand, it is possible that eggshell thickness is less evolutionarily labile than egg colouration and does not differentiate so easily with respect to parasitized hosts.

Another potentially confounding aspect of our study which cannot be properly accounted for is the identity of cuckoo females. Obviously, eggs laid by the same females should be expected to be more similar to each other than to eggs laid by other unrelated females. This issue would be particularly important in our study if specific females laid eggs in nests of different host species. Cuckoo females are believed to be specific with respect to chosen host-species although there are rare cases of females laying eggs in nests of multiple host species [1]. Unfortunately, genetic material gathered in this study does not allow for robust and confident identification of cuckoo eggs coming from specific females: the uncertainty of assigning parents deteriorates rapidly if neither of the parental genotypes is known. However, GPS coordinates collected for all sampled nests indicate that the average distance between sampled host nests in this study was 9 km. According to Davies (2011) [1] (see also [43]), the (mostly non-overlapping) breeding areas of female cuckoos range from 0.21 to 1.68 km^2, resulting in area radii ranging from 250 to 730 m. Even after assuming 730 m as a conservatively large territory size and excluding from the analysis all nests involved in pairs with nest-to-nest distances less than $2 \cdot 730 = 1460$ m (9 nests), we did not find evidence for the presence of significant differentiation in egg phenotypes of cuckoos laying eggs in nests of different hosts. Thus, we assume that our analyses are robust despite the lack of female identities.

Finally, potential bias in our analyses may also be related to egg collection procedure. It is well recognized that hosts tend to reject more dissimilar eggs at higher rates [44,45], and thus dissimilar eggs might be underrepresented in our data set if such cuckoo eggs disappeared prior to egg collection. This problem should be accounted for while interpreting the data, but it is extremely difficult to validate this source of bias. This would be particularly important if we find that the host species showing higher rejection rates matches cuckoo eggs better that the host species showing lower rejection rates. In fact, our data show a rather opposite pattern, so this potential bias is unlikely to alter our interpretation.

In conclusion, our data do not support the hypothesis of specialization of the common cuckoo in parasitizing two species of reed warblers. In the studied population, cuckoos parasitizing two studied hosts seem to lay phenotypically indistinguishable eggs. Our results directly contrasts with other published studies looking at cuckoos exploiting sets of sympatric hosts [5,36]. We failed to find significant differences in egg morphology between cuckoos parasitizing these two species despite clear differentiation between host species. Instead, observed colouration patterns strongly depend on the colour components studied. Chromatic components clearly cluster cuckoo eggs close to the reed warbler eggs, supporting the expectation that cuckoos should match most efficiently the most prevalent host species in a given area. However, achromatically, cuckoo eggs exhibit much more variation that spreads continuously between achromatic components of egg morphology of both host species. Finally, as indicated by the subspaces occupied by cuckoo egg phenotypes in the tetrachromatic colour space, ample amount of chromatic variation in cuckoo eggs spreads in a direction opposite to either of the two available host species. The cuckoo from this gens may exploit other host species present in the study area, and this may explain

this additional variation in cuckoo egg morphology. Further studies should focus on analysing cuckoo egg phenotypes found in other sympatrically occurring host species. It may be particularly interesting to explore selection regimes that may lead to homogenising of cuckoo egg phenotypes even in the presence of clearly distinct host species egg phenotypes. The studied population is a dynamic system, where host species frequencies undergo continuous changes. Our study demonstrates that in colouration studies, a multivariate approach is the one that yields more realistic results. Ignoring size and brightness in our study would clearly favour greater similarity of cuckoo eggs to those of the reed warbler. However, after including brightness and physical dimensions, the variation in egg phenotypes spreads much wider and may overlap phenotypic values of the other host species. Future studies should look more closely at such patterns. In particular, it seems important to establish the relative importance of different aspects of egg morphology (size, brightness, saturation, hue) in the evolution of host egg discrimination and parasite egg mimicry. The amount of observed differentiation may depend on the choice and assembly of characteristics used in such studies. Different aspects of egg morphology may evolve at different rates in brood-parasitic species, substantially biasing simple univariate comparisons of specific traits.

Supporting Information

Table S1 Detailed results of shell-thickness measurement showing differences between host-species tested with a Student t test.

Table S2 Loadings on the first two principal components from all colour and size variables.

Table S3 Correlations of raw variables used in the PCA analysis for hosts (a) and cuckoos (b).

Acknowledgments

We thank two anonymous reviewers for their valuable comments that helped to improve the manuscript. We are also thankful to Kimberly S. Sheldon for useful comments and English editing. Finally, we would like to thank two filed assistants – Monika Czuchra and Wanda Zdunek – for help in searching for new reed warblers' nests.

Author Contributions

Conceived and designed the experiments: AD MC SMD JS. Performed the experiments: AD JS. Analyzed the data: SMD MC JS. Contributed reagents/materials/analysis tools: AD MC SMD. Wrote the paper: AD MC JS SMD.

References

1. Davies N (2011) Cuckoos, cowbirds and other cheats. Poyser Monographs. New York: Princeton University Press. 328p.

2. De L. Brooke M, Davies NB (1988) Egg mimicry by cuckoos *Cuculus canorus* in relation to discrimination by hosts. Nature 335: 630–632. doi:10.1038/335630a0.

3. Moksnes A, Røskaft E (1995) Egg-morphs and host preference in the common cuckoo (*Cuculus canorus*): an analysis of cuckoo and host eggs from European museum collections. J Zool 236: 625–648. doi:10.1111/j.1469-7998.1995.tb02736.x.

4. Stoddard MC, Stevens M (2011) Avian vision and the evolution of egg color mimicry in the common cuckoo. Evol Int J Org Evol 65: 2004–2013. doi:10.1111/j.1558-5646.2011.01262.x.

5. Honza M, Šulc M, Jelínek V, Požgayová M, Procházka P (2014) Brood parasites lay eggs matching the appearance of host clutches. Proc Biol Sci 281: 20132665. doi:10.1098/rspb.2013.2665.

6. Gibbs HL, Sorenson MD, Marchetti K, Brooke MD, Davies NB, et al. (2000) Genetic evidence for female host-specific races of the common cuckoo. Nature 407: 183–186. doi:10.1038/35025058.

7. Fossøy F, Antonov A, Moksnes A, Røskaft E, Vikan JR, et al. (2011) Genetic differentiation among sympatric cuckoo host races: males matter. Proc Biol Sci 278: 1639–1645. doi:10.1098/rspb.2010.2090.

8. Edvardsen E, Moksnes A, Røskaft E, Øien IJ, Honza M (2001) Egg Mimicry in Cuckoos Parasitizing Four Sympatric Species of *Acrocephalus* Warblers. The Condor 103: 829. doi:10.1650/0010-5422(2001)103[0829:EMICPF]2.0.CO;2.

9. Honza M, Taborsky B, Taborsky M, Teuschl Y, Vogl W, et al. (2002) Behaviour of female common cuckoos, *Cuculus canorus*, in the vicinity of host nests before and during egg laying: a radiotelemetry study. Anim Behav 64: 861–868. doi:10.1006/anbe.2002.1969.

10. Takasu F, Moskát C, Muñoz AR, Imanishi S, Nakamura H (2009) Adaptations in the common cuckoo (*Cuculus canorus*) to host eggs in a multiple-hosts system

of brood parasitism. Biol J Linn Soc 98: 291–300. doi:10.1111/j.1095-8312.2009.01288.x.

11. Takasu F (2003) Co-evolutionary dynamics of egg appearance in avian brood parasitism. Evol Ecol Res 5: 345–362.

12. Yang C, Liang W, Cai Y, Shi S, Takasu F, et al. (2010) Coevolution in Action: Disruptive Selection on Egg Colour in an Avian Brood Parasite and Its Host. PLoS ONE 5: e10816. doi:10.1371/journal.pone.0010816.

13. Takasu F (2005) A theoretical consideration on co-evolutionary interactions between avian brood parasites and their hosts. Ornithol Sci 4: 65–72. doi:10.2326/osj.4.65.

14. Kleven O, Moksnes A, Røskaft E, Rudolfsen G, Stokke BG, et al. (2004) Breeding success of common cuckoos Cuculus canorus parasitising four sympatric species of Acrocephalus warblers. J Avian Biol 35: 394–398. doi:10.1111/j.0908-8857.2004.03359.x.

15. Moskat C, Takasu F, Munoz AR, Nakamura H, Ban M, et al. (2012) Cuckoo parasitism on two closely-related Acrocephalus warblers in distant areas: a case of parallel coevolution. Chin Birds 3: 320–329.

16. Honza M, Moksnes A, Røskaft E, Stokke BG (2001) How are different Common Cuckoo Cuculus canorus egg morphs maintained? An evaluation of different hypotheses. Ardea 89: 341–352.

17. Dyrcz A, Hałupka L (2006) Great reed warbler Acrocephalus arundinaceus and reed warbler Acrocephalus scirpaceus respond differently to cuckoo dummy at the nest. J Ornithol 147: 649–652. doi:10.1007/s10336-006-0097-x.

18. Dyrcz A, Hałupka K (2007) Why Does the Frequency of Nest Parasitism by the Cuckoo Differ Considerably Between Two Populations of Warblers Living in the Same Habitat? Ethology 113: 209–213. doi:10.1111/j.1439-0310.2006.01308.x.

19. Cassey P, Thomas GH, Portugal SJ, Maurer G, Hauber ME, et al. (2012) Why are birds' eggs colourful? Eggshell pigments co-vary with life-history and nesting ecology among British breeding non-passerine birds. Biol J Linn Soc 106: 657–672. doi:10.1111/j.1095-8312.2012.01877.x.

20. Antonov A, Stokke B rd G, Moksnes A, Kleven O, Honza M, et al. (2006) Eggshell strength of an obligate brood parasite: a test of the puncture resistance hypothesis. Behav Ecol Sociobiol 60: 11–18. doi:10.1007/s00265-005-0132-6.

21. Honza M, Picman J, Grim T, Novak V, Capek Jr M, et al. (2001) How to hatch from an egg of great structural strength. A study of the Common Cuckoo. J Avian Biol 32: 249–255. doi:10.1111/j.0908-8857.2001.320307.x.

22. Dyrcz A (1981) Breeding ecology of Great Reed Warbler Acrocephalus arundinaceus and Reed Warbler Acrocephalus scirpaceus at fish-ponds in SW Poland and lakes in NW Switzerland. Acta Ornithol 18: 307–334.

23. Cassey P, Ewen JG, Blackburn TM, Hauber ME, Vorobyev M, et al. (2008) Eggshell colour does not predict measures of maternal investment in eggs of Turdus thrushes. Naturwissenschaften 95: 713–721. doi:10.1007/s00114-008-0376-x.

24. Maia R, Eliason CM, Bitton P-P, Doucet SM, Shawkey MD (2013) pavo: an R package for the analysis, visualization and organization of spectral data. Methods Ecol Evol 4: 906–913. doi:10.1111/2041-210X.12069.

25. R Development Core Team (2012) R: A Language and Environment for Statistical Computing.

26. Stoddard MC, Prum RO (2008) Evolution of avian plumage color in a tetrahedral color space: a phylogenetic analysis of new world buntings. Am Nat 171: 755–776. doi:10.1086/587526.

27. Vorobyev M, Osorio D (1998) Receptor noise as a determinant of colour thresholds. Proc Biol Sci 265: 351–358. doi:10.1098/rspb.1998.0302.

28. Rasband WS (2012) ImageJ: image processing and analysis in Java. Bethesda: National Institutes of Health.

29. Hoyt D (1979) Practical methods of estimating volume and fresh weight of bird eggs. The Auk 96: 73–77. Available: http://www.jstor.org/stable/4085401.

30. Moksnes A, Røskaft E, Bičík V, Honza M, Øien IJ (1993) Cuckoo Cuclus canorus parasitism on Acrocephalus Warblers in Southern Moravia in The Czech Republic. J Ornithol 134: 425–434. doi:10.1007/BF01639833.

31. Moskát C, Honza M (2002) European Cuckoo Cuculus canorus parasitism and host's rejection behaviour in a heavily parasitized Great Reed Warbler Acrocephalus arundinaceus population. Ibis 144: 614–622. doi:10.1046/j.1474-919X.2002.00085.x.

32. Liang W, Yang C, Stokke BG, Antonov A, Fossøy F, et al. (2012) Modelling the maintenance of egg polymorphism in avian brood parasites and their hosts. J Evol Biol 25: 916–929. doi:10.1111/j.1420-9101.2012.02484.x.

33. Stoddard MC, Kilner RM, Town C (2014) Pattern recognition algorithm reveals how birds evolve individual egg pattern signatures. Nat Commun 5: 4117. Available: http://www.nature.com/ncomms/2014/140618/ncomms5117/full/ncomms5117.html. Accessed 2014 Aug 2.

34. Spottiswoode CN, Stevens M (2011) How to evade a coevolving brood parasite: egg discrimination versus egg variability as host defences. Proc R Soc B Biol Sci 278: 3566–3573. doi:10.1098/rspb.2011.0401.

35. Lahti DC (2005) Evolution of bird eggs in the absence of cuckoo parasitism. Proc Natl Acad Sci U S A 102: 18057–18062. doi:10.1073/pnas.0508930102.

36. Avilés JM, Stokke BG, Moksnes A, Røskaft E, Asmul M, et al. (2006) Rapid increase in cuckoo egg matching in a recently parasitized reed warbler population. J Evol Biol 19: 1901–1910. doi:10.1111/j.1420-9101.2006.01166.x.

37. Spottiswoode CN, Stevens M (2010) Visual modeling shows that avian host parents use multiple visual cues in rejecting parasitic eggs. Proc Natl Acad Sci U S A 107: 8672–8676. doi:10.1073/pnas.0910486107.

38. Avilés JM, Stokke BG, Moksnes A, Røskaft E, Møller AP (2006) Environmental conditions influence egg color of reed warblers Acrocephalus scirpaceus and their parasite, the common cuckoo Cuculus canorus. Behav Ecol Sociobiol 61: 475–485. doi:10.1007/s00265-006-0275-0.

39. Antonov A, Stokke BG, Vikan JR, Fossøy F, Ranke PS, et al. (2010) Egg phenotype differentiation in sympatric cuckoo Cuculus canorus gentes. J Evol Biol 23: 1170–1182. doi:10.1111/j.1420-9101.2010.01982.x.

40. Rothstein SI (1975) An Experimental and Teleonomic Investigation of Avian Brood Parasitism. The Condor 77: 250–271.

41. Igic B, Hauber ME, Galbraith JA, Grim T, Dearborn DC, et al. (2010) Comparison of micrometer- and scanning electron microscope-based measurements of avian eggshell thickness. J Field Ornithol 81: 402–410. doi:10.1111/j.1557-9263.2010.00296.x.

42. Hargitai R, Moskat C, Ban M, Gil D, Lopez-Rull I, et al. (2010) Eggshell characteristics and yolk composition in the common cuckoo Cuculus canorus: are they adapted to brood parasitism? J Avian Biol 41: 177–185.

43. Nakamura H, Miyazawa Y, Kashiwagi K (2005) Behavior of radio-tracked Common Cuckoo females during the breeding season in Japan. Ornithol Sci 4: 31–41.

44. Cherry MI, Bennett ATD, Moskát C (2007) Host intra-clutch variation, cuckoo egg matching and egg rejection by great reed warblers. Naturwissenschaften 94: 441–447. doi:10.1007/s00114-007-0216-4.

45. Bártol I, Karcza Z, Moskát C, Røskaft E, Kisbenedek T (2002) Responses of great reed warblers Acrocephalus arundinaceus to experimental brood parasitism: the effects of a cuckoo Cuculus canorus dummy and egg mimicry. J Avian Biol 33: 420–425. doi:10.1034/j.1600-048X.2002.02945.x.

7

Influence of the Biotope on the Tick Infestation of Cattle and on the Tick-Borne Pathogen Repertoire of Cattle Ticks in Ethiopia

Sándor Hornok[1]*, Getachew Abichu[2], Marina L. Meli[3], Balázs Tánczos[1], Kinga M. Sulyok[4], Miklós Gyuranecz[4], Enikő Gönczi[3], Róbert Farkas[1], Regina Hofmann-Lehmann[3]

1 Department of Parasitology and Zoology, Faculty of Veterinary Science, Szent István University, Budapest, Hungary, 2 National Research Center, Department of Parasitology, Arachnoentomology Unit, Sebeta, Ethiopia, 3 Clinical Laboratory and Center for Clinical Studies, Vetsuisse Faculty, University of Zurich, Zurich, Switzerland, 4 Institute for Veterinary Medical Research, Centre for Agricultural Research, Hungarian Academy of Sciences, Budapest, Hungary

Abstract

Background: The majority of vector-borne infections occur in the tropics, including Africa, but molecular eco-epidemiological studies are seldom reported from these regions. In particular, most previously published data on ticks in Ethiopia focus on species distribution, and only a few molecular studies on the occurrence of tick-borne pathogens or on ecological factors influencing these. The present study was undertaken to evaluate, if ticks collected from cattle in different Ethiopian biotopes harbour (had access to) different pathogens.

Methods: In South-Western Ethiopia 1032 hard ticks were removed from cattle grazing in three kinds of tick biotopes. DNA was individually extracted from one specimen of both sexes of each tick species per cattle. These samples were molecularly analysed for the presence of tick-borne pathogens.

Results: *Amblyomma variegatum* was significantly more abundant on mid highland, than on moist highland. *Rhipicephalus decoloratus* was absent from savannah lowland, where virtually only *A. cohaerens* was found. In the ticks *Coxiella burnetii* had the highest prevalence on savannah lowland. PCR positivity to *Theileria* spp. did not appear to depend on the biotope, but some genotypes were unique to certain tick species. Significantly more *A. variegatum* specimens were rickettsia-positive, than those of other tick species. The presence of rickettsiae (*R. africae*) appeared to be associated with mid highland in case of *A. variegatum* and *A. cohaerens*. The low level of haemoplasma positivity seemed to be equally distributed among the tick species, but was restricted to one biotope type.

Conclusions: The tick biotope, in which cattle are grazed, will influence not only the tick burden of these hosts, but also the spectrum of pathogens in their ticks. Thus, the presence of pathogens with alternative (non-tick-borne) transmission routes, with transstadial or with transovarial transmission by ticks appeared to be associated with the biotope type, with the tick species, or both, respectively.

Editor: Brian Stevenson, University of Kentucky College of Medicine, United States of America

Funding: MG, KMS and the study in part were supported by the Lendület program (LP2012-22) of the Hungarian Academy of Sciences. Additional funding was provided by the Research Faculty budget of the SZIU-FVS (KK-UK-12006). The funders had no role in study design, data collection and analysis, decision to publish, or preparation of the manuscript.

Competing Interests: The authors have declared that no competing interests exist.

* Email: Hornok.Sandor@aotk.szie.hu

Introduction

Due to the veterinary-medical importance of hard ticks (Acari: Ixodidae) and costs of their control, the transmission of tick-borne diseases remains a challenge for cattle industry in the tropical and subtropical areas of the world, and it is a priority concern for many countries in these regions [1]. On the other hand, while modern molecular biological methods allow more effective detection of pathogens in tick species, these are expensive and require sophisticated laboratory instruments. As a consequence, although vector-borne (tick-borne) infections occur more frequently in the tropics [2], these are less frequently studied with up-to-date methods and relevant data are scarce in the literature.

Ethiopia has the largest livestock population in Africa [3], including 52 million cattle. Because ticks are widely distributed in the country [4], with more than 40 species of 10 genera [5], they severely affect cattle. Major tick-borne diseases in Ethiopia include anaplasmosis, babesiosis and theileriosis [6]. In such a scenario it is utterly important to know those epidemiological factors, which may increase or decrease tick burdens of animals and thus the risks of tick-borne pathogen transmission. However, most regional studies report only the occurrence of tick species on Ethiopian cattle [7,8], and molecular investigations are very few [9].

The farming system, the local cattle breed and the human population show marked differences throughout the various landscapes of Ethiopia. Human and livestock settlements have concentrated in the moist highland areas, whereas dry lowlands allow conditions for traditional nomadic life [3,10]. In the face of such historical settings limited financial resources may explain why less attention was paid to compare epidemiological factors of tick-borne diseases between these highly divergent regions.

Thus the primary aim of the present study was to assess if (1) the tick infestation of cattle (i.e. tick species and/or their abundance), and (2) the presence of certain tick-borne pathogens/groups in ticks is influenced by the regional biotope used for grazing. In this context it was less important to consider here if PCR positivity implies a potential vector role for the relevant tick species in transmitting the detected pathogen. Instead, results focus on and are compared according to biotopes in which ticks could have acquired the evaluated pathogen(s), either from the current or one of their previous hosts.

Materials and Methods

Sample collection

The study area is situated in South-Western Ethiopia, along the Didessa valley (in the region between Nekemte and Jima, coordinates: 09°05′N, 36°33′E–7°40′N, 36°50E). Three types of habitats (biotopes) were selected for tick collection. These biotopes have different altitude, rainfall, relative humidity, temperature, vegetation coverage, wildlife, cultivated crops and livestock animals. Ticks were collected from cattle between June-July of 2012 in the following biotopes: (A) moist highland (above 1500 m altitude, in excess of 900 mm rain annually, temperature 18–20°C, with dense forest vegetation); (B) mid highland (less moist and cool, with mixed vegetation coverage showing altitudinal change); (C) savannah lowland (500–1500 m altitude, annual rainfall below 900 mm, temperature 18–24°C, with shrubs, gallery forests around rivers, woodland). Ticks were removed with strong pointed forceps from the skin of 109 cattle (35, 56 and 18 animals according to the above three biotope types, respectively) in 18 herds. Because this was part of the regular veterinary care and the field studies did not involve endangered or protected species, no specific permissions were required for these activities. All specimens were put into 70% ethanol in a separate vial according to host animal, and stored consequently at around room temperature.

DNA extraction

Amblyomma specimens were prepared by mechanical removal of host tissues around their mouthparts, which are long and firmly cemented into the skin. All ticks were mechanically cleaned and their species identified according to [11,12]. DNA was individually extracted from one specimen of both sexes (and if available, one nymph) of each tick species per cattle. Prior to DNA extraction these 295 specimens were taken out from the 70% ethanol, air dried, and individually washed sequentially in detergent-containing water, in tapwater and in distilled water. Air-dried ticks were then minced with pointed scissors at the bottom of Eppendorf-tubes, in 100 μl of phosphate-buffered saline (PBS). DNA was extracted using the QIAamp DNA blood mini kit (QIAGEN, Hilden, Germany) following the manufacturer's instructions and including an overnight digestion step (incubation at 56°C for at least 8 h) with tissue lysis buffer and Proteinase-K (QIAGEN,

Hilden, Germany). Extractions included an extraction control to monitor cross-contamination of samples.

Conventional PCR and sequencing for piroplasms

A 450 bp long portion of the *18S rRNA* gene of piroplasms was amplified with the primers PIRO-A1 [13] 5′-AGG GAG CCT GAG AGA CGG CTA CC-3′ and PIRO-B [14] 5′-TTA AAT ACG AAT GCC CCC AAC-3′. The reaction mixture contained 1×concentration of Coralload PCR Buffer, 1.5 mM of MgCl$_2$, 0.2 mM of each dNTP, 25 pmol of each primer and 1 U of HotStarTaq Plus DNA Polymerase (QIAgen GmbH, Hilden, Germany) in a final volume of 25 μl. The reaction was run in a T-personal thermocycler (Biometra GmbH, Göttingen, Germany) according to the following program: initial denaturation for 5 min at 95°C was followed by a cycle 94°C for 30 s, 60°C for 30 s and 72°C for 40 s, repeated 35 times and finished with 72°C for 10 min. PCR products were visualized in 1.5% agarose gel prestained with ethidium-bromide. Strongly positive samples were selected for sequencing done by the Macrogen Inc. (Seoul, South Korea). Representative sequences were submitted to the GenBank (accession numbers KJ941104-12). In all PCR procedures positive (*Babesia canis* DNA) and negative controls (sterile deionized water) were included.

Real-time PCR for *Coxiella burnetii*

The samples were screened using a sensitive and specific TaqMan real-time PCR assay for the IS*1111* element of *C. burnetii* [15]. The assay amplifies the superoxide dismutase gene and IS*1111* transposable element of *C. burnetii* with the primers 5′-CCG ATC ATT TGG GCG CT-3′ (forward) and 5′-CGGCGGTGTTTAGGC-3′ (reverse) and the probe 5′-6FAM-TTA ACA CGC CAA GAA ACG TAT CGC TGT G-MGB-3′, at concentrations of 1600, 800 and 200 nM, respectively. The reaction was started with 95°C for 10 min, followed by 40 cycles at 95°C for 15 s and 60°C for 60 s.

PCRs and sequencing for rickettsiae

The presence of the members of spotted fever (SFG) and typhus groups (TG) rickettsiae was detected by using a previously published real-time TaqMan PCR assay specific for a 74-bp fragment of the citrate synthase (*gltA*) gene [16]. The PCR mixture contained a final concentration of 0.2 μM of primers (forward: 5′-TCG CAA ATG TTC ACG GTA CTT T-3′ and reverse: 5′-TCG TGC ATT TCT TTC CAT TGT G-3′) and probe (5′-FAM-TGC AAT AGC AAG AAC CGT AGG CTG GAT G-BHQ-3′), 12.5 μl of 2×qPCR MasterMix Plus Low ROX (Eurogentec), and 5 μl or 2.5 μl of template in a final volume of 25 μl. The *gltA* assay was performed using 45 cycles on a ABI 7500 Fast Real-Time PCR system (Applied Biosystems), with an initial denaturation step of 20 s at 95°C, which was followed by 45 cycles of 95°C for 3 s and 60°C for 30 s [16]. In addition, from 21 samples with low threshold cycle (Ct) values the amplification of an approx. 750 bp fragment of the *gltA* gene and direct sequencing of the PCR product was attempted [17]. Representative sequences were submitted to the GenBank (accession numbers KJ941095-103). Sequences were edited and aligned with a consensus sequence using Geneious Version 7.1.5. For phylogenetic analysis, the sequences were aligned with known rickettsiae sequences from GenBank using ClustalW [18] and, if necessary, manually adjusted. Only the nucleotides available for all included sequences were used in the phylogenetic analysis. A bootstrap phylogenetic tree demonstrating the relationship between the isolates was created by the Neighbor-Joining method [19] using a distance matrix corrected for nucleotide substitutions based on the

Maximum Composite Likelihood model. The dataset was resampled 1,000 times to generate bootstrap values. Phylogenetic and molecular evolutionary analyses were conducted using MEGA version 6 [20].

Real-time PCR for haemotropic mycoplasmas (haemoplasmas)

The presence of haemotropic mycoplasmas or haemoplasmas was evaluated by using a universal screening assay based on the SYBR principle [21]. The primers were designed to fit the *16S rRNA* gene of haemotropic *Mycoplasma* species with known sequences. The reaction volume was 25μl, consisting of 12.5 μl of KAPA SYBR FAST qPCR Kit Master Mix (2x) Universal (KAPA Biosystems), a final concentration of 200 nM of forward primer (5′-AGC AAT RCC ATG TGA ACG ATG AA-3′), and an equimolar mixture (200 nM) of two reverse primers (5′-TGG CAC ATA GTT TGC TGT CAC TT-3′ and 5′-GCT GGC ACA TAG TTA GCT GTC ACT-3′), and 5μl of template DNA. Assays were performed using an ABI 7500 Fast Real-Time PCR system (Applied Biosystems). The SYBR green PCR protocol included an initial step at 95°C for 3 min, followed by 40 cycles of 95°C for 3 s and 60°C for 30 s. After the PCR run, dissociation was performed with the following thermal profile: 95°C for 15 s, 60°C for 1 min, a temperature increase from 60°C to 95°C with 1% gradient (for about 20 min), followed by 95°C for 15 s, and finally 60°C for 15 s.

Statistical analysis

Abundance rates were calculated from the number of individuals of one species, expressed as the percentage of the number of all ticks. Sample prevalence data were analysed by using Fisher's exact test. Differences were regarded significant when P<0.05. Because of the large number of abundance/prevalence rates provided in the text below, exact confidence intervals are not shown.

Results

Abundance and habitat-dependent occurrence of ticks

Altogether 1032 ticks of seven species were collected: 1028 adults and four nymphs (Table 1). Concerning the adults, the most abundant was *Amblyomma variegatum* (558 out of 1028: 54.3%), followed by *A. cohaerens* (307 out of 1028: 29.9%), *Rhipicephalus decoloratus* (135 out of 1028: 13.1%), *Rh. evertsi* (17 out of 1028: 1,7%), *Rh. praetextatus* (8 out of 1028: 0.8%), *A. lepidum* (2 out of 1028: 0.2%) and *Hyalomma rufipes* (1 out of 1028: 0.1%). Among specimens of *Amblyomma* spp. the males predominated over females (752 vs. 115), whereas in case of *Rh. decoloratus* the contrary was true (male vs. female ratio was 9:126).

According to biotopes, *A. variegatum* was significantly more abundant on mid highland (445 out of 686 ticks: 64.9%), than on moist highland (111 out of 248 ticks: 44.8%) (P<0.001). On the other hand, *A. cohaerens* was significantly more abundant on savannah lowland (93 out of 98 ticks: 94.9%), than on either moist highland (82 out of 248 ticks: 33.1%) or mid highland (132 out of 686 ticks: 19.2%) (P<0.001). No *Rh. decoloratus* was found on savannah lowland. Pathogens detected in ticks are summarized in Table 1.

Piroplasms in ticks

Theileria mutans, *T. velifera* and *T. orientalis* were detected with prevalence rates of 8.5% (25 out of 295), 5.4% (16 out of 295) and 5.8% (17 out of 295), respectively. In two *A. variegatum* males *Babesia caballi* was also demonstrated. This strain had the highest,

but only 97% sequence similarity to a *B. caballi* isolate from South Africa (EU642514).

The presence and prevalence of *Theileria* spp. in ticks (and most likely in the cattle from which they had been removed) was apparently not related to the biotopes, as *Rh. decoloratus* was significantly (P<0.001) more frequently *Theileria*-positive, than *A. variegatum* (or *A. cohaerens*) on both moist highland (12 out of 21 vs. 2 out of 35) and mid highland (16 out of 29 vs. 7 out of 81) (Table 1).

However, *Theileria* positivity seemed to be related to tick species, because the overall prevalence was significantly (P<0.001) higher in *Rh. decoloratus* (28 out of 50: 56%), than in *Amblyomma* specimens (25 out of 220: 11.4%). Concerning *Rhipicephalus* spp, *T. velifera* was identified only in *Rh. decoloratus*. Apart from this tick species, *T. orientalis* was detected only in *Rh. praetextatus*, and *T. mutans* in both *Rh. praetextatus* and *Rh. evertsi*.

There was a considerable sequence variation of all three above *Theileria* spp, with altogether eight genotypes, with differences in up to 16 nucleotide positions (Tables 2–4). The highest degree of sequence polymorphism was observed in the case of *T. mutans*, but *T. velifera* and *T. orientalis* had more genotypes in the samples. The numbers of ticks with the two genotypes (A and B) of *T. mutans* were more equilibrated, than for *T. velifera* and *T. orientalis* with the dominance of genotype-A in comparison with B and C (Table 1, bottom).

In addition, it was noted, that if compared not on the species, but on the genotype level of the same *Theileria* spp, the B and C genotypes of *T. orientalis* were exclusively found in *Rh. decoloratus* (Table 1). Similarly, as B and C genotypes of *T. velifera* were unique to *A. cohaerens* (Table 1.), these could be detected only in association with savannah lowland. The genotypes of *T. mutans* were more evenly distributed among ticks species and biotopes.

Coxiella burnetii in ticks

10.8% of ticks (32 out of 295) were positive for the Q fever agent (Table 1). PCR positivity was significantly (P<0.001) more frequently detected in ticks (collected from cattle) on savannah lowland (14 out of 18: 77.8%), than on mid highland (6 out of 173: 3.5%) or moist highland (12 out of 104: 11.5%). These data also imply, that *C. burnetii* was significantly more prevalent in ticks from moist highland, than in those from mid highland (P = 0.011). The biotope-related occurrence of *C. burnetii* is confirmed by the comparison of the same tick species in different biotopes, i.e., there was significantly higher PCR positivity among *A. cohaerens* ticks in savannah lowland (12 out of 15: 80%), than in either mid highland (2 out of 51: 3.9%) or moist highland (2 out of 34: 5.9%).

Rickettsiae in ticks

The overall prevalence of rickettsiae in ticks of the present study was 64.1% (189 out of 295). In the rate of rickettsia prevalence there was a significant (P<0.001) difference between the tick species: more *A. variegatum* (114 out of 118: 96.6%) and *A. lepidum* specimens (2 out of 2: 100%) were PCR positive, than those of *A. cohaerens* (57 out of 100: 57%); but PCR positivity was also significantly more frequently detected among individuals of the latter species, than among those of *Rh. decoloratus* (11 out of 50: 22%) or other *Rhipicephalus* spp. (5 out of 25: 20%) (Table 1). The presence of rickettsiae in ticks was also associated with biotopes in case of *A. cohaerens*: the prevalence was significantly lower on savannah lowland (4 out of 15: 26.7%), than on mid highland (33 out of 51: 64.7%) (P = 0.016) or on mid highland and moist highland taken together (53 out of 85: 62.4%) (P = 0.021). Similarly, *Rh. decoloratus* ticks significantly (P = 0.016) more

Table 1. Distribution of tick species collected from cattle in three biotopes, and results of their molecular analyses.

biotope type	tick species[1]	all male/female specimens	number of ticks for PCR[2]	number of Theileria and Babesia positive samples[3]				number of PCR positive samples		
				T. mutans	T. velifera	T. orientalis	B. caballi	Coxiella	Rickettsia	Haemoplasma
moist highland	A. variegatum	100/11	**35***	-	1^A	1^A	-	5	34	-
	A. cohaerens	68/14	**34**	-	-	2^A	-	2	20	-
	Rh. decoloratus	5/35	21	3^A 2^B	2^A	4^A 1$^{\underline{C}}$	-	2	1	-
	other Rh. spp.	7/7	14	3^B	-	1^A	-	3	3	-
mid highland	A. variegatum	410/35	81	2^A 2^B	2^A	1^A	2	4	80	1
	A. cohaerens	102/30	**51**	2^A	2^A	3^A	-	2	33	1
	A. lepidum	2/0	2	-	-	-	-	-	2	-
	Rh. decoloratus	4/91	**29***	8^A 1^B	4^A	1^A 1$^{\underline{B}}$ 1$^{\underline{C}}$	-	-	10	2
	other Rh. spp.	5/5	10	-	-	1^A	-	-	2	1
savannah lowland	A. variegatum	0/2	**2****	1^B	-	-	-	2	-	-
	A. cohaerens	70/23	15	1^B	1$^{\underline{B}}$ 4$^{\underline{C}}$	-	-	12	4	-
	other Rh. spp.	0/1	1	-	-	-	-	-	-	-
	in toto	773/254	**295**	15^A 10^B	11^A 1^B 4$^{\underline{C}}$	14^A 1^B 2$^{\underline{C}}$	2	32	189	5

[1]Tick species significantly more abundant in a biotope type, than in other(s), are marked with bold character. Abbreviations are "A." for Amblyomma, "Rh." for Rhipicephalus. Other Rhipicephalus spp. imply Rh. praetextatus and Rh. evertsi.

[2]The number of asterisks (* or **) indicates the number of nymphs included in a sample number.

[3]Upper index capital letters (A, B, C) on the numbers of Theileria positive samples indicate genotypes (for legend see Tables 2–4). Theileria genotypes unique to a tick species are marked with underlined superscript. The Hyalomma rufipes male was PCR negative in all tests, therefore not shown.

Table 2. Sequence differences of *18S rRNA* gene of *Theileria mutans* genotypes identified in this study, compared to GenBank reference sequences.

designation	nucleotid position in reference sequence (AF078815)															
	620	621	625	633	634	643	644	646	649	664	665	666	668	671	675	677
reference (AF078815)	A	T	G	T	C	A	G	G	T	A	C	T	G	T	T	T
genotype TM-A	•	•	•	•	•	•	•	–	•	•	•	•	•	•	•	•
genotype TM-B	C	C	C	–	–	G	A	–	C	C	G	A	–	C	C	G

Table 3. Sequence differences of *18S rRNA* gene of *Theileria vellifera* genotypes identified in this study, compared to GenBank reference sequences.

designation	nucleotid position in reference sequence (AF097993)											
	620	627	628	629	637	644	646	648	650	652	653	677_8
reference (AF097993)	A	C	T	A	T	T	T	G	T	T	T	–
genotype TV-A	•	•	•	•	•	•	•	•	•	•	•	G
genotype TV-B	G	•	C	T	A	C	•	•	–	–	C	G
genotype TV-C	G	T	•	C	A	•	–	A	–	–	C	G

Table 4. Sequence differences of *18S rRNA* gene of *Theileria orientalis* genotypes identified in this study, compared to GenBank reference sequences.

designation	nucleotid position in reference sequence (AF236094)						
	626	630	637	639	654	676	677
reference (AF236094)	A	A	G	T	T	–	–
genotype TO-A	•	•	•	•	•	•	•
genotype TO-B	•	•	•	•	•	T	A
genotype TO-C	T	T	–	C	C	–	–

frequently contained rickettsiae on mid highland (10 out of 29: 34.5%), than on moist highland (1 out of 21: 4.8%) (Table 1). In all 21 samples processed for sequencing *R. africae* was identified, with one to eight nucleotide differences to a reference sequence (U59733). The phylogenetic relationships of these isolates with each other and with other rickettsia sequences from the GenBank are shown on Figure 1. In summary, *R. africae* was detected in *A. variegatum, A. cohaerens* and *A. lepidum.*

Haemoplasmas in ticks

The low level of haemoplasma positivity seemed to be equally distributed among the tick species (Table 1). The haemotropic *Mycoplasma* sp(p). in question could not be identified because of low bacterial loads (reflected by high Ct values). The presence of these pathogens was only detected in one type of tick biotope, i.e. on mid highland, but this association was not significant ($P = 0.16$) due to the small number of positive samples.

Discussion

Hard ticks (Acari: Ixodidae) feed on the blood of vertebrates. Although they adversely affect domestic animals in several ways, their economically most important effect on domestic animals is connected to their vector role, as they are able to transmit a broad spectrum of tick-borne pathogens. The productivity losses attributable to related diseases are estimated to be highest in the tropical parts of the world, including Africa [22].

In order to infect new hosts, all tick-borne agents need the availability of certain ixodid species, i.e. competent vectors, which thus determine their geographical distribution. At a local scale, the epidemiology of tick-borne infections is strongly interrelated with the ecology of relevant ticks. In this way tick-borne diseases are usually endemic, implying their focal occurrence according to the suitable habitats of competent vectors.

Results on the abundance of tick species on cattle in the present study are similar to those reported earlier from South-Western Ethiopia, i.e. *A. variegatum, A. cohaerens, Rh. decoloratus* and *Rh. evertsi* being the most important ixodid species [7,8]. The male-biased sex ratio of *Amblyomma* spp. and female-biased sex ratio of *Rh. decoloratus* is also consistent with previous findings [23]. However, here it was also demonstrated that the two most abundant tick species, *A. variegatum* and *A. cohaerens* are associated with mid-highland and savannah lowland, respectively, relevant to the endemicity and epidemiology of tick-borne pathogens they may carry.

Theileria spp. detected in ticks of cattle, i.e. *T. mutans, T. velifera* and *T. orientalis* are widespread in Africa and are usually regarded as mildly pathogenic. However, they were also shown to cause severe anaemia, icterus, even deaths [24–27]. Especially *T. mutans* have been associated with disease in cattle: invasion of brain capillaries by this piroplasm may result in a form of bovine theileriosis known as turning sickness [28].

In Africa these *Theileria* spp. are transmitted by ticks of the genus *Amblyomma* [28,29]. This is consistent with the present findings, i.e. the relatively high prevalence of these piroplasms in ticks of cattle, particularly in *Amblyomma* spp. However, the fact that according to the results of this study *Rh. decoloratus* specimens also contained (had access to) all three above mentioned *Theileria* spp, it may deserve future attention to assess the vector competence of this tick species too. In addition, the prevalence of piroplasms (*Theileria* spp.) in cattle ticks was significantly higher (up to 56%) in the present study, than the 0.5–4% reported in Western Ethiopia recently [9], suggesting that their tick-borne transmission is more likely than previously thought.

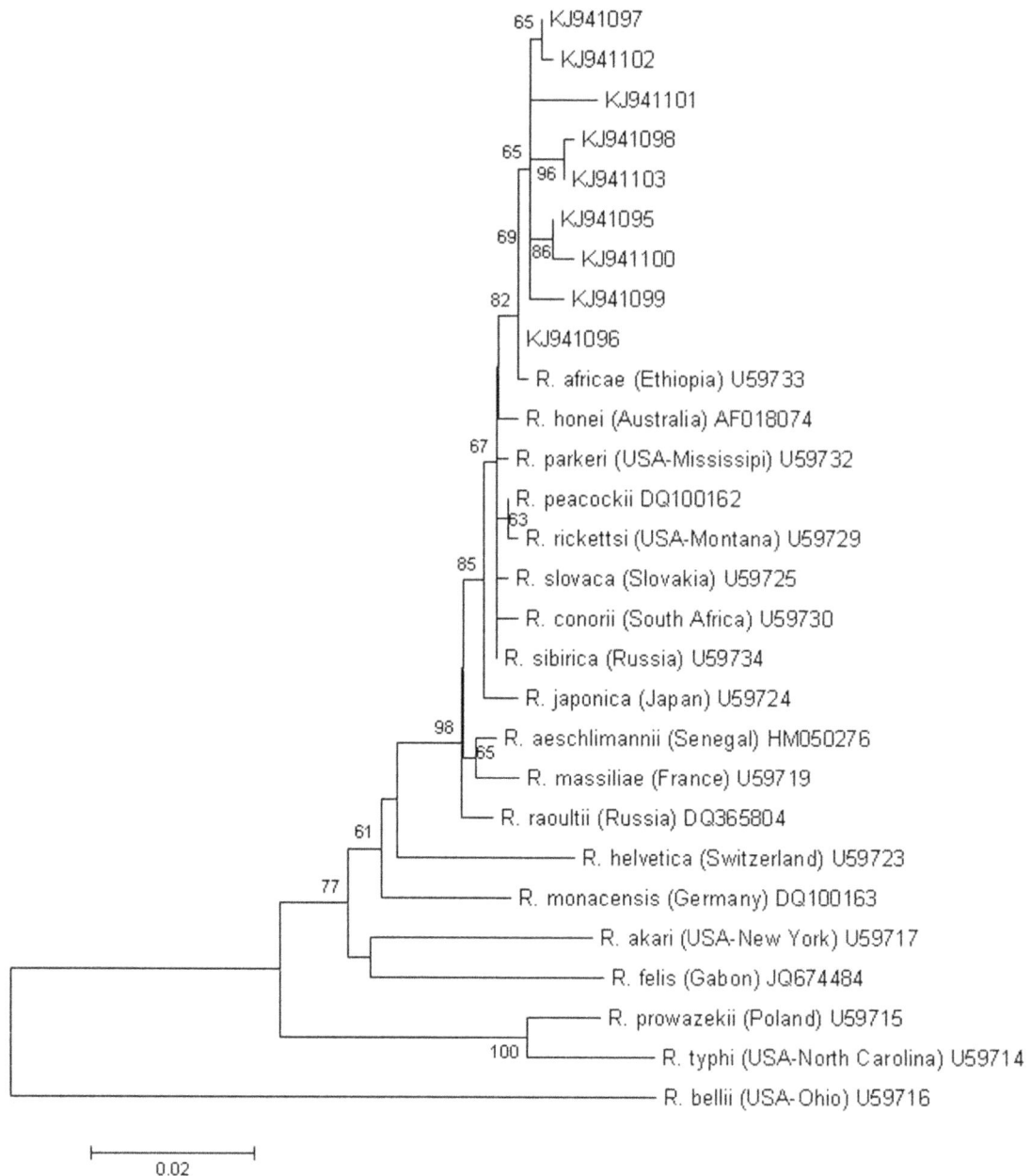

Figure 1. Phylogenetic comparison of *Rickettsia africae* isolates identified in the present study and other rickettsiae based on *gltA* gene sequences. Branch lengths correlate to the number of substitutions inferred according to the scale shown.

A high degree of *18S rRNA* gene sequence polymorphism of these piroplasms was noted here, similarly to that reported in blood samples of the African buffalo (*Syncerus caffer*: the original hosts of *T. mutans*, *T. velifera* and *T. orientalis*) in South Africa [28]. Relevant to the epidemiological significance of the present findings, these genotypes are thought to circulate between some buffalo and cattle populations [28]. The present results also confirm that in comparison with *T. mutans* the sequence variation is less evident in the *18S rRNA* gene of *T. velifera* [28].

According to the data shown here the occurrence of (and risks associated with) these *Theileria* spp. and their genetic variants in South-Western Ethiopia are primarily dependent on the tick species (because they have specific competent vectors), and not on the type of tick-biotope where cattle are grazing. The latter may be

explained by the absence of transovarial maintenance of *Theileria* spp. by ticks in nature [29], unlike in case of babesiae.

Babesia caballi is the large babesia of the horse, causing mainly anaemia. This piroplasm was detected in *A. variegatum* ticks, for the first in East Africa: a finding similar to the identification of the same piroplasm in cattle ticks in West Africa [30]. The sequence was new and highly (3% ≤)divergent from already reported ones. Since cattle were an unlikely source of this piroplasm in the present study, the potential vector role of *A. variegatum* in transmitting *B. caballi* should be further evaluated.

Coxiella burnetii is a tick-borne zoonotic bacterium. Reports are scarce on its presence in ticks in East or West Africa. The Q fever agent is known to occur in Ethiopia for nearly half a century [31]. In a more recent study from West Africa [32] the infection rate with *C. burnetii* showed significant variations in ticks, e.g. in *A. variegatum* it was 0% in one region, and 37.6% in another, which phenomenon could not be completely explained by the authors, and particularly not with differences between tick biotopes. Ticks usually carry *C. burnetii* as reservoirs, and may transfer it between stages (transstadially) and even may inherit it to the next generation (transovarially). Nevertheless, ticks play a subordinate, most likely reservoir role in the epidemiology of Q fever [32], the most significant sources of infection remaining environmental: contact with animal faeces or products, or inhalation of similar substances following aerosolisation. As conditions (prerequisites) for the latter (i.e. dry weather, open area, wind) appear to be more available in savannah woodland, these alternative transmission routes may explain why ticks significantly more often carried (or had access to) *Coxiella* in this kind of biotope (taking into account that cattle were the most likely source of infection of ticks in the present study). Taken together, results presented here may be the first indications that the reservoir role of ticks is related to their habitats, i.e. could be more significant in savannah woodland, than in mid highland or moist highland.

In another West African study [33], also evaluating ticks from cattle, the prevalence of *C. burnetii* was higher (14% vs. 10.8%) and that of rickettsiae was significantly lower (12.5% vs. 64.1%) compared to findings in the present study. In an Ethiopian survey [34] the infection rate of ticks with rickettsiae was also considerably lower (4.1%). Therefore data obtained here, based on highly sensitive real-time PCR analysis, seem to attest for the first time, that rickettsiae may be present in the majority of *A. variegatum* and *A. cohaerens* ticks. In all samples of *A. variegatum* processed for sequencing *Rickettsia africae*, the causative agent of African tick-bite fever was found. This is in line with the known vectorial competence of *A. variegatum* in the transmission of *R. africae* [35]. However, *R. africae* was also identified in a few specimens of *A. cohaerens* and *A. lepidum*, justifying their future evaluation for vector role in the epidemiological cycle. Interestingly, the occurrence of rickettsiae in ticks of cattle, as shown here for the first time, may not only depend on the tick species (i.e. on their competent, specific vector), but also on the biotope type. This may be related to the transovarial maintenance of rickettsiae by ticks (bound to endemic foci) in nature [36].

Haemotropic mycoplasmas (haemoplasmas) are epi-erythrocytic bacteria that may cause anaemia and unthriftiness in various domestic animals or even humans, and were recently reclassified (removed from the order Rickettsiales). Despite *Eperythrozoon* (renamed as *Mycoplasma*) *wenyonii* was reported for the first time in Africa [37], more recent data are not available on the occurrence of haemotropic mycoplasmas in cattle or in ticks in Africa/Ethiopia. The present results not only attest for the first time the occurrence of cattle-related haemoplasmas in ticks collected in Africa, but also show that ticks preferring a more humid environment (in association with mid highland biotopes) may be more exposed to these agents. This biotope-related occurrence may also be partly explained by certain features of haemoplasmas on the group level, because the majority of species have non-tick-borne and other, alternative routes of transmissions [38], thus rendering them more dependent on environmental factors.

In conclusion, the tick biotope, in which cattle were grazed in the evaluated period, influenced not only the tick burden of these hosts, but also the spectrum of pathogens in their ticks. The biotope appeared to be an important limiting factor in case of tick-borne *Coxiella burnetii* and haemoplasmas, which are known to have alternative (non-tick-borne) transmission routes, but no specific tick vectors. The presence of rickettsiae in ticks was influenced by both the biotope type and the tick species, whereas that of *Theileria* spp. only by the latter. This may reflect, that representatives of these two pathogen groups have specific tick vectors, and in these rickettsiae also have the means (i.e. transovarial transmission) of long-term maintenance in nature.

Acknowledgments

Molecular biology work was partially performed using the logistics of the Center for Clinical Studies at the Vetsuisse Faculty of the University of Zurich.

Author Contributions

Conceived and designed the experiments: SH GA MLM BT KMS MG EG RF RHL. Performed the experiments: SH GA MLM BT KMS MG EG RF RHL. Analyzed the data: SH GA MLM BT KMS MG EG RF RHL. Contributed reagents/materials/analysis tools: SH GA MLM MG RF RHL. Contributed to the writing of the manuscript: SH RHL.

References

1. Lodos J, Boue O, de la Fuente J (2000) A model to simulate the effect of vaccination against *Boophilus* ticks on cattle. Vet Parasitol 87: 315–326.

2. Levin BR, Svanborg Edén C (1990) Selection and evolution of virulence in bacteria: an ecumenical excursion and modest suggestion. Parasitology 100: S103–115.

3. Benin S, Ehui S, Pender J (2006) Policies for livestock development in the Ethiopian highlands. In: Pender J, Place F, Ehui S (Eds.): Strategies for Sustainable Land Management in the East African Highlands. IFPRI, Washington DC.

4. Pegram G, Hoogstraal H, Wassef HP (1981) Ticks (Acari Ixodidea) of Ethiopia Distribution, Ecology and Host relationship of species infecting livestock. Bull Entomol Res 71: 339–359.

5. Mekonnen S, Pegram RG, Gebre S, Mekonnen A, Jobre Y, et al. (2007) Zewdie S: A synthetic review of ixodid (Acari: Ixodidae) and argasid (Acari: Argasidae) ticks in Ethiopia and their possible roles in disease transmission. Ethiop Vet J 11: 1–24.

6. Mekonnen S, de Castro J, Gebre S, Hussein I, Regassa A (1992) Ticks, tick-borne diseases and their control in Western Ethiopia. Int J Trop Ins Sci 13: 661–664.

7. Abera M, Mohammed T, Abebe R, Aragaw K, Bekele J (2010) Survey of ixodid ticks in domestic ruminants in Bedelle district, Southwestern Ethiopia. Trop Anim Health Prod 42: 1677–1683.

8. Asrate S, Yalew A (2012) Prevalence of cattle tick infestation in and around Haramaya district, Eastern Ethiopia. J Vet Med Anim Health 4: 84–88.

9. Kumsa B, Signorini M, Teshale S, Tessarin C, Duguma R, et al. (2014) Molecular detection of piroplasms in ixodid ticks infesting cattle and sheep in western Oromia, Ethiopia. Trop Anim Health Prod 46: 27–31.

10. Mekonnen S (1994) Tick and tick-borne disease control in Ethiopia. In: Tick and tick-borne disease control in East, Central and Southern Africa, 1991–1994 (Eds. Musisi FL, Dolan TT). Proceedings of a joint OAU, FAO and ILRAD workshop held in Lilongwe, Malawi 25–28 April 1994. pp.19–21.

11. Hoogstraal H (1956) African Ixodoidea. 1. Ticks of the Sudan with special reference to Equatoria province and with preliminary review of genera *Rhipicephalus*, *Margaropus* and *Hyalomma*. US Navy, Washington D.C.

12. Walker AR, Bouattour A, Camicas J-L, Estrada-Peña A, Horak IG, et al. (2003) Ticks of domestic animals in Africa: a guide to identification of species. Bioscience Reports, Edinburgh.

13. Muhlnickel CJ, Jefferies R, Morgan-Ryan UM, Irwin PJ (2002) *Babesia gibsoni* infection in three dogs in Victoria. Aust Vet J 80: 606–610.

14. Olmeda AS, Armstrong PM, Rosenthal BM, Valladares B, del Castillo A, et al. (1997) A subtropical case of human babesiosis. Acta Trop 67: 229–234.

15. Loftis AD, Reeves WK, Szumlas DE, Abbassy MM, Helmy IM, et al. (2006) Rickettsial agents in Egyptian ticks collected from domestic animals. Exp Appl Acarol 40: 67–81.

16. Boretti FS, Perreten A, Meli ML, Cattori V, Willi B, et al. (2009) Molecular investigations of *Rickettsia helvetica* infection in dogs, foxes, humans and *Ixodes* ticks. Appl Environ Microbiol 75: 3230–3237.

17. Stenos J, Graves SR, Unsworth NB (2005) A highly sensetive and specific real-time PCR assay for the detection of spotted fever and typhus group Rickettsiae. Am J Trop Med Hyg 73: 1083–1085.

18. Thompson JD, Higgins DG, Gibson T J (1994) CLUSTAL W: improving the sensitivity of progressive multiple sequence alignment through sequence weighting, position-specific gap penalties and weight matrix choice. Nuc Acids Res 22: 4673–4680.

19. Saitou N, Nei M (1987) The neighbor-joining method: a new method for reconstructing phylogenetic trees. Mol Biol Evol 4: 406–425.

20. Tamura K, Stecher G, Peterson D, Filipski A, Kumar S (2013) MEGA6: Molecular Evolutionary Genetics Analysis Version 6.0. Mol Biol Evol 30: 2725–2729.

21. Willi B, Meli ML, Lüthy R, Honegger H, Wengi N, et al. (2009) Development and application of universal Hemoplasma screening assay based on the SYBR green PCR principle. J Clin Microbiol 47: 4049–4054.

22. Jongejan F, Uilenberg G (2004) The global importance of ticks. Parasitology 129: S3–S14.

23. Mekonnen S, Hussein I, Bedane B (2001) The distribution of ixodid ticks (Acari: Ixodidae) in central Ethiopia. Onderstepoort J Vet Res 68: 243–251.

24. Rogers RJ, Callow LL (1996) Three fatal cases of *Theileria mutans* infection. Aust Vet J 42: 42–46.

25. Ceci L, Kirvar E, Carelli G, Brown D, Sasanelli M, et al. (1997) Evidence of *Theileria buffeli* infection in cattle in southern Italy. Vet Rec 140: 581–583.

26. Stockham SL, Kjemtrup AM, Conrad PA, Schmidt DA, Scott MA, et al. (2000) Theileriosis in a Missouri beef herd caused by *Theileria buffeli*: case report, herd investigation, ultrastructure, phylogenetic analysis, and experimental transmission. Vet Pathol 37: 11–21.

27. Kamau J, de Vos AJ, Playford M, Salim B, Kinyanjui P, et al. (2011) Emergence of new types of *Theileria orientalis* in Australian cattle and possible cause of theileriosis outbreaks. Parasit Vectors 4: 22.

28. Chaisi ME, Collins NE, Potgieter FT, Oosthuizen MC (2013) Sequence variation identified in the 18S rRNA gene of *Theileria mutans* and *Theileria velifera* from the African buffalo (*Syncerus caffer*). Vet Parasitol 191: 132–137.

29. Bishop R, Musoke A, Morzaria S, Gardner M, Nene V (2004) *Theileria*: intracellular protozoan parasites of wild and domestic ruminants transmitted by ixodid ticks. Parasitology 129: S271–83.

30. Tomassone L, Pagani P, De Meneghi D (2005) Detection of *Babesia caballi* in *Amblyomma variegatum* ticks (Acari: Ixodidae) collected from cattle in the Republic of Guinea. Parassitologia 47: 247–251.

31. Philip CB, Hoogstraal H, Reiss-Gutfreund R, Clifford CM (1966) Evidence of rickettsial disease agents in ticks from Ethiopian cattle. Bull World Health Organ 35: 127–131.

32. Mediannikov O, Fenollar F, Socolovschi C, Diatta G, Bassene H, et al. (2010) *Coxiella burnetii* in humans and ticks in rural Senegal. Plos Negl Trop Dis 4: e654.

33. Reye AL, Arinola OG, Hübschen JM, Muller CP (2012) Pathogen prevalence in ticks collected from the vegetation and livestock in Nigeria. Appl Environ Microbiol 78: 2562–2568.

34. Mura A, Socolovschi C, Ginesta J, Lafrance B, Magnan S, et al. (2008) Molecular detection of spotted fever group rickettsiae in ticks from Ethiopia and Chad. Trans R Soc Trop Med Hyg 102: 945–949.

35. Socolovschi C, Huynh TP, Davoust B, Gomez J, Raoult D, et al. (2009) Transovarial and trans-stadial transmission of *Rickettsiae africae* in *Amblyomma variegatum* ticks. Clin Microbiol Infect 15 (Suppl. 2): 317–318.

36. Perlman SJ, Hunter MS, Zchori-Fein E (2006) The emerging diversity of Rickettsia. Proc Royal Soc B Biol Sci 273:2097–2106.

37. Neitz WO (1940) Eperythrozoonosis in cattle. Onderstepoort J Vet Sci 14: 9–28.

38. Yang Z, Yan C, Yu F, Hua X (2007) Haemotrophic mycoplasma: Review of aetiology and prevalence. Rev Med Microbiol 18: 1–3.

8

A Country on the Verge of Malaria Elimination – The Kingdom of Saudi Arabia

Michael Coleman[1]*, Mohammed H. Al-Zahrani[2], Marlize Coleman[1], Janet Hemingway[1], Abdiasiis Omar[1], Michelle C. Stanton[1], Eddie K. Thomsen[1], Adel A. Alsheikh[2], Raafat F. Alhakeem[2], Phillip J. McCall[1], Abdullah A. Al Rabeeah[1,2], Ziad A. Memish[1,2,3]

1 Vector Biology Department, Liverpool School of Tropical Medicine, Liverpool, United Kingdom, 2 Public Health Directorate, Ministry of Health, Riyadh, Kingdom of Saudi Arabia, 3 College of Medicine, Alfaisal University, Riyadh, Kingdom of Saudi Arabia

Abstract

Significant headway has been made in the global fight against malaria in the past decade and as more countries enter the elimination phase, attention is now focused on identifying effective strategies to shrink the malaria map. Saudi Arabia experienced an outbreak of malaria in 1998, but is now on the brink of malaria elimination, with just 82 autochthonous cases reported in 2012. A review of published and grey literature was performed to identify the control strategies that have contributed to this achievement. The number of autochthonous malaria cases in Saudi Arabia decreased by 99.8% between 1998 and 2012. The initial steep decline in malaria cases coincided with a rapid scaling up of vector control measures. Incidence continued to be reported at low levels (between 0.01 and 0.1 per 1,000 of the population) until the adoption of artesunate plus sulfadoxine-pyrimethamine as first line treatment and the establishment of a regional partnership for a malaria-free Arabian Peninsula, both of which occurred in 2007. Since 2007, incidence has decreased by nearly an order of magnitude. Malaria incidence is now very low, but a high proportion of imported cases, continued potential for autochthonous transmission, and an increased proportion of cases attributable to *Plasmodium vivax* all present challenges to Saudi Arabia as they work toward elimination by 2015.

Editor: Joseph A. Keating, Tulane University School of Public Health and Tropical Medicine, United States of America

Funding: This work was funded by the Kingdom of Saudi Arabia Malaria Ministry of Health who also allowed access to datasets that have not been published and the Innovative Vector Control Consortium for supporting staff from LSTM, as well as unpublished PhD student work of LSTM. The funders had no role in study design, data collection and analysis, decision to publish, or preparation of the manuscript.

Competing Interests: The authors have declared that no competing interests exist.

* Email: mcoleman@lstmed.ac.uk

Introduction

The recent global increase in malaria control efforts has contributed to major reductions in the burden of the disease [1,2]. Since 2007, four countries have eliminated malaria and been certified by the World Health Organization (WHO) as malaria free [3]. Today, 34 countries, including the Kingdom of Saudi Arabia, are actively attempting to eliminate malaria[4].

Malaria control in Saudi Arabia was initiated in 1948 by the Arabian American Oil Company (ARAMCO) in the Eastern province, primarily to protect employees living around the oases [5]. This programme was used by the Saudi Arabian government as the template for a national malaria programme in 1952, [5], which targeted malarious districts across the kingdom and was designed to protect pilgrims en route to the holy sites of Mecca and Medina. Saudi Arabia joined the WHO global malaria eradication effort in 1963 and, by the early 1970s, transmission was arrested in the Eastern and Northern provinces, eliminating malaria in the Palaearctic ecozone [6]. Despite this success, Saudi Arabia switched from malaria eradication to control in 1977 [7], following the worldwide abandonment of the goal of global malaria eradication.

Today in Saudi Arabia, malaria persists in the provinces of Aseer and Jazan [8], both bordering the Republic of Yemen. Following a series of outbreaks (of which the worst was in 1998), malaria control was intensified and the goal of malaria elimination in Saudi Arabia was re-established in 2004, with an elimination target of 2015. After nearly a decade of activity, progress towards this goal is reviewed in this report. Successful strategies and continued challenges are highlighted and discussed so that other countries may benefit from lessons learned in Saudi Arabia.

Methods

Data Sources

Data for this review were identified by literature searches of PubMed and general searches using the Google and WHO search engines. Search term strings included 'malaria' AND 'Saudi Arabia'. The bibliographies of selected documents were used to identify additional data and information sources. The National Malaria Control Programme provided access to archived annual reports and statistical data. Malaria data was obtained from the Ministry Of Health annual statistical reports (1980–2012) [9]. Population data was obtained from the Central Statistical Office.

Due to the limited number of data sources available, no formal exclusion criteria were used.

Analysis

Population figures were extracted from the national censuses in 1992, 2004, and 2010. Population numbers for the interim years were estimated by calculating the average annual population growth between censuses and applying this accordingly. Malaria incidence was calculated as the total number of autochthonous cases divided by the total population.

Findings

Vectors and Epidemiology of Malaria

Records of vectors and epidemiology of malaria in Saudi Arabia are limited. A total of 15 Anopheles species have been recorded, five of which are known to be competent malaria vectors: *An. arabiensis, An. sergentii, An. stephensi, An. superpictus* and *An. culicifacies* [8–14]. *An. arabiensis* is currently the only known vector of malaria in Saudi Arabia [10,11], with low reported sporozoite rates (<1%) [11,12]. Abundance of *An. arabiensis* typically peaks following the rains, when multiple breeding sites appear [10], although there is some evidence from Jazan to suggest that irrigation may contribute to malaria risk [12]. With only 40–42% of blood meals in *An. arabiensis* females being of human origin [11,12], vector populations are clearly supported primarily by other host species, most likely domestic cattle, a preferred host of *An. arabiensis* throughout its range. As *An. arabiensis* rests and feeds outdoors as well as indoors, vector control methods targeting the human home, such as indoor residual spraying (IRS) and insecticide treated nets (ITNs, or long-lasting insecticidal nets, LLINs), are not sufficient to control this vector, necessitating the use of additional approaches.

With high aridity and low population density, much of the environment of the Arabian Peninsula is unsuitable for the malaria parasite. However, perennial mountain streams support numerous fertile valleys and oases in the southern provinces of Saudi Arabia, where human communities and vector populations coincide, and from where most malaria cases are reported.

Throughout the 1980s, there were a relatively consistent number of confirmed malaria cases in the country, ranging from approximately 5,000 to 17,000 annually. Cases were largely confirmed by microscopy. In 1990, recognising the importance of imported malaria, Saudi Arabia began to distinguish between autochthonous acquired and imported malaria cases by classifying patients according to their travel history, which revealed that imported malaria constituted the vast majority of annual cases.

Prior to 1998, national incidence ranged from 0.33/1,000 to 0.96/1,000. In 1998, a major outbreak of autochthonous acquired malaria occurred, in which the total number of confirmed cases in the country reached 36,139. This doubled the total annual incidence to 1.87/1,000 compared to0.92/1,000 in the previous year (Figure 1). Certain areas of the country were disproportionately affected by the outbreak, and incidence in the regions of Jazan and Qunfudah increased from 10/1,000 to 20/1,000 during the outbreak and from 12/1,000 to 44/1,000, respectively [9].

The 1998 outbreak triggered an aggressive centrally coordinated control campaign, the result of which was a steep decline in autochthonous transmitted malaria until 2004. During the 2007–2008 season, a further decrease of nearly an order of magnitude occurred from 0.02/1,000 to 0.002/1,000 (Figure 2).

As the incidence of autochthonous cases has fallen, the proportion of total imported malaria cases has steadily increased from 11% in 1998 to 98% in 2011. By 2010, only two health regions, Aseer (incidence of 0.001/1,000) and Jazan (incidence of 0.013/1,000), recorded any autochthonous malaria cases, and only two malaria-attributable deaths resulting from autochthonous transmission were recorded nationally in 2011. The latest region in Saudi Arabia to eliminate malaria was Qunfudah, which has reported no autochthonous transmission since 2009 (Figure 3) [13]. Additionally, a shift in the predominant parasite species has occurred in the last seven years, from predominantly *P. falciparum* to a majority of the imported *P.vivax* species (Figure 4) [9].

Control and elimination activities

The current elimination strategy in Saudi Arabia focuses mainly on: a) targeting high risk areas for sustained preventative measures such as long lasting insecticide treated nets (LLINs) and indoor residual spraying (IRS); b) management of infection through rapid confirmed diagnosis and treatment; c) individual case follow up and reactive surveillance with appropriate treatment and vector control and, d) active case detection at borders with screening and treatment. The details of this strategy are discussed below.

A. Sustained vector control targeting high-risk areas. Due to the low incidence of malaria, comprehensive vector control is not considered to be cost effective, and is instead guided by case detection for targeted control measures [14,15]. This maintains coverage of interventions to interrupt transmission in the populations at risk [16,17]. A total of 68 active transmission foci were recorded in Saudi Arabia in 2011 [13]. The vector control programme in these areas used IRS, distribution of LLINs, larviciding and space spraying in an integrated manner.

DDT-based IRS was introduced in the Eastern Province in 1948 by ARAMCO to protect its workforce. DDT use in the ARAMCO and national programmes continued until 1954, when resistance was detected in *An. stephensi*. This prompted a switch in insecticide to dieldrin for three years before dieldrin resistance was detected in 1957 [18]. During this period, malaria cases decreased from 2000 in 1947 to 54 in 1956 [5] and the ARAMCO IRS campaign became the model for a national programme extended to all malarious areas of the country. In 1963, the government and WHO jointly started a pre-eradication campaign, and by the mid 1970's, IRS with DDT or dieldrin was used as the primary strategy to eliminate foci of transmission throughout the eastern and central regions of the country. Despite this, active transmission persisted in southern regions of the country and along the Red Sea coast. Following a decline in IRS efforts in the late 1970's, targeted IRS with DDT was re-introduced in Jazan and other southern regions in the early1980's [19]. IRS was then scaled back in 1987 when the organophosphate, fenitrothion, replaced DDT as the number of autochthonous cases declined.

Following the 1998 epidemic, IRS activities were scaled up again in southern Saudi Arabia. Focused IRS with DDT was carried out in 1999 at the beginning of peak malaria transmission (April to October). However, it became apparent that DDT resistance still existed, necessitating a switch to the pyrethroid lambda-cyhalothrin (10% EC) in 1999 and deltamethrin (25%WB) in 2004. By 2011, IRS with either deltamethrin (Kaothrine 25 WG) or lambda-cyhalothrin (ICON or Demand 10CS) protected 83% of the 3.15 million persons at risk [20] (Table 1).

Treatment of vector breeding sites with insecticide was introduced as a component of integrated vector control in 1971, initially with Paris Green and then with temephos (EC50). Combined with source reduction, weekly larviciding of 10–15 km sections of the affected wadis (seasonal river valleys) with temephos was performed [6]. Following detection of resistance in 2000, temephos was replaced by larvicides using insect growth

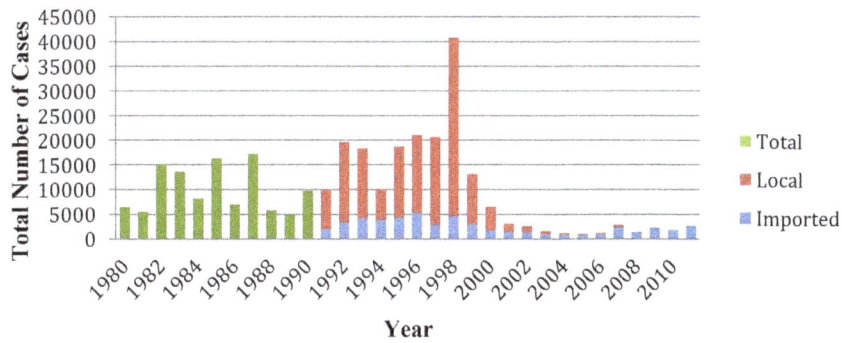

Figure 1. Number of malaria cases in Saudi Arabia 1980 to 2011.

regulators, predominantly difubenzuron, methoperene and pyr-iproxyfen. Larviciding remains one of the primary vector control strategies used in the active transmission foci in Saudi Arabia. Space spraying was used in the early control campaigns to rapidly reduce adult mosquito vector densities around wadis [6]. As this was the only breeding site, this was a successful strategy. Today, space spraying using the pyrethroid insecticides bifenthrin, bioallethrin and cyphenothrin remains a key component of vector control [21]. Currently it is used in response to the detection of a malaria case, when spraying in and around the index home and six nearest neighbours is performed. Following a successful trial of permethrin insecticide treated nets [22], LLINs were added to the national control programme in 1999, with free distribution to all age groups in focal areas. Since 2008, over 680,000 alpha-cypermethrin LLINs (Interceptor, BASF, Germany) have been distributed during the scale up for elimination (Table 1).

Post 1998, the efforts and resources dedicated to malaria control and elimination are shown in the sustained quantities of insecticide used (Figure 5). The current control programme delivers a range of insecticides from different classes, although pyrethroids are predominantly used against the adult insects. Although the importance of insecticide resistance as a major barrier to the success in malaria control and elimination is well known, [23] no routine insecticide susceptibility was undertaken on the malaria vectors in Saudi Arabia.

B. Rapid confirmed diagnosis and treatment. Diagnosis and treatment of malaria has been provided free in Saudi Arabia since the inception of the malaria programme [24,25]. All cases are confirmed positive by microscopic examination [26]. However, microscopy facilities are not always available (especially with active case detection). In these situations, a rapid diagnostic test (RDT) is used to confirm cases. A blood smear is always taken and read when conditions permit as a measure of quality assurance.

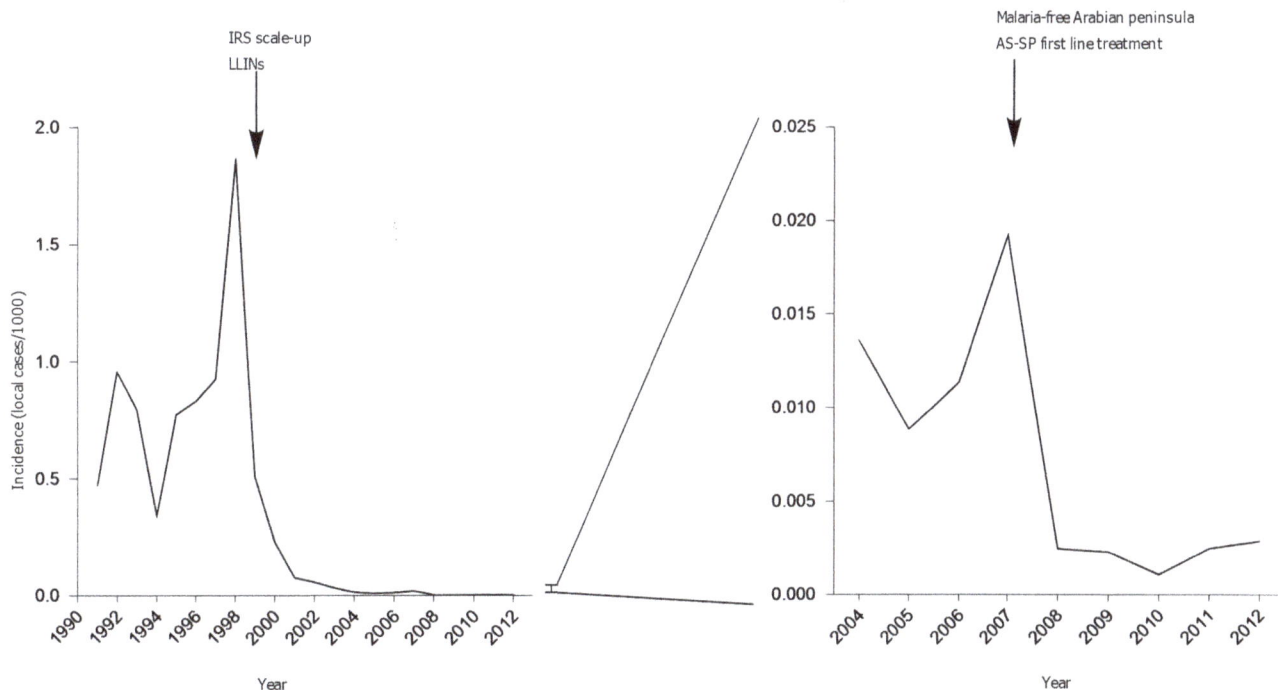

Figure 2. Annual parasite incidence of autochthonous malaria cases with timing of major control milestones indicated. IRS: indoor residual spraying, LLIN: long-lasting insecticidal nets, AS-SP: sulfadoxine pyrimethamine. Although the plan for a malaria-free Arabian peninsula was established in 2001, it was not funded until 2007.

Malaria incidence by Health Region, 1998 and 2012

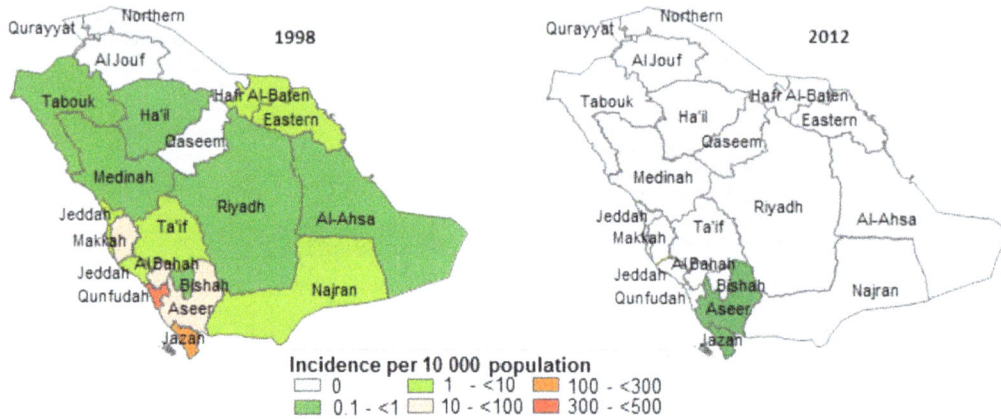

Incidence per 10 000 population
- 0
- 0.1 - <1
- 1 - <10
- 10 - <100
- 100 - <300
- 300 - <500

Total malaria case numbers reported, 1998 and 2012

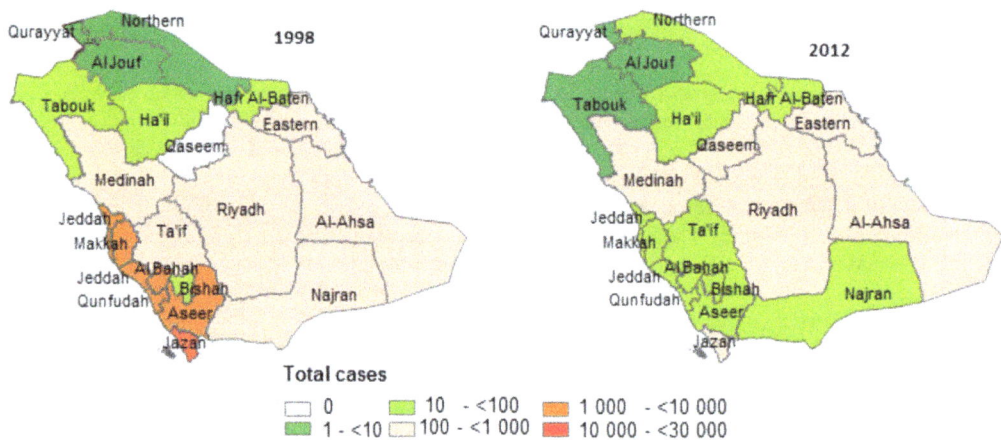

Total cases
- 0
- 1 - <10
- 10 - <100
- 100 - <1 000
- 1 000 - <10 000
- 10 000 - <30 000

Percentage imported cases reported, 1998 and 2012

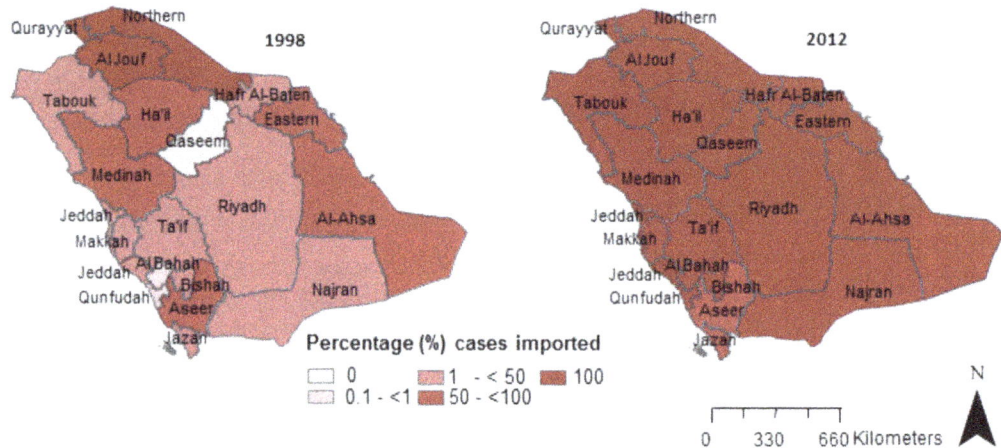

Percentage (%) cases imported
- 0
- 0.1 - <1
- 1 - < 50
- 50 - <100
- 100

0   330   660 Kilometers

Figure 3. Maps showing the changes in incidence of autochthonous cases, the total number of cases and percentage of imported cases.

With the detection of chloroquine resistance in *P. falciparum* [27], artesunate plus sulfadoxine-pyrimethamine was made the first line treatment in 2007, with artemether-lumefanthrine being used for complicated malaria and quinine used for severe complicated malaria. Chloroquine plus primaquine are used for *P. vivax*. As ACTs kill developing gametocytes and reduce malaria transmission [28,29], this treatment strategy is of benefit in areas of low transmission, such as Saudi Arabia, where mature gametocytes may account for residual transmission and outbreaks [30].

C. Reactive case detection. Malaria is a notifiable disease in Saudi Arabia and diagnosis is predominantly based on microscopy [25]. Most malaria is identified through passive case detection at

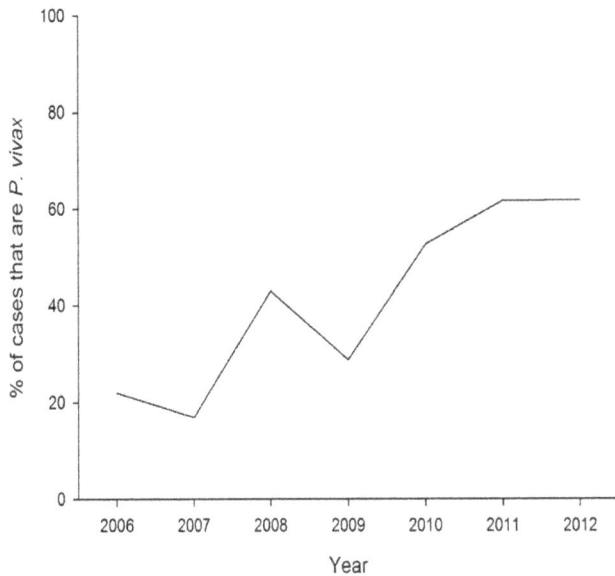

Figure 4. Proportion of confirmed cases caused by *Plasmodium vivax* **from 2006–2012.**

health facilities. Since 1991, all cases are reported within 24 hours to the local malaria centre, using a standard form that triggers a follow-up analysis to determine the probable source of infection. Based on the concept that malaria cases tend to cluster [31–33], reactive case detection focuses on other members of the household and neighbouring houses. Occupants are tested and treated for malaria, and appropriate space spraying and/or larviciding vector control is performed. This rapid action minimises the potential for any onward transmission from an index case. A similar system has been applied successfully in South Africa [34].

D. Active case detection at the border. As autochthonous endemicity is reduced or eliminated, controlling imported malaria becomes more important [4,35]. While there has been a marked reduction in the actual number of imported malaria cases in recent years (4,657 and 2,719 cases in 1998 and 2011, respectively), the proportion of imported cases has increased dramatically (11.4% and 97.5% of malaria cases in 1998 and 2011 were imported, respectively) [9]. Therefore, there is a clear risk of human movement reintroducing malaria into areas of elimination and creating outbreaks in Saudi Arabia as seen elsewhere in other countries [36–38].

There are three main sources of imported malaria in Saudi Arabia. First, Saudi Arabia relies on a large expatriate work force, many of which originate from malaria endemic countries in the Middle East, Africa or Asia. Second, Saudi Arabia contains the most important Islamic holy sites, visited by many millions of pilgrims from every country worldwide. The public health importance of imported malaria and other diseases from pilgrims attending the Hajj has long been recognised [39,40], and reactive case detection is used as the primary method of limiting transmission in these groups. Lastly, and perhaps most importantly, Saudi Arabia shares a southern border with a malaria endemic country, the Republic of Yemen [20].

Yemen has the second highest incidence of malaria in the Mediterranean region, with an estimated 40% of paediatric admissions due to malaria in some settings [20,41]. The highest incidences of malaria in Yemen occur in the Tihama coastal plain, a hot and humid geographical region where *An. arabiensis* is common [11,42], which extends north into Saudi Arabia's Jazan province. High levels of human movement occur between both countries with an estimated 3,000 illegal immigrants crossing from Yemen to Saudi Arabia on a daily basis and some 20,000 Saudi Arabians spending weekends in Yemen [43,44].

To limit the risk of imported malaria, plans were established in 2001 for a malaria-free Arabian peninsula by 2020 [38]. The plan was endorsed in 2005 by the Eastern Mediterranean Regional Office of WHO [45] and by the Health Ministers of the Gulf Cooperation Council in 2007, with financial support of $42.7 million [41]. Elimination in the Arabian Peninsula has been largely successful, and only Saudi Arabia and Yemen have yet to achieve malaria free status [46].

Cross border collaboration between Saudi Arabia and Yemen has included the establishment of malaria centres offering free screening and treatment, mostly for Yemenis living in border villages. These are supported by mobile teams that carry out active case detection to target high risk populations living in the border villages [47]. These mobile teams detected ~50% of imported cases in 2010–11 in Jazan. Malaria control in Yemen includes joint Saudi-Yemeni vector control teams, which are responsible for IRS and for space-spraying a 10 km wide buffer zone inside Yemen. Unfortunately, recent political unrest and a deteriorating political situation in Yemen have halted these efforts.

Discussion

Since the outbreak in 1998, the efforts to control and eliminate malaria in the Kingdom of Saudi Arabia have produced significant gains. The total number of cases has remained consistently low

Table 1. Coverage of ITNs and the number of persons protected by IRS.

Year	Number ITNs distributed	Persons protected by IRS
2004	460,000	unknown
2005	81,364	unknown
2006	0	94,350
2007	0	unknown
2008	250,000	unknown
2009	250,000	2,457,965
2010	81,050	2,500,000
2011	100,000	2,600,000

Source: World malaria reports [20,48].

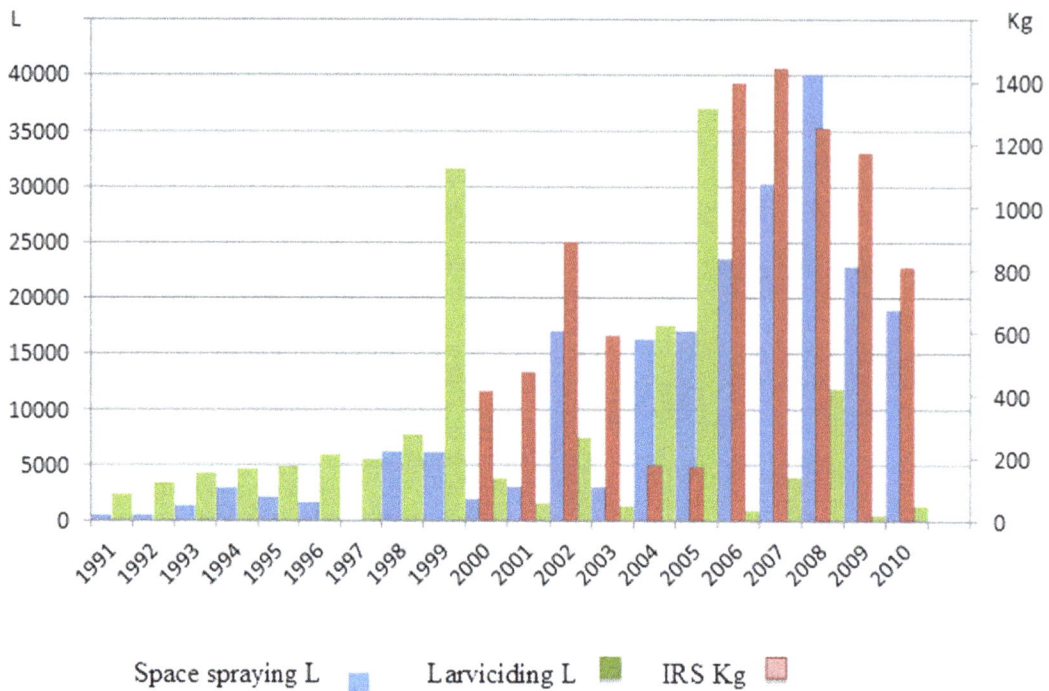

Figure 5. Amount of insecticide used for vector control from 1991 to 2011.

since 2001 and the number of autochthonous cases has decreased from 467 in 2007 [48] to just 82 in 2012 [13]. With incidence already well below 0.01 cases per 1,000 persons, elimination of malaria in Saudi Arabia is a realistic goal. However, a number of considerable challenges must be overcome in order to achieve this target.

Migration of humans (and parasites) has been shown to rapidly undermine gains made in malaria control efforts [49,50], and in Saudi Arabia, imported malaria still remains a major problem. The border with Yemen is an area of central concern, and controlling malaria in this region is a major challenge for Saudi Arabia in achieving and maintaining a malaria free status [46]. Autochthonous transmission is limited due to an effective response by the malaria control programme and/or unfavourable conditions for an outbreak to occur. However, the number of autochthonous cases increased to from 29 in 2010 to 69 and 82 in 2011 and 2012, respectively. This, in part, is probably due to the disruption of malaria control activities on the Yemeni side of the border and an increase in both legal and illegal immigrants into Saudi Arabia. Continued importation combined with the presence of vectors means that autochthonous transmission will inevitably occur. Saudi Arabia must find ways to mitigate this risk and ensure that endemic transmission is halted.

Saudi Arabia is likely to experience further difficulties as the proportion of cases attributable to *P. vivax* increases. *P. vivax* is generally less responsive to control than *P. falciparum* for a number of reasons. First, dormant liver stages can lead to relapse. Second, the extrinsic incubation period is shorter. Third, infective gametocytes are produced concurrently with asexual stages. Fourth, parasite densities are often lower than the detection threshold of diagnostic tests. Lastly, adherence to the 7–14 day primaquine regimen to treat liver stages is often poor. All of these factors make *P. vivax* residual transmission more difficult to detect and control.

The way forward

Regional efforts such as the Lubombo Spatial Development Initiative [35] and the Asia Pacific Malaria Elimination Network [51,52] are excellent examples of where cross border collaborations have succeeded in addressing the challenge of imported malaria. Today, seven countries (Bahrain, Kuwait, Oman, Qatar, Saudi Arabia, the United Arab Emirates and Yemen) remain committed to supporting intensification of malaria control efforts on the Arabian Peninsula [20]. It is clear that reinvigoration of the border collaboration with Yemen will be necessary to achieve a malaria-free Saudi Arabia by 2015. Novel strategies of quickly identifying imported cases should also be considered. Recent work in Swaziland, a country also targeting elimination by 2015, has shown that snowball and time-location sampling can quickly identify networks of individuals at high-risk of bringing malaria into the country [53]. Adoption of this approach in Saudi Arabia would allow for targeted screening and treatment, preventing onward transmission. Additional strategies may include parasite genotyping, which would allow more accurate determination of common sources of infection.

Preventing the growing proportion of P.vivax cases will require implementing new but currently available strategies to quickly identify and treat those infected with the parasite. This may include sensitive, point-of-care diagnostics that are compatible with field conditions, such as a loop-attenuated isothermal amplification assay (LAMP). This assay is as sensitive as PCR detection, and suitable for deployment in remote settings has been shown to be equally as sensitive as PCR, but able to be deployed in remote settings [54].

Preventing autochthonous transmission requires a robust and sophisticated information system that enables timely response to outbreaks. Saudi Arabia aims to strengthen their control programme using emerging information technologies that can inform elimination efforts and maintain the disease free status. It is currently enhancing its surveillance systems through the adoption of a Disease Data Management System, a spatial decision support tool [55] that has a focus on malaria clusters and outbreaks [34,56] with mapping functionality required to support elimination and ongoing surveillance by mapping down to the village and even household level as recommended by WHO [57]. This will enhance the control programme's ability to investigate disease clusters and adjust operational plans accordingly. An independent review recently highlighted the advantages of such a system [58]. The failure to maintain high quality surveillance after near-elimination was a key factor that resulted in the disease returning in India [59] and Madagascar highlands [60].

Malaria elimination is a surprisingly stable state [61]. However, sustained elimination is unlikely to be attributable to continuous vector control. Rather, economic development that reduces vector-human contact and improved health systems are the major drivers in transitioning from malaria-endemic to malaria-free. As

such, while improved management of imported malaria, more rapid case detection, and highly targeted interventions should be encouraged in the short-term, investment in health system infrastructure and overall socioeconomic development should be envisioned as the driving force behind malaria elimination in-country and regionally. Given the continued political and financial support from the government, the prospects to attain and sustain elimination in Saudi Arabia are positive.

Author Contributions

Conceived and designed the experiments: Michael Coleman MHA JH PM ZM. Performed the experiments: Marlize Coleman AO AA AR. Analyzed the data: Michael Coleman MS ET. Contributed reagents/materials/analysis tools: Michael Coleman JH ZM. Wrote the paper: Michael Coleman JH ET ZM.

References

1. Hay SI, Guerra CA, Gething PW, Patil AP, Tatem AJ, et al. (2009) A world malaria map: Plasmodium falciparum endemicity in 2007. PLoSMed 6: e1000048.
2. Hay SI, Guerra CA, Tatem AJ, Noor AM, Snow RW (2004) The global distribution and population at risk of malaria: past, present, and future. The Lancet Infectious Diseases 4: 327–336.
3. World Health Organization (2013) World Malaria Report 2013. Geneva: WHO.
4. Cotter C, Sturrock HJ, Hsiang MS, Liu J, Phillips AA, et al. (2013) The changing epidemiology of malaria elimination: new strategies for new challenges. Lancet: 10–6736.
5. Daggy RH (1959) Malaria in oases of Eastern Saudi Arabia. AmJTropMedHyg 8: 223–291.
6. Ministry Of Health, Saudi Arabia (1984) Malaria Control Programme in the Kingdom of Saudi Arabia 1983. Malaria Control Services, report.
7. World Health Organization office for the Eastern Mediteranian Region (1986) Joint Government/WHO malaria review misison to Saudi Arabia. WHO Regional Office for the Eastern Mediterranean Region.
8. Malaria Atlas Project (2013) http://www.map ox ac k/explore/countries/SAU/ #!mosquito–vectors. Accessed 2013.
9. Kingdom of Saudi Arabia, Ministry of Health (2012) Health Statistcal Year Books 1980 to 2012. Saudi Arabia: Ministry of Health.
10. Abdoon AM, Alshahrani AM (2003) Prevalence and distribution of anopheline mosquitoes in malaria endemic areas of Asir region, Saudi Arabia. East MediterrHealth J 9: 240–247.
11. Al-Sheikh AAH (2004) Studies on the Ecology, Vectorial Role and Population Structure of Anopheles arabiensis in the Tihama Region of Saudi Arabia and Yemen: University of Liverpool, UK.
12. Al-Sheik AA (2011) Larval habitat, ecology, seasonal abundance and vectorial role in malaria transmission of Anopheles arabiensis in Jazan Region of Saudi Arabia. JEgyptSocParasitol 41: 615–634.
13. World Health Organization (2012) World Malaria Report: 2012. WHO Geneva.
14. Smith DL, McKenzie FE, Snow RW, Hay SI (2007) Revisiting the basic reproductive number for malaria and its implications for malaria control. PLoSBiol 5: e42.
15. World Health Organization (2008) Global Malaria Control and Elimination:report of a technical review. WHO Geneva.
16. Muller P, Chouaibou M, Pignatelli P, Etang J, Walker ED, et al. (2008) Pyrethroid tolerance is associated with elevated expression of antioxidants and agricultural practice in Anopheles arabiensis sampled from an area of cotton fields in Northern Cameroon. MolEcol 17: 1145–1155.
17. Pampana E (1969) A testbook for malaria eradication. London: Oxford University Press.
18. Peffly RL (1959) Insecticide resistance in anophelines in eastern Saudi Arabia. BullWorld Health Organ 20: 757–76.: 757–776.
19. World Health Organization (1986) Joint Government/WHO Malaria Review Mission to Saudi Arabia 5–23 July 1986.
20. World Health Organization (2011) World Malaria Report: 2011. Geneva: WHO.
21. Sebai ZA (1988) Malaria in Saudi Arabia. TropDoct 18: 183–188.
22. Jamjoom GA, Mahfouz AA, Badawi IA, Omar MS, al-Zoghaibi OS, et al. (1994) Acceptability and usage of permethrin-impregnated mosquito bed nets in rural southwestern Saudi Arabia. TropGeogrMed 46: 355–357.
23. World Health Organization (2012) Global Plan for Insecticide Resistance Management, in Malaria Vectors. WHO Geneva.
24. Ministry of Health, Saudi Arabia (2008) The National Policy of Malaria Case Management in The Kingdom of Saudi Arabia. Saudi Arabia.
25. Ministry of Health, Saudi Arabia (2012) The National Policy of Malaria Case Management in The Kingdom of Saudi Arabia. Saudi Arabia.
26. Nzila A, Al-Zahrani I (2013) Drugs for the treatment of malaria in the Kingdom of Saudi Arabia. Saudi Med J 34: 569–578.
27. Malik GM, Seidi O, El-Taher A, Mohammed AS (1998) Clinical aspects of malaria in the Asir Region, Saudi Arabia. AnnSaudiMed 18: 15–17.
28. Ding Y, Ortelli F, Rossiter LC, Hemingway J, Ranson H (2003) The Anopheles gambiae glutathione transferase supergene family: annotation, phylogeny and expression profiles. BMCGenomics 4: 35.
29. Craig A, Kyes S, Ranson H, Hemingway J (2003) Malaria parasite and vector genomes: partners in crime. Trends Parasitol 19: 356–362.
30. White NJ (2008) The role of anti-malarial drugs in eliminating malaria. MalarJ 7 Suppl 1:S8. doi: 10.1186/1475-2875-7-S1-S8.: S8-7.
31. Dicko A, Mantel C, Kouriba B, Sagara I, Thera MA, et al. (2005) Season, fever prevalence and pyrogenic threshold for malaria disease definition in an endemic area of Mali. TropMedIntHealth 10: 550–556.
32. Branch O, Casapia WM, Gamboa DV, Hernandez JN, Alava FF, et al. (2005) Clustered local transmission and asymptomatic Plasmodium falciparum and Plasmodium vivax malaria infections in a recently emerged, hypoendemic Peruvian Amazon community. MalarJ 4: 27.: 27.
33. Brooker S, Clarke S, Njagi JK, Polack S, Mugo B, et al. (2004) Spatial clustering of malaria and associated risk factors during an epidemic in a highland area of western Kenya. TropMedIntHealth 9: 757–766.
34. Coleman M, Coleman M, Mabuza AM, Kok G, Coetzee M, et al. (2008) Evaluation of an operational malaria outbreak identification and response system in Mpumalanga Province, South Africa. Malar J 7: 69.: 69.
35. Sharp BL, Kleinschmidt I, Streat E, Maharaj R, Barnes KI, et al. (2007) Seven years of regional malaria control collaboration—Mozambique, South Africa, and Swaziland. AmJTropMedHyg 76: 42–47.
36. Cohen JM, Smith DL, Cotter C, Ward A, Yamey G, et al. (2012) Malaria resurgence: a systematic review and assessment of its causes. MalarJ 11: 122. doi: 10.1186/1475-2875-11-122.: 122–111.
37. Maltezou HC, Tsolia M, Polymerou I, Theodoridou M (2013) Paediatric malaria in Greece in the era of global population mobility. TravelMedInfectDis: 10.
38. Craig MH, Kleinschmidt I, Le SD, Sharp BL (2004) Exploring 30 years of malaria case data in KwaZulu-Natal, South Africa: part II. The impact of non-climatic factors. TropMedIntHealth 9: 1258–1266.
39. Farid MA (1956) The implications of Anopheles sergenti for malaria eradication programmes east of the Mediterranean. BullWorld Health Organ 15: 821–828.
40. Khan AS, Qureshi F, Shah AH, Malik SA (2002) Spectrum of malaria in Hajj pilgrims in the year 2000. JAyubMedCollAbbottabad 14: 19–21.
41. Al-Taiar A, Jaffar S, Assabri A, Al-Habori M, Azazy A, et al. (2006) Severe malaria in children in Yemen: two site observational study. BMJ 333: 827.
42. Alzahrani MH (2006) Impact of Irrigation Systems on Malaria and Rift Valley Fever Transmission in Jizan, Saudi Arabia: University of Liverpool, UK.
43. Meleigy M (2007) Arabian Peninsula states launch plan to eradicate malaria. BMJ 20;334: 117.
44. Meleigy M (2007) The quest to be free of malaria. BullWorld Health Organ 85: 507–508.
45. Atta H, Zamani G (2008) The progress of Roll Back Malaria in the Eastern Mediterranean Region over the past decade. East MediterrHealth J 14 Suppl: S82–9.: S82–S89.
46. Snow RW, Amratia P, Zamani G, Mmundia CW, Noor AM, et al. (2013) The Malaria Transition on the Arabian Peninsula: Progress toward a Malaria-Free Region between 1960–2010. AdvParasitol 82: 205–51. doi: 10.1016/B978-0-12-407706-5.00003-4.: 205–251.
47. Macauley C (2005) Aggressive active case detection: a malaria control strategy based on the Brazilian model. SocSciMed 60: 563–573.
48. World Health Organization (2007) World Malaria Report: 2007. WHO Geneva.

49. Wickramage K, Galappaththy GN (2013) Malaria burden in irregular migrants returning to Sri Lanka from human smuggling operations in West Africa and implications for a country reaching malaria elimination. TransRSocTropMed-Hyg 107: 337–340.

50. Vakali A, Patsoula E, Spanakos G, Danis K, Vassalou E, et al. (2012) Malaria in Greece, 1975 to 2010. EuroSurveill 17: 20322.

51. Alilio MS, Kitua A, Njunwa K, Medina M, Ronn AM, et al. (2004) Malaria control at the district level in Africa: the case of the muheza district in northeastern Tanzania. AmJTropMedHyg 71: 205–213.

52. Hsiang MS, Abeyasinghe R, Whittaker M, Feachem RG (2010) Malaria elimination in Asia-Pacific: an under-told story. Lancet 375: 1586–1587.

53. Koita K, Novotny J, Kunene S, Zulu Z, Ntshalintshali N, et al. (2013) Targeting imported malaria through social networks: a potential strategy for malaria elimination in Swaziland. Malar J 12: 219.

54. Hopkins H, Gonzalez IJ, Polley SD, Angutoko P, Ategeka J, et al. (2013) Highly sensitive detection of malaria parasitemia in a malaria-endemic setting: performance of a new loop-mediated isothermal amplification kit in a remote clinic in Uganda. J Infect Dis 208: 645–652.

55. Eisen L, Coleman M, Lozano-Fuentes S, McEachen N, Orlans M, et al. (2011) Multi-Disease Data Managment System Platform for Vector-Borne Diseases. PLoSNeglTropDis 5: e1016.

56. Coleman M, Coleman M, Mabuza A, Kok G, Coetzee M, et al. (2008) Using the SaTScan method to detect local malaria clusters for guiding malaria control. Malar J Submitted.

57. World Health Organization (2007) Malaria Elimination. A field manual for low and moderate endemic countries. WHO Geneva.

58. Mabuza A, Kok G, Groepe MA, Misiani E, Shandukani M, et al. (2012) Active Case Detection Towards Malaria Elimination: Lessons Learned From Mpumlanga Province, South Africa. Abstract Book 2012. The American Journal of Tropical Medicine and Hygine.

59. Sharma VP (1996) Re-emergence of malaria in India. Indian JMedRes 103: 26–45.: 26–45.

60. Mouchet J, Laventure S, Blanchy S, Fioramonti R, Rakotonjanabelo A, et al. (1997) The reconquest of the Madagascar highlands by malaria. BullSocPatho-lExot 90: 162–168.

61. Chiyaka C, Tatem AJ, Cohen JM, Gething PW, Johnston G, et al. (2013) Infectious disease. The stability of malaria elimination. Science 339: 909–910.

9

A Systematic Health Assessment of Indian Ocean Bottlenose (*Tursiops aduncus*) and Indo-Pacific Humpback (*Sousa plumbea*) Dolphins Incidentally Caught in Shark Nets off the KwaZulu-Natal Coast, South Africa

Emily P. Lane[1]*, Morné de Wet[2], Peter Thompson[2], Ursula Siebert[3], Peter Wohlsein[4], Stephanie Plön[5¤]

1 Department of Research and Scientific Services, National Zoological Gardens of South Africa, Pretoria, South Africa, **2** Epidemiology Section, Department of Production Animal Studies, Faculty of Veterinary Science, University of Pretoria, Pretoria, South Africa, **3** Institute for Terrestrial and Aquatic Wildlife Research, University of Veterinary Medicine, Hannover, Foundation, Germany, **4** Department of Pathology, University of Veterinary Medicine, Hannover, Foundation, Germany, **5** South African Institute for Aquatic Biodiversity, c/o Port Elizabeth Museum/Bayworld, Port Elizabeth, South Africa

Abstract

Coastal dolphins are regarded as indicators of changes in coastal marine ecosystem health that could impact humans utilizing the marine environment for food or recreation. Necropsy and histology examinations were performed on 35 Indian Ocean bottlenose dolphins (*Tursiops aduncus*) and five Indo-Pacific humpback dolphins (*Sousa plumbea*) incidentally caught in shark nets off the KwaZulu-Natal coast, South Africa, between 2010 and 2012. Parasitic lesions included pneumonia (85%), abdominal and thoracic serositis (75%), gastroenteritis (70%), hepatitis (62%), and endometritis (42%). Parasitic species identified were *Halocercus* sp. (lung), *Crassicauda* sp. (skeletal muscle) and *Xenobalanus globicipitis* (skin). Additional findings included bronchiolar epithelial mineralisation (83%), splenic filamentous tags (45%), non-suppurative meningoencephalitis (39%), and myocardial fibrosis (26%). No immunohistochemically positive reaction was present in lesions suggestive of dolphin morbillivirus, *Toxoplasma gondii* and *Brucella* spp. The first confirmed cases of lobomycosis and sarcocystosis in South African dolphins were documented. Most lesions were mild, and all animals were considered to be in good nutritional condition, based on blubber thickness and muscle mass. Apparent temporal changes in parasitic disease prevalence may indicate a change in the host/parasite interface. This study provided valuable baseline information on conditions affecting coastal dolphin populations in South Africa and, to our knowledge, constitutes the first reported systematic health assessment in incidentally caught dolphins in the Southern Hemisphere. Further research on temporal disease trends as well as disease pathophysiology and anthropogenic factors affecting these populations is needed.

Editor: Lloyd Vaughan, Veterinary Pathology, Switzerland

Funding: Pathological investigations on cetaceans caught in shark nets in South Africa was funded by the German Science Foundation (SI 1542/4-1) as part of a Research Cooperation Programme with the South African National Research Foundation (Grant number 707140), as well as by a National Research Foundation SEAChange grant (Grant number 74241). The funders had no role in study design, data collection and analysis, decision to publish, or preparation of the manuscript.

Competing Interests: The authors have declared that no competing interests exist.

* Email: emily@nzg.ac.za

¤ Current address: Coastal and Marine Research Unit, Nelson Mandela Metropolitan University, Port Elizabeth, South Africa

Introduction

Surveillance and research on diseases in wildlife populations present many challenges but are important tools to identify changes in ecosystem health and emerging threats to human and animal health [1]. Health assessments in coastal cetaceans can be used to indirectly monitor marine ecosystem health, investigate the effects of human activities on animal health, and identify risks to humans utilizing the same habitat for food or recreation [2,3]. Marine mammal researchers over the past 40 years have raised concerns about deteriorating ocean health. Although increased surveillance and improved diagnostic techniques may account for a portion of the recent proliferation of disease reports [4], mortality events due to harmful algal blooms and morbillivirus outbreaks are thought to be increasingly common in the North Atlantic [4–6]. However, lack of baseline data precludes accurate recording of temporal changes in the prevalence of many diseases [4,7,8]. Expected increasing effects of climate change, inter- and intra-specific competition and habitat degradation as well as exposure to pollutants, lend new urgency to understanding the causes of marine mammal disease outbreaks [3,7–9].

Coastal cetaceans are particularly vulnerable to anthropogenic impacts including net entanglement [10], boat strike [11], disturbances due to boat traffic [10], pollution [7], nutrient enrichment [10], novel pathogens [12], habitat degradation [10], and prey depletion through fishing [10,12]. Dolphins have long life spans [12,13], feed at a high trophic level [13], and their fat stores accumulate chemical pollutants [13–15]. Increased mortalities in polluted waters during morbillivirus epidemics suggest that pollutants may impair disease defense mechanisms [12]. Habitat destruction and prey depletion increase inter- and intra-species competition and stress that further undermine host defense mechanisms [7,12]. Nutrient enrichment with sewage and fertilizers has been implicated in an increase in the occurrence of devastating toxic algal blooms [16,17]. River runoff from urban areas may be responsible for the introduction of new marine pathogens such as *T. gondii* [18,19].

Both *Tursiops aduncus* (Indian Ocean bottlenose dolphin) and *Sousa plumbea* (Indo-Pacific humpback dolphin) occur along the Southern African coast within 10 km of the shore, [20–23]. Gill nets are deployed off the South African east coast by the KwaZulu-Natal Sharks Board (KZNSB) to reduce the risk of shark-human interactions [22,24]. Approximately 20 dolphins, mainly *T. aduncus* and *S. plumbea*, are incidentally caught (by-caught) annually in the shark nets [25]. This paper reports the results of the first systematic health assessment of incidentally caught coastal dolphins, based on 40 animals examined between 2010 and 2012. Pathological findings are analyzed in relation to species, catch location, age, sex, and body condition. This survey provides valuable baseline data for assessing the health status of these dolphin populations and for future monitoring of temporal and spatial health trends.

Materials And Methods

Ethics Statement

Evaluation of dolphins incidentally caught in the shark nets was performed under research permits issued to the Port Elizabeth Museum/Bayworld (PEM) by the South African Departments of Environmental Affairs and Agriculture, Forestry and Fisheries (RES2012/40 and RES2013/19). The protocol for this study was approved by the Research Committee of the Faculty of Veterinary Science; the Animal Use and Care Committee of the University of Pretoria (Protocol V011/12) and the Ethics and Scientific Committee of the National Zoological Gardens of South Africa (P10/23). Formalin-fixed tissues are stored at the PEM; paraffin embedded tissues and glass slides are stored at the National Zoological Gardens of South Africa.

From April 2010 to April 2012, dead dolphins were retrieved from the shark nets, weighed and frozen at -20°C by the KZNSB. Every 6–8 months, carcasses were defrosted and morphological measurements taken [26]. Of the 46 dolphins retrieved, 35 *T. aduncus* and five *S. plumbea* were deemed sufficiently fresh for necropsy and histopathological examination [27]. Age was estimated by total body length in *T. aduncus* [21] and by counting the annual growth layers in a mandibular tooth in *S. plumbea*. Animals were classified as unweaned calves (<2 years), juveniles (2–12 years), or sexually mature adults (>12 years) [21,28]. Blubber thickness measurements were used (ventral, lateral and dorsal midline cranial to the dorsal fin) to assess nutritional condition [29].

Using a standard necropsy and sampling protocol [30], all organs were examined macroscopically and representative samples fixed in 10% buffered formalin. Paraffin wax embedded tissues were sectioned (5 μm) and stained with haematoxylin and eosin (HE). Selected tissues were also stained with Gram, Von Kossa (VK), Stamps, Masson's Trichrome (MT), Ziehl-Neelsen (ZN), Gomori's methenamine silver (GMS), Perl's prussian blue, Hall's bile, periodic acid-Schiff (PAS), Fontana Masson's and Bielschowsky's modified silver stains [31]. Immunohistochemical reactions for

Toxoplasma gondii (Department of Pathology, University of Pretoria) and dolphin morbillivirus (Department of Pathology, University of Veterinary Medicine, Hannover) [33] were performed on sections where lymphoplasmacytic inflammation was present in the brain, lung, muscle or heart.

Parasites found during necropsy were preserved in 70% ethanol and identified according to published methods [34]. Lung tissue samples from all 40 dolphins were frozen, until the end of the collection period, thawed in the laboratory and cultured using standard bacteriological methods.

Statistical analyses

Animals were divided into two groups based on capture location region: North and South of Ifafa beach (Figure 1), since population and genetic studies of *T. aduncus* indicate that these are different subpopulations [35,36]. Too few *S. plumbea* were sampled for statistical analysis; all statistical comparisons are for *T. aduncus* only, unless otherwise stated. Blubber thickness was compared between age classes and sample sites using a linear mixed model adjusted for sex and region with Bonferroni correction for multiple comparisons. Occurrence of selected lesions with possible biological significance was compared between species, and for *T. aduncus*, between age classes, sexes and capture location region using Fisher's exact test. For univariable associations with $p<0.25$, adjustment for possible confounding between age class, sex and region was done using multivariable exact logistic regression models. Associations between the occurrence of selected lesions within the same animals was tested using McNemar's test. Due to the exploratory nature of the analysis and the relatively small sample size, significance was assessed at $p<0.1$. Statistical analysis was done using Stata 12.1 (StataCorp, College Station, TX, U.S.A.).

Results

More *T. aduncus* (35; 88%) were caught in the nets than *S. plumbea* (5; 12%) (Figure 1). Most *T. aduncus* (25; 71%) and all five *S. plumbea* were sampled from the northern region nets; and seven of the 35 *T. aduncus* (20%) were from the nets off Durban. Most *T. aduncus* in all age classes were females (24; 69%); and more juveniles (16; 46%) and calves (11; 31%) were caught than adults (8; 22%) of both sexes.

Blubber was thicker at the dorsal and thinner at the lateral sampling site for each age class ($p<0.05$; Figure 2). Blubber thickness did not differ between the sexes or between dolphins from different regions. Blubber was thicker in juveniles and adults compared to calves, at the dorsal ($p<0.001$) and ventral ($p<0.05$) sites.

Moderate to severe autolysis, putrefaction and freezing artefact were present histologically in most organs, particularly in the respiratory and intestinal mucosae, pancreas, brain and eye. Eosinophils were relatively well preserved compared to other inflammatory cells. Freezing distorted tissue architecture and caused lysis of erythrocytes. In addition, variable numbers of variably sized, round to oval, vacuoles (<0.1 cm diameter) with no associated nuclei or saprophytic bacteria were found in blood vessel lumina and the parenchyma of various organs. Mild to severe, acute congestion was present in most organs in all the dolphins.

Dolphin number, species, sex, age, sampling region, lesion severity and health status for *T. aduncus* and *S. plumbea* are listed in Table S1. Common and newly reported lesions and lesions that may have affected organ function are described below, along with their prevalence in *T. aduncus* and *S. plumbea* (Table 1). Exact

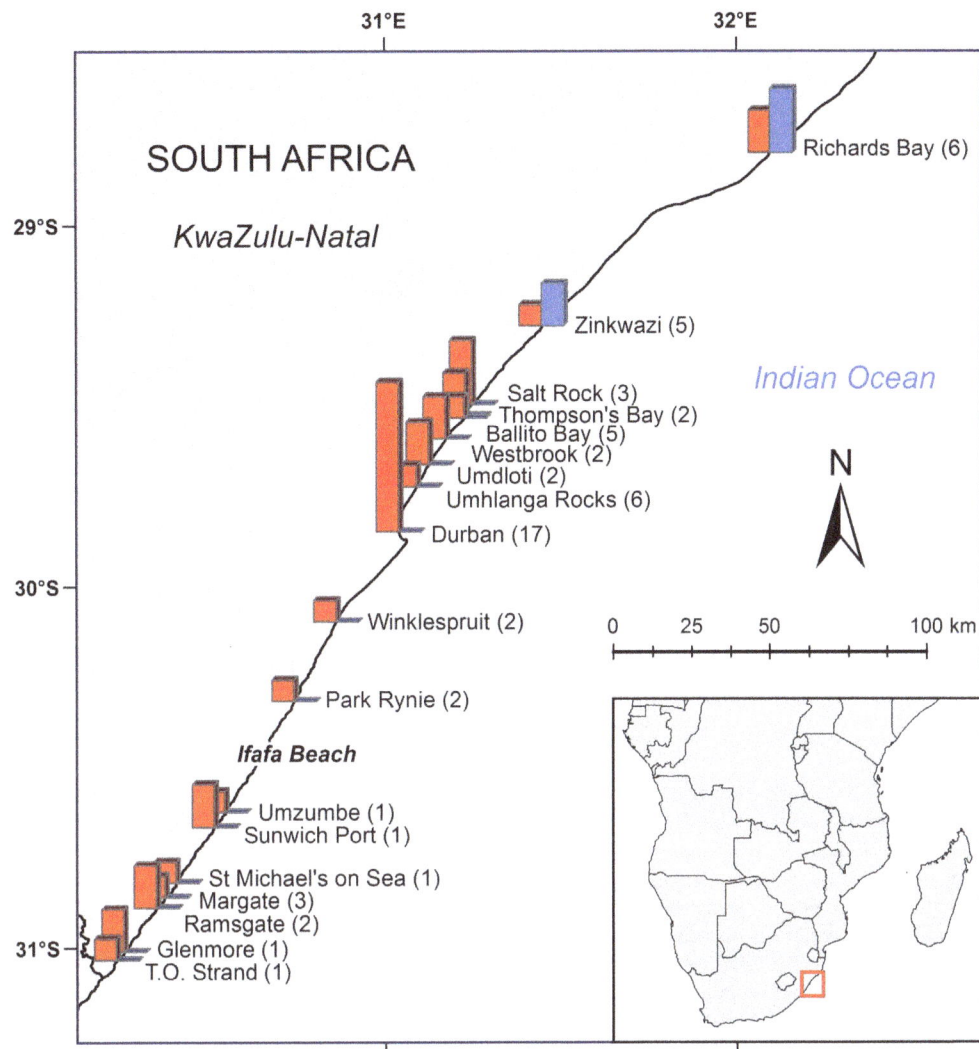

Figure 1. Location (beach name), number of shark nets per beach (in parenthesis) and number of _T. aduncus_ (red) and _S. plumbea_ (blue) sampled along the KwaZulu-Natal coast, South Africa. Gill nets are 110 m long and 10 m deep. Adapted from [87].

logistic regression models for lesions significantly associated (p< 0.1) with age class, sex and region are given in Table 2. Supplementary materials include a complete list, with prevalence by species, age class and region, of all pathological findings (Table S2) and common pathology observed in _T. aduncus_ by age class, sex and region (Table S3).

Mild to severe, multifocal to diffuse, acute pulmonary congestion, oedema and emphysema were common, characterized by lungs that were heavy, poorly collapsed, mottled pink to deep red and contained air-filled bullae (1–4 mm diameter) beneath the pleura and throughout the lung parenchyma. White foam filled airways of affected lungs. Variable numbers of fine white round helminths (<50×1×1 mm, _Halocercus_ sp.) were present in multiple firm, white to tan, unencapsulated pulmonary nodules (<2 cm diameter) and ectatic bronchi (<8 cm diameter) in 37 animals (93%), in all ages and both sexes and species (Figure 3). Affected bronchi were lined by discontinuous attenuated epithelium, with large amounts of necrotic cellular and inflammatory debris and medium number of filarial larvae. Similar inflammation often extended into and disrupted the architecture of adjacent pulmonary parenchyma. Nematode adults, with (#5, 11, 16) or

without (#6, 8, 10, 37, 40) microfilaria were present in these inflammatory lung lesions in eight (20%) animals. In addition, mild, multifocal lymphoplasmacytic and variably eosinophilic bronchointerstitial pneumonia was present in dolphins of all age classes, both sexes and species. Pneumonia was also frequently accompanied by follicular lymphoid hyperplasia of bronchus associated lymphoid tissue (18 animals; 45%).

Clustered or scattered connective tissue nodules enclosing variably mineralized necrotic debris, mixed with eosinophils, lymphocytes and plasma cells and, in some cases sections of nematodes, occurred throughout the lung parenchyma (<4 cm diameter), often close to bronchioles (16 animals, 40%). Mild to moderate, multifocal, subacute lymphoplasmacytic and variably eosinophilic tracheobronchitis, with no apparent relationship to areas of bronchiectasis or parasites, was present in 12 (44%) _T. aduncus_ calves and juveniles.

Small numbers of firm, white, pleural or subpleural plaques or nodules (<5 mm diameter), occasionally containing caseous material, were seen in 16 (40%) animals. These consisted histologically of chronic pleuritis characterized by variably thick fibrous connective tissue foci containing variably mineralized

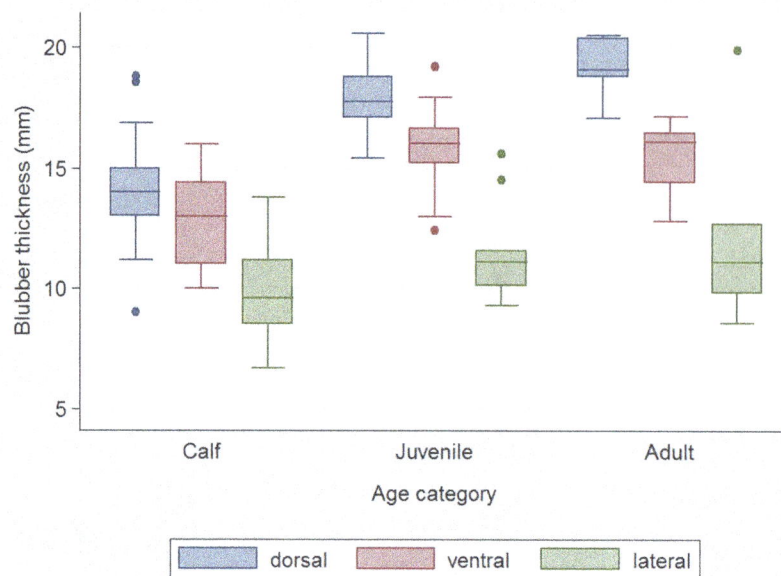

Figure 2. Blubber thickness (mm) of *T. aduncus* in three age classes. Box extends from 25th to 75th percentile, horizontal line represents the median, whiskers extend to the smallest and largest observations that are <1.5 times removed from the interquartile range (IQR), and dots represent outliers.

necrotic inflammatory and cellular debris with moderate lymphoid follicular hyperplasia and mild pleural and interstitial fibrosis in the adjacent tissue. Mild, multifocal lymphoplasmacytic and variably eosinophilic pleuritis that was not detected on gross examination was found in 12 calves and juveniles (30%) of both species. Pleural arterioles were prominent on the visceral pleura. One male *T. aduncus* calf (#14) had a large subpleural focus of bronchiectasis (8 cm diameter) lined by compressed lung tissue (2–3 mm thick) and bronchiolar epithelium which contained a few fine filamentous white helminths (<1 mm thick, 3–5 cm long). Thick white firmly attached adhesions between the parietal and visceral pleura and the diaphragm were present in two female *T. aduncus* (#1, 23). Histologically, these consisted of bands of mature fibrous connective tissue infiltrated with small foci of lymphocytes and plasma cells. The pleural surfaces of one juvenile and one adult male *S. plumbea* (#38, 40) were covered in small fibrovascular tags (<1 cm long) with variably plasmacytic and eosinophilic pleuritis and moderate pleural and interstitial fibrosis. In *T. aduncus* no association was found between pneumonia and pleuritis (p = 0.653).

Autolysis and freezing artefact precluded detailed assessment of lymphoid tissue, however, mild to moderate follicular and paracortical lymphoid hyperplasia were seen in ten animals with respiratory tract inflammation (#6, 16, 17, 19, 22–25, 34, 36, 38) and six with lung marginal lymph node serositis characterized by aggregates of small numbers of eosinophils, lymphocytes, macrophages and plasma cells in the lung marginal lymph node connective tissue capsule (# 9, 17, 19, 22, 36, 38). Inflammation also often extended to the connective tissue between the lung and the lung marginal lymph node. Lymphoid tissue appeared depleted in two female juvenile *T aduncus* (#27, 28). Mild, focal, neutrophilic and histiocytic, necrotising lung marginal lymph node lymphadenitis was seen in association with suspected fungal hyphae in a juvenile male *T. aduncus* (#25), although the lesion was not present on serial sections stained with GMS. While 12 lung sections contained small to large numbers of mixed bacteria

in blood vessels, interstitium and alveoli (H&E and Gram stains), these were not associated with necrosis or neutrophilic inflammation. A variety of bacteria were isolated on routine lung cultures, including *Pantoea agglomerans*, *Enterococcus solitarius*, *Enterobacter gergoviae*, *Shewanella algae* and *S. putrefaciens*, *Photobacterium damselae*, *Aeromonas media*, *Lactococcus garviae*, *Clostridium tertium*, *Streptococcus* from the *viridians* group, *Psychrobacter* sp, *Enterococcus* sp., *Micrococcus* sp., *Lactobacillus* sp., *Brevundimonas* sp., *Bacillus sp.*, *Acinetobacter* sp. *and Proteus* sp. Lung samples from 16 animals tested by immunohistochemistry contained no dolphin morbillivirus or *Toxoplasma* antigen.

Multiple variably mineralized deposits were common, occurring beneath or replacing the bronchial and bronchiolar mucosae. Unfortunately, details of the lesions in these animals were obscured by autolysis of the bronchiolar epithelium. Both affected and unaffected dolphins originated from both regions, were from all age classes, and of both sexes and species.

In *T. aduncus*, all three gastric compartments contained raised, firm tan nodules with central pores (<1 cm diameter); lesions were more common in the 3rd compartment (p = 0.004). Moderate to severe, multifocal, chronic lymphoplasmacytic and eosinophilic pyloric gastritis with variable calcification of the adjacent mucosa was associated with trematodes of the subfamily Brachycladiinae (Figure 4). Prevalence increased with age (p = 0.097), although this was not statistically significant in the multivariable model (p = 0.123). Eosinophilic and lymphoplasmacytic gastritis of variable severity and chronicity that was not detected on gross examination affected all three gastric compartments. The prevalence of this gastritis also increased with age (p = 0.034), as did the prevalence of similar enteritis (p = 0.002). Adult nematodes (*Anisakidae*) were found in gastro-intestinal tract of two *T. aduncus* (#26, 33). Lingual myocytes contained sarcocysts, without associated inflammation, in one *T. aduncus* calf (#13, Figure 5).

Although the livers were macroscopically unremarkable, eosinophilic and variably lymphoplasmacytic, and occasionally necro-

Table 1. Common pathology observed in Indian Ocean bottlenose (*Tursiops aduncus*) and Indo-Pacific humpback (*Sousa plumbea*) dolphins incidentally caught in shark nets, and bivariable association with species.

Lesion/abnormality	Total (%)	Species (*n*)		
		T. aduncus	*S. plumbea*	p*
Combined pneumonia	93	32/35	5/5	1.000
Bronchopneumonia	18	7/35	0/5	0.565
Interstitial pneumonia	63	22/35	3/5	1.000
Broncho-interstitial pneumonia	30	9/35	3/5	0.149
Pulmonary parasites	15	6/35	0/5	1.000
Pleuritis	30	10/35	2/5	0.627
Bronchiolar mucosal calcification	83	29/35	4/5	1.000
Pulmonary anthracosis	8	2/35	1/5	0.338
Gastritis all compartments	68	24/34	2/4	0.577
First and second compartment gastritis	63	23/34	1/4	0.132
Third compartment gastritis	65	14/21	1/2	1.000
Parasitic nodules all compartments	32	12/34	0/4	0.556
Parasitic nodules in the first and second gastric compartments	8	3/34	0/4	1.000
Parasitic nodules in the third gastric compartment	43	10/21	0/2	0.486
Pyloric mucosal calcification	26	5/21	1/2	0.462
Enteritis	68	25/35	2/5	0.307
Periportal hepatitis	54	21/35	0/4	**0.037**
Hepatic serositis	23	9/35	0/4	0.556
Periportal fibrosis	26	9/35	1/4	1.000
Hepatic trematode eggs	8	3/35	0/4	1.000
Bile ductular hyperplasia	44	15/35	2/4	1.000
Splenic filamentous peritonitis	45	17/35	1/5	0.355
Splenic serositis	28	11/35	0/5	0.298
Cervical lymph node serositis	26	10/34	0/5	0.302
Mesenteric lymphnode serositis	46	15/34	3/5	0.647
Marginal lymph node serositis	43	11/27	2/3	0.565
Marginal lymph node anthracosis	10	3/27	0/3	1.000
Endometritis	42	10/24	1/2	1.000
Metritis	23	5/24	1/2	0.415
Oophoritis	19	4/24	1/2	0.354
Mastitis	43	3/7	-	-
Mammary corpora amylacea	43	3/7	-	-
Testicular serositis	38	3/10	2/3	0.510
Endo-, myo- and epicarditis	51	20/35	0/4	**0.047**
Cardiac fibrosis	26	9/35	1/4	1.000
Meningoencephalitis	39	7/16	0/2	0.497
Myositis	19	6/32	1/5	1.000
Combined serositis	75	26/35	4/5	1.000
Abdominal serositis	60	20/35	4/5	0.631
Thoracic serositis	20	18/35	2/5	1.000

*Fisher's exact test; statistically significant results (p<0.100) in bold.

tizing, periportal hepatitis and cholangitis of variable severity and chronicity were present in 21 (60%) *T. aduncus*. Adults were more often affected than calves (p = 0.044), although the association was not significant on multivariable analysis (p = 0.112). Green-brown, triangular trematode eggs (Figure 6) were found in the portal triads of three *T. aduncus* (#13, 26, 32). Moderate to marked

hyperplasia of the bile duct epithelium was present in a *T. aduncus* (#21) and two *S. plumbea* (#36, 38) with cholangitis. Significantly, although two *S. plumbea* had cholangitis, no animals of this species had hepatitis (p = 0.037). Mild to severe, multifocal to diffuse increases in periportal mature fibrous connective tissue was observed with age in *T. aduncus* (p = 0.020). The presence of

Table 2. Associations of age, sex and region with presence of various lesions in *T. aduncus*: results of multivariable exact logistic regression models.

Variable and level		Age class			Sex	Region
		Calf (<2 y)	Juvenile (2–12 y)	Adult (>12 y)	male vs. female	south vs. north
Pleuritis	OR[1]	1*	1.54	0.26	**6.50**	1.17
	95% C.I.[2]	–	0.19, 13.65	0.00, 2.53	**0.98, 59.17**	0.00, 11.33
	p*	–	0.952	0.270	**0.053**	1.000
Pulmonary pneumoconiosis	OR	1*	1.00	**9.52**	1.50	3.00
	95% C.I.	–	0.00, ∞	**0.72, ∞**	0.04, ∞	0.08, ∞
	p	–	–	**0.085**	0.800	0.500
Enteritis	OR	1*	**15.26***	6.55	0.33	0.17
	95% C.I.	–	**1.95, ∞**	0.82, ∞	0.02, 3.78	0.00, 2.48
	p	–	**0.006**	0.080	0.573	0.303
Gastritis	OR	1*	5.66	**6.21**	1.13	1.97
	95% C.I.	–	0.57, 291.4	**0.78, ∞**	0.14, 9.89	0.23, 26.05
	p	–	0.201	**0.090**	0.141	0.785
Gastritis (compartments 1&2)	OR	1*	7.38	**7.02**	1.36	1.09
	95% C.I.	–	0.76, 376.8	**0.89, ∞**	0.17, 10.91	0.12, 10.17
	p	–	0.104	**0.066**	1.000	1.000
Periportal fibrosis	OR	1*	3.02	**12.64**	1.21	1.71
	95% C.I.	–	0.30, 41.55	**1.17, 223.9**	0.13, 9.89	0.18, 15.94
	p	–	0.482	**0.033**	1.000	0.884
Splenic tags	OR	1*	2.20	4.33	2.01	**7.75**
	95% C.I.	–	0.29, 18.27	0.42, 67.24	0.33, 14.21	**1.10, 99.82**
	p	–	0.607	0.300	0.621	**0.037**
Splenic serositis	OR	1*	2.81	5.41	**11.07**	3.3
	95% C.I.	–	0.27, 40.96	0.37, 117.1	**1.51, 152.0**	0.36, 45.61
	p	–	0.553	0.307	**0.012**	0.408
Cervical lymph node serositis	OR	1*	1.42	4.77	**7.42**	3.90
	95% C.I.	–	0.13, 15.20	0.36, 90.18	**1.04, 95.28**	0.45, 54.29
	p	–	1.000	0.327	**0.045**	0.297
Mesenteric lymph node serositis	OR	1*	2.85	**16.82**	3.56	0.94
	95% C.I.	–	0.42, 23.36	**1.92, ∞**	0.53, 29.82	0.10, 7.54
	p	–	0.377	**0.009**	0.247	1.000
Endometritis	OR	1*	0.92	**8.10**	–	0.92
	95% C.I.	–	0.07, 9.14	**0.87, ∞**	–	0.07, 9.14
	p	–	1.000	**0.067**	–	1.000
Cardiac fibrosis	OR	1*	**13.97**	**51.63**	4.29	0.71
	95% C.I.	–	**1.54, ∞**	**5.35, ∞**	0.26, 280.2	0.04, 13.09
	p	–	**0.017**	**0.001**	0.498	1.000
Myositis	OR	1*	5.73	**14.31**	0.26	0.33
	95% C.I.	–	0.20, 470.3	**1.31, ∞**	0.00, 2.31	0.00, 3.55
	p	–	0.473	**0.029**	0.246	0.381
Abdominal serositis	OR	1*	4.05	**11.18**	3.00	0.73
	95% C.I.	–	0.63, 35.27	**1.37, ∞**	0.44, 26.62	0.08, 5.42
	p	–	0.177	**0.022**	0.362	1.000

[1] OR = Odds ratio.
[2] 95% C.I. = 95% confidence interval.
*statistically significant results (p<0.100) in bold.

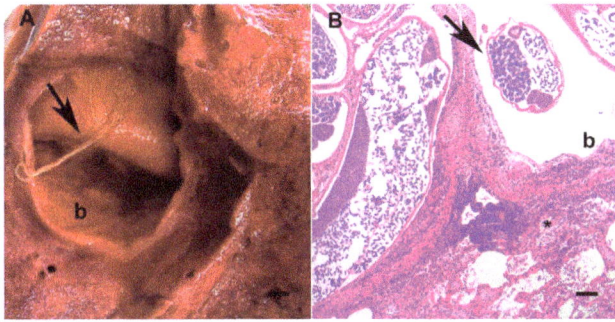

Figure 3. Parasitic pneumonia. A: Ectatic bronchus (b) containing thin (1–2 mm diameter), long, white helminths identified as *Halocercus* sp. (arrow). Bar = 5 mm. B: Pulmonary helminths (arrow) in an ectatic bronchiole (b) with eosinophilic and lymphoplasmacytic interstitial pneumonia (*) and an adjacent follicle of mildly hyperplastic bronchiolar-associated lymphoid tissue (HE, bar = 250 μm).

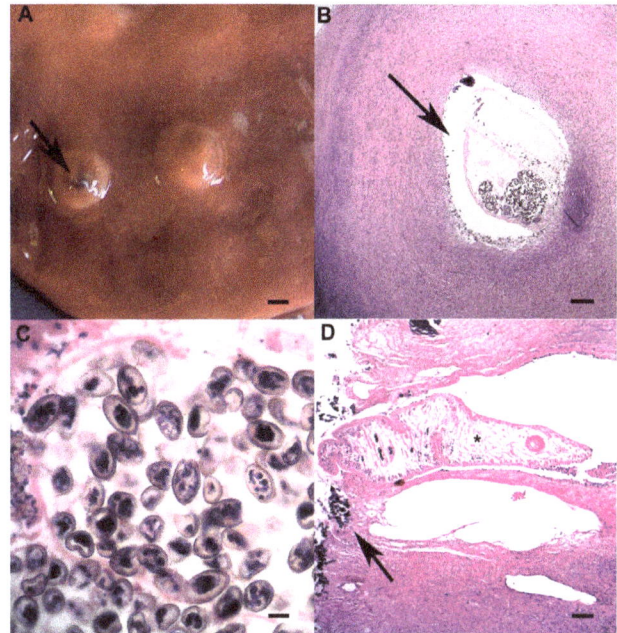

Figure 4. Gastric trematode associated lesions. A: Firm, round parasitic nodules (<1 cm diameter) with a small pore opening to the gastric lumen (arrow). Bar = 0.4 cm. B: Adult trematode (arrow) in the center of a focus of extensive fibrosis (HE, bar = 0.5 mm). C: Embryonated trematode eggs (280×160 μm, HE, bar = 150 μm). D: Parasitic nodule with adult trematode blocking the pore and irregular mineralized foci (arrow) in the adjacent superficial gastric epithelium (HE, bar = 500 μm).

increased portal connective tissues was positively associated with the presence of trematode eggs (p = 0.013). Mildly to moderately increased numbers of small bile ductules in the portal triads and under the hepatic capsule were interpreted as mild to moderate bile ductular hyperplasia in 17 (42.5%) animals of all ages and both sexes. Portal connective tissue was positively associated with bile ductular hyperplasia (p = 0.009) but not with portal hepatitis (p = 0.468).

Subjectively, increased numbers of eosinophilic cell lines were present in the rib bone marrow in 22 animals of both species (75% of *T. aduncus* and 33% of *S. plumbea*) and from both regions (80% north and 63% south). Mild to moderate, multifocal, variably eosinophilic and lymphoplasmacytic oophoritis that was not detected on gross examination was found in 21% of *T. aduncus* females and one *S. plumbea* female (#37). Endometritis was more common in adults (100%) than in calves (31%) and juveniles (29%), (p = 0.044) and consisted of small clusters of lymphocytes, plasma cells and variable numbers of eosinophils and neutrophils in the endometrium. A single adult *T. aduncus* (#32) had a trematode egg associated with the endometritis. Mild to moderate, multifocal, variably eosinophilic and lymphoplasmacytic metritis that was not detected on gross examination, was found in five *T. aduncus* (#13, 30, 31, 35, 36) and a single *S. plumbea* (#37). A positive association with age was found (p = 0.019), although this association was not significant on multivariable analysis (p = 0.107). Stamps stain for *Brucella* bacteria was negative in 12 females and all five males tested.

Mild to moderate, focal to multifocal, lymphoplasmacytic epicarditis, endocarditis and myocarditis (Figure 7A), that were not detected on gross examination, were seen in *T. aduncus* (20; 51%) but not in *S. plumbea* (p = 0.047). The highest prevalence was in juveniles (80%) (p = 0.060), although this was not significant in the multivariable model (p = 0.451). Immunohistochemistry of affected histologic sections did not demonstrate *T. gondii* antigen. Mild, focal to multifocal myocardial fibrosis (Figure 7B) was found in ten (51%) animals of both species for which heart was examined (#21, 24, 26, 29–34, 38). Prevalence increased with age (p = 0.001) and was positively associated with adrenal cortical hyperplasia (p = 0.043) but not correlated with epi-, endo-, or myocarditis (p = 0.393).

Mild, multifocal, lymphocytic meningoencephalitis was found in only seven (39%) *T. aduncus* (#7, 9, 18, 21, 25, 27, 29). Stamps and Gram histologic stains and immunohistochemistry of affected

sections did not demonstrate *Brucella*, other bacteria, *T. gondii* or dolphin morbillivirus antigen.

Multiple slightly raised, firm, white serosal nodules (<1 cm diameter) were present on various abdominal organs, mainly in *T. aduncus*. Animals from both regions and all age classes were affected (Figure 8). Histologically, these corresponded to mild, variably eosinophilic lymphoplasmacytic and necrotizing serositis

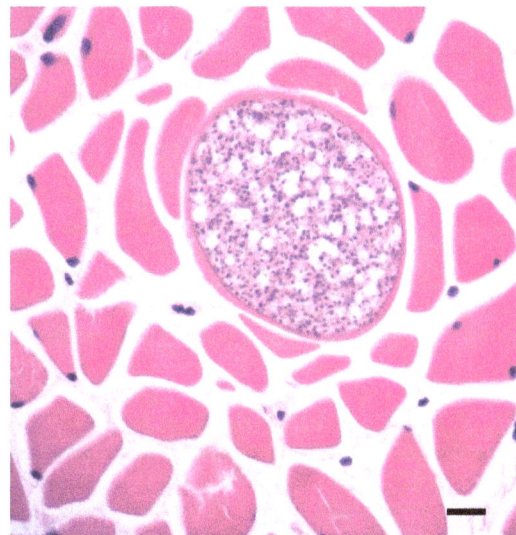

Figure 5. Lingual Sarcocystis. Sarcocyst containing myriad metrocytes in a muscle fiber of the tongue (HE, bar = 150 μm).

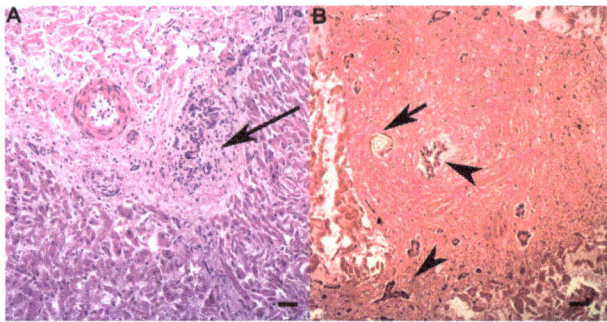

Figure 6. Hepatic lesions in *T. aduncus*. A: Mild proliferation (hyperplasia) of small portal bile ductules (arrow) (HE, bar = 100 μm): B: Severe hepatic periportal fibrosis associated with a trematode egg (arrow, 100 μm diameter). Note the bile ductules with hyperplastic epithelium (arrowheads, HE, bar = 100 μm).

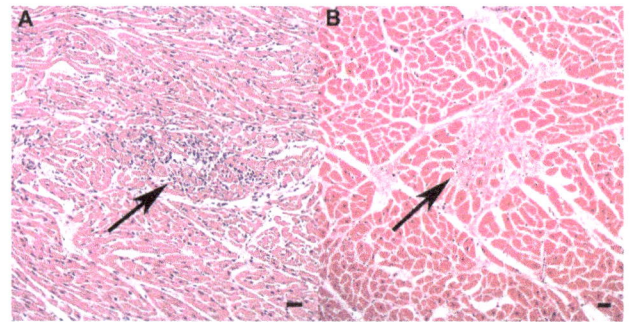

Figure 7. Myocardial lesions. A: Mild focal lymphoplasmacytic myocarditis (arrow) (HE, bar = 50 μm). B: Mild focal myocardial fibrosis (arrows) (HE, bar = 120 μm).

affecting the fibrous capsule of the mesenteric lymph node (#2, 9, 13, 15, 17, 18, 20, 21, 23–26, 29, 30, 34, 35, 37, 38, 40), spleen (#9, 13, 17, 18, 23–26, 29, 32, 33, 35), liver (#4, 9, 15, 17, 21, 24, 25, 33), testis (#18, 24, 33, 38, 39), kidney (#5, 32, 36, 39), diaphragm (#7, 26, 30), and epididymis (#40), as well as adipose tissue adjacent to the mesenteric lymph node (#7, 8). Multifocal to diffuse, lymphoplasmacytic and eosinophilic inflammation was present in the mesenteric lymph node in five animals (#24, 30, 31, 34, 35), the testis in a *T. aduncus* calf (#18) and the spermatic cord in a juvenile *T. aduncus* (#38). Nematode larvae were associated with the mesenteric lymph node serositis in two juvenile male *T. aduncus* (#25, 29). These lesions were variably associated with mesenteric lymph node lymphoid hyperplasia (Table S1). The prevalence of the mesenteric lymph node serositis increased significantly with age in *T. aduncus* (p = 0.009). Male *T. aduncus* were more often affected with splenic serositis than females (p = 0.015). Renal serositis was not associated with the mild, multifocal, mainly lymphoplasmacytic, renal interstitial nephritis seen in 11 animals (Table S1).

Long, slender, splenic tags occurred in a higher proportion of *T. aduncus* (49%) than *S. plumbea* (20%) (Figure 9). Histologically, these filamentous projections of the splenic capsule consisted of fibrovascular connective tissue with minimal or mild, multifocal, lymphoplasmacytic and eosinophilic inflammation. Splenic tags were significantly more common in dolphins from the southern coast (80%) than the northern coast (36%) (p = 0.027) and were associated with splenic serositis (p = 0.034).

Mild, multifocal, lymphoplasmacytic interstitial skeletal myositis was present in ten (27%) dolphins of both species and sexes from the northern region (#9, 21, 22, 23, 30, 31, 32, 33, 27, 38). Prevalence increased with age (p = 0.007). Immunohistochemistry of affected histologic sections did not demonstrate *T. gondii*. Multiple raised, pale pink cystic lesions (<1 cm diameter) containing adult *Crassicauda* sp. were associated with moderate, locally extensive, chronic, eosinophilic myositis in the musculature next to the mammary gland in one *T. aduncus* adult female (#22), which also had round basophilic crystalline structures with variable mineralized cores (interpreted as *corpora amylacea*) in the adjacent otherwise unremarkable mammary gland. Mild, multifocal, interstitial mammary gland inflammation with pleocellular infiltrates was present in two *T. aduncus* calves (#3, 6) and one juvenile (#21). Sarcocysts, without associated inflammation, were found in neck and intercostal muscle of one *T. aduncus* calf (#13).

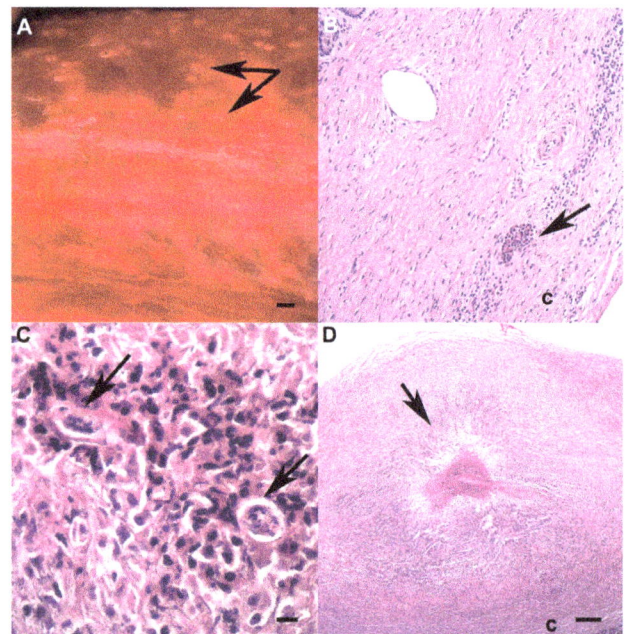

Figure 8. Abdominal serositis. A: Peritoneum overlying the testis contains multiple, slightly raised, firm white nodules, some of which contain depressed red centers (arrows). Bar = 5 mm. B: Eosinophil aggregate (arrow) and lymphoplasmacytic serositis in the testicular capsule (c). Note the seminiferous tubule in the upper left corner (HE, bar = 100 μm). C: Mesenteric lymph node serositis with intra-lesional nematode larvae in the capsule (60 μm diameter, arrows) (HE, bar = 30 μm). D: Severe focal granulomatous testicular serositis in the testicular capsule (c) with a central area of necrosis (arrow) resembling a helminth migration tract. Note seminiferous tubules at bottom right (HE, bar = 500 μm).

All animals had superficial cutaneous linear abrasions (net marks), particularly over the thorax, flippers, flukes and head, associated with subcutaneous congestion or haemorrhage in some cases (#1, 9, 20, 30, 32). An adult male *S. plumbea* (#40) had two flat, pale-tan, lobular, cutaneous soft masses below the dorsal fin (10 mm diameter) with a light brown exudate on the cut surface. Histologically, large numbers of large foamy macrophages and rare multinucleate giant cells infiltrated the skin and subcutis with a large number of intra-lesional round yeasts (7–10 μm diameter) that stained positive on both GMS and PAS, consistent with

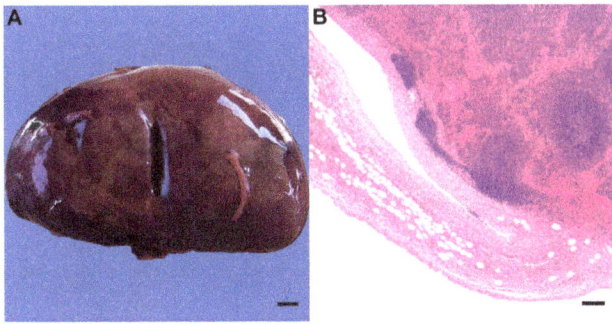

Figure 9. Splenic filamentous peritonitis. A: Fine long filamentous tags (1×2×30 mm) on the splenic capsule. Bar = 10 mm. B: Splenic tag consisting of mature fibrovascular connective tissue (HE, bar = 500 μm).

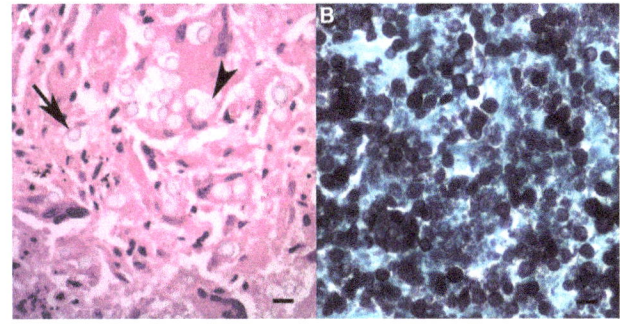

Figure 10. Cutaneous lobomycosis. A: Moderate numbers of round to oval refractile yeasts occur free in the subcutis (arrow) or within multinucleate giant cells (arrowhead, HE, bar = 10 μm). B: Large numbers of deep blue-black staining yeasts (GMS, bar = 10 μm).

lobomycosis (Figure 10). Small aggregates of lymphocytes, plasma cells, neutrophils or eosinophils occurred in the mammary gland interstitium of two *T. aduncus* calves (#3, 6) and one *T. aduncus* juvenile (# 8).

Mild, focal, lymphoplasmacytic and eosinophilic steatitis affecting the adipose tissue around the cervical lymph node was present in four animals (#11, 15, 20, 26). Mild, multifocal, lymphoplasmacytic and histiocytic inflammation of the capsule of the cervical lymph node and or surrounding adipose tissue was present in nine animals (#13, 17, 18, 22, 24, 26, 33, 34, 35), affecting more males (55%) than females (17%) (p = 0.045). This finding had no association with mild to moderate follicular and paracortical lymphoid hyperplasia present in this lymph node in 17 animals (Table S1).

Discussion

This study is the first reported systematic health assessment of incidentally caught dolphins in the Southern Hemisphere. This valuable information on the current prevalence of disease in the coastal dolphin populations of South Africa can be used as a baseline for future monitoring projects.

The degree of autolysis and freezing artefact varied between animals and organs, and likely masked subtle histological features such as necrosis and tissue and inflammatory cellular detail. The presence and patterns of inflammation and parasites could, however, be confidently diagnosed, as has been documented in harbour porpoises (*Phocoena phocoena*) [38,40,41] and fur seals (*Arctocephalus forsteri*) [42].

Correct interpretation of tissue changes as pathological was hampered by the small sample size and the lack of standardized descriptions of tissue anatomy in dolphins. Also, in contrast to regularly dewormed domestic species, establishing normal tissue parameters is complex in free-ranging mammals which often harbour large numbers of internal parasites that may vary with age, geographical location and season. Focal (#16, 19, 25, 26, 32), multifocal (#33, 39) or diffuse (#6, 35) increases in the amounts of mature connective tissue spatially unrelated to pneumonia were compared to pulmonary connective tissue amounts in the remaining animals and subjectively diagnosed as pulmonary fibrosis. Similarly, increased amounts of periportal mature connective tissue was positively associated with age, but not with the presence of periportal hepatitis (Table S3). This may therefore be an age-related change in *T. aduncus*, although it is not clear whether this is related to trematode infections which are more numerous in older animals (Table S2). Increased numbers of small bile ducts in the portal triads and under the hepatic capsule were

noted in 17 animals (Table S1); this change was subjectively associated with increased amounts of mature connective tissue, based on comparison between livers in other animals in this series and on our knowledge of similar lesions in terrestrial mammals. Too few animals were examined to assess whether the number of small bile ducts in the hepatic portal zone is variable in these species or is related to inflammatory changes. Documentation of the amount of connective tissue in well preserved tissues from newborn animals and any age-related increases, in the absence of pathological changes, would facilitate correct interpretation of the amount of pulmonary and hepatic connective tissue in these species.

Widespread tissue congestion and pulmonary emphysema and oedema are described in other net-captured cetaceans and are likely due to terminal heart failure and or drowning [37–39]. Clear, round vacuoles in various tissues and air emboli in blood vessels in a wide range of tissues were possibly a result of either drowning or supersaturation [84,88]; however without more detailed studies regarding the pathophysiology of drowning in cetaceans the distinction between these two possibilities is uncertain. The histological location, absence of nuclei, and variable size of tissue and intravascular vacuoles excluded adipocytes; the absence of bacteria associated with the vacuoles (HE and Gram) make gas produced by saprophytic bacteria unlikely. However, only bubble content analysis would confirm supersaturation [89].

As expected, most of the lesions noted in these incidentally caught dolphins were mild to moderate and severe lesions were mostly focal (Table S1) and the dolphins were judged to be healthy. Blubber thickness measurements were within previously published ranges for *T. aduncus* from the KwaZulu-Natal coast [29]. No reference ranges or prior data are available for blubber thickness in *S. plumbea* from the KwaZulu-Natal coast. None of the animals with the thinnest blubber had major or multiple significant lesions and no statistical association between thinner blubber and pathology could be demonstrated. Therefore, we concluded that all animals were at least in fair nutritional condition. Although parasite levels in free-ranging animals generally have little effect on the host, factors such as stress, altered nutrition, anthropogenic factors, pollutants or concurrent disease may compromise the host's immune system and increase the severity and prevalence of parasitic infections [40]. Parasite burdens may then be used as indicators for the overall health status of an individual [40]. This assumption should, however, be made with caution, as environmental factors such as pollution may also negatively affect parasite populations [43]. Pollutant analysis on stored tissues from these dolphins would be valuable.

However, myocardial inflammation and fibrosis as well as meningoencephalitis may affect organ function and therefore be significant for the individual dolphin. Fertility and therefore population dynamics could also be affected by oophoritis, endometritis and orchitis but since one pregnant female had mild metritis, this lesion alone may not impair fertility.

Although autolysis and freezing artefact likely obscured subtle lesions, visible lesions in the respiratory and gastro-intestinal tracts were largely parasitic, as expected in incidentally caught free-ranging animals. Lesions were generally mild compared to those described in other health investigations [38,39,41,44]. The presence of lungworms was less common (20%) than has been reported for stranded *T. truncatus* (77%) and *S. coeruleoalba* (76.5%) from the Northern Hemisphere [45,46]. Eosinophilic pneumonia, even in the absence of visible parasites, was likely parasitic [47,48].

Halocercus spp. are common in the lungs of many dolphin species, although the complete life cycle remains unknown [44,49]. They are generally considered to be of no clinical importance in *T. truncatus* from Florida [45]. Since parasites were recovered more often from calves than from juveniles, and no parasites recovered from adults, the infestation is likely established *in utero* or through milk ingestion [45,49]. Adult animals more often showed only chronic or resolving infections; however, heavily infested adults that died due to parasitism would have been missed in this survey. The variable lymphoplasmacytic inflammation and accompanying follicular lymphoid hyperplasia may indicate the presence of persistent foreign antigen and activation of the adaptive immune response despite clearance of the infestation in older animals [50]. As has been described previously [45], pulmonary interstitial fibrosis was significantly more common in older animals. Interstitial pulmonary fibrosis is a sequel to repetitive, persistent, or severe damage to the endothelial or epithelial cells, inflammation of the alveolar septa, or chronic pulmonary hypertension [51]. In dolphins it has commonly been reported in chronic morbillivirus [52–54] and parasitic infections [44,45]. However, no association between fibrosis and pneumonia or pulmonary verminosis could be demonstrated in this study.

Gastric parasitic nodules due to the trematode *Pholeter gastrophilus* infestation are a common incidental finding in dolphins [49,55,56]. As described previously, nodules were mainly in the pyloric compartment. Nematodes belonging to the family Anisakidae have an indirect life cycle, with animals ingesting infective larvae in infected fish and squid [49]. This likely explains the higher prevalence in juveniles and adults, since calves only become infected once they start consuming fish. Observed species differences in the prevalence of parasitic lesions in the liver, stomach, spleen, lung and lymph nodes may be a result of the small sample size of *S. plumbea*. Alternatively, the parasites that cause these lesions could be host specific due to consumption of different fish and squid species that act as intermediate or paratenic hosts [49]. Of the 94 prey species recorded in *T. aduncus* and 54 prey species in *S. plumbea*, only 25 species are eaten by both *T. aduncus* and *S. plumbea* [57,58]. Changing diet due to changes in prey population dynamics, climate change and or anthropogenic influences may affect parasite loads and is a key topic for future research.

Parasites, including the trematodes *Campula*, *Oschmarinella*, and *Brachycladium* (formerly *Zalophotrema*) which have been found in hepatic ducts, were the most likely cause of the hepatitis and periportal hepatitis in *T. aduncus* [44,49,56]. The life cycle of these brachycladiids is not known [49]. The eosinophilic

oophoritis, endometritis, metritis and orchitis were also probably caused by parasites, supported by the trematode egg present in one case. The positive association with age (up to 100% of adult animals) suggests an indirect life cycle. Small sample size, bias towards younger animals and autolysis precludes a definitive diagnosis of increased bone marrow eosinophilic myelopoiesis; however, a predominance of eosinophilic bone marrow cell lines could reflect the widespread parasitism in these dolphins. Sarcocysts have not previously been reported in dolphins from South African waters, although they have been reported in other cetacean populations [18,49,59–62]

Widespread serosal eosinophilic or fibrotic abdominal serosal lesions were reported to have increased in prevalence in 2009 (*pers. comm.* S. Plön). Similar lesions are described in in domestic horses with *Strongylus* spp migrations, and in domestic pigs due to chronic bacterial serositis. Most of the lesions were chronic with no definitive indication of aetiology. However, parasite larvae were found in the capsules of two mesenteric lymph nodes, and a necrotic tract suggestive of a migration tract was found in another mesenteric lymph node. Lack of association between serosal lesions and pulmonary verminosis, hepatic trematode eggs, or gastric trematodes may be due to the fact that these parasites were not the cause of the lesions, or perhaps due to temporal changes in lesion location and severity over the life cycle of the parasite. Changes in the ecology of food species acting as parasite intermediate hosts could explain the apparent changes in the prevalence of these lesions. Further research is needed on the identity of the parasite, its life cycle and the possible changes in host, environment and prey factors that may influence parasitic loads. Although the inflammatory nature of the splenic serositis resembles that in other abdominal organs, the aetiology of the splenic tags remains uncertain and further research is needed to determine their significance and explain why they are more common in *T. aduncus*, particularly from the southern region.

No histological or immunohistochemical evidence of dolphin morbillivirus infection, brucellosis or toxoplasmosis was found. However, cetacean morbillivirus antibodies were previously found in a *D. delphinus* that stranded approximately 350 km south of the study area [65]. Regrettably, no pathological information is available for this animal and paired serum samples could not be taken to confirm active infection. This population of dolphins may be less susceptible to these diseases than other populations. Alternately, the prevalence of these diseases may have been too low to detect in our study. However, the absence of histological or immunohistochemically stained antigen in the tissues from these dolphins may also have occurred due to poor tissue preservation or loss of antigen integrity due to formalin fixation. The antibody used to detect morbillivirus antigen was a pan-morbillivirus antibody and has been used with success in *Phoca vitulina* (harbour seal) [33], and *S. coeruleoalba* [38]. Commercially available immunohistochemical stains used in this study have been used effectively to detect *T. gondii* in dolphins [66]. The modified ZN (Stamps) stain is an accepted method of demonstrating *Brucella* spp. organisms in tissues [67,68]. This is a crucial area for future research, given the presence of inflammatory lesions compatible with these diseases and their worldwide distribution. Continued monitoring of these dolphin populations is needed as reliable detection of infectious agents present at low prevalence can only be accomplished by testing larger numbers of animals but access to live free-ranging coastal dolphins is limited [65,69]. Microbiological culture and biotyping of brain, spleen and reproductive tract isolates will be conducted in future. Serological and molecular diagnostic tests for *Brucella* spp. and *T. gondii* are also needed. If these dolphin populations are in fact naïve to these pathogens, their introduction could have devastating consequences, as has been documented previously in other populations elsewhere during morbillivirus epidemics [5,52,54,70–73].

No animals had lesions consistent with bacterial pneumonia and no primary bacterial pathogens were isolated from the lung. However, autolysis and freezing may have compromised culture success. Isolation of opportunistic bacteria such as *Aeromonas media* and *Photobacterium damselae* is consistent with previous reports [72]. *Shewanella algae* is commonly isolated from marine environments, and is an opportunistic human pathogen [74]. Remaining bacteria were considered contaminants or normal commensals.

Granulomatous dermatitis associated with fungi is consistent with the zoonotic disease lobomycosis [5,75,76]. This is, to our knowledge, the first confirmed report of lobomycosis in South African waters, although macroscopic lobomycosis-like disease has been documented in other Indian Ocean populations of *T. aduncus* [77]. Impaired adaptive immunity was found in endemically affected *T. truncatus* from the Indian River Lagoon, Florida [78]. The exact aetiology of the immunosuppression in dolphins has not yet been determined, but both environmental contaminants, such as mercury and polychlorinated biphenyls, and chronic stress as result of anthropogenic factors have been suggested [76,78]. No evidence of immunosuppression was found histologically in the dolphins in this study, although differential white cell counts, determination of lymphocyte subpopulations, phagocytic activity and lysozyme activity, amongst other tests [78], were not possible in incidentally caught animals.

While some variation in the width of the adrenal cortex and occasional cortical nodules were seen in the cortex or medulla in these animals, such variation could have been due to differing planes of section. Blood and faecal adrenocortical hormone assays, adrenal weights and objective measurement of adrenal cortico-medullary ratios by point-counting techniques [32] as well as systematic evaluation of the pituitary are needed to evaluate the possibility of stress in this dolphin population. Adrenal hyperplasia has been attributed to chronic stress from long-term debilitating disease or injury in *T. truncatus* in the Gulf of Mexico [32,79]; however, the animals in this study had relatively mild pathology. Environmental stressors, such as competition for resources, and anthropogenic factors, such as boat traffic, seismic or military activities warrant evaluation. Myocardial fibrosis is a non-specific indication of prior tissue damage due to inflammation or necrosis. Myocardial necrosis and fibrosis in stranded and incidentally caught *T. truncatus* and *S. coeruleoalba* from the Gulf of Mexico were attributed to the acute and chronic effects, respectively, of high catecholamine levels [79]. The association of cardiac fibrosis with age may indicate that the effects are cumulative. Cardiac fibrosis was not associated with myocarditis in *T. aduncus*; however, the small sample size precludes definitive conclusions on the aetiology of either lesion. Similarly, the small sample size, including only one adult *S plumbea*, may account for the absence of epicarditis, endocarditis or myocarditis seen in this species. Although mild cutaneous depigmentation (#1), lacerations (#6, 26, 31), and barnacles (#3) were documented, inter and intra-specific aggression could not be reliably distinguished from boat strike or other anthropogenic injury.

The higher numbers of *T. aduncus*, caught in the nets all along the coast likely reflects the relative population size and more widespread distribution of this species [21,22]. All five *S. plumbea* were caught on two adjacent beaches in the northern region (Figure 1), where they occur in higher numbers than in the south [80,81]. The fact that calves and juveniles are more inquisitive and inexperienced may explain why *T. aduncus* calves and juveniles

were caught more often than adults [82]. Females with calves also feed closer to shore, and therefore to the nets, which results in higher capture rates of adult females and calves [64,83].

Mineralization of the bronchiolar epithelium has previously been attributed to lungworm infection [85,86]. Bronchiolar mineralization is not a common feature of verminous pneumonia in cetaceans [38,44], but is occasionally seen in harbour porpoises from the North Sea (P. Wohlsein, *pers. comm.*). Foreign particles are thought to accumulate in the lung due to the inability of dolphins to cough. These particles become inspissated, undergo calcification and are later incorporated into the bronchial wall [85]. Additional investigations are underway to determine the distribution and exact location of the material. Small foci of mineralisation were present in 24 dolphins in a wide range of tissues, in addition to the airways (Table S1). In mammals, metastatic tissue mineralisation due to disturbed calcium and phosphorus metabolism typically occurs on the intercostal pleura, pulmonary and renal cortical basement membrane, and the middle and deep gastric mucosa [90]. Since these sites were not involved, and no indication of renal failure, neoplasia, or granulomatous inflammation that could result in secondary hyperparathyroidism were present, the mineralisation seen in these dolphins was assumed to be dystrophic changes due to minor tissue damage. However, since neither pituitary nor parathyroid glands were routinely sampled we cannot rule out the possibility of altered calcium homeostasis in these dolphins. We consider that nutritional hyperparathyroidism (due to altered calcium, phosphate or Vitamin D metabolism) is unlikely to be common in free-ranging animals; and cannot rule out the possibility of emerging secondary marine plant intoxication through ingestion of herbivorous fish.

Conclusion

In the first systematic health assessment of incidentally caught coastal dolphins in the Southern Hemisphere, we report the first confirmed cases of lobomycosis and sarcocystosis in dolphins from the South African coast. While optimum samples are not provided by frozen, incidentally caught animals, this study still yielded valuable information on the current prevalence of disease in the two dolphin populations, which can be used as a baseline for future monitoring projects, not only of the health status of the population, but also that of the environment. This may prove particularly important for *S. plumbea*, whose coastal habitat, restricted distribution range, and small population size make it prone to a number of threats, including anthropogenic impacts. These findings further highlight the importance of disease investigation in marine mammals.

Supporting Information

Table S1 Summary of mild, moderate and severe lesions and overall health status for each of 35 Indian Ocean bottlenose (T. aduncus) and five Indo-Pacific humpback (S. plumbea) dolphins incidentally caught in shark nets along the KwaZulu-Natal coast, South Africa, 2010-2012.

Table S2 Complete pathological findings for indicating occurrence (lesion/number of organ evaluated) and percentage per species, age group, and region (for both species combined).

Table S3 Common pathology observed in Tursiops aduncus and associations with sex, age and region.

Acknowledgments

The authors would also like to thank the staff and students of the Port Elizabeth Museum, in particular Dr. Greg Hofmeyr; staff of the National Zoological Gardens of South Africa; staff of the KwaZulu-Natal Sharks Board, in particular Geremy Cliff; laboratory staff of the Department of Pathology, Faculty of Veterinary Science, University of Pretoria, and the Department of Pathology, University of Veterinary Medicine, Hannover, Foundation, Germany. Particular thanks go to Dr. David Zimmerman for necropsy data and sampling in 2010; Dr. Maryke Henton (IDEXX South Africa) for culture and identification of the bacteria; Drs. Kerstin Junker (Agricultural Research Council – Onderstepoort Veterinary Institute, Pretoria, South Africa) and Kristina Lehnert (Institute of Terrestrial and Aquatic Wildlife, University of Veterinary Medicine, Hannover, Foundation, Germany) for parasite identification; and Dr. Ingrid de Wet for help during dissections.

Disclaimer

Any opinion, findings and conclusions or recommendations expressed in this material are those of the author(s) and therefore the NRF does not accept any liability in regard thereto.

Author Contributions

Conceived and designed the experiments: MdW EPL US PW PT SP. Performed the experiments: MdW EPL US PW SP. Analyzed the data: MdW EPL PW PT. Contributed reagents/materials/analysis tools: EPL PT SP. Wrote the paper: MdW EPL US PW PT SP.

References

1. Ryser-Degiorgis M (2013) Wildlife health investigations: Needs, challenges and recommendations. BMC Vet Res 9: 223–240.
2. Bossart GD (2006) Marine mammals as sentinel species for oceans and human health. Oceanography 19: 134–137.
3. Harvell CD, Mitchell CE, Ward JR, Altizer S, Dobson AP, et al. (2002) Climate warming and disease risks for terrestrial and marine biota. Science 296: 2158–2162.
4. Gulland FMD, Hall AJ (2007) Is marine mammal health deteriorating? Trends in the global reporting of marine mammal disease. Ecohealth 4: 135–150.
5. Van Bressem MF, Raga JA, Guardo G, Jepson PD, Duignan PJ, et al. (2009) Emerging infectious diseases in cetaceans worldwide and the possible role of environmental stressors. Dis Aquat Org 86: 143–157.
6. Raga JA, Banyard A, Domingo M, Corteyn M, Van Bressem MF, et al. (2008) Dolphin morbillivirus epizootic resurgence, Mediterranean Sea. Emerg Infect Dis 14: 471–473.
7. Harvell CD, Kim K, Burkholder JM, Colwell RR, Epstein PR, et al. (1999) Emerging marine diseases - climate links and anthropogenic factors. Science 285: 1505–1510.
8. Ward JR, Lafferty KD (2004) The elusive baseline of marine disease: Are diseases in ocean ecosystems increasing? PLoSBiol 2, e120 2: 542–546.
9. Epstein RP, Sherman D, Spanger-Siegfried E, Langston A, Prasad S (1998) Marine ecosystems: Emerging diseases as indicators of change. Boston: Harvard Medical School. MA. 85 p.
10. Geraci JR, Lounsbury VJ (2009) Health. In: Perrin WF, Würsig B, Thewissen JGM, editors. Encyclopedia of Marine Mammals (second edition). London: Academic Press. pp.546–553.
11. Bar K, Slooten E (1999) Effects of tourism on dusky dolphins at Kaikoura. Conservation Advisory Science Notes 229: 5–10.
12. Lafferty KD, Porter JW, Ford SE (2004) Are diseases increasing in the ocean? Annu Rev Ecol Evol Syst 35: 31–54.
13. Wells RS, Rhinehart HL, Hansen LJ, Sweeney JC, Townsend FI, et al. (2004) Bottlenose dolphins as marine ecosystem sentinels: Developing a health monitoring system. EcoHealth 1: 246–254.
14. Reddy ML, Dierauf LA, Gulland FMD (2001) Marine mammals as sentinels of ocean health. In: Dierauf LA, Gulland FMD, editors. CRC Handbook of marine mammal medicine. Boca Raton: CRC Press Inc. pp.3–13.
15. O'Shea TJ, Bossart GD, Fournier M, Vos JG (2003) Conclusions and perspectives for the future. In: Vos JG, Bossart GD, Fournier M, O'Shea TJ, editors. Toxicology of Marine Mammals. New York: Taylor & Francis. pp.595–613.
16. Riva GT, Johnson CK, Gulland FMD, Langlois GW, Heyning JE, et al. (2009) Association of an unusual marine mammal mortality event with *Pseudo-nitzschia* spp. blooms along the southern California coastline. J Wildl Dis 45: 109–121.
17. Flewelling LJ, Naar JP, Abbott JP, Baden GD, Barros NB, et al. (2005) Red tides and marine mammal mortalities. Nature 435: 755–756.
18. Dubey JP, Zarnke R, Thomas NJ, Wong SK, Bonn WV, et al. (2003) *Toxoplasma gondii*, *Neospora caninum*, *Sarcocystis neurona*, and *Sarcocystis canis*-like infections in marine mammals. Vet Parasitol 116: 275–296.
19. Miller MA, Gardner IA, Kreuder C, Paradies DM, Worcester KR, et al. (2002) Coastal freshwater runoff is a risk factor for *Toxoplasma gondii* infection of southern sea otters (*Enhydra lutris nereis*). Int J Parasitol 32: 997–1006.
20. Best PB (2007) Whales and dolphins of the Southern African subregion. Cape Town: Cambridge University Press. 352 p.
21. Cockcroft VG, Ross GJB (1990) Age, growth and reproduction in bottlenose dolphins (*Tursiops truncates*) from the east coast of Southern Africa. Fish B-NOAA 88: 289–302.
22. Cockcroft VG, Ross GJB, Peddemors VM (1990) Bottlenose dolphin *Tursiops truncatus* distribution in Natal's coastal waters. S Afr J Marine Sci 9: 1–10.
23. Karczmarski L, Cockcroft VG, Mclachlan A (2000) Habitat use and preferences of Indo-Pacific humpback dolphins *Sousa chinensis* in Algoa Bay, South Africa. Mar Mamm Sci 16: 65–79.
24. Cockcroft VG (1994) Is there common cause for dolphin captures? A review of dolphin catches in shark nets off Natal, South Africa. Report of the international whaling commission Special issue 15: 541–547.
25. KwaZulu-Natal Sharks Board (2009) Catch statistics. http://shark.co.za, last accessed 15 January 2013.
26. Norris KS (1961) Standardized methods for measuring and recording data on the smaller cetaceans. J Mammal 42: 471–476.
27. Geraci JR, Lounsbury VJ (2005) Specimen and data collection. In: Geraci JR, Lounsbury VJ, editors. Marine mammals ashore: A field guide for strandings (second edition). Baltimore, Maryland: National Aquarium in Baltimore. pp.167–251.
28. Jefferson TA, Hung SK, Robertson KM, Archer FI (2012) Life history of the Indo-Pacific humpback dolphin in the pearl river estuary, southern china. Mar Mamm Sci 28: 84–104. Adapted by Z. Nolte, Rhodes University.
29. Young DD (1998) Aspects of condition in captive and free-ranging dolphins. PhD thesis. Grahamstown, Rhodes University. 436 p.
30. De Wet M (2013) A systematic health assessment of two dolphin species by-caught in shark nets off the KwaZulu-Natal coast, South Africa. MSc Thesis. Pretoria: University of Pretoria. 152 pp.
31. Böck P, Romeis B (1989) Romeis mikroskopische technik. Munchen, Germany: Publisher Urban and Schwarzenberg. 697 p.
32. Clark LS, Cowan DF, Pfeiffer DC (2006) Morphological changes in the Atlantic bottlenose dolphin (*Tursiops truncatus*) adrenal gland associated with chronic stress. J Comp Pathol 135: 208–216.
33. Stimmer L, Siebert U, Wohlsein P, Fontaine JJ, Baumgärtner W, et al. (2010) Viral protein expression and phenotyping of inflammatory responses in the central nervous system of phocine distemper virus-infected harbour seals (*Phoca vitulina*). Vet Microbiol 145: 23–33.
34. Lehnert K, Raga JA, Siebert U (2007) Parasites in harbour seals (*Phoca vitulina*) from the German Wadden Sea between two phocine distemper virus epidemics. Helgol Mar Res 61: 239–245.
35. Peddemors VM (1999) Delphinids of Southern Africa: A review of their distribution, status and life history. J Cetac Res Manage 1: 157–165.
36. Natoli A, Peddemors VM, Hoelzel AR (2008) Population structure of bottlenose dolphins (*Tursiops aduncus*) impacted by by-catch along the east coast of South Africa. Conserv Genet 9: 627–636.
37. Kuiken T, Simpson VR, Allchin CR, Bennett PM, Codd GA, et al. (1994) Mass mortality of common dolphins (*Delphinus delphis*) in south-west England due to incidental capture in fishing gear. Vet Rec 134: 81–89.
38. Siebert U, Wünschmann A, Weiss R, Frank H, Benke H, et al. (2001) Post-mortem findings in harbour porpoises (*Phocoena phocoena*) from the German North and Baltic Seas. J Comp Pathol 124: 102–114.
39. Duignan PJ (2003) Disease investigations in stranded marine mammals, 1999–2002. Department of Conservation, Science Internal Series. 32 p.
40. Siebert U, Joiris C, Holsbeek L, Benke H, Failing K, et al. (1999) Potential relation between mercury concentrations and necropsy findings in cetaceans from Gerrman waters of the North and Baltic Seas. Mar Pollut Bull 38: 285–295.
41. Siebert U, Tolley K, Vikingsson GA, Ólafsdottir D, Lehnert K, et al. (2006) Pathological findings in harbour porpoises (*Phocoena phocoena*) from Norwegian and Icelandic waters. J Comp Pathol 134: 134–142.
42. Roe WD, Gartrell BD, Gartrell BD, Hunter SA (2012) Freezing and thawing of pinniped carcasses results in artefacts that resemble traumatic lesions. Vet J 194: 326–331.
43. Torchin ME, Lafferty KD, Kuris AM (2002) Parasites and marine invasions. Parasitology 124: S137–S151.
44. Jauniaux T, Petitjean D, Brenez C, Borrens M, Borrens L, et al. (2002) Postmortem findings and causes of death of harbour porpoises (*Phocoena phocoena*) stranded from 1990 to 2000 along the coastlines of Belgium and northern France. J Comp Pathol 126: 243–253.

45. Fauquier DA, Kinsel MJ, Dailey MD, Sutton GE, Stolen MK, et al. (2010) Prevalence and pathology of lungworm infection in bottlenose dolphins Tursiops truncatus from southwest Florida. Dis Aquat Org 88: 85–90.

46. Cornaglia E, Rebora L, Gili C, Guardo G (2000) Histopathological and immunohistochemical studies on cetaceans found stranded on the coast of Italy between 1990 and 1997. J Vet Med A 47: 129–142.

47. Van Dijk JE, Gruys E, Mauwen JMVM (2007) Colour atlas of veterinary pathology. Spain: Sauderns Elsevier. 200 p.

48. Bossart GD, Reidarson TH, Dierauf LA, Duffield DA (2001) Clinical pathology. In: Dierauf LA, Gulland FMD, editors. CRC Handbook of marine mammal medicine. Boca Raton: CRC Press Inc. pp.383–436.

49. Raga JA, Fernández M, Balbuena JA, Aznar FJ (2009) Parasites. In: Perrin WF, Würsig B, Thewissen JGM, editors. Encyclopedia of Marine Mammals (second edition). London: Academic Press. pp.821–830.

50. King DP, Aldridge BM, Kennedy-Stoskopf S, Scott JT (2001) Immunology. In: Dierauf LA, Gulland FMD, editors. CRC Handbook of marine mammal medicine. Boca Raton: CRC Press Inc. pp.237–252.

51. Caswell JL, Williams KJ (2007) Lungs. In: Maxie MG, editor. Jubb, Kennedy & Palmer's pathology of domestic animals (fifth edition). Edinburgh: Elsevier. pp.540–575.

52. Domingo M, Visa J, Pumarola M, Marco AJ, Ferrer L, et al. (1992) Pathologic and immunocytochemical studies of morbillivirus infection in striped dolphins (Stenella coeruleoalba). Vet Pathol 29: 1–10.

53. Kennedy S (1998) Morbillivirus infections in aquatic mammals. J Comp Pathol 119: 201–225.

54. Lipscomb TP, Kennedy S, Moffett D, Krafft A, Klaunberg BA, et al. (1996) Morbilliviral epizootic in bottlenose dolphins of the Gulf of Mexico. J Vet Diagn Invest 8: 283–290.

55. Aznar FJ, Fognani P, Balbuena JA, Pietrobelli M, Raga JA (2006) Distribution of Pholeter gastrophilus (digenea) within the stomach of four odontocete species: The role of the diet and digestive physiology of hosts. Parasitology 133: 369–380.

56. Geraci JR, St. Aubin DJ (1987) Effects of parasites on marine mammals. Int J Parasitol 17: 407–414.

57. Venter K (2009) Diet of humpback dolphins (Sousa plumbea) along the southeastern coast of South Africa.

58. Kaiser SML (2012) Feeding ecology and dietary patterns of the Indo-Pacific bottlenose dolphin (Tursiops aduncus) incidentally caught in the shark nets off KwaZulu-Natal, South Africa. MSc thesis. Port Elizabeth: Nelson Mandela Metropolitan University. 68 p.

59. Daily M, Stroud R (1978) Parasites and associated pathology observed in cetaceans stranded along the Oregon coast. J Wildl Dis 14: 503–511.

60. Munday BL, Mason RW, Hartley WJ, Presidente PJ, Obendorf D (1978) Sarcocystis and related organisms in Australian wildlife: survey findings in mammals. J Wildl Dis 14: 417–433.

61. Resendes AR, Juan-Salls C, Almeria S, Maj N, Domingo M, et al. (2002) Hepatic sarcocystosis in a striped dolphin (Stenella coeruleoalba) from the Spanish Mediterranean coast. J Parasitol 88: 206–209.

62. Lehnert K, Seibel H, Hasselmeier IWP, Iversen M, Nielsen NH, et al. Change in parasite burden and associated pathology in harbour porpoises (Phocoena phocoena) in west Greenland. In Press.

63. Brown CC, Baker DC, Barker IK (2007) Peritoneum and retroperitoneum. In: Maxie MG, editor. Jubb, Kennedy & Palmer's pathology of domestic animals (fifth edition). Edinburgh: Elsevier. pp.279–296.

64. Cockcroft VG, Ross GJB (1990) Food and feeding of the Indian Ocean bottlenose dolphin off southern Natal, South Africa. In: Leatherwood S, Reeves RR, editors. The bottlenose dolphin.San Diego : Academic Press. pp.295–308.

65. Van Bressem MF, Waerebeek K, Jepson PD, Raga JA, Duignan PJ, et al. (2001) An insight into the epidemiology of dolphin morbillivirus worldwide. Vet Microbiol 81: 287–304.

66. Di Guardo G, Proietto U, Francesco CE, Marsilio F, Zaccaroni A, et al. (2010) Cerebral toxoplasmosis in striped dolphins (Stenella coeruleoalba) stranded along the Ligurian Sea coast of Italy. Vet Pathol 47: 245–253.

67. Alton GG, Jones LM, Pietz DE (1975) Laboratory techniques in brucellosis. Geneva: World Health Organization. 80 p.

68. Foster G, MacMillan AP, Godfroid J, Howie F, Ross HM, et al. (2002) A review of Brucella sp. infection of sea mammals with particular emphasis on isolates from Scotland. Vet Microbiol 90: 563–580.

69. Dubey JP, Fair PA, Bossart GD, Hill DFR, Sreekumar C, et al. (2005) A comparison of several serologic tests to detect antibodies to Toxoplasma gondii in naturally exposed bottlenose dolphins (Tursiops truncatus). J Parasitol 91: 1074–1081.

70. Calzada N, Lockyer C, Aguilar A (1994) Age and sex composition of the striped dolphin die-off in the western Mediterranean. Mar Mamm Sci 10: 299–310.

71. Di Guardo G, Marruchella G, Agrimi U, Kennedy S (2005) Morbillivirus infections in aquatic mammals: a brief overview. J Vet Med A 52: 88–93.

72. Keck N, Kwiatek O, Dhermain F, Dupraz F, Boulet H, et al. (2010) Resurgence of morbillivirus infection in Mediterranean dolphins off the French coast. Vet Rec 166: 654–655.

73. Lipscomb TP, Schulman FY, Moffett D, Kennedy S (1994) Morbilliviral disease in Atlantic bottlenose dolphins (Tursiops truncatus) from the 1987–1988 epizootic. J Wildl Dis 30: 567–571.

74. Tsai M, You H, Tang Y, Liu J (2008) Shewanella soft tissue infection: case report and literature review. Int J Infect Dis 12: e119–124.

75. Higgens R (2000) Bacteria and fungi of marine mammals: a review. Canadian Vet J 41: 105–116.

76. Reif JS, Mazzoil MS, McCulloch SD, Varela RA, Goldstein JD, et al. (2006) Lobomycosis in Atlantic bottlenose dolphins from the Indian River Lagoon, Florida. J Am Vet Med Assoc 228: 104–108.

77. Kiszka J, Van Bressem MF, Pusineri C (2009) Lobomycosis-like disease and other skin conditions in Indo-Pacific bottlenose dolphins, Tursiops aduncus, from the Indian Ocean. Dis Aquat Org 84: 151–157.

78. Reif JS, Peden-Adams M, Romano TA, Rice CD, Fair PA, et al. (2009) Immune dysfunction in Atlantic bottlenose dolphins (Tursiops truncatus) with lobomycosis. Med Mycol 47: 125–135.

79. Turnbull BS, Cowan DF (1998) Myocardial contraction band necrosis in stranded cetaceans. J Comp Pathol 118: 317–327.

80. Atkins S, Cliff G, Pillay N. (2013) Humpback dolphin by-catch in the shark nets in KwaZulu-Natal, South Africa. Biol Conserv 159: 442–449.

81. Durham B (1994) The distribution and abundance of humpback dolphin (Sousa chinensis) along the Natal coast, South Africa. MSc thesis. Durban: University of Natal. 32 p.

82. Peddemors VM (1995) The aetiology of bottlenose dolphin capture in shark nets off Natal, South Africa. PhD thesis. Port Elizabeth: University of Port Elizabeth. 83 p.

83. Cockcroft VG (1992) Incidental capture of bottlenose dolphins (Tursiops truncatus) in shark nets: an assessment of some possible causes. J Zool 226: 123–134.

84. Moore MJ, Bogomolni AL, Dennison SE, Early G, Garner MM, et al. (2009) Gas bubbles in seals, dolphins, and porpoises entangled and drowned at depth in gillnets. Vet Pathol 46: 536–547.

85. Woodard JC, Zam SG, Caldwell DK, Caldwell MC (1969) Some parasitic diseases of dolphins. Vet Pathol 6: 257–272.

86. Zappulli V, Mazzariol S, Cavicchioli L, Petterino C, Bargelloni L, et al. (2005) Fatal necrotizing fasciitis and myositis in a captive common bottlenose dolphin (Tursiops truncatus) associated with Streptococcus agalactiae. J Vet Diagn Invest 17: 617–622.

87. KwaZulu-Natal Sharks Board (2011) Shark nets, drumlins and safe swimming. Available: http://shark.co.za. Accessed 2013 Jan 15.

88. Bernaldo de Quirós Y, González-Diaz O, Arbelo M, Sierra E, Sacchini S, et al. (2012) Decompression vs. decomposition: distribution, amount, and gas composition of bubbles in stranded marine mammals. Frontiers in Physiology 3 (177): 1–19.

89. Bernaldo de Quirós Y, González-Diaz O, Saavedra P, Arbelo M, Sierra E, et al. (2011) Methodology for in situ gas sampling, transport and laboratory analysis of gases from stranded cetaceans. Scientific Reports 1 (193): 1–10.

90. Maxie MG, Newman SJ (2007) Urinary System. In: Maxie MG, editor. Jubb, Kennedy & Palmer's pathology of domestic animals (fifth edition). Edinburgh: Elsevier. pp.433–436.

10

Identification of Paralogous Life-Cycle Stage Specific Cytoskeletal Proteins in the Parasite *Trypanosoma brucei*

Neil Portman[1,2]*, Keith Gull[1]

1 Sir William Dunn School of Pathology, University of Oxford, Oxford, United Kingdom, **2** Faculty of Veterinary Science, University of Sydney, Sydney, Australia

Abstract

The life cycle of the African trypanosome *Trypanosoma brucei*, is characterised by a transition between insect and mammalian hosts representing very different environments that present the parasite with very different challenges. These challenges are met by the expression of life-cycle stage-specific cohorts of proteins, which function in systems such as metabolism and immune evasion. These life-cycle transitions are also accompanied by morphological rearrangements orchestrated by microtubule dynamics and associated proteins of the subpellicular microtubule array. Here we employed a gel-based comparative proteomic technique, Difference Gel Electrophoresis, to identify cytoskeletal proteins that are expressed differentially in mammalian infective and insect form trypanosomes. From this analysis we identified a pair of novel, paralogous proteins, one of which is expressed in the procyclic form and the other in the bloodstream form. We show that these proteins, CAP51 and CAP51V, localise to the subpellicular corset of microtubules and are essential for correct organisation of the cytoskeleton and successful cytokinesis in their respective life cycle stages. We demonstrate for the first time redundancy of function between life-cycle stage specific paralogous sets in the cytoskeleton and reveal modification of cytoskeletal components *in situ* prior to their removal during differentiation from the bloodstream form to the insect form. These specific results emphasise a more generic concept that the trypanosome genome encodes a cohort of cytoskeletal components that are present in at least two forms with life-cycle stage-specific expression.

Editor: Ziyin Li, University of Texas Medical School at Houston, United States of America

Funding: This work was funded through a Wellcome Trust Principal Fellowship and a Wellcome Trust programme grant to KG. The funders had no role in study design, data collection and analysis, decision to publish, or preparation of the manuscript.

Competing Interests: The authors have declared that no competing interests exist.

* Email: neil.portman@sydney.edu.au

Introduction

The African trypanosome, *Trypanosoma brucei* is a single-celled obligate parasite that causes African sleeping sickness. It has a complex lifecycle with multiple distinct phases encompassing a passage through an insect vector - the Tsetse fly - and a mammalian host [1]. From the midgut of the insect vector, parasites migrate to and colonise the salivary glands. When the fly takes a bloodmeal, parasites with pre-adaptations to the mammalian bloodstream are transferred to the mammalian host. An extracellular bloodstream infection is established in the host which eventually results in the production of parasites competent for transfer back to a Tsetse fly. This life-cycle involves both proliferative and non-proliferative stages, each with distinct morphology that reflect the very different environments encountered by the parasite [1]. The most widely studied life cycle stages are the procyclic form from the Tsetse midgut and the long slender form that colonises the mammalian bloodstream.

The various lifecycle morphologies range from extremely long, slender cells to relatively short, broad cells and life-cycle transitions are, in many cases, accompanied by rearrangements of the relative position of nuclei and kinetoplast (the concatenated mitochondrial DNA). Such drastic changes in morphology are accomplished through asymmetric cytokinesis [2] or through differentiation of growth arrested cells [3] but all must be accomplished within the

continuous presence of the persistent microtubule cytoskeleton – a highly ordered array of subpellicular microtubules that underlies the plasma membrane.

The transition from the long slender bloodstream form to the procyclic form occurs via a growth arrested short stumpy bloodstream form [4]. This differentiation involves a range of metabolic and morphological changes including mitochondrial elaboration and exchange of the protective surface glycoprotein coat used in immune evasion. Although both cell types are morphologically trypomastigotes (i.e. the kinetoplast is positioned to the posterior of the nucleus and the flagellum is attached to the cell body for much of its length) bloodstream and procyclic forms differ in terms of the relative positions of the kinetoplast and the organisation of the nuclei and kinetoplasts during the cell cycle [1,5–9]. The kinetoplast is positioned much closer to the posterior pole of the cell in the bloodstream form and in dividing cells the nuclei and kinetoplasts are arranged in a posterior-K-K-N-N-anterior morphology as opposed to the posterior-K-N-K-N-anterior morphology adopted by procyclic form dividing cells.

Notwithstanding the differences in morphology between bloodstream and procyclic forms, at the ultrastructural level the organisation and appearance of the microtubules of the subpellicular corset are indistinguishable. However, a growing number of cytoskeletal components have been identified that have undergone gene duplication with the resultant paralogous genes showing life-cycle stage-specific expression in *T. brucei* [10–15]. These include components of the subpellicular corset such as CAP5.5 and CAP5.5V as well as proteins involved in the Flagellum Attachment Zone (FAZ), a specialised domain of the cytoskeleton that connects the flagellum to the cell body.

Recent studies have suggested that the *T. brucei* lifecycle is accompanied by changes to the cell proteome [16,17], but the extent and significance of the changes for certain cytoskeletal components remains unclear. We have performed a comparative proteomic analysis of isolated cytoskeletons from bloodstream and procyclic form cultures. We identified a novel pair of paralogous cytoskeletal proteins that show life-cycle stage specificity and have roles in the organisation of microtubules in the subpellicular corset in their respective life-cycle stages. We show that these two proteins exhibit redundancy of function when expressed in the exogenous life-cycle stage; the first demonstration of this phenomenon for *T. brucei* stage specific cytoskeletal components. Finally we provide evidence of protein modification prior to removal of one of these proteins during differentiation from the bloodstream form to the procyclic form.

Results

Candidate cytoskeletal proteins identified using comparative proteomics

We compared the protein composition of detergent extracted bloodstream form cells to that of detergent extracted procyclic form cells using Difference Gel Electrophoresis (DiGE) as previously described [18] (Figure 1A, Figure S1). A total of 36 spots showing a difference in density between the two samples were cut from the gel for mass spectrometric analysis. This yielded 71 identifications based on at least two unique peptides and with a confidence interval greater than 99%, forming a non-redundant candidate set of 49 proteins. Both CAP5.5 and CAP5.5V were present in this set and showed greater abundance in samples from procyclic forms and bloodstream forms respectively. A total of 18 proteins were identified that are annotated as hypothetical at TriTrypDb.

Two of the hypothetical proteins identified - Tb927.7.2640 and Tb927.7.2650 - are encoded by open reading frames that are adjacent to one another on chromosome 7 and that are related but not identical.

Gel spots corresponding to Tb927.7.2640 appeared most strongly in procyclic form samples and spots corresponding to Tb927.7.2650 showed a greater density in bloodstream form samples. Tb927.7.2640 (procyclic form) has an apparent molecular weight of 51 kDa and hence we have named these proteins CAP51 (Tb927.7.2640) and CAP51V (Tb927.7.2650) in accordance with previous nomenclatures [11,12].

Two-way alignment of the protein sequences showed high similarity for most of the length but with a region of very low similarity towards the N terminus (Figure 1B). This region is around 120 residues long in CAP51 and 195 residues long in CAP51V and in the latter case includes a complex lysine rich repetitive element. The 5′UTRs of the two genes are identical for around 200 bp upstream of the open reading frames but no significant similarity could be found between the 3′UTRs. Alignment to the NCBI non-redundant protein database by BLASTP showed that homologues to these proteins are restricted to kinetoplastids, with both *Leishmania major* (LmjF.22.0730) and *Trypanosoma cruzi* (Tc00.1047053506859.170) genomes encoding a single homologue. Querying either CAP51 or CAP51V protein sequences against the Pfam database revealed no known domains

or motifs. Both proteins are predicted to form coiled-coils in the C-terminal conserved regions (via COILS, http://embnet.vital-it.ch/software/COILS_form.html).

CAP51 proteins localise to the subpellicular microtubule corset

We introduced a C terminal YFP::Ty tag into one of the endogenous alleles for both proteins individually in both bloodstream and procyclic form cells, preserving the endogenous 3′UTRs of the genes and any regulatory signals contained therein. Western blot analysis of whole cell extracts from each of these cell lines using the anti-Ty tag monoclonal antibody BB2 (Figure 1C) confirmed that CAP51 is expressed in procyclic forms but is not expressed above the level of detection in bloodstream forms. CAP51V is expressed in bloodstream forms but is not detected in procyclic forms. Analysis of native YFP fluorescence in whole cells and detergent extracted cells with co-staining for the marker protein WCB [19] showed that both proteins localise to the subpellicular corset in their respective life cycle stages (Figure 1D, E). The YFP signal in both lifecycle stages decorated the entire subpellicular corset evenly, with the exception of the extreme posterior end of procyclic form cells, and persisted throughout the cell cycle. No YFP signal was detected in the flagellum in either life cycle stage.

CAPs during differentiation

To examine the expression of CAP proteins during differentiation, we incubated the CAP51V::YFP::Ty and CAP51::YFP::Ty monomorphic bloodstream form cell lines at 28°C in the presence of 6 mM cis-aconitate, a treatment that has been shown to induce differentiation to procyclic forms in culture conditions [20].

CAP51V::YFP::Ty showed uniform fluorescence across the microtubule corset - as described above - which persisted through 24 and 48 hours after exposure to cis-aconitate but at a steadily decreasing intensity (Figure 2A, all images captured and processed using identical parameters). Western blot analysis of detergent extracted cells with BB2 (Figure 2D) showed that CAP51-V::YFP::Ty was present as a single band of the correct apparent molecular weight in undifferentiating bloodstream forms. Surprisingly, by 24 hours after induction a second, larger band appeared and by 48 hours only the higher molecular weight band was detectable. By 72 hours no CAP51V::YFP::Ty was detectable. The apparent difference in molecular weight between the two CAP51V bands was approximately 10 kDa.

As before, CAP51::YFP::Ty was not detected before exposure to differentiation conditions. However, by 24 hours after induction of differentiation, fluorescence at the posterior end of cells was observed (Figure 2B). By 48 hours after induction of differentiation, cells with uniform fluorescence over the whole cytoskeleton with the exception of the posterior end, comparable to that seen in cultured procyclic forms, were observed (Figure 2C). CAP51::YFP::Ty was undetectable by Western blot analysis using BB2 at early time points but appeared between 16 and 24 hours after differentiation was induced (Figure 2E). This was slightly later than the appearance of CAP5.5 that was first detected 16 hours after induction of differentiation.

Ablation of CAP proteins leads to aberrant morphology and compromised cytokinesis

We generated cell lines in both bloodstream and procyclic form cells with doxycycline inducible RNAi against each CAP mRNA individually, targeting the region of low similarity near the 5′ end of each coding sequence. Population doubling times were

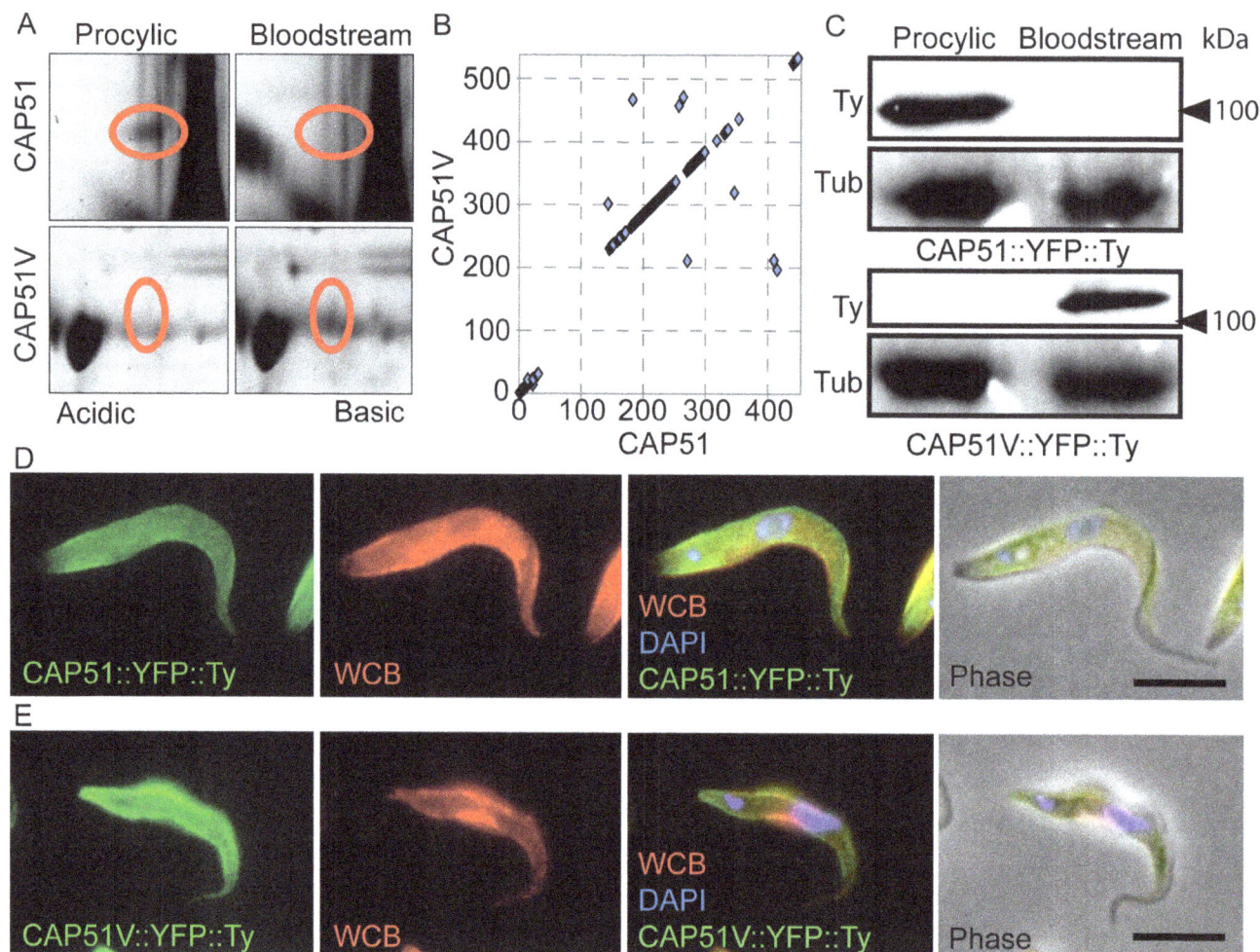

Figure 1. CAP51 and CAP51V are paralogous lifecycle stage-regulated cytoskeletal components. (A) Sections from a 2D-DiGE comparison of detergent extracted Bloodstream forms and Procyclic forms (Full gels in Figure S1) showing views of the gel regions corresponding to CAP51 and CAP51V. The spot corresponding to CAP51 is present in samples from procyclic form only and the spot corresponding to CAP51V has greater density in bloodstream form samples. (B) Dotplot alignment of CAP51 (x axis) and CAP51V (y axis) protein sequences using a 4 residue window. CAP51V contains a 75 residue lysine rich insert that is not present in CAP51. (C) Western blot of detergent extracted bloodstream and procyclic form cells expressing YFP::Ty tagged CAP51V or CAP51. CAP51V::YFP::Ty is detected only in the bloodstream form and CAP51::YFP::Ty only in the procyclic form. A ponceau stain of the membrane showing the tubulin (Tub) region is included as an indication of relative loading. (D) Procyclic form cells expressing CAP51::YFP::Ty. CAP51 localises to the whole of the subpellicular corset apart from the extreme posterior end of the cell and is not present on the flagellum. (E) Bloodstream form cells expressing CAP51V::YFP::Ty. CAP51V also localises to the subpellicular corset with no signal detected on the flagellum. Red = whole cell body (WCB), blue = DAPI, bar = 5 μm.

determined in both life-cycle stages in the presence and absence of RNAi against each protein. RNAi against either CAP resulted in a decrease in population growth rate compared to non-induced cells in the corresponding life cycle stage (Figure 3A, B). Cells with abnormal numbers of nuclei and kinetoplasts, including monstrous multinucleate cells (greater than 2N) and anucleate zoids, began to accumulate by 72 hours after induction of RNAi against CAP51 in procyclic forms (Figure 3C&D). In bloodstream forms, cells undergoing aberrant cytokinesis (Figure 3E) were evident by 24 hours after induction of RNAi against CAP51V and cytoplasts with no detectable DNA content appeared. In the reciprocal experiments, RNAi against CAP51V in the procyclic form or CAP51 in the bloodstream form had no effect on population growth rates or cell morphologies (not shown). Addition of a YFP::Ty tag to CAP51 in the CAP51 RNAi background showed that by 48 hours after RNAi induction, protein was lost preferentially in the posterior portion of the cell (Figure 3F, G).

Loss of CAP protein did not affect the distribution of the cytoskeletal markers CAP5.5 in procyclic form or WCB in either life cycle stage. CAP51 ablation in procyclic forms resulted in cells exhibiting distorted morphology such that the diameter of the cell at the midpoint was abnormally large by 48 hours after induction compared to the rest of the cell body.

Examination of bloodstream form cells 24 hours after induction of RNAi against CAP51V by thin section electron microscopy showed that the tight organisation of the subpellicular array of microtubules was lost as a result of RNAi (Figure 4). In wild-type cells the subpellicular microtubules form a regularly spaced single layer array closely apposed to the plasma membrane. In CAP51V-depleted bloodstream form cells, sections that contained a nucleus (i.e. around the mid-point of the cell, n = 72) showed disruptions to the subpellicular array with individual microtubules displaced along an axis orthogonal to the plane of the array (Figure 4, A–D). These perturbations in the microtubule array occurred in short

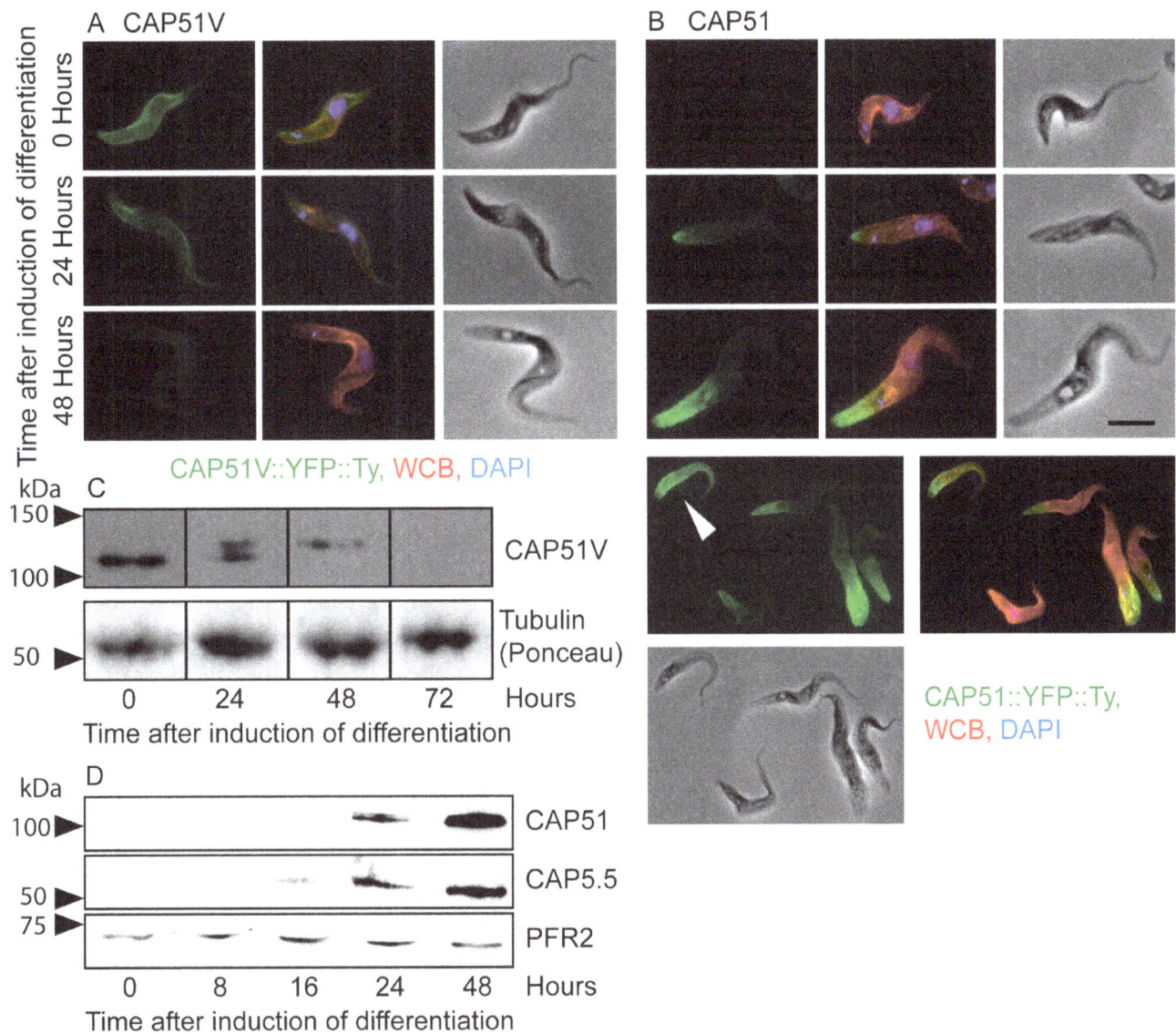

Figure 2. CAP proteins during differentiation Monomorphic bloodstream form cells with CAP51V::YFP::Ty or CAP51::YFP::Ty expressed from an endogenous locus were induced to differentiate into procyclic forms by the addition of cis-aconitate and incubation at 28°c. (A) CAP51V signal (green) disappears uniformly across the cell body during differentiation. (B) CAP51 signal (green) intrudes from the posterior end of the cell during the course of the differentiation and (C) cells with uniform fluorescence are detected by 48 hours after induction (white arrowhead). Red = WCB, blue = DAPI, bar = 5 µm. (D) Western blot of CAP51V::YTFP::Ty on detergent extracted cells using BB2. A single band is visible in bloodstream forms but over the time course a second band approximately 10 kDa larger than the first appears. Neither form of CAP51V is detectable by 72 hours after induction. Tubulin (ponceau) is shown as an indication of relative loading. (E) Western blot of CAP51::YFP::Ty on detergent extracted cells using BB2. CAP51 is undetectable in bloodstream form cells. A band of the correct predicted molecular mass appears by 24 hours, slightly later than the first detection of CAP5.5 (detected with the CAP5.5 antibody) at 16 hours. PFR2 (L8C4) is shown as an indication of relative loading.

runs with other areas of the array in the same cross-section appearing normal. In sections with an associated flagellar profile but no nucleus (i.e. an anterior position in the cell, n = 56), the disorganisation of the microtubule array was more pronounced with multiple layered sheets of microtubules apparent (Figure 4, E–H). Disorganisation of the microtubule array was observed rarely in cross sections without an associated flagellar profile (i.e. posterior to the flagellar pocket, n = 76).

CAP functions are complementary

Phenotypes associated with RNAi ablation of CAP proteins were similar in the two life-cycle stages suggesting that the two proteins have similar functions. We expressed Ty::GFP::CAP51V in procyclic forms and Ty::GFP::CAP51 in bloodstream forms from an inducible promoter. In both cases cells grew normally after induction and the ectopic protein localised to the sub-pellicular array (Figure 5A). When RNAi of the endogenous protein and ectopic expression of the exogenous protein were induced simultaneously, the growth rate/cell division defects were rescued in both life cycle stages (Figure 5B, C) and the subpellicular array appeared normal by TEM analysis (not shown). Simultaneous induction of RNAi against the endogenous protein and ectopic expression of the RNAi target protein failed to rescue the phenotype in either life-cycle stage (not shown).

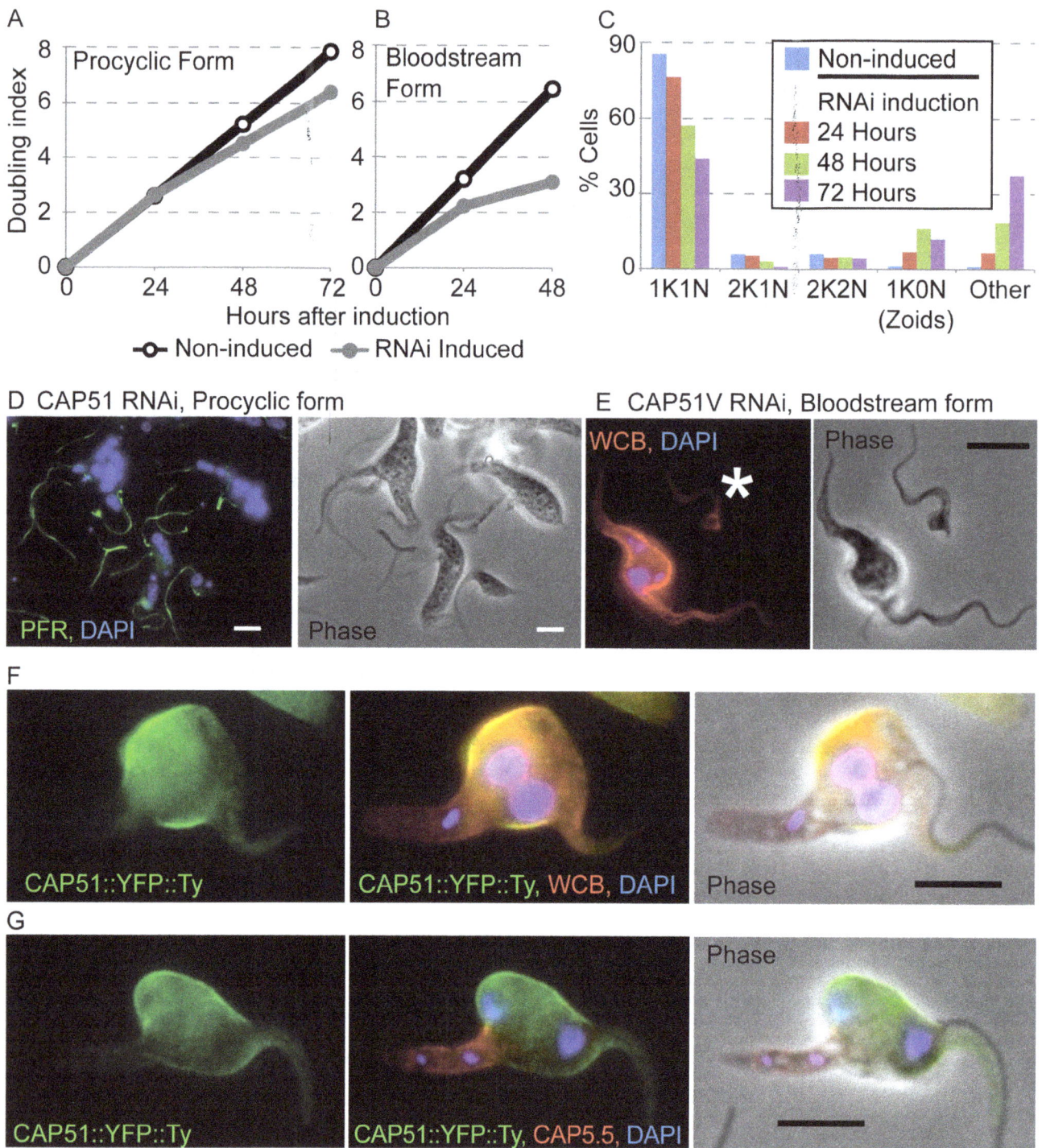

Figure 3. Ablation of CAP51 and CAP51V leads to reduced growth rate and aberrant morphology. (A, B) RNAi mediated ablation of CAP51 in procyclic forms (A) and CAP51V in bloodstream forms (B) results in reduced growth rate. Representative plots from three replicates are shown. Grey, RNAi induced; black non-induced. (C) Ablation of CAP51 in procyclic forms results in an accumulation of cells with aberrant nucleus/kinetoplast numbers including the production of 1K0N zoids. Cells with multiple nuclei (i.e. greater than 2) are classified as "Other". Representative counts from three replicates are shown. (D) 72 hours after induction of RNAi against CAP51 in procyclic forms, multinucleate cells with multiple kinetoplasts and flagella and 1K0N zoids can readily be observed, green = PFR (L8C4). (E) RNAi against CAP51V in the bloodstream form. Cells undergo aberrant cytokinesis and cytoplasts that contain no nuclear or kinetoplast DNA (white asterisk) are also observed. Red = WCB. (F, G) RNAi against CAP51 in the procyclic form. CAP51 (green) is lost preferentially in the posterior portion of the cell which remains positive for other markers of the subpellicular corset (F) WCB (red) and (G) CAP5.5 (red). Cell morphology is distorted with the diameter of the midpoint of the cell being abnormally large compared to the anterior and posterior. (D–G) Blue = DAPI, bar = 5 μm.

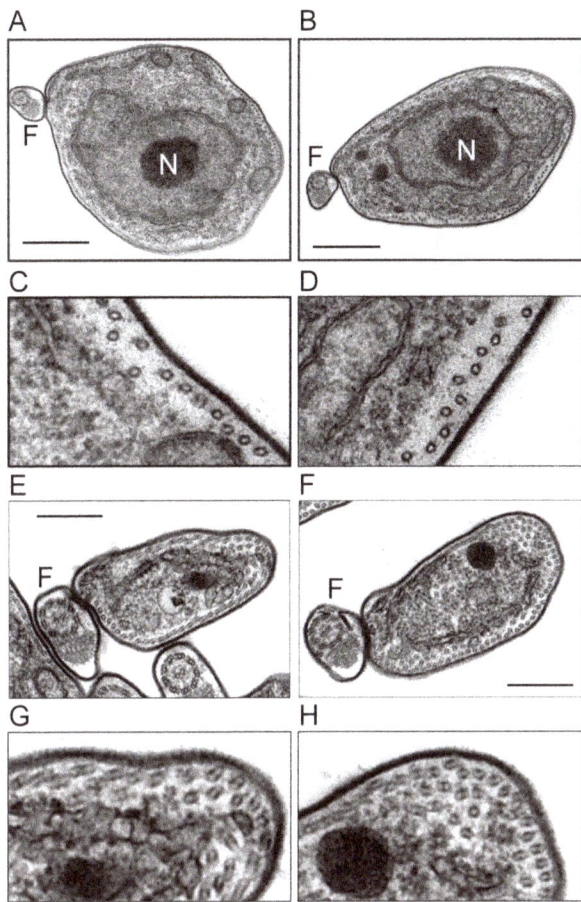

Figure 4. Ablation of CAP proteins disrupts the organisation of the microtubule corset. Thin section TEM of transverse sections through bloodstream form trypanosomes 24 hours after induction of RNAi against CAP51V. (A, B) The microtubules of the subpellicular array become disordered at the midpoint of the cell. Bar 800 nm (C, D) higher magnification view of part of the subpellicular array of A and B respectively. (E, F) Multiple layers of microtubules are present at the anterior end of the cell. Bar 400 nm (G, H) higher magnification view of part of the subpellicular array of E and F respectively. Similar disruptions to the organisation of the subpellicular corset can be seen in procyclic form sections 48 hours after induction of RNAi (not shown). N = Nucleus, F = flagellum.

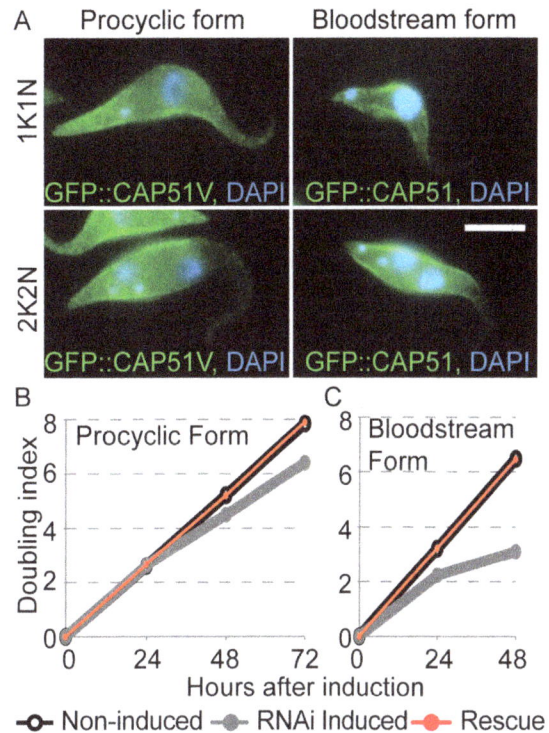

Figure 5. Localisation of CAP proteins to the subpellicular microtubule array is not life-cycle stage dependant. A) Ty::GFP::CAP51V and Ty::GFP::CAP51 expressed from inducible ectopic loci for 24 hours localise to the subpellicular array of procyclic form and bloodstream form cells respectively. Expression of the ectopic protein does not disrupt positioning of the nuclei and kinetoplasts of 2K2N cells in either life-cycle stage. Bar = 5 µm. (C, D) Simultaneous induction of RNAi against endogenous protein and expression of ectopic protein rescues the parental growth defect. (C) Ablation of CAP51 in procyclic forms reduces growth rate (grey) that is rescued by simultaneous expression of CAP51V (red). (D) Ablation of CAP51V in the bloodstream form reduces growth rate (grey) that can be rescued by simultaneous expression of CAP51 (red). Representative plots from three independent replicates are shown. Growth rates of non-induced cells (black) are comparable to that of rescued cells and expression of the endogenous protein with an N terminal GFP tag from an inducible ectopic locus did not rescue the growth defect (not shown).

Discussion

Paralogous cytoskeletal proteins with life-cycle stage-specificity

A growing number of cytoskeleton-associated proteins have been identified that show life-cycle stage-specific expression in *T. brucei*. Expression of CAP5.5 [10] and the related CAP5.5V [11] is restricted to the procyclic form and the bloodstream form respectively. Ablation of either protein in its endogenous life-cycle stage resulted in the accumulation of cells with abnormal numbers of nuclei and/or kinetoplasts, particularly the 1K0N cytoplasts known as zoids. The similarity in phenotypes suggests that CAP5.5 and CAP5.5V play analogous roles in their respective life-cycle stages. Another pair of related cytoskeleton associated proteins, CAP15 and CAP17, have been shown to stabilise microtubules and exhibit regulated expression in the life cycle [12]. Both proteins have analogous localisations at the anterior end of the subpellicular corset and overexpression of either in procyclic forms

resulted in a similar organelle positioning/cytokinesis defect phenotype.

There is now evidence that several components of the FAZ are present as paralogues in the genome and exhibit life-cycle regulated expression profiles. FLA1, an essential component of the FAZ, is significantly up-regulated in the procyclic form [21] whereas the recently identified FLA2, which is highly similar to FLA1 but contains a 44 residue proline rich insert [13], is significantly up-regulated in the bloodstream form. The transmembrane protein FLA3 also has a crucial role in flagellum attachment but only appears to be expressed in the bloodstream form [14]. A so-far unnamed protein, related to FLA3 and with similar domain architecture appears to be significantly upregulated in procyclic forms [14]. Our own data support both procyclic form specific expression and FAZ localisation for this protein [15]. Similarly as with FLA1/FLA2, a distinguishing feature of this pair is a 30 amino acid insert in the bloodstream form protein.

Using DiGE we identified numerous proteins that showed greater abundance in one or other of the two lifecycle stages analysed. These included both CAP5.5 and CAP5.5V that showed greater abundance in samples from the appropriate life-cycle stage (procyclic form and bloodstream form respectively [11]), hence validating our approach. Two pairs of hypothetical proteins in the dataset shared an orthologue group and of these CAP51 and CAP51V exhibited a pattern of life-cycle stage regulated expression. This pattern is supported by data from two recent global comparisons of protein expression in the two life-cycle stages [16,17]. The genomes of *L. major* and *T. cruzi* both encode a single CAP orthologue and BLASTP analysis detected no related proteins outside the *Kinetoplastida*.

Both proteins localise to the subpellicular microtubule array evenly throughout the cell cycle, with the exception of the extreme posterior end of procyclic form cells where no CAP51 was detected. Several studies have shown that there are differences in the regulation of microtubules at the posterior end of procyclic and bloodstream form cells, strikingly displayed by the nozzle phenotype in the procyclic form [22–24]. We speculate that this specific regulation of microtubule extension in procyclic forms requires the exclusion of certain cytoskeletal components from the posterior end of the cell.

CAP51V is modified during differentiation

During differentiation of monomorphic bloodstream form cells to procyclic form, CAP51 first appeared at the posterior end of cells between 16 and 24 hours after induction of differentiation and spread towards the anterior end of the cell to give a uniform distribution similar to that seen in cultured procyclic forms - a pattern consistent with that observed previously for CAP5.5 [25]. By contrast, CAP51V disappeared evenly across the whole cell body during differentiation. Analysis of the cells by Western blot showed that this was accompanied by an apparent increase in the molecular weight of a proportion of the protein by 24 hours with all remaining protein present at the higher molecular weight by 48 hours. This shift in mobility on the gel corresponds to an increase in molecular weight of approximately 10 kDa which is consistent with the approximate shift that would result from the addition of ubiquitin. Ubiquitination is associated with marking proteins for degradation or for relocation in the cell and this would certainly correlate with the observed behaviour of CAP51V during differentiation. Interestingly, the samples used for Western blot analysis were treated with detergent prior to electrophoresis of the insoluble fraction so the apparent modification of CAP51V must take place whilst it is still associated with the cytoskeleton.

CAP proteins function in the organisation of subpellicular microtubules

RNAi-mediated knockdown of both CAP proteins in their native life-cycle stages resulted in the accumulation of cells with aberrant morphologies. CAP proteins were lost first from the posterior end of cells after RNAi induction. During the cell cycle, cells become both longer and wider prior to cell division [9,26,27]. Length is increased by the extension of the subpellicular microtubules into the posterior end of the cell. With the exception of a specialised quartet of microtubules associated with the FAZ, all of the subpellicular microtubules are orientated with their more dynamic plus ends towards the posterior of the cell [28]. Width is increased by the intercalation of very short microtubules into the array which then extend between the existing microtubules [5,9]. In order for this to occur the existing intermicrotubule connections must be severed. These mechanisms may go some way to explain the pattern of loss of CAP proteins - within a single cell cycle, microtubules at the posterior of the cell extend in the absence of new protein. The apparent loss of protein at the posterior of the cell would therefore more properly be considered a loss of gain of protein. Although the loss of gain would also presumably be

occurring at the sites (or potential sites) of new microtubule intercalation [9,24,29], this may be disguised at the level of immunofluorescence analysis by the presence of old protein associated with the existing microtubule array. An opposing model would be that during the extensive remodelling associated with cell growth and division, cytoskeletal proteins are actively stripped from microtubules during microtubule elongation and intercalation and rapidly replaced. In this way an absence of new protein would be most readily observed at the dynamic posterior end of the cell. This pattern of protein loss is consistent with that previously described for RNAi against other cytoskeletal proteins such as CAP5.5 and WCB [11,19].

At time points prior to the appearance of large numbers of multi-nucleate cells, 2N cells exhibited a particular morphology where the diameter of the mid-portion of the cell appeared abnormally large. Our analysis of mutant cells by TEM showed that this morphology is accompanied by disruptions to the organisation of the subpellicular microtubule array, with a loss of the regular, ordered arrangement of microtubules into a continuous sheet seen in wild type cells. This loss of organisation is similar to that previously reported for RNAi against CAP5.5 and CAP5.5V [10,11]. Here it was suggested that the phenotype observed may be due to defects in the process of either severing or establishing the inter-microtubule-connections necessary for the intercalation and extension of new microtubules into the sheet during the cell cycle [5]. We hypothesise that CAP51/51V also play a role in inter-microtubule connections. The increase in cell width in the midportion of cells depleted for CAP51 is consistent with a model whereby short nucleating microtubules are incorporated into the array as normal but further intercalation of the growing ends as they extend away from the point of insertion is prevented. Our TEM observations of disruptions to the microtubular array supports the view that microtubules are forced to extend outside the plane of the array but do seem to maintain some connection to the existing microtubules. The appearance of layered sheets of microtubules at the anterior ends of the cell can then be explained as microtubules extending in different planes being constricted into the smaller cell body diameter at the anterior end of the cell.

Why does the cell require independently regulated versions of some cytoskeletal components?

Trypanosomes are unusual amongst eukaryotes in that genes are transcribed as polycistronic units [30]. In general genes do not contain introns and regulation of gene expression is accomplished post-transcriptionally. This means that all protein diversity must be encoded as discreet open reading frames (there is no opportunity for alternative splicing). We therefore hypothesised that two versions of CAP51 were present in the genome to address different functional requirements between life cycle stages. However, we found that expressing CAP51V in procyclic forms or CAP51 in bloodstream forms with simultaneous ablation of the endogenous protein completely rescued the RNAi phenotype. The ectopic protein localised to the subpellicular array and rescued cells exhibited normal morphologies. It is possible that over time small, initially undetectable aberrant phenotypes might accumulate although rescued cells have been kept in culture under induction for several weeks with no apparent detrimental effects. It is also possible that residual expression of the endogenous protein, bolstered by the presence of ectopic protein, is sufficient for normal cell functions. However, our data suggest that as well as being complementary when expressed ectopically, the proteins perform equivalent functions in their respective life-cycle stages. It appears that in terms of function within the cytoskeleton, these two proteins are essentially identical.

So why else might the cell require two independently regulated versions of the same protein? Differential CAP functions may be required under conditions to which the cells are not subjected in culture, or in life-cycle stages that were not examined in this study. Alternatively it may be that the regulation of expression itself that is important, i.e. it is the difference between the 3′UTRs that is critical. Being able to regulate each open reading frame individually may give the cell more fine control over gene expression than could be afforded by regulatory elements associated with a single open reading frame. Perhaps it is during the restructuring processes that occur in the life-cycle rather than those that occur in the individual cell cycles that such fine control becomes critical.

Conclusion

The organisation and regulation of the cytoskeleton are key factors in the success of *T. brucei*, driving the morphological adaptations to each new environment encountered by the parasite through its life cycle. The extent to which the molecular composition of the cytoskeleton is life-cycle stage dependant is gradually being explored. There is now evidence for a number of stage-specific proteins, encompassing both the microtubule array and even more specialised cytoskeletal structures such as the FAZ. That many of these proteins exist as paralogous sets likely reflects the unusual arrangement of the trypanosome genome with its almost complete lack of introns and alternative splicing. However, the reasons for encoding similar proteins with similar functions for stage specific expression remain unclear. As our knowledge of these paralogous sets improves, our understanding of the regulation of the cytoskeleton through the life cycle will also increase.

Materials and Methods

Plasmid construction

Primers used in this study are presented in Table S1. RNAi constructs were generated by amplification of 300–700 bp of the target open reading frame from genomic DNA and insertion between the XbaI sites of p2T7-177 [31]. For ectopic expression of recombinant proteins with an N terminal Ty::GFP tag from a constitutive rDNA promoter, the entire open reading frame of the gene of interest was amplified and inserted in frame between the XbaI and BamHI restriction sites of pDex577-G [32]. For the expression of chimeric proteins with a C terminal YFP::Ty tag from endogenous loci, 150–250 bp of in-frame sequence up to but not including the stop codon and 150–250 bp of the 3′UTR of genes of interest were amplified and inserted in this order between the HindIII and SpeI restriction sites of pEnT6B-Ty::YFP::Ty, separated by an XhoI restriction site. The entire intergenic region downstream of the gene of interest was amplified and inserted between the BamHI and SphI restriction sites to preserve any regulatory elements in the 3′UTR.

Trypanosome cell culture

Procyclic forms (427) were cultured at 28°C in SDM-79 medium supplemented with 10% v/v foetal bovine serum (Gibco) [33]. Bloodstream forms (927) were cultured at 37°C with 5% CO_2 in HMI-9 medium supplemented with 15% v/v foetal bovine serum (Gibco). Cells were diluted as necessary to maintain the culture in log-phase. Cell density was measured using a Casy-counter (Model TT; Sharfe systems).

Induction of RNAi or ectopic gene expression was achieved by the addition of doxycycline to the culture medium to a final concentration of 1 μg ml^{-1}.

Differentiation of monomorphic bloodstream forms to procyclic forms was induced by the addition of 6 mM cis-aconitate to the culture medium and transfer of the culture to 28°C.

Transmission Electron Microscopy

For TEM cells were fixed in culture as previously described [34]. Following embedding and sectioning, samples were viewed in an FEI Tecnai-F12 electron microscope (FEI Company Ltd.) operating at 80 kV.

Preparation of cells for light microscopy

Cells were harvested by centrifugation washed once in PBS and resuspended in PBS to a density of 1×10^7 cells ml^{-1}. Cells were then settled onto glass slides and washed in PBS.

For whole cells: cells were fixed with 3.7% w/v paraformalde-hyde for 20 minutes before permeabilisation by incubation in methanol for 20 minutes at −20°C.

For cytoskeletons: cells were washed once with 1% v/v Nonidet P40 (Sigma) in PEME and fixed by incubation in methanol for 20 minutes at −20°C.

Immunoglobulin isotype-specific secondary antibodies conjugated to either fluorophore 488 or fluorophore 594 (Alexafluor, Invitrogen) were used to visualise antibodies. Samples were mounted in Vectashield mounting medium with 4′, 6′-diamino-2-phenylindole (Vector Laboratories Inc) for visualisation of DNA and examined on a Leica DM5500B. Primary antibodies used in this study are as follows: BB2 – Ty epitope [35]; L8C4 – PFR2 [36]; WCB – Whole Cell Body [37]; CAP5.5 – CAP5.5 [10]. All image processing and analysis was performed in ImageJ (http://rsbweb.nih.gov/ij) using built in functions or custom scripts.

Protein electrophoresis

Protein samples were prepared as previously described [34]. For Western blot analysis either 5×10^6 (whole cell) or 1×10^7 (cytoskeleton) cell equivalents were separated by SDS PAGE.

DiGE comparison of protein samples and mass spectrometric identification of peptides were conducted as previously described [18].

Supporting Information

Figure S1   2D-DiGE comparison of detergent extracted (A) Bloodstream forms and (B) Procyclic forms. Spots that showed a twofold or greater change between the samples are circled. (C) and (D) show close up views of the gel regions corresponding to CAP51V and CAP51 respectively.

Table S1   PCR Primers used in this study.

Acknowledgments

The authors would like to thank Mike Shaw for technical assistance with electron microscopy.

Author Contributions

Conceived and designed the experiments: NP. Performed the experiments: NP. Analyzed the data: NP KG. Contributed reagents/materials/analysis tools: NP KG. Contributed to the writing of the manuscript: NP KG.

References

1. Vickerman K (1985) Developmental cycles and biology of pathogenic trypanosomes. Br. Med. Bull. 41: 105–114.
2. Sharma R, Peacock L, Gluenz E, Gull K, Gibson W, et al. (2008) Asymmetric cell division as a route to reduction in cell length and change in cell morphology in trypanosomes. Protist 159: 137–151.
3. Matthews KR (2005) The developmental cell biology of Trypanosoma brucei. J. Cell Sci. 118: 283–290.
4. Fenn K, Matthews KR (2007) The cell biology of Trypanosoma brucei differentiation. Curr. Opin. Microbiol. 10: 539–546.
5. Sherwin T, Gull K (1989) Visualization of detyrosination along single microtubules reveals novel mechanisms of assembly during cytoskeletal duplication in trypanosomes. Cell 57: 211–221.
6. Woodward R, Gull K (1990) Timing of nuclear and kinetoplast DNA replication and early morphological events in the cell cycle of Trypanosoma brucei. J. Cell Sci. 95 (Pt 1): 49–57.
7. Tyler KM, Matthews KR, Gull K (2001) Anisomorphic cell division by African trypanosomes. Protist 152: 367–378.
8. Vickerman K (1965) Polymorphism and mitochondrial activity in sleeping sickness trypanosomes. Nature 208: 762–766.
9. Wheeler RJ, Scheumann N, Wickstead B, Gull K, Vaughan S (2013) Cytokinesis in Trypanosoma brucei differs between bloodstream and tsetse trypomastigote forms: implications for microtubule-based morphogenesis and mutant analysis. Mol. Microbiol. 90: 1339–1355.
10. Hertz-Fowler C, Ersfeld K, Gull K (2001) CAP5.5, a life-cycle-regulated, cytoskeleton-associated protein is a member of a novel family of calpain-related proteins in Trypanosoma brucei. Mol. Biochem. Parasitol. 116: 25–34.
11. Olego-Fernandez S, Vaughan S, Shaw MK, Gull K, Ginger ML (2009) Cell morphogenesis of Trypanosoma brucei requires the paralogous, differentially expressed calpain-related proteins CAP5.5 and CAP5.5V. Protist 160: 576–590.
12. Vedrenne C, Giroud C, Robinson DR, Besteiro S, Bosc C, et al. (2002) Two related subpellicular cytoskeleton-associated proteins in Trypanosoma brucei stabilize microtubules. Mol. Biol. Cell 13: 1058–1070.
13. LaCount DJ, Barrett B, Donelson JE (2002) Trypanosoma brucei FLA1 is required for flagellum attachment and cytokinesis. J. Biol. Chem. 277: 17580–17588.
14. Woods K, Nic a'Bhaird N, Dooley C, Perez-Morga D, Nolan DP (2013) Identification and characterization of a stage specific membrane protein involved in flagellar attachment in Trypanosoma brucei. PLoS One 8: e52846 doi: 52810.51371/journal.pone.0052846.
15. Portman N (2011) Thesis: University of Oxford, Oxford. UK.
16. Butter F, Bucerius F, Michel M, Cicova Z, Mann M, et al. (2013) Comparative proteomics of two life cycle stages of stable isotope-labeled Trypanosoma brucei reveals novel components of the parasite's host adaptation machinery. Mol. Cell. Proteomics 12: 172–179.
17. Urbaniak MD, Guther ML, Ferguson MA (2012) Comparative SILAC proteomic analysis of Trypanosoma brucei bloodstream and procyclic lifecycle stages. PLoS One 7: e36619 doi: 36610.31371/journal.pone.0036619.
18. Portman N, Lacomble S, Thomas B, McKean PG, Gull K (2009) Combining RNA interference mutants and comparative proteomics to identify protein components and dependences in a eukaryotic flagellum. J. Biol. Chem. 284: 5610–5619.
19. Baines A, Gull K (2008) WCB is a C2 domain protein defining the plasma membrane - sub-pellicular microtubule corset of kinetoplastid parasites. Protist 159: 115–125.
20. Brun R, Schonenberger M (1981) Stimulating effect of citrate and cis-Aconitate on the transformation of Trypanosoma brucei bloodstream forms to procyclic forms in vitro. Z. Parasitenkd. 66: 17–24.
21. Koumandou VL, Natesan SK, Sergeenko T, Field MC (2008) The trypanosome transcriptome is remodelled during differentiation but displays limited responsiveness within life stages. BMC Genomics 9: 298.
22. Hendriks EF, Robinson DR, Hinkins M, Matthews KR (2001) A novel CCCH protein which modulates differentiation of Trypanosoma brucei to its procyclic form. EMBO J. 20: 6700–6711.
23. Hammarton TC, Engstler M, Mottram JC (2004) The Trypanosoma brucei cyclin, CYC2, is required for cell cycle progression through G1 phase and for maintenance of procyclic form cell morphology. J. Biol. Chem. 279: 24757–24764.
24. Sheriff O, Lim LF, He CY (2014) Tracking the Biogenesis and Inheritance of Subpellicular Microtubule in Trypanosoma brucei with Inducible YFP- alpha - Tubulin. Biomed. Res. Int. 2014: 893272.
25. Matthews KR, Gull K (1994) Evidence for an interplay between cell cycle progression and the initiation of differentiation between life cycle forms of African trypanosomes. J. Cell Biol. 125: 1147–1156.
26. Rotureau B, Subota I, Bastin P (2011) Molecular bases of cytoskeleton plasticity during the Trypanosoma brucei parasite cycle. Cell. Microbiol. 13: 705–716.
27. Farr H, Gull K (2012) Cytokinesis in trypanosomes. Cytoskeleton (Hoboken) 69: 931–941.
28. Robinson DR, Sherwin T, Ploubidou A, Byard EH, Gull K (1995) Microtubule polarity and dynamics in the control of organelle positioning, segregation, and cytokinesis in the trypanosome cell cycle. J. Cell Biol. 128: 1163–1172.
29. Sherwin T, Gull K (1989) The cell division cycle of Trypanosoma brucei brucei: timing of event markers and cytoskeletal modulations. Philos. Trans. R. Soc. Lond. B Biol. Sci. 323: 573–588.
30. Siegel TN, Gunasekera K, Cross GA, Ochsenreiter T (2011) Gene expression in Trypanosoma brucei: lessons from high-throughput RNA sequencing. Trends Parasitol. 27: 434–441.
31. Wickstead B, Ersfeld K, Gull K (2002) Targeting of a tetracycline-inducible expression system to the transcriptionally silent minichromosomes of Trypanosoma brucei. Mol. Biochem. Parasitol. 125: 211–216.
32. Kelly S, Reed J, Kramer S, Ellis L, Webb H, et al. (2007) Functional genomics in Trypanosoma brucei: a collection of vectors for the expression of tagged proteins from endogenous and ectopic gene loci. Mol. Biochem. Parasitol. 154: 103–109.
33. Brun R, Jenni L (1977) A new semi-defined medium for Trypanosoma brucei sspp. Acta Trop. 34: 21–33.
34. Broadhead R, Dawe HR, Farr H, Griffiths S, Hart SR, et al. (2006) Flagellar motility is required for the viability of the bloodstream trypanosome. Nature 440: 224–227.
35. Bastin P, Bagherzadeh Z, Matthews KR, Gull K (1996) A novel epitope tag system to study protein targeting and organelle biogenesis in Trypanosoma brucei. Mol. Biochem. Parasitol. 77: 235–239.
36. Kohl L, Sherwin T, Gull K (1999) Assembly of the paraflagellar rod and the flagellum attachment zone complex during the Trypanosoma brucei cell cycle. J. Eukaryot. Microbiol. 46: 105–109.
37. Woods A, Baines AJ, Gull K (1992) A high molecular mass phosphoprotein defined by a novel monoclonal antibody is closely associated with the intermicrotubule cross bridges in the Trypanosoma brucei cytoskeleton. J. Cell Sci. 103 (Pt 3): 665–675.

11

Nuclear Glycolytic Enzyme Enolase of *Toxoplasma gondii* Functions as a Transcriptional Regulator

Thomas Mouveaux[1], Gabrielle Oria[1], Elisabeth Werkmeister[1], Christian Slomianny[2], Barbara A. Fox[3], David J. Bzik[3], Stanislas Tomavo[1]*

1 Center for Infection and Immunity of Lille, CNRS UMR 8204, INSERM U 1019, Institut Pasteur de Lille, Université Lille Nord de France, Lille, France, **2** Laboratory of Cell Physiology, INSERM U 1003, Université Lille Nord de France, Villeneuve d'Ascq, France, **3** Department of Microbiology and Immunology, The Geisel School of Medicine at Dartmouth, Lebanon, New Hampshire, United States of America

Abstract

Apicomplexan parasites including *Toxoplasma gondii* have complex life cycles within different hosts and their infectivity relies on their capacity to regulate gene expression. However, little is known about the nuclear factors that regulate gene expression in these pathogens. Here, we report that *T. gondii* enolase TgENO2 is targeted to the nucleus of actively replicating parasites, where it specifically binds to nuclear chromatin *in vivo*. Using a ChIP-Seq technique, we provide evidence for TgENO2 enrichment at the 5' untranslated gene regions containing the putative promoters of 241 nuclear genes. Ectopic expression of HA-tagged TgENO1 or TgENO2 led to changes in transcript levels of numerous gene targets. Targeted disruption of TgENO1 gene results in a decrease in brain cyst burden of chronically infected mice and in changes in transcript levels of several nuclear genes. Complementation of this knockout mutant with ectopic TgENO1-HA fully restored normal transcript levels. Our findings reveal that enolase functions extend beyond glycolytic activity and include a direct role in coordinating gene regulation in *T. gondii*.

Editor: Laura J. Knoll, University of Wisconsin Medical School, United States of America

Funding: This work was supported by grants from the Centre National de la Recherche Scientifique (CNRS), Institut Médicale de la Recherche Scientifique (INSERM) and the Institut Pasteur de Lille (IPL) and Laboratoire d'Excellence (LabEx) ParaFrap from the National Agency for Research [PIA, ANR-11-LABX-0024] to ST. The funding organizations had no role in study design, data collection and analysis, decision to publish, or preparation of the manuscript.

* Email: Stan.Tomavo@pasteur-lille.fr

Introduction

Apicomplexan parasites are important pathogens of humans and domestic animals that cause diseases with a widespread impact on global health. The life cycles of these obligate intracellular parasites are complex, involving multiple proliferative and non-growing stages that ensure successful parasite transmission. Pathogenesis, virulence, and disease severity are critically influenced by asexual stage growth rates that can lead to increased parasite biomass and significant tissue destruction and inflammation. *Toxoplasma gondii* is distinct from nearly all other members of the phylum Apicomplexa in its exceptionally large host range, which includes all warm-blooded animals. Although the advent of acquired immune deficiency syndrome (AIDS) has drawn attention to *T. gondii* as a serious opportunistic parasite, it has long been a major medical and veterinary problem responsible for causing abortion or congenital birth defects in both humans and livestock [1]. The infection is incurable because of the parasite's ability to differentiate from rapidly replicating tachyzoite stages into latent cysts containing the bradyzoite stages that are impervious to the host's immune system and current therapeutic drugs. *T. gondii* cysts and dormant bradyzoites persist in the brain of the infected host and play key roles in pathogenesis because they can convert to virulent tachyzoites in immune-compromised individuals with AIDS and transplant patients. This stage conversion is triggered by the host immune response and impairment of the immune system in HIV-infected individuals can lead to lethal toxoplasmic encephalitis. The ability of *T. gondii* to cycle between one parasitic stage and another, a process known as interconversion, is central to its pathogenesis. However, very little is known about the mechanisms involved in stage interconversion, and the key nuclear factors that control *T. gondii* differentiation remains to be discovered. Even more intriguing, completion of sequencing and annotation of the genomes of *T. gondii* and other apicomplexan parasites revealed a relatively low number of genes encoding transcription factors [2–7]. In contrast, the basal core transcriptional machinery and the protein-coding genes involved in nucleosome assembly and chromatin remodelling machinery were found to be well-conserved, leading to the proposal that gene regulation in *T. gondii* and other apicomplexan parasites is controlled mainly by epigenetic mechanisms [8–10]. However, the complexity of the parasite life cycle suggests that other nuclear factors are likely to be involved in both basic and stage-specific regulation of gene expression in the apicomplexan parasites. Recently, bioinformatics searches for DNA-binding domains have uncovered a family of proteins homologous to the plant transcription factor Apetala2, named ApiAP2 for apicomplexan AP2-like factor [11–17]. In *T. gondii*, two recent studies have described the transcriptional control of genes encoding key factors involved in translation (ribosomal proteins), host cell invasion, and modulators of immune responses (kinases, pseudo kinases, or ROP proteins of the apical secretory organelles named rhoptries) by nuclear ApiAP2 factors [14,15]. Nevertheless, the balance between regulatory mechanisms and the relative roles of epigenetic and transcriptional machineries remain uncertain.

Our laboratory and others have previously established that *T. gondii* stage conversion is accompanied by the expression and nuclear localization of two stage-specific glycolytic enolases ENO1 (TgENO1) and ENO2 (TgENO2), also named 2-phospho-D-glycerate hydrolase (EC 4.2.1.11), that convert 2-phosphoglycerate to phosphoenolpyruvate in the glycolytic pathway of *T. gondii* [18–22]. TgENO1 is bradyzoite-specific enzyme only detected in the dormant encysted forms present in the brains of chronically infected mice while TgENO2 is exclusively expressed in the rapidly dividing and virulent tachyzoites [18–20]. During early intracellular proliferation and development, tachyzoites and bradyzoites exhibited very strong nuclear labelling for enolase, but in mature parasites this was markedly reduced to levels below those seen in the cytoplasm [21]. In addition, specific nuclear localization of TgENO2 has also been described in actively developing coccidian (both asexual and sexual) stages in a manner similar to the tachyzoites [21]. In mammalian cells, part of the enolase protein has been shown to function as a transcription factor that represses expression of the *c-myc* gene promoter [23,24]. In the plant *Arabidopsis thaliana*, which has no bona fide *c-myc* homolog, cold-responsive gene transcription is controlled by a bi-functional enolase that binds to the promoter of the gene encoding the zinc finger protein STZ/ZAT10 [25]. In both animal and plant cells, binding of nuclear enolases to the TATA motif is required for the control of transcription regulation [23–25]. However, in contrast to mammalian and plant cells, a survey of the whole genome of *T. gondii* or other apicomplexan parasites reveals no functional TATA and CCAAT boxes in their gene promoters. Therefore, the functions associated with nuclear localization of *T. gondii* enolases remain to be elucidated.

Here, we report that TgENO1 and TgENO2 are preferentially targeted to the nucleus of intracellular and actively dividing *T. gondii*, where they regulate gene expression. Targeted disruption of the TgENO1 gene resulted in a decrease in brain cyst burden in chronically infected mice and in changes of transcript levels of several nuclear gene targets in bradyzoites. Complementation of this knockout mutant with ectopic TgENO1-HA restored these transcripts to normal levels. Our data show that TgENO1 and TgENO2 are nuclear factor that share the capacity for binding to putative gene promoters and to control gene expression during intracellular proliferation of *T. gondii*.

Materials and Methods

Parasite culture and nuclear extract preparation

Tachyzoites of the avirulent 76 K (type II) strain [19,20] and the virulent *ΔKu80* (type I) strain [26], which was generously provided by Dr Vern Carruthers (University of Michigan, USA), were used in this study. These parasite strains were maintained *in vitro* by serial passage in confluent monolayers of human foreskin fibroblasts (HFF, from the American Type Culture Collection [ATCC]) grown in Dulbecco's modified Eagle's medium (Bio Whittaker) supplemented with 10% fetal calf serum (Gibco, BRL), 2 mM glutamine, and 0.05% gentamycin. Tachyzoites were allowed to grow until they lysed HFF cells spontaneously and were harvested by filtration through a glass wool column and a 3-μm pore filter. Encysted bradyzoites were purified from brains of

chronically infected OlaHsd mice by pepsin digestion (0.05 mg/ml pepsin in 170 mM NaCl, 60 mM HCl) for 30 minutes at 37°C. Nuclear extracts were obtained from approximately 4×10^8 purified tachyzoites of *T. gondii* 76K strain as previously described [22]. The quality of the nuclear extract was checked by Western blot analysis to confirm the absence of cytoplasmic contaminant (the glycolytic lactate dehydrogenase [27]) and enrichment of a *T. gondii* nuclear protein (TgDRE, a nuclear repair enzyme [28]).

Experimental infection in mice and ethics statement

All animal experiments were performed following the guidelines of the Pasteur Institute Pasteur of Lille animal study board, which conforms to the **A**msterdam **P**rotocol on animal protection and welfare, and **D**irective 86/609/EEC on the **P**rotection of **A**nimals **U**sed for **E**xperimental and **O**ther **S**cientific **P**urposes, updated in the **C**ouncil of **E**urope's **A**ppendix A (http://conventions.coe.int/Treaty/EN/Treaties/PDF/123-Arev.pdf). The animal work also complied with the French law (n°87-848 dated 19-10-1987) and the **E**uropean **C**ommunities **A**mendment of **C**ruelty to **A**nimals **A**ct 1976. All animals were fed with regular diet and all procedures were in accordance with national regulations on animal experimentation and welfare authorized by the French Ministry of Agriculture and Veterinary committee (Permit number: 59-009145). The Pasteur Institute of Lille and the CNRS Committee on the Ethics of Animal Experiments specifically approved this study. Purified tachyzoites (10^3 tachyzoites) from the parental 76K strain and the transgenic derivative ectopically expressing TgENO1-HA or TgENO2-HA were inoculated into groups of seven female 6- to 8-week-old CBA/J mice. For phenotypic studies of the TgENO1 knockout mutants, 5×10^2 tachyzoites from the Pru*Δku80ΔTgeno1* mutant and parental Pru*Δku80* strain were purified as described above and inoculated into a group of six female 6- to 8-week-old Balb/C mice. *In vivo* cyst formation was determined by harvesting mouse brain 6–8 weeks after infection. Cysts were purified using Percoll gradients as described above and washed with PBS. The cyst wall was stained by FITC-labelled *dolichol biflorus* lectin and counted under observation by inverted phase microscopy.

DNA manipulation and expression vector cloning

All primers used during this study and their purposes are described in Tables S1, S2 and S3. The open reading frames corresponding to TgENO1-HA and TgENO2-HA were amplified by polymerase chain reaction (PCR) using the forward and reverse primers indicated in Table S1. The forward primer contained a NsiI restriction site upstream of the coding sequence and the reverse primer contained a PacI restriction site downstream of the nucleotide sequence coding for the 12 amino acids of the HA epitope tag (CYPYDVPDYASL). PCR was performed as described previously [19] using the *T. gondii* 76K strain as a template, 50 pmol of each primer, and Pfu polymerase (Promega). The reaction conditions were 39 cycles of denaturation at 95°C for 1 minute, annealing at 66°C for 1 minute, and elongation at 72°C for 3 minutes, with a final elongation for 10 minutes at 72°C. PCR-amplified DNA fragments encoding TgENO1-HA and TgENO2-HA were double-digested with NsiI and PacI, gel-purified using the Gene clean spin kit (Q-Biogene) and cloned into the pSAG1-Bleo and pTUB5-Bleo vectors, respectively.

T. gondii transfection and stable transformation

For ectopic expression of *T. gondii* enolases, 100 μg of the TgENO1-HA and TgENO2-HA DNA constructs was linearized by *Kpn*I, purified by phenol/chloroform extraction, resuspended in 100 μl cytomix and transfected into 76K strain tachyzoites using a BTX electroporation system as previously described [22]. Stable transformants were selected and cloned in the presence of 5 µg/ml bleomycin. Phenotypic studies of transgenic parasites over-expressing TgENO1-HA or TgENO2-HA were performed by plaque assays and direct staining and counting of intracellular parasites. Briefly, 10^3 extracellular wild type or transgenic parasites were inoculated onto confluent HFF monolayers in 24-well plates for 2 hours at 37°C. Plaques were stained with crystal violet after 5–7 days in normal culture conditions. Experiments were repeated two or three times with triplicate wells. For the proliferation assay, 10^5 parasites per well were incubated in a 24-well plate for 18, 24, or 36 hours. Fixation and staining was carried out using the RAL555 kit (RAL Diagnostics, France). The number of parasites per vacuole was counted in 15 fields per slide with a minimum of three slides per condition.

Promoter studies

For promoter-driven expression, a plasmid containing the reporter luciferase gene and 787 bp of the 5′-untranslated region (UTR) containing the putative promoter of the TgMag1 gene (TGME49_070240) was constructed using the pair of primers described in Table S1. Two successive mutations of the parental TgMag1 plasmid were performed to generate $Δ_1$TgMag1 and $Δ_2$TgMag2 plasmids using the Quick-change Multi Site-Directed Mutagenesis Kit (Agilent). Briefly, the reaction mix was composed of 2.5 µl of 10× Quick-change Multi reaction buffer, 0.75 µl of Quick Solution, 50 ng of ds-DNA template, 100 ng of forward mutagenic primers, 100 ng of reverse mutagenic primers, 1 µl of dNTP mix, and 1 µl of Quick-Change Multi enzyme with double-distilled H_2O to a final volume of 25 µl. The mutagenesis reaction was performed by 29 cycles of 1 minute at 95°C, 1 minute at 55°C, and 2 minutes/kb of plasmid length at 65°C (13 minutes in this case). The PCR product was kept on ice for 2 minutes to cool the reaction and 1 µl of Dpn I restriction enzyme (10 U/µl) was added to digest the double-stranded DNA template. After digestion at 37°C for 1 hour, the mutated plasmid was used to transform E. coli and positive colonies were isolated.

Generation of knockout mutants using ENO1-targeting plasmid and complementation

The type II bradyzoite-specific ENO1 gene locus is defined by TGME49_068860 (Chromosome VIII, 6,242,505 to 6,244,976) in www.Toxodb.org (version 7.3). The plasmid pΔENO1 was constructed by yeast recombinational cloning using previously described methods [29]. A 1,003-bp 5′ ENO1 target flanking sequence amplified from type II Pru genomic DNA, the HXGPRT cDNA minigene cassette, and a 1,058-bp 3′ ENO1 target flanking sequence amplified from type II Pru genomic DNA were fused into the yeast shuttle plasmid pRS416. The deletion was engineered to remove 366 bp of the ENO1 5′-UTR, the entire ENO1 protein coding region, and 283 bp of the ENO1 3′-UTR. The oligonucleotide primers used to construct pΔENO1 and to validate targeted deletion of the ENO1 gene are shown in Tables S2 and S3, respectively.

For complementation, the ΔENO1 mutant was transfected with 30 µg of plasmid containing the coding region of ENO1 that was tagged with HA and flanked downstream by the heterologous promoter of the T. gondii sortilin (TgSORTLR) gene [30] and the 3′ UTR of the SAG1 gene. The linearized DNA was cleaned by ethanol precipitation and resuspended with 82 µl of Human T Cell Nucleofector solution plus 18 µl of buffer from the Amaxa Human T Cell Nucleofector Kit. A pellet containing $3×10^6$ ΔENO1 mutant cells was resuspended in the DNA solution and transfection was performed using the Amaxa electroporation

system and U33 program. After transfection, 10^5 parasites were loaded onto a cover slide containing a confluent monolayer of HFF for the immunofluorescence assay (IFA) and the remaining parasites were used to infect a 25-cm² flask containing confluent HFF cells, grown for 48 hours, and harvested for Western blot analysis. The intracellular parasites were cultured under tachyzoite growth (culture medium at pH 7) or bradyzoite growth (culture medium at pH 8) conditions.

Electrophoretic mobility shift assays

Electrophoretic mobility shift assays (EMSA) were performed using a band shift assay kit (Thermo Scientific, France) according to the manufacturer's instructions. Gel retardation assays were performed with 27-bp biotinylated primers containing the TTTTTCTTCTC motif present in the TgMag1 promoter (probe A), the human TATA box (probe B), or an irrelevant control sequence (probe C) (Sigma, France). The biotinylated double-stranded oligonucleotides (20 fmol) were incubated for 20 min at room temperature with 1 µg of purified bacterial TgENO2 recombinant protein in binding buffer (10 mM Tris-HCl at pH 7.5, 50 mM NaCl and 0.5 mM DTT) containing 1 µg Poly(dI-dC):Poly(dI-dC) (Thermo Scientific, France), 0.01% NP40, and 10% glycerol. For competition experiments, a 100-fold excess of self or unrelated unlabeled double-stranded oligonucleotides was added to the binding reaction together with the recombinant TgENO2 protein and before addition of the labelled probe. Loading dye (25 mM Tris-HCl pH 7.5, 0.02% bromophenol blue, 0.02% xylene, and 4% glycerol) was added and the complexes were separated on 6% non-denaturing polyacrylamide gels in running buffer (7 mM Tris-HCl at pH 7.5, 3 mM $NaC_2H_2O_2$ and 1 mM EDTA). The DNA was transferred onto a Nylon membrane and detected using the Light Shift Chemiluminescent EMSA Kit (Thermo Scientific, France).

Western blots

Total protein extracts and pulled down proteins were resuspended in 25 µl Laemmli sample buffer, boiled, separated by SDS-PAGE, and transferred to Hybond ECL nitrocellulose (Amersham). Immunoblot analysis was performed using specific monoclonal anti-SAG1 (P30), anti-SAG4 (P36) or polyclonal anti-TgENO1 and anti-TgENO2 antibodies. The specificity of the rabbit polyclonal anti-TgENO1 and anti-TgENO2 antibodies has been previously described [19,20]. The blots were incubated with peroxidase-conjugated secondary antibodies followed by chemiluminescence detection.

Chromatin immunoprecipitation and high-throughput sequencing

Chromatin immunoprecipitation (ChIP) was performed as described previously [31] with slight modifications. Briefly, chromatin from intracellular parasites (wild type 76K or ENO2-HA strain) grown in HFF cells (three 150-cm2 flasks) was cross-linked for 10 minutes with 1% formaldehyde at room temperature and purified as described [31]. After cross-linking, the extracts were sonicated to yield chromatin fragments of 500–1,000 bp. Immunoprecipitations were performed using polyclonal anti-ENO1 and anti-ENO2 antibodies and monoclonal or polyclonal anti-HA antibodies (Invitrogen). The immunoprecipitate was incubated at 4°C overnight and washed as described [31]. DNA was subjected to proteinase K digestion for 2 hours and purified using the Qiagen PCR purification kit. Pre-immune sera were used as a negative control. The primers used for PCR are listed in Table S1. ChIP PCR products were electrophoresed on agarose gels, stained with ethidium bromide, and photographed using a UV-light scanner.

For ChIP-Seq, chromatin was immunoprecipitated by anti-ENO2 antibody as above and amplified using a GenomePlex Amplification of ChIP DNA kit (Sigma-Aldrich). The amplified ChIP products were electrophoresed on agarose gels and stained with ethidium bromide. DNA fragments with a length of 500 bp or less were purified and processed for high-throughput sequencing (Genoscreen, Pasteur Institute of Lille). The GsFLX bead adaptors and specific tag (MID) were introduced to the flanking 5' and 3' end of each purified DNA sample according to the manufacturer's instructions. The ChIP-Seq GsFLX libraries were analyzed on a Bio analyzer 2100 using Agilent RNA 6000 Pico methods and quantified using Quant-iT TM RiboGreen (Invitrogen). Equimolar amounts of the libraries were mixed, fixed on beads, and amplified using the GS FLX Titanium emPCR Kit (454 Life Sciences, Roche Diagnostics). The beads were purified, enriched, counted using Beckman Coulter Z1, and deposited on the GS FLX Titanium Pico Titer Plate (454 Life Sciences, Roche Diagnostics). The pyro sequencing reaction was performed using a GS FLX Titanium Sequencing Kit and Genome Sequencer FLX Instrument (454 Life Sciences, Roche Diagnostics). For bioinformatics analyses, the GSMapper software v 2.3 (Roche) was used to align reads for each sample using the updated *T. gondii* ME49 genome databases downloaded from http://www.toxodb.org. Only sequences of a minimum overlapping length of 40 nucleotides with at least 90% identity were considered for further analyses. The ChIP-Seq data specific to TgENO2 or TgENO2-HA were subtracted from non-specific sequences obtained with the pre-immune sera used as negative control and data collected for specific sequences immunoprecipitated by anti-ENO2 and anti-HA antibodies were grouped into four sub-groups of associated contigs. Signal Map software was used to schematically represent the number of reads per contig and their corresponding positions, which allowed visualization of putative gene promoter occupancy by ENO2 for all chromosomes of *T. gondii*.

Quantitative real-time PCR and RT-PCR

Anti-ENO2 or anti-HA chromatin immunoprecipitates were subjected to qPCR to validate genes identified by ChIP-Seq or to quantify the level of chromatin that was specifically immunoprecipitated. For quantitative reverse transcriptase PCR (qRT-PCR), tachyzoites of wild-type parasites and parasites that over-expressed TgENO1-HA or TgENO2-HA were purified from infected HFF cells and RNA from 10^8 tachyzoites was reverse transcribed for 1 hour at 42°C in a buffer containing 1 M oligo $(dT)_{18}$ primer, 2 mM dNTPs, 40 U rRNasin (Promega) and 25 U AMV reverse transcriptase (Roche) for subsequent PCR amplification using the primers listed in Table S1. The *T. gondii* β-tubulin housekeeping gene was used as a negative control. We confirmed that each primer pair amplified a single product, identified as a single band in an acrylamide gel and as a single peak within the qPCR dissociation curve. The primer pairs displayed amplification efficiency greater than 90%. Quantitative PCR was performed with the Maxima TM SYBR Green qPCR Master Mix Kit (Fermentas) using the Mx3005P TM real-time PCR System (Stratagene). ROX Solution was used as a passive reference for all analyses and the qPCR was repeated three times, each in duplicate. Gene expression in TgENO1 and TgENO2 over-expressing parasites was represented as fold expression relative to wild-type after normalization.

Immunofluorescence assay and confocal imaging

The immunofluorescence assay (IFA) was performed as described previously [32]. Briefly, 2×10^5 intracellular tachyzoites were fixed with 4% paraformaldehyde in PBS for 15 minutes on ice, washed twice with PBS, and dried on Teflon slides. Intracellular parasites were permeabilized with 0.1% Triton X-100 in PBS containing 0.1% glycine for 10 minutes at room temperature. Samples were blocked with 3% BSA in the same buffer and then primary antibodies were added in the same buffer and incubated for 1 hour at 37°C. Secondary antibody coupled to Alexa-488 or Alexa-633 (Molecular Probes; diluted 1:1000) was added together with DAPI to stain the nucleus. The samples were examined with a Zeiss Axioplan microscope or by confocal imaging with a LSM710 microscope (Zeiss) and a Plan Apochromat objective (Plan-Apochromat 63x/1.40 Oil DIC M27, Zeiss) as previously described [31]. Quantification of TgENO2 signal in the nucleus and cytoplasm of the intracellular parasites was performed using 8-bit images (256 grey levels) that were acquired sequentially and averaged four times. Size was adapted to the optimal resolution that could be obtained by the confocal setup (0.13 μm per pixel). UV excitation (laser diode 405 nm) enabled imaging of the DAPI signal of parasite nuclei (blue image corresponding to spectral range 410–495 nm) and excitation at 561 nm allowed imaging of the TgENO2 signal (red fluorescence corresponding to spectral ranges 565–700 nm). Image analysis was performed using ImageJ (NIH) software. A short macro enabled us to automatically determine the relative proportion of TgENO2 signal in the nucleus versus the cytoplasmic signal. The cytoplasm of parasite was delimited by fluorescent background staining of anti-ENO2 while the DAPI enabled imaging parasite nuclei as indicated above. For each image, regions of interest (ROIs) delimiting the nucleus and cytoplasm were determined and the integrated intensity was calculated for each ROI as described (30). The proportion of signal contained in the nuclei corresponds to $100 \times$ (Nucleus intensity)/(Nucleus + Cytoplasm intensity).

Statistical analysis

Statistical differences between groups of mice used in this study were evaluated by the Student's t-test. Mann-Whitney test was used for analysis of intracellular growth of TgENO1-HA and TgENO2-HA *versus* wild type tachyzoites after colorimetric staining and microscopic observation and for mouse survival curves.

Results

TgENO2 is dynamically shuttled from the cytoplasm to the nucleus of actively dividing *T. gondii*

We previously reported that rabbit polyclonal anti-TgENO1 and anti-TgENO2 antibodies that are highly specific to ENO1 and ENO2 protein, respectively, and do not show cross-reactivity to their mammalian counterparts detect these two glycolytic enolase isoenzymes in the nuclei of bradyzoites and tachyzoites, respectively [19,20]. To rule out the possibility of cross-reaction with other unrelated parasitic nuclear proteins, we expressed ectopic Tg-ENO2 tagged with influenza hemagglutinin (HA) in the tachyzoites of 76K (type II) strain under the control of a strong tubulin gene promoter (Fig. 1A). Immunofluorescence assays (IFA) confirmed that the ectopic TgENO2-HA recognized by the rat polyclonal anti-HA antibody perfectly co-localized with endogenous TgENO2 and that the signal was strongly detected in the nuclei of intracellular parasites (Fig. 1A).

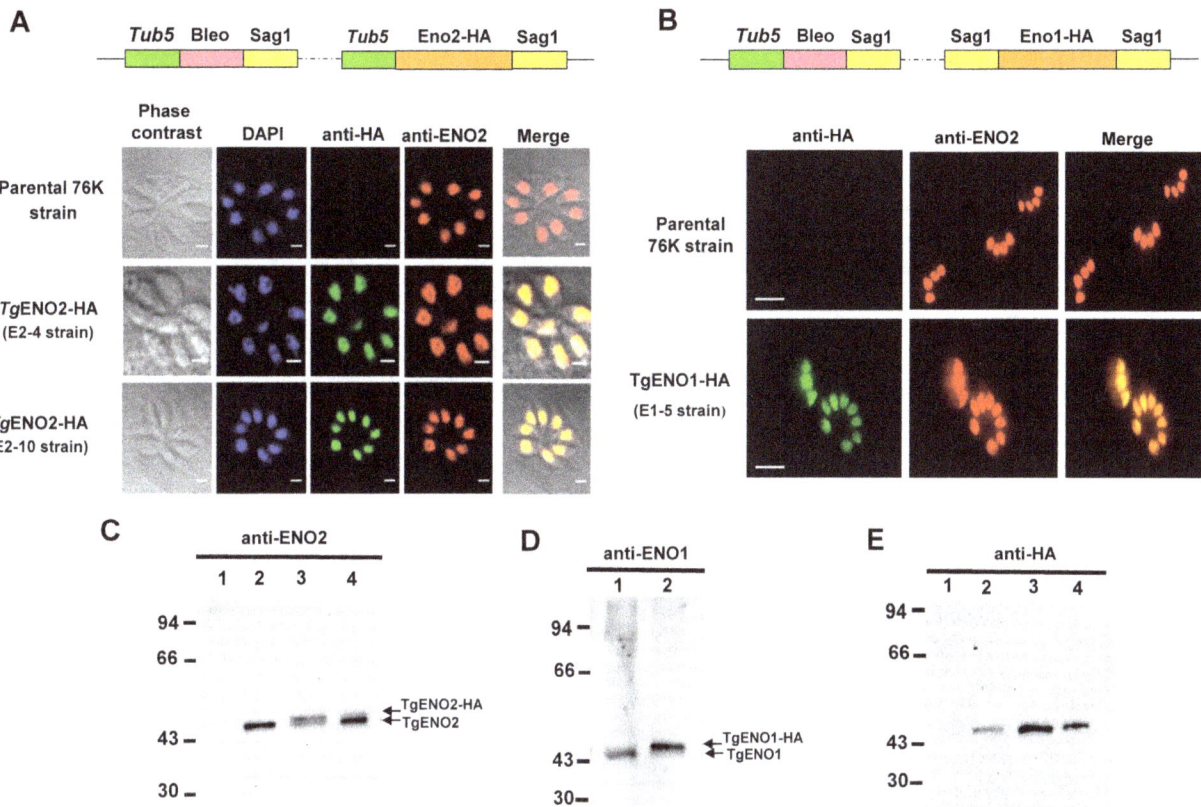

Figure 1. Ectopic expression of TgENO2-HA and TgENO1-HA in wild type *T. gondii* 76K tachyzoites confirms predominant nuclear enolase localization. A) Schematic representation of the vector used for stable ectopic expression of TgENO2-HA in parasite clones. Expression of TgENO2-HA protein driven by the heterologous tubulin gene promoters was demonstrated by co-localization of fluorescent images using anti-HA and anti-ENO2 antibodies. Similar to endogenous TgENO2 protein, ectopic TgENO2-HA was exclusively localized in the nucleus of two stable clones (E2-4 and E2-10) selected after bleomycin treatment. B) Schematic representation of the vector used for stable ectopic expression of TgENO1-HA in parasite clones. Expression of TgENO1-HA protein driven by heterologous tubulin gene promoters was demonstrated by co-localization of fluorescent images obtained using anti-HA and anti-ENO1 antibodies. Similar to endogenous ENO1 protein, TgENO1-HA protein was exclusively localized in the nucleus of one stable clone (E1-5) selected after bleomycin treatment. C) Western blots of transgenic and ectopically expressed TgENO2-HA protein. Total protein extracts from E2-4 and E2-10 tachyzoites were loaded in lanes 2 and 3 and probed with anti-ENO2. Lane 2 contained total protein from wild-type 76K *T. gondii* tachyzoites. Lane 1 corresponds to total protein extract from encysted bradyzoites isolated from the brain of mice chronically infected with wild type *T. gondii* 76K strain. C) Western blots of transgenic and ectopically expressed TgENO1-HA protein. Total SDS-extracted proteins from E2-4 and E2-10 transgenic tachyzoites were loaded in lanes 3 and 4 and probed with anti-ENO2 antibodies. Lane 2 contained total SDS-extracted proteins from wild-type 76 K *T. gondii* tachyzoites and lane 1 corresponds to total SDS-extracted proteins from encysted bradyzoites isolated from the brains of mice chronically infected with wild type *T. gondii* 76K strain. D) Western blot analysis of transgenic and ectopically expressed TgENO1-HA protein. Total SDS-extracted proteins from the E1-5 strain was loaded in lane 2 and probed with anti-ENO1 antibodies. Lane 1 corresponds to total-SDS extracted proteins from encysted bradyzoites as described above. E) Western blots of total SDS-extracted proteins from E1-5, E2-4 and E2-10 tachyzoites were loaded in lanes 2, 3 and 4 and probed with anti-HA antibodies. Lane 1 contained total SDS-extracted proteins from wild-type 76 K *T. gondii* tachyzoites.

Because of the limitations of obtaining sufficient numbers of encysted bradyzoites from brains of chronically infected mice, we engineered transgenic tachyzoites that ectopically expressed TgENO1-HA under the control of the promoter from the major surface antigen 1 (*SAG1*) gene. Figure 1B shows a representative positive clone (E1-5) in which ectopic TgENO1-HA co-localized with nuclear endogenous TgENO2, similar to ectopic TgENO2-HA shown in Figure 1A. The anti-HA fluorescent signals were perfectly superimposed with anti-ENO2 staining (red signals), demonstrating that both transgenic bradyzoite-specific TgENO1-HA and TgENO2 were properly targeted to the nuclei of actively dividing tachyzoites (merged pictures). We confirmed the expression of both ectopic TgENO2-HA and TgENO1-HA by Western blots using polyclonal antibodies specific to HA (Fig. 1E, lanes 2-4), anti-ENO2 (Fig. 1C, lanes 3 and 4) and TgENO1-HA

(Fig. 1D, lane 2). The endogenous TgENO2 protein in tachyzoites (Fig. 1C, lane 2) and TgENO1 protein of encysted bradyzoites isolated from the brains of chronically infected mice (Fig. 1D, lane 1) were used to monitor the protein levels of these positive controls.

The level of nuclear enolases increased in actively dividing tachyzoites of *T. gondii*

We examined the subcellular localization of TgENO2-HA and its endogenous counterpart in intracellular tachyzoites at 0, 6, 12, 18, 24, 30, or 36 hours post-infection using immunofluorescence confocal microscopy (Fig. 2A and 2B). These kinetic studies revealed that the proportion of ectopic TgENO2-HA and endogenous TgENO2 detected in the nuclei of intracellular tachyzoites increased from 20% to 60% during the 12-h period of intracellular growth, reaching 70% by 32 h post-invasion

(Fig. 2C). The increased nuclear fluorescence of TgENO2-HA and TgENO2 correlated with the increase in the number of intracellular tachyzoites from 1 to 32 daughter parasites (Fig. 2C). Conversely, the level of cytoplasmic TgENO2 in the tachyzoites decreased to 20–30% (Fig. 2C). We conclude that a substantial proportion of ectopic and endogenous enolases are dynamically targeted to the nuclei of actively replicating tachyzoites. These data also suggest that the glycolytic enolase of *T. gondii* might have novel nuclear functions that are distinct from its classic role as a cytoplasmic enzyme required for energy production.

Genome-wide TgENO2 occupancy of gene promoters defined by ChIP-Seq

We next investigated the interactions of TgENO2 with 5′-UTR gene sequences that might correspond to putative gene promoters *in vivo* using chromatin immunoprecipitation followed by high-throughput sequencing (ChIP-Seq). Intracellular actively dividing tachyzoites of wild-type *T. gondii* 76K strain or transgenic TgENO2-HA parasites were fixed by formaldehyde and released from host cells. After chromosome fragmentation by sonication, the chromatin was immunoprecipitated using specific anti-TgENO2 or anti-HA antibodies, or with pre-immune or non-relevant anti-ROP1 sera as negative controls. The immunoprecipitates were subjected to high-throughput sequencing and bioinformatics analyses using genome data from http://www.toxodb.org. After comparison of sequences and removal of common genes that were targeted by both pre-immune and specific anti-TgENO2 and anti-HA sera, the 5′-UTRs corre-

sponding to the putative promoters of 241 genes that were commonly pulled down by both antibodies were selected (Table S4). The gene promoters pulled down were expressed in both tachyzoite and bradyzoite stages, suggesting that there is no clear link between genes targeted by nuclear enolase and their stage-specific expression (Table S4). In addition, these antibodies also bound to coding regions of the same genes or other gene targets (Table S5). We have no obvious explanations of the binding of nuclear TgENO2 to coding regions of these target genes identified by ChIP-Seq. We cannot rule out the possibility that binding to gene coding regions may also account for transcription regulation. However, we do not further investigate the role of TgENO2 binding to coding regions because this issue is beyond the scope of this study. Therefore, we focused on TgENO2 produced hits that were localized to promoter regions. Among the gene promoters identified were mostly 44% (107) of encoded hypothetical proteins, 4% encoded metabolic enzymes, 5% encoded translation factors, 5% encoded cytoskeleton, trafficking, and transporter proteins, and 12% encoded other enzymes (Table S4). Interestingly, ChIP-Seq also identified 5 genes corresponding to the novel plant-like AP2 transcription factor AP2 that have been demonstrated to regulate transcription in apicomplexan parasites [13–17]. We conclude that *T. gondii* enolase may be important in nuclear functions linked to the transcription and translation of genes involved in parasite growth and intracellular development. Surprisingly, very few genes coding surface proteins, or microneme and rhoptry components were identified with the stringency set at reads of a minimum of 40 bp.

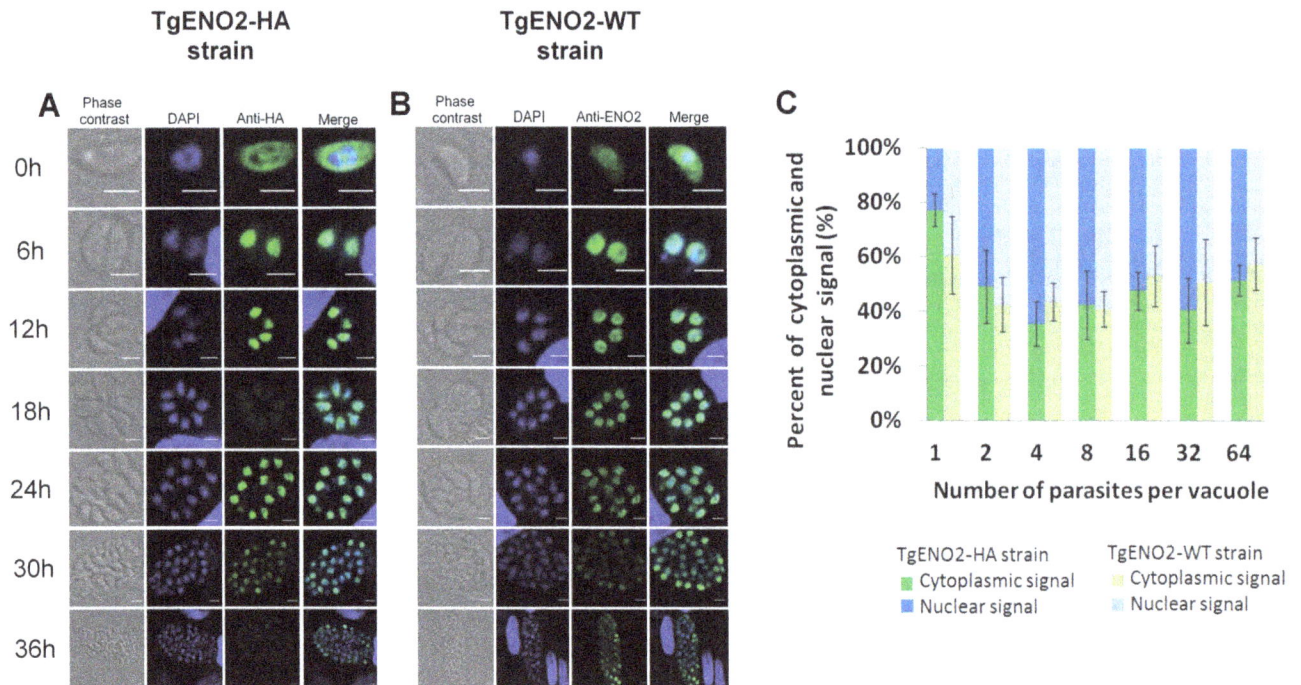

Figure 2. Expression of ectopic TgENO2-HA and native TgENO2 proteins is predominantly nuclear and increases with intracellular replication of *T. gondii*. A) Kinetics of nuclear accumulation of ectopic TgENO2-HA protein in transgenic tachyzoites at different time points (0, 6, 12, 18, 24, 30, and 36 h post-infection) during the intracellular division cycle. Intracellular dividing transgenic tachyzoites were fixed and stained with rabbit polyclonal anti-HA and DAPI followed by confocal imaging. Scale bars, 5 μm. B) Kinetics of nuclear accumulation of native TgENO2 protein in wild type *T. gondii* tachyzoites at different time points (0, 6, 12, 18, 24, 30, and 36 h post-infection) during the intracellular division cycle. Intracellular dividing transgenic tachyzoites were fixed and stained with rabbit polyclonal anti-ENO2 antibodies and DAPI followed by confocal imaging. Scale bars, 5 μm. C) Quantification of cytoplasmic and nuclear levels of ectopic TgENO2-HA and native TgENO2 in intracellular dividing tachyzoites of *T. gondii*. Experiments were repeated three times (n = 3, P<0.001). Quantifications were performed on at least 8–10 independent intracellular vacuoles using ImageJ software and bioinformatics tools as described in Materials and Methods.

Quantitative real-time PCR analysis was used to demonstrate the specificity of chromatin immunoprecipitates from wild-type and TgENO2-HA strains pulled down with anti-TgENO2 (Fig. 3A) and anti-HA (Fig. 3B) antibodies, respectively. We confirmed that eight genes (glucose-6-phosphate isomerase, dense granule protein 1 (GRA1), cyst matrix protein 1 (TgMag1), hexokinase, glucose transporter putative, phosphoglycerate mutase, rhoptry protein 16 (ROP16), and glucose-6-phosphate dehydrogenase) were significantly and specifically targeted by nuclear TgENO2 *in vivo* using anti-ENO2 (Fig. 3A) and anti-HA antibodies (Fig. 3B). As a negative control, a gene (TgME49_0022080) encoding a hypothetical protein that was not present in the list of nuclear TgENO2 target genes (Table S4) was not amplified (Fig. 3A and 3B), confirming specific binding of nuclear TgENO2 *in vivo* on these eight selected genes. These experiments were performed three times from two independent ChIP experiments with reproducible and statistically significant results (P<0.0001). We also pulled down the 5′-UTRs corresponding to ROP16, TgENO1, and TgENO2 with a stringency of nucleotide length less than 40 bases (approximately 20–30 bp), suggesting that the number of gene hits of nuclear TgENO2 may actually be higher than the 241 genes shown in Table S4. Pre-immune (naïve) serum did not bind to these eight genes. We conclude that TgENO2 is targeted to the nuclei of intracellular dividing tachyzoites of *T. gondii* where it specifically binds to the promoter regions of numerous genes.

Evidence for specific *T. gondii* enolase-DNA interactions

Searches for a conserved motif common to the gene promoters targeted by nuclear TgENO2 using the bioinformatics MEME tool identified the motif TTTTCT (Fig. 4A), which is present at least once in 238 out of the 241 putative promoters targeted by TgENO2 and listed in Table S4. The TTTTCT motif was detected in the 2,500-bp sequence upstream of the start codon (ATG) of 99% of the target genes of TgENO2. However, this motif is also present in the promoter of many other genes in *T. gondii* genome that were not pulled down by TgENO2 in addition to its presence in coding regions, suggesting that this motif alone may not be sufficient for the efficient and specific binding to gene promoters *in vivo*. Nevertheless, we established that the TTTTCT motif is directly involved in DNA-protein interactions *in vitro* using bacterial recombinant TgENO2 protein, which was purified to homogeneity (Fig. 4C) and tested for its binding to this motif designed from the 787-bp putative promoter of *TgMAG1*, one of the target genes identified in Table S4. It should be mentioned that TgMAG1 also called cyst matrix antigen is not a bradyzoite specific protein as was originally described [33] but was later shown to be expressed by both tachyzoites and bradyzoites [34]. Figure 4D showed that recombinant TgENO2 protein specifically interacted with the TTTTCT motif. In contrast, TgENO2 protein did not bind to a c-Myc motif from animal or plant [23–25] that contains a canonical TATA box (probe B), or to the unrelated probe C (Fig. 5B). Thus, we conclude that nuclear enolases of *T. gondii* could possibly bind to DNA motif present in the promoter regions of parasite genes, which may allow multi-complex nuclear factors to control gene expression in intracellular proliferating tachyzoites.

The TTTTC motif is involved in promoter activation

Given that TgENO2 clearly targets the TTTTC motif in many putative gene promoters and the bacterially purified recombinant protein also binds specifically to this motif, it seems likely that this motif is responsible, at least in part, for regulating gene expression. To address this hypothesis, the putative promoter region of

TgMag1 was fused to firefly luciferase and the double TTTTC-like motif (TTTTTCTTCTC, see TgMag1 in Figure S1) present in the promoter was first mutated to <u>ATCGA</u>TCTC (Δ_1TgMag1, Figure S2), followed by a second mutation to <u>ATCGAGCGC</u> (Δ_2TgMAg2, Figure S3). A dual luciferase assay was carried out after transient transfection with constructs containing wild-type or mutated promoters of TgMag1 gene in addition to vector encoding *Renilla* luciferase for standardization. The <u>ATCGA</u>TCTC mutation in TgMag1 promoter (Δ_1TgMag1) resulted in a 60% decrease in luciferase expression compared with wild-type (Fig. 4E). The combination of both mutations in Δ_2TgMag2 (<u>ATCGAGCGC</u>) resulted in a 70% reduction in luciferase expression compared with wild-type (Fig. 4E). We conclude that deletion of the TTTTC motif in TgMag1 promoter has a significant negative effect on its activity *in vivo*.

TgENO1 and TgENO2 bind to promoters *in vivo* and modulate gene expression

Next, we showed that both endogenous nuclear TgENO2 or ectopically expressed TgENO2-HA and TgENO1-HA enolases bound to the active gene promoter of TgENO2 in rapidly replicating tachyzoites using ChIP assays (Fig. 5A). We also demonstrated that ectopic TgENO1-HA, which is illegitimately expressed in the transgenic tachyzoites, bound specifically to the *TgENO2* gene promoter *in vivo* similar to the binding of nuclear TgENO2-HA (Fig. 5A). We conclude that nuclear enolases can bind to their own gene promoters and probably modulate gene activation and repression in T. gondii. Additionally, we showed that the RNA profiles of these genes described above were significantly modulated by ectopic nuclear TgENO1-HA and TgENO2-HA (Fig. 5B). The quantitative RT-PCR data shown in Figure 5B reveal that four out of the eight genes tested were negatively regulated in intracellular E1-5 tachyzoites expressing TgENO1-HA (Fig. 5B, red columns). In contrast, six out of eight genes were positively regulated in the intracellular E2-4 tachyzoites expressing TgENO2-HA (Fig. 5B, blue columns). The mRNA level of the housekeeping β-tubulin gene was unchanged, as expected. We conclude that nuclear enolases are likely involved in the transcriptional control of numerous genes through binding to their promoter regions. These data suggest that nuclear TgENO2 and TgENO1 associate with chromatin, leading to either up-regulation or down-regulation of gene expression.

Targeted disruption of the bradyzoite-specific TgENO1 gene

To gain more insight into the biological functions of nuclear enolases during infection, we aimed to knock out the TgENO1 and TgENO2 genes. Despite several attempts, we failed to obtain targeted deletion of the tachyzoite-specific TgENO2 in intracellular tachyzoites using direct or inducible gene disruption strategies. This failure seems to suggest that TgENO2 is essential for intracellular parasite growth, as might be expected, considering its nuclear regulatory roles in addition to its function in glycolysis and ATP production. Alternatively, it may reflect the weaker strength of the conditional promoter compared with the TgENO2 promoter, or the *Tgeno2* locus may be refractory to double homologous recombination. Nevertheless, we did successfully target the *Tgeno1* locus by double homologous recombination for deletion of the bradyzoite-specific TgENO1 gene in intracellular tachyzoites (Figure S4). A PmeI-linearized pΔENO1 targeting plasmid was transfected into *T. gondii* strain PruΔ-ku80Δhxgprt that exhibits highly enhanced homologous recombination [29], and transfected parasites were selected in myco-

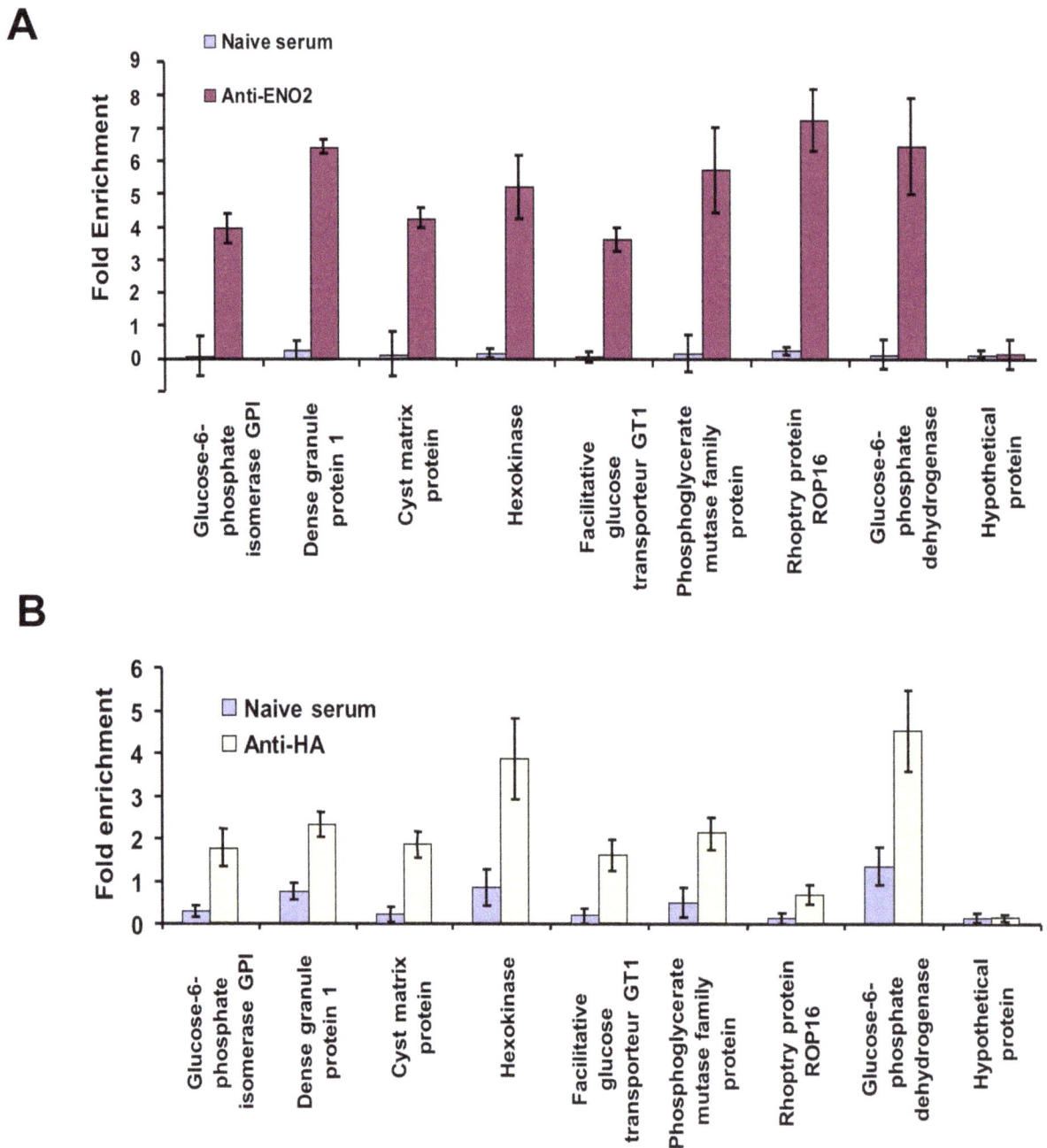

Figure 3. Validation of native TgENO2 and transgenic TgENO2-HA protein bound to several putative gene promoters. A) Quantitative real-time PCR analysis of chromatin immunoprecipitates from three independent experiments (n = 3, P<0.0001) demonstrates specific binding of nuclear TgENO2 *in vivo* to eight selected genes identified by ChIP-Seq (see Table S4). A gene encoding a hypothetical protein that was absent from the gene hits (TgME49_ 0022080) was used as a negative control. B) Quantitative real-time PCR analysis of chromatin immunoprecipitates from three independent experiments (n = 3, P<0.0001) demonstrates specific binding of nuclear transgenic TgENO2-HA *in vivo* to eight selected genes identified by ChIP-Seq (see Table S4). The TgME49_ 0022080 gene was used as a negative control.

phenolic acid (MPA). MPA-resistant clones were isolated and the genotype of the clones was evaluated by PCR (Fig. S4A and S4B). In PCR1, the EXR primer was paired with the CXF primer (Table S2) to verify the presence of the 5′ target flank in all clones and the parental strain. PCR2 was used to identify clones with a deleted *ENO1* gene by the absence of PCR product (clones 1–12, Fig. S4B) compared with the parental strain (lane 13, Fig. S4B). To verify targeted integration of *HXGPRT* into the deleted *ENO1* locus, we used PCR3 and PCR4 to respectively verify 5′ and 3′

integration of the HXGPRT selectable marker in the entire ENO1 knockout (Fig. S4B). For further phenotypic studies, we selected one of these positive *T. gondii* clones and confirmed specific deletion of the entire open reading frame of TgENO1 gene (Fig. S4C, see primers in Table S3).

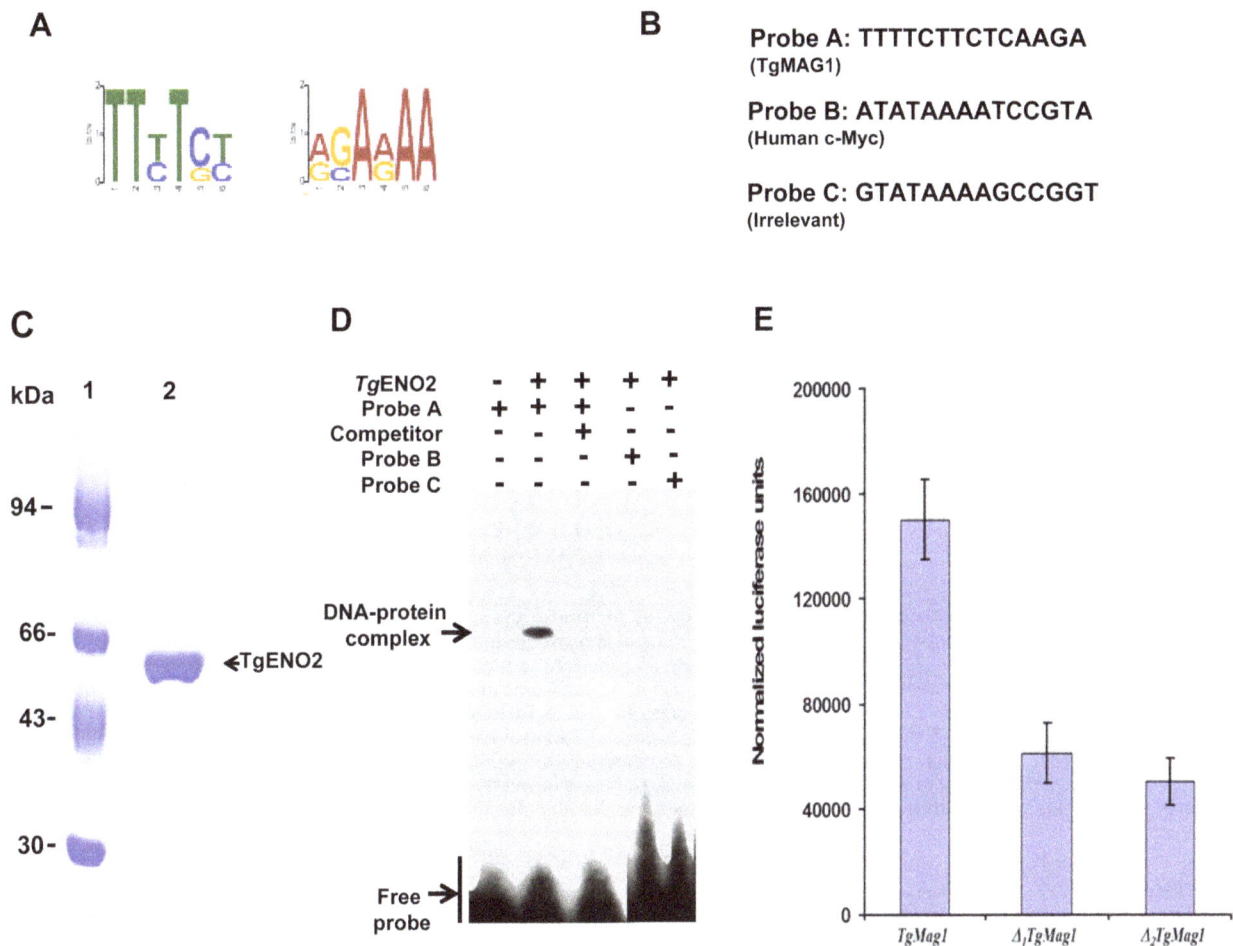

Figure 4. Role of the TTTTCT motif in specific TgENO2-DNA interactions and promoter activity. A) The TTTTCT motif was identified in the putative gene promoters targeted by nuclear TgENO2 using ChIP-Seq (Table S4) and the MEME bioinformatics tool (a motif-based sequence analysis tool). B) Nucleotide sequences of probes corresponding to the TTTTCT motif in the promoter of TgMag1 gene, the TATA box from human c-Myc gene, and a non-relevant motif used as a negative control. C) Expression and purification of recombinant TgENO2 fused to His-Tag. D) Electrophoretic band shift assays using recombinant TgENO2 incubated with or without the probes described in panel A. The unlabeled competitor was present at 100-fold excess. E) The GCTAGC motif is required for efficient transcription of the TgMag1 gene. The putative promoter of TgMag1, corresponding to a 787-bp region upstream of the start codon, was subjected to site-directed mutagenesis resulting in sequential disruption of the single TTTCT motif within the TTTTCTTCTC motif of TgMag1 to $\underline{ATCGA}$TCTC (Δ_1TgMag1) and then to $\underline{ATCGAGCGC}$ (Δ_2TgMAg2). These two mutant promoters and the wild-type promoter were cloned upstream of a reporter luciferase construct and assayed for their ability to drive transcription. The transcriptional potential of mutated promoters was measured as firefly luciferase activity normalized to activity of a vector encoding *Renilla* luciferase. These experiments have been performed three times ($n = 3$, $p < 0.001$).

Targeted deletion of TgENO1 reduces cyst burden in the brain of chronically infected mice

We consistently noticed a severe decrease in the number of brain tissue cysts in mice infected with TgENO1 knockout mutants compared with those infected with parental parasites (Fig. 6A). Specifically, at 4 weeks post-infection, the average of brain tissue cyst burden in mice inoculated with parental Pru$\Delta ku80$ was about 300 cysts per brain compared with 50 cysts per brain in mice infected with the knockout Pru$\Delta ku80\Delta Tgeno1$ mutants. These experiments were independently performed three times using different preparations of freshly harvested extracellular tachzyoites with identical results. Furthermore, we observed a significant over-expression of the tachyzoite-specific TgENO2 in the knockout Pru$\Delta ku80\Delta Tgeno1$ mutants as shown by Western blots (Fig. 6B, left panel, compare lane 2 to lane 1). As compared to the housekeeping actin used for to verify loading of equal protein amounts (Fig. 6B, right panel), it appears that TgENO2 protein

level in Pru$\Delta ku80\Delta Tgeno1$ mutants is approximately two-fold higher (Fig. 6B, left panel, lane 2) than that of the parental Pru$\Delta ku80$ parasites (Fig. 6B, left panel, lane 1). In contrast, the levels of other tachyzoite-specific glycolytic enzymes such as lactate dehydrogenase LDH1, glucose 6-phosphate isomerase (G6PI) were unchanged in $\Delta ku80\Delta Tgeno1$ mutants *versus* the parental Pru$\Delta ku80$ parasites (Fig. 6C, lanes 5 and 6). In addition, the knock-out of TgENO1 does not alter the expression of the bradyzoite-specific LDH2 enzyme (Fig. 6C, LDH2 panel, lanes 5 and 6). Again, the housekeeping actin was used as a loading control (Fig. 6C, Actin panel). In addition, we also showed that the ectopic expression of TgENO1-HA and TgENO2-HA in the E1-4, E2-4 and E2-10 strains used for ChIP-sequencing and qPCR does not have any obvious consequences in the protein levels of other glycolytic enzymes such as LDH1, LDH2 and G6-PI (Fig. 6C, lanes 2, 3 and 4). Taken together, these data indicate that the intracellular Pru$\Delta ku80\Delta Tgeno1$ mutants and the ectopic

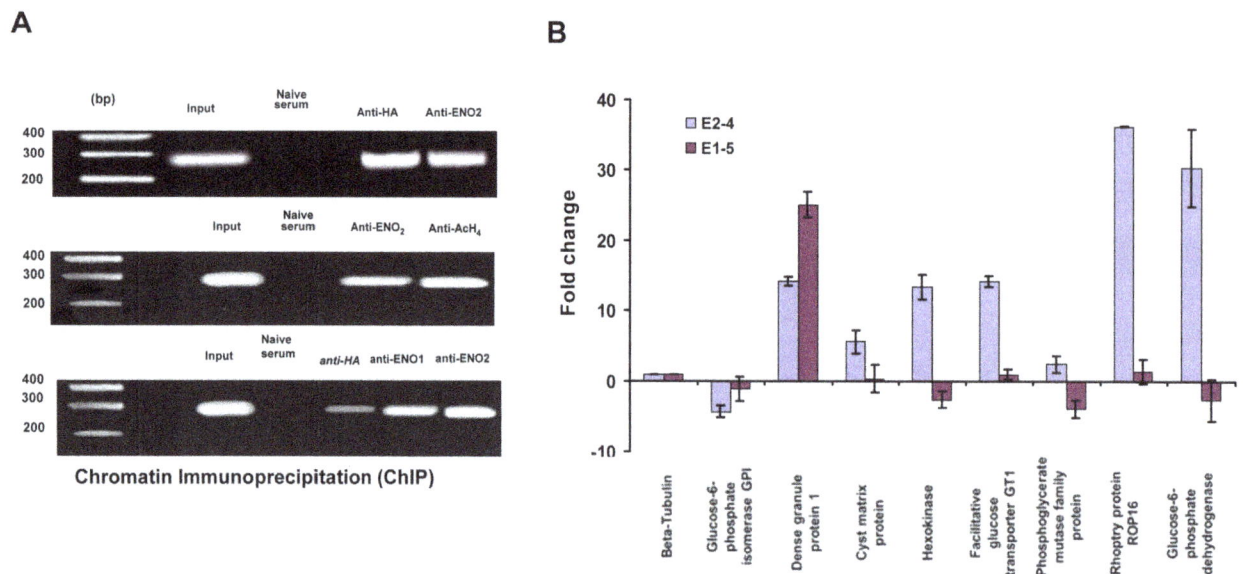

Figure 5. Nuclear enolases bind to promoters of several genes including their own and modulate transcriptional expression of several gene targets. A) The upper panel shows comparative chromatin immunoprecipitation (ChIP) from intracellular tachyzoites of a strain that ectopically overexpressed TgENO2-HA (E2-4) using polyclonal anti-HA and anti-ENO2 antibodies. No chromatin was immunoprecipitated with a non-relevant non-immune serum used as negative control. The middle panel shows chromatin immunoprecipitation (ChIP) of intracellular parasites that ectopically express TgENO2-HA (strain E2-4) with polyclonal antibodies specific to TgENO2 or acetyl histone H4, the epigenetic mark of an active promoter. The lower panel shows ChIP of parasites that ectopically express TgENO1-HA parasites (strain E1-5) with polyclonal antibodies specific to HA, TgENO1, and TgENO2. B) Changes in gene expression were confirmed by qRT-PCR analysis in the intracellular tachyzoites over-expressing TgENO2-HA and TgENO1-HA, compared with the beta-tubulin housekeeping gene and the expression of these genes in wild-type parasites. These experiments were repeated twice with triplicate samples and identical results were obtained n = 3, (P<0.001).

TgENO1 and TgENO2 parasite over-expressers grew normally as tachyzoites under our experimental culture conditions.

Targeted deletion of TgENO1 directly influenced transcriptional regulation

To validate the phenotypic consequences of the Pru*Δku80ΔT-geno1* knockout, we complemented this mutant with a second plasmid carrying the TgENO1 gene that ectopically expressed the enzyme. We verified that TgENO1 protein was not expressed in the tachyzoites of the Pru*Δku80ΔTgeno1* knockout mutants using Western blot analysis, monoclonal anti-P30, which is directed to tachyzoite-specific surface antigen 1 (SAG1), and anti-TgENO1 antibodies (Fig. 7A) and confocal microscopy (Fig. 7C) whereas the complemented mutants clearly expressed ectopic TgENO1-HA protein (Fig. 7B and 7D). The complemented mutants also expressed ectopic TgENO1-HA under *in vitro* bradyzoite-induced conditions (increased alkaline pH), as shown by Western blot analysis using anti-HA (Fig. 8B, lane 3), anti-ENO1 antibodies and monoclonal anti-P36, which is directed to bradyzoite-specific surface antigen SAG4 [37] (Fig. 8B, lane 4). Confocal microscopy showed that TgENO1-HA protein was localized to the nuclei of the alkaline pH-induced bradyzoites (Fig. 8D), with positive staining of the cyst wall by the lectin WGA marker [35,36]. As expected, no endogenous TgENO1 protein was detected in the Pru*Δku80ΔTgeno1* mutant transfected with empty plasmid under alkaline pH bradyzoite-induced conditions (Fig. 8A and 8C).

Having confirmed that complementation of the Pru*Δku80ΔT-geno1* knockout mutant with an ectopically expressing vector resulted in the expression of TgENO1-HA that was localized to the nuclei of tachyzoites and *in vitro*-induced bradyzoites, we could directly compare the effect of the absence or presence of

nuclear TgENO1 protein on the transcriptional regulation of tachyzoite and bradyzoites of *T. gondii*.

We compared transcript levels under experimental conditions that support the intracellular growth of tachyzoites (normal pH 7) or stress-induced bradyzoite differentiation (alkaline pH 8.2), as an *in vitro* experimental model of bradyzoite and cyst formation, by qRT-PCR. We found that expression of the eight nuclear genes targeted by the nuclear TgENO2 was differentially modulated 2- to 5-fold between intracellular tachyzoites of Pru*Δku80ΔTgeno1* mutants and parental Pru*Δku80* (Fig. 9A). Specifically, the transcript levels of five genes were increased in the mutant and those of three genes were unchanged (Fig. 9A). In contrast, transcript levels of seven of these eight genes were consistently decreased in *in vitro*-induced bradyzoites from Pru*Δku80ΔT-geno1* mutants relative to parental Pru*Δku80* (Fig. 9B). We noticed that complementation of Pru*Δku80ΔTgeno1* mutants by ectopic expression of TgENO1-HA did not significantly affect the decrease in transcript levels of these eight target genes in the tachyzoites (Fig. 9C) whereas the normal transcript levels of most gene targets (5 out 7) were restored in *in vitro*-induced bradyzoites, as expected for a functional complementation (Fig. 9D). We did not compare the production of cysts in the brains of chronically infected mice because we were unable to obtain stable complemented parasite lines despite several attempts. In addition, the low number of cysts obtained in mice brains also limited the quantity of RNA that could be isolated for qRT-PCR. Nevertheless, our data indicate that lack of TgENO1 has a direct impact on gene regulation and influences transcript levels during the *in vitro* interconversion between the rapidly replicating tachyzoite and the dormant bradyzoites of *T. gondii*.

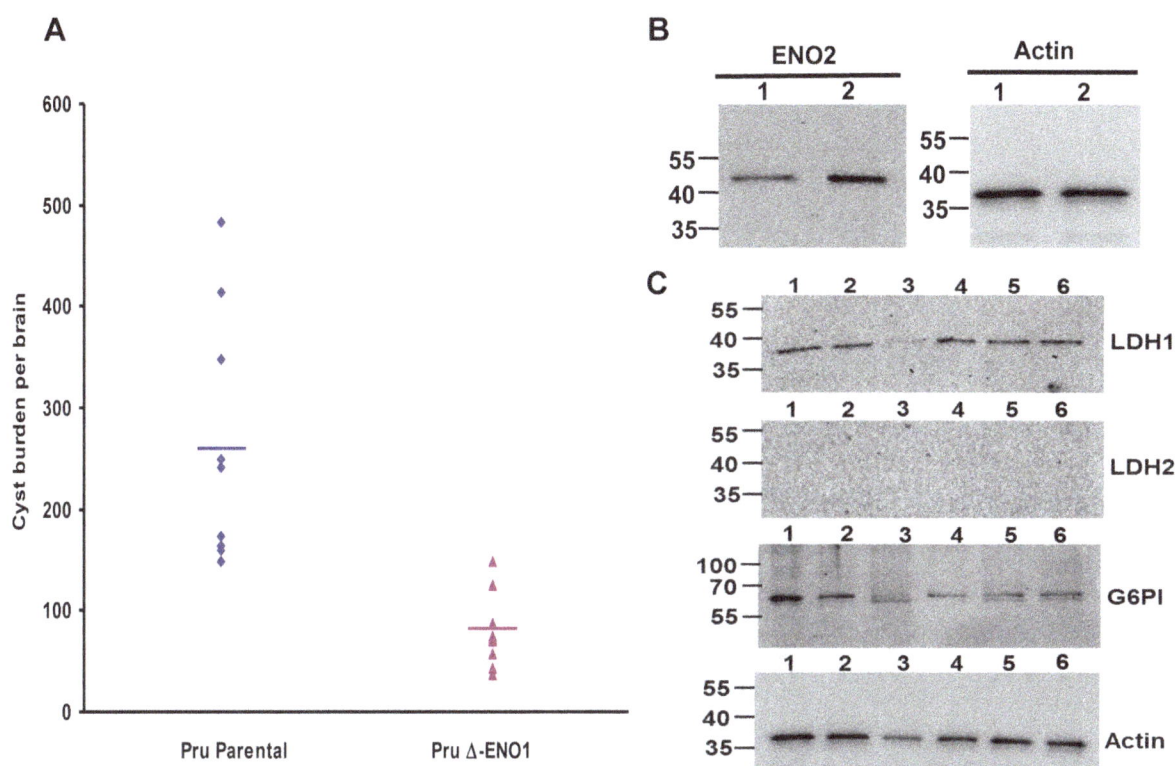

Figure 6. Targeted deletion of TgENO1 reduces cyst burden in the brain of chronically infected mice. A) The total number of cysts per brain of mice infected with 5×10^2 tachyzoites from the PruΔku80ΔTgeno1 mutant or parental PruΔku80 was counted after staining with FITC-labeled *dolichol biflorus* lectin. A group of nine mice was used for each experiment, and the experiment was repeated twice with similar results (n = 2, P< 0.001). Cyst burden (total number of cysts per brain) of the PruΔku80ΔTgeno1 mutant was significantly lower than that of the parental PruΔku80 strain. B) Western blots of total SDS-extracted proteins from knockout PruΔku80ΔTgeno1 mutants (lane 1) and parental PruΔku80 tachyzoites (lane 2). Left panel was probed with the polyclonal anti-ENO2 antibodies while the right panel was stained with the monoclonal anti-actin antibodies. C) Western blots of mutants and wild type parasites. Lane 1, total SDS-protein extracts from wild type 76K tachyzoites. Lane 2, total SDS-extracted proteins from transgenic E1-5 tachyzoites. Lane 3, total SDS-extracted proteins from transgenic E2-4. Lane 4, total SDS-extracted proteins from transgenic E2-10. Lane 5, total SDS-extracted proteins from parental PruΔku80 tachyzoites. Lane 6, total SDS-extracted proteins from knock-out PruΔku80ΔTgeno1 mutants. Blots were stained with polyclonal antibodies specific to LDH1, LDH2, G6PI and monoclonal antibodies specific to actin. The numbers on the left indicate molecular markers in kilodaltons.

Discussion

In the present study we aimed to decipher novel functions of stage-specific glycolytic enolases in *T. gondii* that display dual cytoplasmic and nuclear localization, but are considerably enriched in the nuclei of actively dividing intracellular parasites. It is somewhat intriguing that no nuclear localization sequence (NLS) is present in *T. gondii* enolases, although the presence of a NLS has not been described in any of the human, plant, or parasite enolases that are efficiently transported into the nucleus for repression of gene transcription [23–25]. We cannot rule out the possibility that enolases enter the nucleus through interaction with a partner or chaperone that contains a functional NLS. Additionally, the size of enolase is close to the cut off for the nuclear pore, and it is possible that an NLS is not required but the protein might be trapped in one compartment or another under the appropriate conditions. To date, the mechanism by which enolases and other glycolytic enzymes are transported into the nucleus remains unknown and represents an interesting issue that is beyond the scope of our present investigation. Here, we focused on elucidating the novel nuclear functions of *T. gondii* enolases as an important step towards a better understanding of gene regulation in *T. gondii*. We have previously shown that silencing of tachyzoite enolase 2 (TgENO2) using double-strand RNA

inhibition strategies alters the nuclear targeting of bradyzoite enolase 1 (TgENO1) [38]. These observations suggest that concomitant expression of both isoenzymes in the nuclei of intermediate parasitic forms at the early stages of interconversion might play a role in their nuclear targeting and in the regulation of genes involved in parasite differentiation and cyst formation [36]. Here, we demonstrate that *T. gondii* enolase binds specifically to approximately 241 putative gene promoters *in vivo*, which represents about 3% of the total gene content in the parasite genome. We provide further evidence that this binding positively or negatively influences gene expression in *T. gondii*.

For many years, glycolytic enzymes have been considered to be housekeeping cytoplasmic proteins. However, recent studies have provided evidence that some glycolytic enzymes are multifaceted and perform multiple functions, rather than just being simple components of the glycolytic pathway [39]. For example, glycolytic enzymes have been shown to have additional functions in transcriptional regulation (hexokinase-2, lactate dehydrogenase, glyceraldehyde-3-phosphate dehydrogenase or GAPD), stimulation of cell motility (glucose-6-phosphate isomerase), and the regulation of apoptosis (glucokinase and GAPD) [39]. It is well established that enolase is a multifunctional protein that is involved in gene transcription in other eukaryotes [23–25,39] including the protozoan parasite *Entamoeba histolytica* [40]. Although no non-

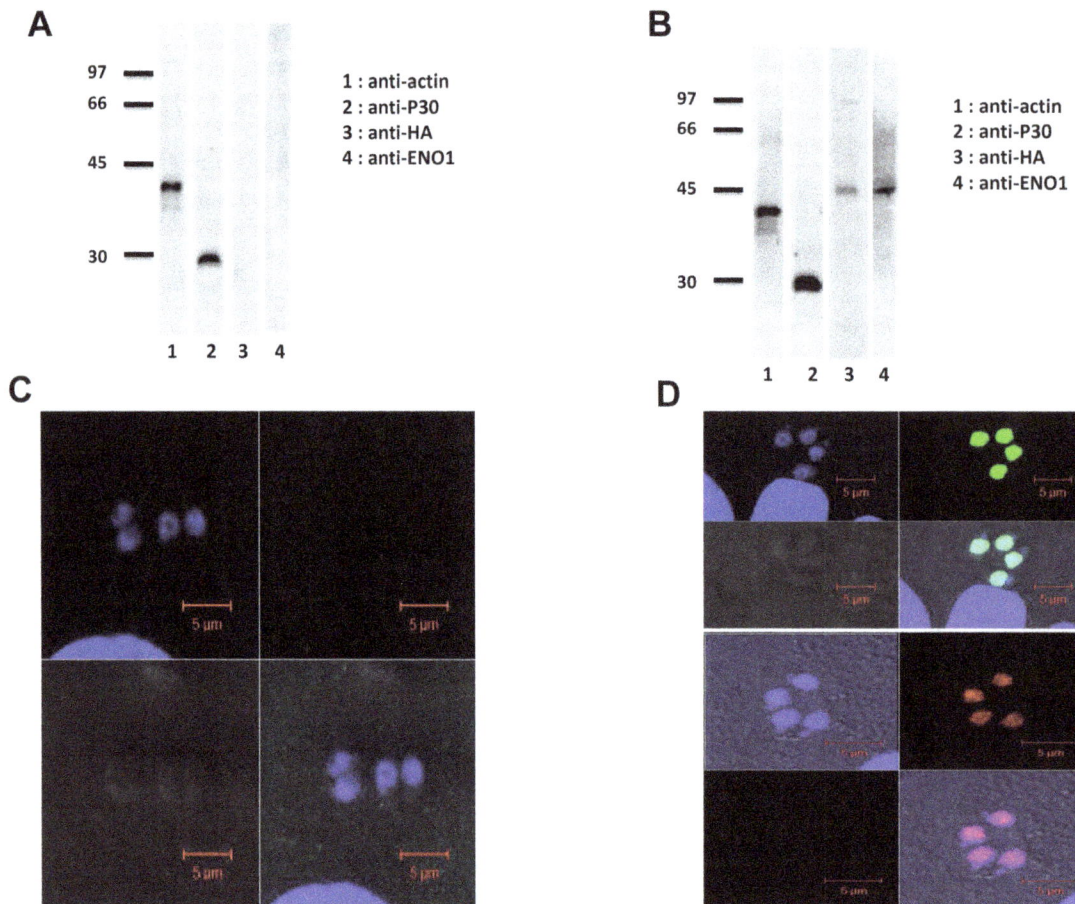

Figure 7. Complementation of Pru∆ku80∆Tgeno1 knockout mutant by ectopic expression of TgENO1-HA under tachyzoite culture conditions. A) Western blot analysis of Pru∆ku80∆Tgeno1 mutants using total SDS-extracted antigens from approximately 2×10⁶ tachyzoites and probed with monoclonal antibodies specific to *T. gondii* actin (a housekeeping protein, lane 1), tachyzoite-specific major surface protein 1 or SAG1 (anti-P30, lane 2), the influenza hemagglutinin (HA)-tag (anti-HA, lane 3) and TgENO1 (lane 4). B) Western blot analysis of total SDS-extracted antigens from approximately 2×10⁶ intracellular tachyzoites from complemented Pru∆ku80∆Tgeno1 mutants and probed with antibodies specific to *T. gondii* actin (lane 1), tachyzoite-specific major surface protein 1 or SAG1 (anti-P30, lane 2), the influenza hemagglutinin (HA)-tag (anti-HA, lane 3) and TgENO1 (lane 4). C) Confocal microscopy of intracellular tachyzoites of Pru∆ku80∆Tgeno1 mutants that were grown in confluent HFF cells and stained with polyclonal anti-HA antibodies. The panels show nuclei stained by DAPI (upper left panel), IFA (upper right panel), phase contrast (lower left panel), and merged image (lower right panel). Scale bars are 5 µm. D) Confocal microscopy of intracellular tachyzoites from Pru∆ku80∆Tgeno1 mutants that were complemented with TgENO1-HA plasmid and grown in confluent HFF cells. The first four upper panels show staining with polyclonal anti-HA antibodies (green). The second four lower panels show double staining with FITC-labeled *dolichol biflorus* lectin (cyst wall marker, green) and polyclonal anti-HA antibodies (red). The panels show nuclei stained with DAPI (upper left panel, blue), IFA (upper right panel, green or red), phase contrast (lower left panel), and merged image (lower right panel). Scale bars are 5 µm.

glycolytic functions have been described for *T. gondii* lactate dehydrogenases (LDH1 and LDH2), which are exclusively localized to the cytosol [21,27], *T. gondii* aldolase is a key component of the actin-myosin motor essential for parasite gliding and host cell invasion [41,42]. The results of this study support the notion that *T. gondii* nuclear enolases also have a non-glycolytic function through binding to gene promoters *in vivo*, leading to a decrease or increase in transcript levels of numerous genes. In mammals and plants, nuclear enolases modulate gene regulation through direct binding to a TATA box in their target gene promoters [23–25]. Interestingly, we found that *T. gondii* enolases are very different from their counterparts in higher eukaryotes because they do not bind to the TATA motif of mammalian and plant gene promoters. This is consistent with the fact that no functional TATA motif has been shown to be involved in transcription regulation of *T. gondii* [43,44] or other apicomplexan parasites. Moreover, *T. gondii* enolases also have striking

differences from higher eukaryote counterparts with regard to their plant-like structural peculiarities [19,20]. We did, however, identify a putative common TTTTCT motif within the promoter sequences of all *T. gondii* enolase target genes identified by ChIP-Seq. Using gel retardation assays; we further demonstrated that bacterial recombinant *T. gondii* ENO1 and ENO2 proteins specifically bound to this TTTTCT motif. Additionally, deletion of this motif TTTCT motif from the putative promoter of gene encoding the cyst matrix antigen, also known as TgMag1 that is present in both bradyzoites and tachyzoites, significantly decreased the promoter activity *in vivo*, suggesting that binding of nuclear enolases to this motif may be involved in gene regulation. We cannot rule out the possibility that the nuclear functions of TgENO2 may also involve epigenetic mechanisms. Future identification of other nuclear partners that specifically interact with TgENO2 will advance our understanding of how nuclear enolases control gene expression in *T. gondii*.

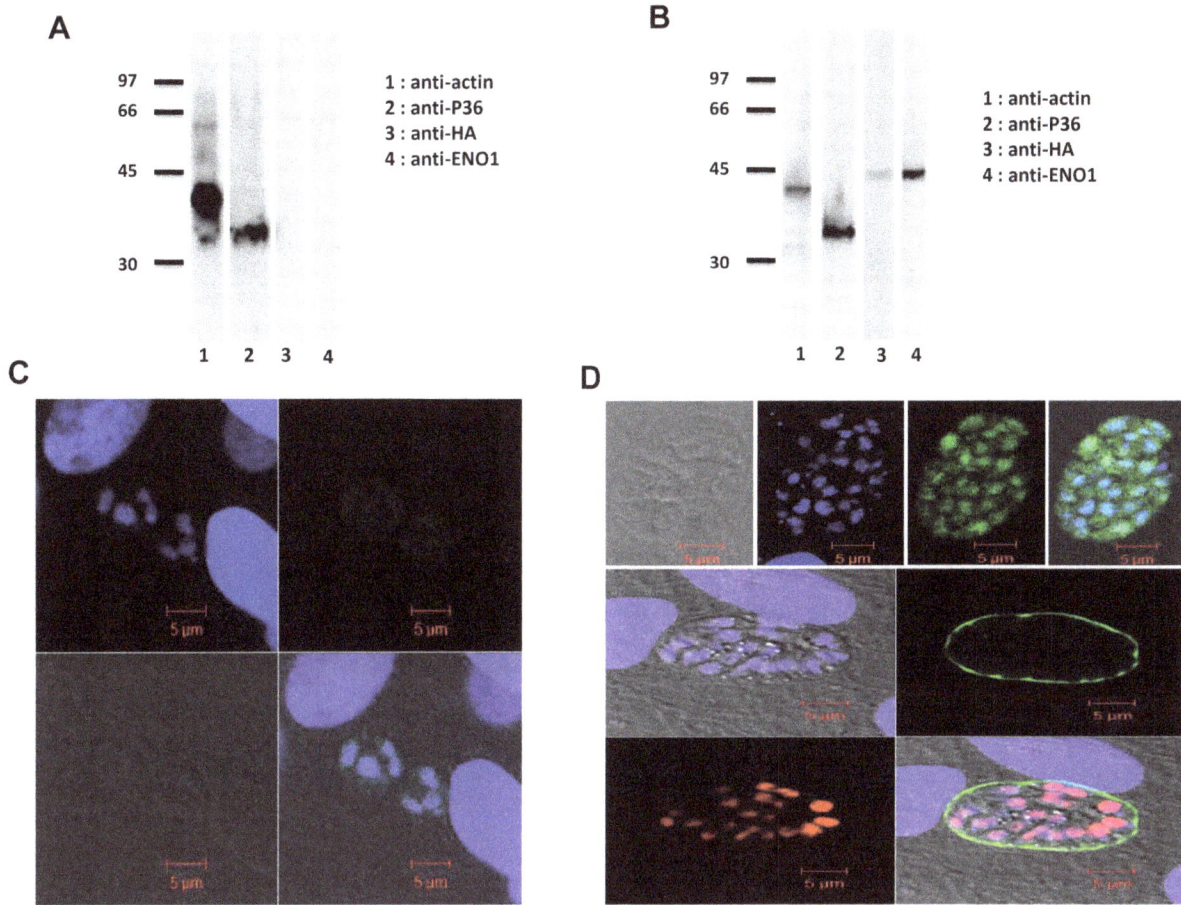

Figure 8. Complementation of Pru*Δku80ΔTgeno1* mutants by ectopic expression of TgENO1-HA and growth under bradyzoite culture conditions. A) Western blot analysis of Pru*Δku80ΔTgeno1* mutants using total SDS-extracted antigens from *in vitro*-induced intracellular bradyzoites and probed with monoclonal antibodies specific to *T. gondii* actin (lane 1), bradyzoite-specific major surface protein 1 (anti-P36, lane 2), the influenza hemagglutinin (HA)-tag (anti-HA, lane 3) and TgENO1 (lane 4). B) Western blot analysis of complemented Pru*Δku80ΔTgeno1* mutants using total SDS-extracted antigens from *in vitro*-induced intracellular bradyzoites and probed with monoclonal antibodies specific to *T. gondii* actin (lane 1), bradyzoite-specific major surface protein (anti-P36, lane 2), the influenza hemagglutinin (HA)-tag (anti-HA, lane 3) and TgENO1 (lane 4). C) Confocal microscopy of *in vitro*-induced intracellular bradyzoites of Pru*Δku80ΔTgeno1* mutants that were grown in confluent HFF cells and stained with polyclonal anti-HA antibodies. The panels show nuclei stained with DAPI (upper left panel), IFA (upper right panel), phase contrast (lower left panel), and merged images (lower right panel). Scale bars are 5 μm. D) Confocal microscopy of *in vitro*-induced intracellular bradyzoites of complemented Pru*Δku80ΔTgeno1* mutants that were grown in confluent HFF cells and stained with anti-HA antibodies (upper panels show phase contrast, DAPI, IFA, and merged images). Complemented Pru*Δku80ΔTgeno1* mutants were grown in confluent HFF cells before double staining with FITC-labeled *dolichol biflorus* lectin (cyst wall marker, green) and polyclonal anti-HA antibodies (red). The nuclei were stained by DAPI (left middle panel, blue). IFA reveals staining of the cyst wall (right middle panel, green) and ectopic TgENO1-HA protein expressed in the complemented Pru*Δku80ΔTgeno1* mutants was stained by anti-HA antibodies (lower left panel, red). Merged images are shown in the lower right panel. Scale bars are 5 μm.

To date, only a few nuclear factors have been characterized in *T. gondii* and other apicomplexan parasites [6,7,11–17,31,45] and little is known about specific *cis*-elements and nuclear factors that regulate gene expression in these parasites [11–17,43,44]. We demonstrate that ectopic expression of *T. gondii* enolases or targeted disruption of the *TgENO1* gene directly impacts transcript levels. Moreover, knockout of the TgENO1 gene results in a significant decrease in the cyst burden in the brains of chronically infected mice, suggesting that nuclear TgENO1 may be important for controlling expression of genes that are required for proper and efficient bradyzoite differentiation and cyst formation. We cannot rule out that the bradyzoites formed by the TgENO1 KO mutants may also have other defects. The limited number of brain tissue cysts and bradyzoites obtained from the KO mutants did not allow us to further examine the

ultrastructure of these mutants by electron microscopy. Nevertheless, our inability to generate knockout mutant of the tachyzoite-specific TgENO2 despite several attempts and various strategies including the conditional KO method indicates that *T. gondii* enolases may be key nuclear factors, and at least one isoform of enolase should always be present in the nuclei of the parasites to ensure their intracellular growth and development. However, we observed no significant difference between the *in vitro* intracellular replication rates of the Pru*Δku80ΔTgeno1* ENO1 knockout and the parental Pru*Δku80* strains. In addition, both KO and parental strains are able to mount acute toxoplasmosis in mice. The presence of TgENO1 protein in the nuclei of the complemented Pru*Δku80ΔTgeno1* mutant and its restorative effect on transcriptional regulation *in vitro* strongly suggests that nuclear targeting of these glycolytic enzymes may represent the key sensor that

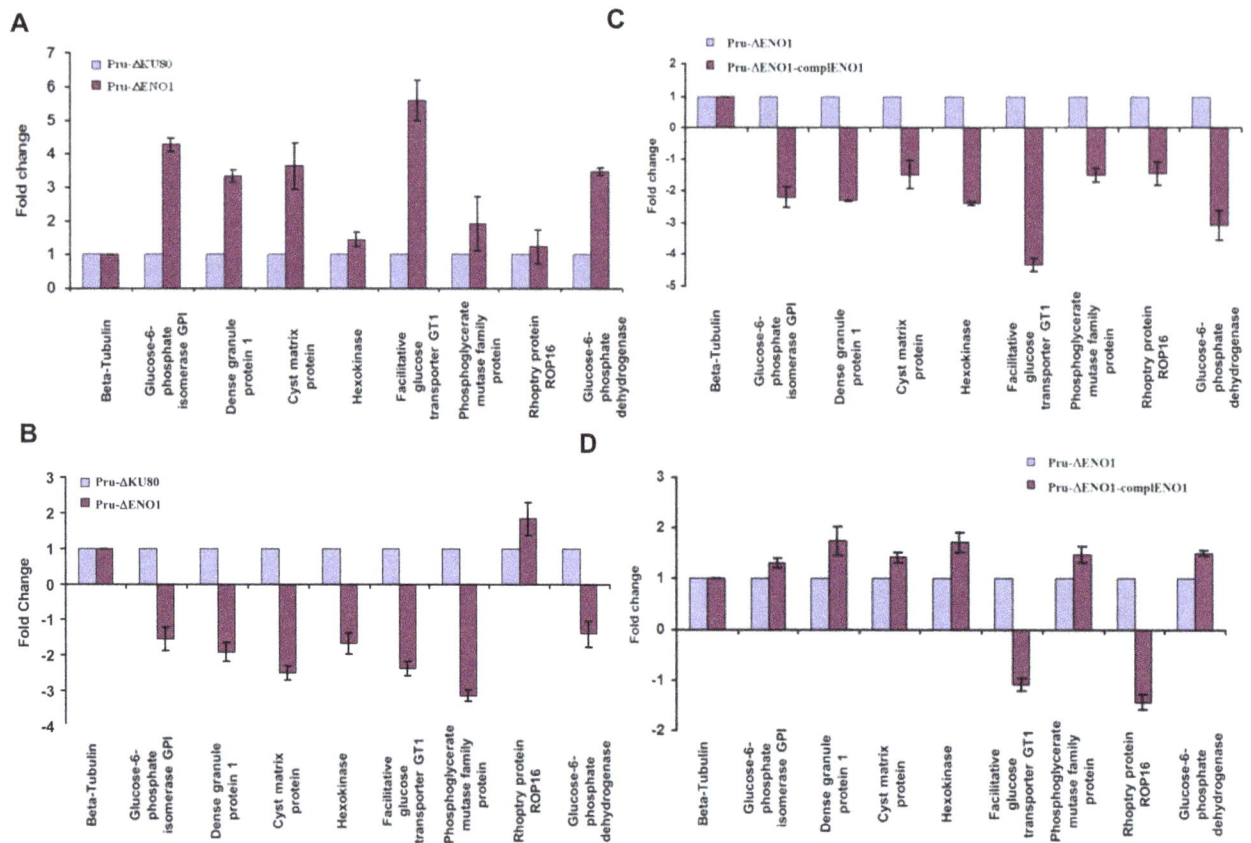

Figure 9. Targeted deletion of TgENO1 and complementation of the knockout demonstrate that nuclear enolase modulates gene transcript levels. A) Total RNA extracted from intracellular tachyzoites of PruΔku80ΔTgeno1 mutants or parental PruΔku80 grown in normal culture conditions was reverse transcribed and used for quantitative RT-PCR. The eight enolase target genes selected by ChIP-Seq and beta tubulin, a housekeeping gene, were analyzed. B) Total RNA was extracted from intracellular tachyzoites of the PruΔku80ΔTgeno1 mutant or parental PruΔku80 after bradyzoite interconversion *in vitro* using alkaline (pH 8) stress culture conditions. After reverse transcription, the cDNA was subjected to quantitative RT-PCR analysis of the eight enolase target genes selected by ChIP-Seq and beta tubulin. C) Total RNA extracted from intracellular tachyzoites of PruΔku80ΔTgeno1 mutants complemented with TgENO1 plasmid or parental PruΔku80 parasites grown in normal culture conditions was reverse transcribed and used for quantitative RT-PCR analysis of the eight enolase target genes and beta tubulin. D) Total RNA was extracted from intracellular tachyzoites from PruΔku80ΔTgeno1 mutant complemented with TgENO1 plasmid or parental PruΔku80 parasites after bradyzoite interconversion *in vitro* using alkaline (pH 8) stress culture conditions. After reverse transcription, the cDNA was subjected to quantitative RT-PCR analysis of the eight enolase target genes selected by ChIP-Seq and beta tubulin. These experiments were performed at least three times with identical results (n = 3, P<0.0001).

regulates gene transcription during stage interconversion. However, we cannot exclude the possibility that the heterologous TgSORTLR gene promoter [30] used to complement the TgENO1 KO mutants may induce an effect on the timing of activity or target occupancy by an already active TgENO1 protein in the tachyzoite. Nevertheless, our complementation with this ectopic TgENO1-HA successfully restored the changes provoked by the deletion of endogenous TgENO1gene, as expected.

In conclusion, our data reveal the existence of new nuclear regulatory functions for nuclear enolases in *T. gondii*. We hypothesize that links may be established between metabolic sensors and transcription through *T. gondii* enolases or other enzymes that participate in cell metabolism. Our previous findings indicated that TgENO1 and TgENO2 display distinct enzymatic properties that correlate with differences in the metabolic needs of the rapidly dividing virulent tachyzoites *versus* the slowly replicating encysted bradyzoites [20,36]. Furthermore, it appears that the presence of both TgENO1 and TgENO2 isoforms is not deleterious, as demonstrated by the simultaneous ectopic expression of these two isoenzymes in intracellular tachyzoites. More-

over, the absence of TgENO1 also leads to positive or negative changes in transcript levels of numerous target genes in tachyzoites or in bradyzoites. Thus, we propose a model for the functions of TgENO2 and TgENO1 in which they accumulate in the nucleus of actively dividing tachzyoites or bradyzoites and bind to gene promoters, leading to either repression or activation of transcription depending on other interacting partners, the promoter context, the intracellular niche, and differentiation status of the parasite. These novel nuclear functions of *T. gondii* enolases may involve changes in chromatin structure that control gene expression during parasite proliferation, virulence, differentiation, and cyst formation.

Supporting Information

Figure S1 Nucleotide sequence of the putative promoter of TgMag1 (cyst matrix gene 1, TGME49_070240), a 787-bp region upstream of the start codon containing the double TTTCT-like motif TTTTTCTTCTC sequence (red). This region was subcloned into the reporter luciferase vector for promoter activity studies.

Figure S2 The first site-directed mutagenesis reaction of the putative TgMag1 promoter resulting in the disruption of a single TTTCT motif within the TTTTTCTTCTC motif of the promoter to <u>ATCGA</u>TCTC to give $\varDelta_1 Tg$Mag1 plasmid, which was used in promoter activity assays.

Figure S3 The first mutagenized vector described above ($\varDelta_1 Tg$Mag1) was subjected to a second round of mutagenesis to generate <u>ATCGAGCGC</u> ($\varDelta_2 Tg$Mag2). This mutant promoter was also cloned upstream of the reporter luciferase construct for promoter activity assays.

Figure S4 Targeted deletion of the *ENO1* gene. A) Strategy for deleting the *TgENO1* gene in the Pru$\varDelta ku80\varDelta hxgprt$ strain using MPA selection. PCR1-4, locations of PCR products used to verify the genotype (not to scale; see Table S2). B) Validation of *TgENO1*-deleted clones based on the products of PCR1 (655 bp), PCR2 (356 bp), PCR3 (1,181 bp), and PCR4 (1,304 bp). Top panel: 12 randomly selected MPA-resistant clones were assayed by PCR1 and PCR2 (lanes 1–12). Lane 13 corresponds to parental Pru$\varDelta ku80$ DNA assayed by PCR1 and PCR2. Lane 14, no-template control. Clones 1–12 exhibited perfect deletion of the *TgENO1* gene. Bottom panel: clones 1–6 with deletion of *TgENO1* were assayed by PCR4 (lanes 1–6) and PCR3 (lanes 11–16). Parental Pru$\varDelta ku80$ DNA was also assayed by PCR4 (lane 7) and PCR3 (lane 17). Lanes 8 and 18 show no-template controls. DNA size ladder is shown in lanes 9 and 15. Clones 1–6 were validated as *TgENO1* knockouts with the genotype Pru$\varDelta ku80\varDelta T$-*geno1*. C) One knockout mutant was checked for perfect allelic integration and double homologous recombination using two primers (forward and reverse) specific for the open reading frame of TgENO1 and two other primers in the ORF (reverse) and the *TgENO1* promoter (forward). Superoxide dismutase (SOD) was used as a PCR control. The sequences of the primers are indicated in Table S3. All *in vitro* and *in vivo* phenotypic studies were performed using this *TgENO1* knockout mutant.

Table S1 Primers used in this study. The names and sequences of all primers used in this study are listed together with

the associated gene targets and experimental applications. Underlined regions of primer sequences indicate an additional HA Tag, no gene-specific pLIC regions were required for either cloning. F = forward primer, R = reverse primer.

Table S2 Oligonucleotide primers used for construction of *ENO1* targeting vector.

Table S3 Oligonucleotide primers used for validation of *ENO1* deletion.

Table S4 Identification of **T. gondii** genes and promoters defined by genome-wide TgENO2 occupancy and ChIP-Seq. The list of gene targets was obtained using bioinformatics analyses and genome data from http://www.toxodb.org. After comparison of data from three independent experiments only genes that were identified in ChIP using anti-TgENO2 and anti-HA are shown. Genes that were non-specifically pulled down by the naïve sera used as a negative control were removed.

Table S5 Identification of **T. gondii** genes and ORFs defined by genome-wide TgENO2 occupancy and ChIP-Seq. The list of gene ORF targets was obtained using bioinformatics analyses and genome data from http://www.toxodb.org. After comparison of data from three independent experiments only genes that were identified in ChIP using anti-TgENO2 and anti-HA are shown.

Acknowledgments

We thank E. Dewailly and J. Duflot for excellent technical assistance. Special thanks to all members of the lab for fruitful discussions. We also thank Dr D. Ferguson for providing the anti-LDH1 and anti-LDH2 antibodies.

Author Contributions

Conceived and designed the experiments: DJB ST. Performed the experiments: TM GO EW CS BAF DJB ST. Analyzed the data: EW CS DJB ST. Contributed reagents/materials/analysis tools: BAF DJB ST. Contributed to the writing of the manuscript: DJB ST.

References

1. Kim K, Weiss LM (2008) *Toxoplasma*: the next 100 years. Microbes Infect 10: 978–984.
2. Aravind L, Iyer LM, Wellems TE, Miller LH (2003) *Plasmodium* biology: genomic gleanings. Cell 115: 771–785.
3. Gardner MJ, Hall N, Fung E, White O, Berriman M, et al. (2002) Genome sequence of the human malaria parasite *Plasmodium falciparum*. Nature 419: 498–511.
4. Templeton TJ, Iyer LM, Anantharaman V, Enomoto S, Abrahante JE, et al. (2004) Comparative analysis of apicomplexa and genomic diversity in eukaryotes. Genome Res 14: 1686–1695.
5. Abrahamsen MS, Templeton TJ, Enomoto S, Abrahante JE, Zhu G, et al. (2004) Complete genome sequence of the apicomplexan, *Cryptosporidium parvum*. Science 304: 441–445.
6. Callebaut I, Prat K, Meurice E, Mornon JP, Tomavo S (2005) Prediction of the general transcription factors associated with RNA polymerase II in *Plasmodium falciparum*: conserved features and differences relative to other eukaryotes. BMC Genomics 6: 100.
7. Meissner M, Soldati D (2005) The transcription machinery and the molecular toolbox to control gene expression in *Toxoplasma gondii* and other protozoan parasites. Microbes Infect 7: 1376–1384.
8. Saksouk N, Bhatti MM, Kieffer S, Smith AT, Musset K, et al. (2005) Histone-modifying complexes regulate gene expression pertinent to the differentiation of the protozoan parasite *Toxoplasma gondii*. Mol Cell Biol 25: 10301–10314.
9. Gissot M, Kelly KA, Ajioka JW, Greally JM, Kim K (2007) Epigenomic modifications predict active promoters and gene structure in *Toxoplasma gondii*. PLoS Pathog 3: e77.
10. Hakimi MA, Deitsch KW (2007) Epigenetics in Apicomplexa: control of gene expression during cell cycle progression, differentiation and antigenic variation. Current Opinion in Microbiology 10: 357–362.
11. Balaji S, Babu MM, Iyer LM, Aravind L (2005) Discovery of the principal specific transcription factors of Apicomplexa and their implication for the evolution of the AP2-integrase DNA binding domains. Nucleic Acids Res 33: 3994–4006.
12. De Silva EK, Gehrke AR, Olszewski K, Leon I, Chahal JS, et al. (2008) Specific DNA-binding by Apicomplexan AP2 transcription factors. Proc Natl Acad Sci USA 105: 8393–8398.
13. Yuda M, Iwanaga S, Shigenobu S, Mair GR, Janse CJ, et al. (2009) Identification of a transcription factor in the mosquito-invasive stage of malaria parasites. Mol Microbiol 71: 1402–1414.
14. Hutson SL, Mui E, Kinsley K, Witola WH, Behnke MS, et al. (2010) *T. gondii* RP promoters & knockdown reveal molecular pathways associated with proliferation and cell-cycle arrest. PLoS One 5: e14057.
15. Behnke MS, Wootton JC, Lehmann MM, Radke JB, Lucas O, et al. (2010) Coordinated progression through two subtranscriptomes underlies the tachyzoite cycle of *Toxoplasma gondii*. PLoS One 5:e12354.

16. Walker R, Gissot M, Croken MM, Huot L, Hot D, et al. (2012) The *Toxoplasma* nuclear factor TgAP2XI-4 controls bradyzoite gene expression and cyst formation. Mol Microbiol 87: 641–655.

17. Walker R, Gissot M, Huot L, Alayi TD, Hot D, et al. (2013) *Toxoplasma* transcription factor TgAP2XI-5 regulates the expression of genes involved in parasite virulence and host invasion. J Biol Chem 288: 31127–31138.

18. Manger ID, Hehl A, Parmley S, Sibley LD, Marra M, et al. (1998) Expressed sequence tag analysis of the bradyzoite stage of *Toxoplasma gondii*: identification of developmentally regulated genes. Infect Immun 66: 1632–1637.

19. Dzierszinski F, Popescu O, Toursel C, Slomianny C, Yahiaoui B, et al. (1999) The protozoan parasite *Toxoplasma gondii* expresses two functional plant-like glycolytic enzymes. Implications for evolutionary origin of apicomplexans. J Biol Chem 274: 24888–24895.

20. Dzierszinski F, Mortuaire M, Dendouga N, Popescu O, Tomavo S (2001) Differential expression of two plant-like enolases with distinct enzymatic and antigenic properties during stage conversion of the protozoan parasite *Toxoplasma gondii*. J Mol Biol 309: 1017–1027.

21. Ferguson DJ, Parmley SF, Tomavo S (2002) Evidence for nuclear localisation of two stage-specific isoenzymes of enolase in *Toxoplasma gondii* correlates with active parasite replication. Int J Parasitol 32: 1399–1410.

22. Kibe M, Coppin A, Dendouga N, Oria G, Meurice E, et al. (2005) Transcriptional regulation of two stage-specifically expressed genes in the protozoan parasite *Toxoplasma gondii*. Nucleic Acids Res 33: 1722–1736.

23. Feo S, Arcuri D, Piddini E, Passantino R, Giallongo A (2000) ENO1 gene product binds to the c-myc promoter and acts as a transcriptional repressor: relationship with Myc promoter-binding protein 1 (MBP-1). FEBS Lett 473: 47–52.

24. Subramanian A, Miller DM (2000) Structural analysis of α-enolase. Mapping the functional domains involved in down-regulation of the *c-myc* protooncogene. J Biol Chem 271: 5958–5965.

25. Lee H, Guo Y, Ohta M, Xiong L, Stevenson B, et al. (2002) LOS2, a genetic locus required for cold-responsive gene transcription encodes a bi-functional enolase. EMBO J 21: 2692–2702.

26. Huynh MH, Carruthers VB (2009) Tagging of endogenous genes in a *Toxoplasma gondii* strain lacking Ku80. Eukaryot Cell 8: 530–539.

27. Yang S, Parmley SF (1997) *Toxoplasma gondii* expresses two distinct lactate dehydrogenase homologous genes during its life cycle in intermediate hosts. Gene 184: 1–12.

28. Frénal K, Callebaut I, Wecker K, Prochnicka-Chalufour A, Dendouga N, et al. (2006) Structural and functional characterization of the TgDRE multidomain protein, a DNA repair enzyme from *Toxoplasma gondii*. Biochemistry 45: 4867–4874.

29. Fox BA, Falla A, Rommereim LM, Tomita T, Gigley JP, et al. (2011) Type II *Toxoplasma gondii* KU80 knockout strains enable functional analysis of genes required for cyst development and latent infection. Eukaryot Cell 10: 1193–1206.

30. Sloves PJ, Delhaye S, Mouveaux T, Werkmeister E, Slomianny C, et al. (2012) *Toxoplasma* sortilin-like receptor regulates protein transport and is essential for apical secretory organelle biogenesis and host infection. Cell Host Microbe 17: 515–527.

31. Olguin-Lamas A, Madec E, Hovasse A, Werkmeister E, Callebaut I, et al. (2011) A novel *Toxoplasma gondii* nuclear factor TgNF3 is a dynamic chromatin-associated component, modulator of nucleolar architecture and parasite virulence. PLoS Pathog 7:e1001328.

32. Fauquenoy S, Morelle W, Hovasse A, Bednarczyk A, Slomianny C, et al. (2008) Proteomics and glycomics analyses of N-glycosylated structures involved in *Toxoplasma gondii*-host cell interactions. Mol Cell Proteomics 7: 891–910.

33. Parmley SF, Yang S, Harth G, Sibley LD, Sucharczuk A, et al. (1994) Molecular characterization of a 65-kilodalton *Toxoplasma gondii* antigen expressed abundantly in the matrix of tissue cysts. Mol Biochem Parasitol. 66: 283–296.

34. Ferguson DJ, Parmley SF (2002) *Toxoplasma gondii* MAG1 protein expression. Trends Parasitol. 18: 482

35. Knoll LJ, Boothroyd JC (1998) Isolation of developmentally regulated genes from *Toxoplasma gondii* by a gene trap with the positive and negative selectable marker hypoxanthine-xanthine-guanine phosphoribosyltransferase. Mol Cell Biol 18: 807–814.

36. Tomavo S (2001) The differential expression of multiple isoenzyme forms during stage conversion of *Toxoplasma gondii*: an adaptive developmental strategy. http://www.ncbi.nlm.nih.gov/pubmed/11429165 Int J Parasitol 31: 1023–1031.

37. Tomavo S, Fortier B, Soete M, Ansel C, Camus D, et al. (1991) Characterization of bradyzoite-specific antigens of *Toxoplasma gondii*. Infection and Immunity 59: 3750–3753.

38. Holmes M, Liwak U, Pricop I, Wang X, Tomavo S, et al. (2010) Silencing of tachyzoite enolase 2 alters nuclear targeting of bradyzoite enolase 1 in *Toxoplasma gondii*. Microbes Infect 12: 19–27.

39. Kim JW, Dang CV (2005) Multifaceted roles of glycolytic enzymes. Trends Biochem Sci 30: 142–150.

40. Tovy A, Siman Tov R, Gaentzsch R, Helm M, Ankri S (2010) A new nuclear function of the *Entamoeba histolytica* glycolytic enzyme enolase: the metabolic regulation of cytosine-5 methyltransferase 2 (Dnmt2) activity. PLoS Pathog 6:e1000775.

41. Jewett TJ, Sibley LD (2003) Aldolase forms a bridge between cell surface adhesins and the actin cytoskeleton in apicomplexan parasites. Mol Cell 11: 885–894.

42. Starnes GL, Coincon M, Sygusch J, Sibley LD (2009) Aldolase is essential for energy production and bridging adhesin-actin cytoskeletal interactions during parasite invasion of host cells. Cell Host Microbe 5: 353–364.

43. Soldati D, Boothroyd JC (1995) A selector of transcription initiation in the protozoan parasite *Toxoplasma gondii*. Mol Cell Biol 15: 87–93.

44. Mercier C, Lefebvre-Van Hende S, Garber GE, Lecordier L, Capron A, et al. (1996) Common cis-acting elements critical for the expression of several genes of *Toxoplasma gondii*. Mol Microbiol 21: 421–428.

45. Vanchinathan P, Brewer JL, Harb OS, Boothroyd JC, Singh U (2005) Disruption of a locus encoding a nucleolar zinc finger protein decreases tachyzoite-to-bradyzoite differentiation in *Toxoplasma gondii*. Infect Immun 73: 6680–6688.

12

A Full-Length *Plasmodium falciparum* Recombinant Circumsporozoite Protein Expressed by *Pseudomonas fluorescens* Platform as a Malaria Vaccine Candidate

Amy R. Noe[1,9], Diego Espinosa[2,9], Xiangming Li[3], Jordana G. A. Coelho-dos-Reis[3], Ryota Funakoshi[3], Steve Giardina[1], Hongfan Jin[4], Diane M. Retallack[4], Ryan Haverstock[4], Jeffrey R. Allen[4], Thomas S. Vedvick[5], Christopher B. Fox[5], Steven G. Reed[5], Ramses Ayala[1], Brian Roberts[1], Scott B. Winram[1], John Sacci[6], Moriya Tsuji[3], Fidel Zavala[2], Gabriel M. Gutierrez[1]*

1 Leidos Inc., Frederick, Maryland, United States of America, 2 Johns Hopkins Malaria Research Institute and Department of Molecular Microbiology and Immunology, Johns Hopkins Bloomberg School of Public Health, Johns Hopkins University, Baltimore, Maryland, United States of America, 3 HIV and Malaria Vaccine Program, Aaron Diamond AIDS Research Center, Affiliate of The Rockefeller University, New York, New York, United States of America, 4 Pfenex Inc., San Diego, California, United States of America, 5 Infectious Disease Research Institute, Seattle, Washington, United States of America, 6 Department of Microbiology and Immunology, University of Maryland School of Medicine, Baltimore, Maryland, United States of America

Abstract

The circumsporozoite protein (CSP) of *Plasmodium falciparum* is a major surface protein, which forms a dense coat on the sporozoite's surface. Preclinical research on CSP and clinical evaluation of a CSP fragment-based RTS, S/AS01 vaccine have demonstrated a modest degree of protection against *P. falciparum*, mediated in part by humoral immunity and in part by cell-mediated immunity. Given the partial protective efficacy of the RTS, S/AS01 vaccine in a recent Phase 3 trial, further improvement of CSP-based vaccines is crucial. In this report, we describe the preclinical development of a full-length, recombinant CSP (rCSP)-based vaccine candidate against *P. falciparum* malaria suitable for current Good Manufacturing Practice (cGMP) production. Utilizing a novel high-throughput *Pseudomonas fluorescens* expression platform, we demonstrated greater efficacy of full-length rCSP as compared to N-terminally truncated versions, rapidly down-selected a promising lead vaccine candidate, and developed a high-yield purification process to express immunologically active, intact antigen for clinical trial material production. The rCSP, when formulated with various adjuvants, induced antigen-specific antibody responses as measured by enzyme-linked immunosorbent assay (ELISA) and immunofluorescence assay (IFA), as well as CD4+ T-cell responses as determined by ELISpot. The adjuvanted rCSP vaccine conferred protection in mice when challenged with transgenic *P. berghei* sporozoites containing the *P. falciparum* repeat region of CSP. Furthermore, heterologous prime/boost regimens with adjuvanted rCSP and an adenovirus type 35-vectored CSP (Ad35CS) showed modest improvements in eliciting CSP-specific T-cell responses and anti-malarial protection, depending on the order of vaccine delivery. Collectively, these data support the importance of further clinical development of adjuvanted rCSP, either as a stand-alone product or as one of the components in a heterologous prime/boost strategy, ultimately acting as an effective vaccine candidate for the mitigation of *P. falciparum*-induced malaria.

Editor: Takafumi Tsuboi, Ehime University, Japan

Funding: This work was funded by NIAID/DMID Malaria Vaccine Production Support Services (MVPSS) contract number AI-N01-054210 to Leidos Health Inc. and executed through subcontracts to Johns Hopkins University, Aaron Diamond AIDS Research Center, Pfenex Inc., and the University of Maryland as well as a material transfer agreement with the Infectious Disease Research Institute. Co-authors Amy R. Noe, Steve Giardina, Ramses Ayala, Brian Roberts, Scott B. Winram and Gabriel M. Gutierrez are employed by Leidos Inc. Leidos Inc., as directed by the MVPSS prime contract, provided support in the form of salaries for authors, ARN, SG, RA, BR, SBW and GMG and Leidos Inc. planned the work, designed the experiments, conducted data analysis, made the decision to publish, and played a major role in manuscript preparation but had no role in data collection. The specific roles of these authors are articulated in the Author Contributions section. Co-authors Hongfan Jin, Diane Retallack, Ryan Haverstock and Jeff Allen are employed by Pfenex Inc. Pfenex Inc., as directed by an MVPSS subcontract, provided support in the form of salaries for authors HJ, DR, RH and JA, and Pfenex Inc. performed study design, conducted data collection and analysis, and assisted in preparation of the manuscript regarding expression and analytical testing of recombinant CSP. Pfenex, Inc. did not play an additional role in planning the work or the decision to publish. The specific roles of these authors are articulated in the Author Contributions section.

* Email: gabriel.m.gutierrez@leidos.com

9 These authors contributed equally to this work.

Introduction

World-wide, mortality from malaria is estimated in the hundreds of thousands each year, with more than 200 million cases estimated for 2012 [1]. Given the incidence of this disease, development of a malaria vaccine is a priority [2]. However, the complex lifecycle of the malaria parasite has, in part, hampered efforts to achieve this goal. Briefly, parasites are first delivered into the human host as sporozoites by the bite of an infected mosquito, and then rapidly travel to the liver where they invade hepatocytes. Once in the liver, they develop into merozoites and are subsequently released into the blood stream, where they invade red blood cells. Merozoite multiplication results in many thousands of parasite-infected erythrocytes in the host bloodstream, leading to clinical disease. Therefore, targeting a vaccine that prevents the parasite from entering and/or exiting the liver would provide the most effective prevention of disease. For children younger than age 5 residing in endemic regions of Africa and other developing nations, sterile immunity is the most desirable outcome of a vaccine [1]. The strategy of immunizing with high doses of irradiated sporozoites has, in fact, achieved immunity to the pre-erythrocytic stages of Plasmodium [3,4]. Importantly, immune response to circumsporozoite protein (CSP) has been shown to be a necessary component of the protective immune response induced by irradiated sporozoite, since in the absence of an immune response to CSP, protection in mice achieved upon vaccination with irradiated sporozoites was greatly diminished compared to that in mice mounting an active immune response to CSP [5,6].

CSP is the major surface protein of the sporozoite and forms a dense coat on the parasite's surface. Structurally, CSP is divided into three regions: (1) the NH_2-terminal region, which includes the ligand-binding domain [7], multiple human HLA-restricted epitopes [8,9], a proteolytic processing site [10], and a region I; (2) a centrally-located repeat region; and (3) the COOH-terminus, which contains a type I thrombospondin repeat region. Since its discovery more than 30 years ago, much attention has been given to developing a vaccine against CSP [11]. Antibody response to sporozoites is largely to the immuno-dominant central repeat region of CSP, and previous studies have shown that high titers of antibodies to this region can confer protection from malaria infection in rodent models [12,13,14]; however, translating these results to humans has been difficult. Various CSP configurations have only generated a sufficient immune response to confer partial protection, and the precise specificity and function of protective antibodies has not been determined. Considerable evidence from studies in animals and humans suggests that both antibody and T-cell immune responses are likely needed to control malaria infection [15,16,17,18,19,20,21,22,23].

Phase 2 clinical trials of an RTS, S/AS01 vaccine, a virus-like particle (VLP) consisting of a fragment (central repeat and C-terminal regions) of the CSP fused to a Hepatitis B Virus Surface Antigen (HBVsAg) (developed by GlaxoSmithKline), provided evidence for a high level of anti-CSP antibody response that correlated with reduced clinical malaria episodes. However, analysis of the Phase 3 trial of the RTS, S/AS01 vaccine showed that only one third of infants were protected by the CSP-fragment-based vaccine candidate [24]. The partial protective efficacy of the RTS, S/AS01 vaccine has demonstrated viability of a CSP-based malaria vaccine and suggests that further improvement of a CSP-based vaccine may provide an increase in clinical efficacy. Considering the nature of domains contained within the N-terminus of the protein, one key approach by which to improve the efficacy of a CSP-based subunit vaccine would be to utilize the full-length protein. In fact, data indicate that antibody responses to the N-terminus of CSP are associated with protection [25,26]. Therefore, it is reasonable to postulate that a properly folded, full-length CSP will enhance the quality, magnitude, and breadth of protective antibody and T-cell responses. Indeed, a recent report showed that a full-length, recombinant P. falciparum CSP, expressed in E. coli, provided significant protection when administered with adjuvants in a malaria mouse challenge model [27]. However, manufacture of adequate amounts of clinical-grade full-length rCSP has been problematic [28]. Specifically, current Good Manufacturing Practice (cGMP) production of the protein at a large economical scale suitable for clinical evaluation and advanced product development has yet to be reported. This lack of clinical grade, full-length CSP may be due in large part to inherent unique properties of the P. falciparum parasite, which include an extremely A/T-rich genome with many lysine and arginine repeats, and proteins that contain multiple disulfide bonds. Expression of malaria proteins in bacterial systems, such as E. coli, often results in insoluble expression that requires purification from inclusion bodies and steps to refold the protein. Furthermore, malaria parasites lack N-linked glycosylation machinery, thereby making common eukaryotic expression platforms less effective. Therefore, as part of our CSP vaccine development strategy, we sought novel expression platforms to overcome the inherent obstacles associated with expression of malaria antigens listed above.

Here, we describe the successful expression and pre-clinical testing of a recombinant, full-length P. falciparum CSP that is immunogenic and biologically active. Briefly, we employed a high-throughput process to rapidly screen hundreds of Pseudomonas fluorescens expression strains starting from plasmid construction with varying promoters, secretion leaders, and translation initiation sequences. Strains producing soluble, high-yield, and full-length P. falciparum rCSP were identified and characterized. The purified full-length recombinant CSP (rCSP) was tested in animal studies for immunogenicity and efficacy in combination with several adjuvant formulations. Initially, incomplete Freund's adjuvant was used to establish baseline activity across a panel of biological assays. Other adjuvants more appropriate for use in humans were later evaluated with rCSP, including Alhydrogel, AdjuPhos, Glucopyranosyl Lipid Adjuvant-Stable Emulsion (GLA-SE), and a GLA-liposome-QS21 formulation (GLA-LSQ). When adjuvanted with GLA formulations, rCSP produced robust antibody titers that neutralized invasion of malarial sporozoites into hepatocytes in vitro. Furthermore, immunization with GLA rCSP formulations was effective at inhibiting liver stage development in mice (>90%) using a transgenic parasite model of infection. Although a detectable CD4+ T-cell response was seen with GLA formulations in mice, CD8+ T-cell response was low. In an attempt to improve the magnitude of T-cell response, a heterologous prime-boost strategy with rCSP and Ad35CS was employed. Ad35CS was used because of its low seroprevalence in endemic regions and its safety profile in the clinic [29]. Interestingly, we found that the order in which the various components of the prime-boost strategy were administered played a significant role in eliciting immunologically potent effects.

This study corroborates the hypothesis that immunization with full-length CSP can effectively induce both humoral and cell-mediated immune responses. Such a vaccine is more likely to achieve sufficient levels of sterilizing immunity by antibody neutralization of sporozoites prior to hepatocyte infection and/or through T-cell-mediated clearing of malaria-infected hepatocytes. Collectively, this is the first report of a process for generating clinical-grade full-length recombinant CSP that is not only manufacturable at a robust scale, but also is highly immunogenic, thus allowing us to swiftly move the rCSP forward for clinical evaluation.

Results

Design and Selection of rCSP Expression Strains

Twenty *P. fluorescens* host strains, including protease deletion mutants and strains expressing protein folding modulators, were each transformed with sixteen different expression plasmids comprised of varying secretion leader sequences and ribosome binding site combinations. The resulting 320 *P. fluorescens* expression strains were screened for rCSP expression at the 0.5 mL scale. Five strains with the highest level of expression were selected from the initial screening and evaluated at the 4 mL fermentation scale, under 9 different fermentation conditions such as varied combinations of inducer concentration, induction pH, and temperature (data not shown). Three strains expressing high levels of rCSP (CSP-1, CSP-2, and CSP-3) were selected for further characterization (Figure 1a). Conditions were then scaled up to a high cell density fermentation process in 1 L conventional bioreactors where yields ranged up to 4 g/L depending on strain and fermentation condition tested (data not shown). Purification was initially performed using a 2-column method with anion-exchange and hydrophobic interaction chromatography to achieve material of sufficient purity (>90%) for analytical assessment and in vivo functional tests. SDS-PAGE analysis showed a single protein band run at apparent molecular weight ~55 kDa for all three strains (Figure 1b). Western blot analysis using an anti-CSP mAb confirmed the band as CSP (Figure 1c). A faint band at approximately 120 kDa was also visible by western blot and represents a dimeric form of rCSP not reduced under the sample preparation conditions used for the western blot assay.

Purification and Analytical Assessment of rCSP

In addition to the initial assessments by gel electrophoresis, the purity and the extent of rCSP multimerization for all lots were analyzed by size exclusion (SE) and reversed phase (RP) high-performance liquid chromatography (HPLC). During development of the purification process, a strong tendency of the protein to multimerize was alleviated by addition of a disaggregant (urea) and a reducing agent to process buffers. A representative SE-HPLC chromatogram shows a minor peak at approximately 15 minutes corresponding to the rCSP dimer, and a major peak at approximately 17 minutes corresponding to the rCSP monomer (Figure 2a). A representative RP-HPLC chromatogram shows a major peak at approximately 19.2 minutes corresponding to the rCSP monomer with a detectable shoulder that was identified as a pyroglutamic acid derivative of rCSP (data not shown) as well as a minor peak at approximately 20.6 minutes corresponding to the rCSP dimer (Figure 2b). Based on HPLC results, the percentage of dimer was estimated as <10% after the 2-column purification process and final buffer exchange.

To confirm identity and integrity of the expression products, N-terminal sequencing and intact mass analysis were performed. Sequencing the N-terminus of each expression product confirmed the proteins as CSP for all three strains. In addition, it was determined that rCSP from strain CSP-1 was 100% intact, rCSP from strain CSP-2 was only 46% intact with clipping occurring after amino acid 13 or amino acid 17, and rCSP from strain CSP-3 was 68% intact with the first 5 amino acids being trimmed from the clipped protein (Figure 2c). Intact mass analysis confirmed these results as the transformed and integrated maximum entropy spectra revealed the presence of a principal mass component at 38,724–38,726 Da for all three samples; the theoretical molecular weight for full-length CSP matches at 38,725 Da. A single full-length expression product was found for strain CSP-1 (Figure 2d), while the other two strains had N-terminal truncated variants present. For the CSP-2 strain, two truncated rCSP species were found, one starting at Val14 and the other at Leu18 with actual

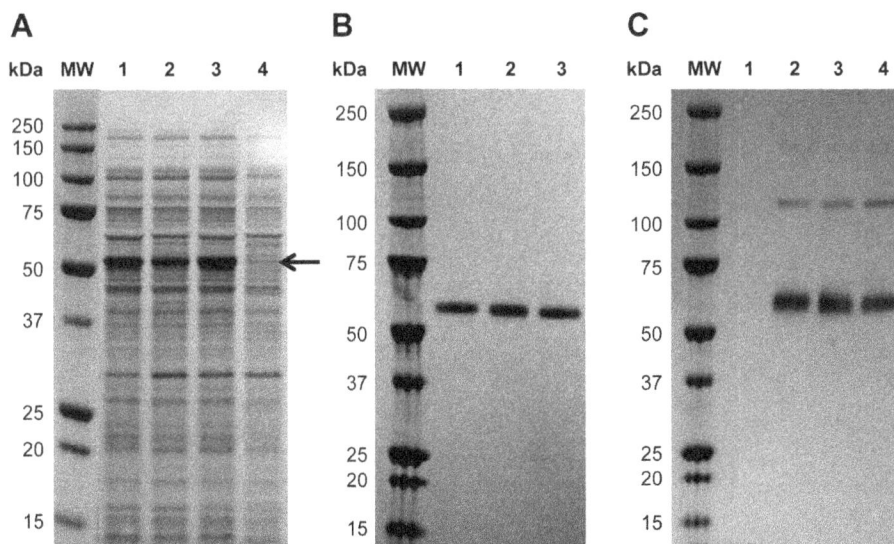

Figure 1. rCSP Strain Selection and Identification. (**A**) Reduced SDS-PAGE analysis of lysate samples prior to purification. Lanes 1–3 represent samples from strain CSP-1, CSP-3, and CSP-2 fermentations, respectively. Lane 4 represents a sample from a culture of *P. fluorescens* transformed with plasmid vector containing no insert. Arrow denotes rCSP in lysate samples. (**B**) Reduced SDS-PAGE analysis of purified proteins from a 2-column purification process. Lanes 1–3 represent samples from CSP-1, CSP-2, and CSP-3 productions, respectively. (**C**) Western blot of non-reduced purified CSP-1, CSP-2, and CSP-3 proteins (lanes 2–4, respectively). Lane 1 represents a sample from a culture of *P. fluorescens* transformed with plasmid vector containing no insert.

A

B

C

D

E

F

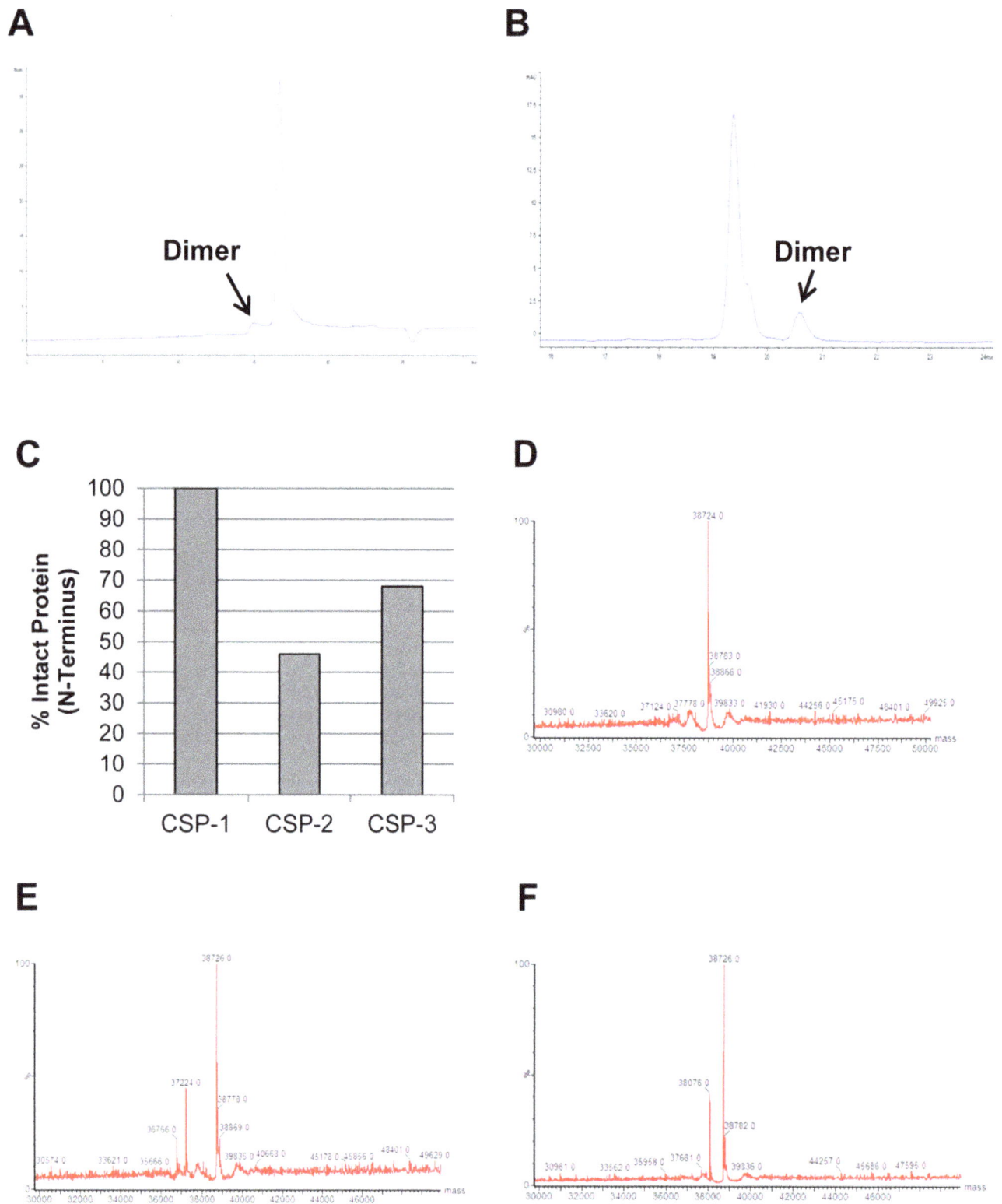

Figure 2. Analytical Assessment of rCSP Purity. Expression products from the three strains (CSP-1, CSP-2, and CSP-3) were characterized using a panel of analytical assays. (**A**) A SE-HPLC chromatogram of rCSP from the CSP-1 strain purified with the 2-column process shows a major peak representing monomeric rCSP and minor peak representing dimeric rCSP. The dimer represents ~9% of the rCSP present. (**B**) A RP-HPLC chromatogram of rCSP from the CSP-1 strain purified with the 2-column process shows a major peak representing monomeric rCSP and minor peak representing dimeric rCSP. The dimer represents ~9.5% of the rCSP present. (**C**) N-terminal sequencing demonstrated that the CSP-1 protein was fully intact; however, CSP-2 and CSP-3 showed clipping at the N-terminus. (**D**) Intact mass analysis of rCSP from the CSP-1 strain showed a single peak at 38724 Da. (**E**) Intact mass analysis of rCSP from the CSP-2 strain showed one peak at 38,726 Da and two truncated forms with minor peaks at 37,224 Da and 36,766 Da. (**F**) Intact mass analysis of rCSP from the CSP-3 strain showed one peak at 38726 Da and one truncated form with a minor peak at 38,076 Da.

molecular weights of 36,766 Da and 37,224 Da, and theoretical molecular weights of 37,220 Da and 36,765 Da, respectively (Figure 2e). For the CSP-3 strain, one truncated species was found starting at Tyr6 with an actual molecular weight of 38,076 Da and a theoretical molecular weight of 38,073 Da (Figure 2f).

Biological and Functional Evaluation of rCSP Strains

To address whether the CSP strains have biological activity and whether the N-terminus of the CSP protein impacts its biological function, we tested all three expression products for their ability to elicit an immune response in vivo. Furthermore, the ability of the CSP strains to induce protective anti-plasmodial response was assessed by quantifying the in vitro parasite neutralization activity of the generated antisera and the in vivo inhibition of liver stage development in a mouse model. For this initial study, CSP-1, CSP-2, and CSP-3 derived proteins were formulated in incomplete Freund's adjuvant. ELISA titers were similar among pooled sera from mice immunized with protein generated from the three strains, as measured against a CSP repeat region peptide [NANP]$_6$ by ELISA (Figure 3a). Optical density (OD) in sera from mice injected only with incomplete Freund's adjuvant (the adjuvant control) was less than 0.1. In addition to ELISA, immunogenicity was also measured via immunofluorescence assay IFA using transgenic P. berghei sporozoites containing the P. falciparum repeat region and portion of the N-terminal region of CSP (Pb-CS[Pf]) (Figure 3a). As with ELISA, the IFA titers were similar among sera from mice immunized with the three recombinant proteins. The IFA also confirmed reactivity of all the anti-rCSP sera to the surface of Pb-CS(Pf) sporozoites (Figure 3b, only reactivity of anti-CSP-1 sera is shown). There was no specific sporozoite surface fluorescence seen with the sera from the adjuvant control mouse cohort at the highest concentration tested (data not shown). To determine whether ELISA and IFA titers were associated with functional antibodies, the same sera pools were assessed in an inhibition of sporozoite invasion (ISI) assay with a cultured hepatocytoma cell line. More pronounced differences among the three sera pools were seen in the ISI assay (Figure 3c). Indeed, a 1:100 dilution of anti-CSP-1 serum achieved an 89% reduction in the level of hepatocyte invasion by P. falciparum sporozoites (as compared to naïve mouse sera); however, similar dilutions of anti-CSP-2 or anti-CSP-3 sera were only able to attain 77% and 70% invasion reduction, respectively. To further compare protective efficacy of the three rCSP expression products, immunized mice were challenged with Pb-CS(Pf) sporozoites to determine the degree of the inhibition of liver stage (LS) parasite development (Figure 3c). The reduction in parasite liver load in mice immunized with CSP-1 or CSP-3 proteins was statistically significant compared to naïve mice and mice administered adjuvant alone. However, this was not the case with mice immunized with CSP-2. The reduction in liver load in mice administered CSP-1, CSP-2, or CSP-3 proteins was 98% (p = 0.008), 76% (p = 0.151), and 97% (p = 0.008), respectively, when standardized to the naïve mouse control group. Liver parasite load in mice administered adjuvant alone was similar to that of the naïve control group with only a non-significant decrease (16%) seen in the adjuvant alone group as compared to the naïve control group. The results from the ISI and LS inhibition assays suggest that even small deletions at the N terminal can have a deleterious effect on immunogenic efficacy; therefore, based on the percentage of intact sequence and the optimal performance in the protective efficacy assays, CSP-1 was selected as the lead candidate strain for further rCSP production and rCSP-based vaccine development.

Adjuvant Formulation Screening for rCSP Vaccine

While inclusion of the disaggregant (urea) and a reducing agent in process buffers resulted in low levels of rCSP multimerization, ultimately, a third column (utilizing hydrophobic interaction chromatography) was added to the purification scheme as an additional polishing step and to further reduce the level of rCSP dimer in the final product. The successful expression and production of rCSP allowed us to further advance product development by moving forward in search of appropriate formulations suitable for clinical testing. Along with two common alum-based adjuvant systems, we also selected two TLR4 agonist-based adjuvants, GLA-SE and GLA-LSQ, for testing. Both formulations consist of a highly pure [30], synthetic hexaacylated monophosphorylated lipid A-like structure that has been shown to be an effective adjuvant for vaccine candidates against other infectious diseases [31,32,33]. Another important consideration for optimal adjuvant effectiveness is to improve formulation by using appropriate systems to increase uptake by cells, particularly antigen-presenting cells, and to provide sustained release of the antigen. In the case of GLA-SE, the GLA is formulated into a squalene-in-water nanoemulsion. In contrast, GLA-LSQ is a nanoliposomal formulation that includes QS21, a saponin. Each adjuvant formulation was mixed with rCSP prior to immuniza-tion. Protein formulated in incomplete Freund's adjuvant was also included in several experiments to bridge these data with the previously detailed down-selection data. Immunization with rCSP in the context of GLA-SE, GLA-LSQ, and incomplete Freund's adjuvant formulations resulted in similar immunogenicity of the different sera when evaluated by ELISA and IFA, whereas the titers of the sera obtained from mice immunized with both alum-based adjuvants were considerably lower (Figure 4a and 4b). ELISA titer was calculated at OD = 1.0; sera from mice injected with adjuvant only (each adjuvant was tested) did not achieve ODs of 1.0 by ELISA (ODs were <0.15 with these sera pools). The pattern of Pb-CS(Pf) sporozoite surface fluorescence with sera from all formulations evaluated in IFA was characteristic of reactivity to CSP and similar to that shown in Figure 3b (data not shown).

To further investigate whether the vaccine-induced immune responses are sustainable, the longevity of the antigen-specific humoral immune response to each of the rCSP formulations was measured in a time-course study throughout the entire injection schedule and for several months following the final injection (Figure 4c). Peak antibody response was seen 2 or 8 weeks after the final injection, depending on the formulation, with the GLA-LSQ formulation demonstrating the highest antibody response overall, followed by GLA-SE and then the alum-based formulations. Sera from mice injected with adjuvant-only did not achieve an OD of 1.0 by ELISA (ODs were <0.15 with these sera). The overall titers achieved with the GLA-SE and GLA-LSQ formulations were significantly higher compared to the alum-based formulations. In addition, the overall titers achieved with the GLA-LSQ formulation were significantly higher compared to the GLA-SE formulation (repeated measures ANOVA and post hoc comparisons). At the final collection time point (approximately 22 weeks after the final immunization), the titers achieved with the GLA-LSQ continued to be significantly higher compared to the GLA-SE and alhydrogel formulations (p values = 0.003 and 0.00003, respectively using significant comparisons [Tukey] from a one-way ANOVA).

The ability of the GLA-SE and GLA-LSQ formulations to induce protective immunity was also comparable to that of the incomplete Freund's adjuvant formulation. Upon challenge of the immunized mice with Pb-CS(Pf) sporozoites, all three formula-

A

B

C

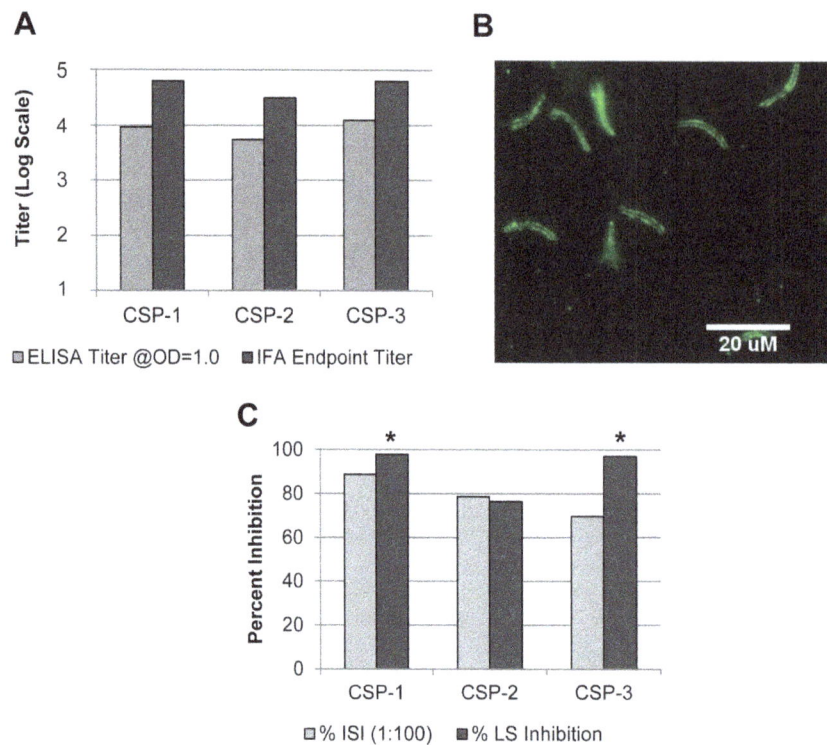

Figure 3. Immunogenicity and Efficacy of rCSP Immunization in Mice. Five C57BL/6 mice per cohort were immunized 3x with 25 µg of rCSP from 3 production strains (CSP-1, CSP-2, and CSP-3) formulated in incomplete Freund's adjuvant. Sera were collected 2 weeks after the last immunization and pooled for analysis by ELISA, IFA, and ISI. Mice were challenged with Pb-CS(Pf) parasites 3 weeks after the last immunization. (**A**) Pooled sera from mice immunized with the 3 recombinant proteins showed similar ELISA and IFA titers. ELISA titers to a [NANP]$_6$ peptide were calculated at OD = 1.0 based on 4-parameter logistic curve fits, and IFA titers were determined based on the lowest serum dilution to give sporozoites-specific fluorescence above background level shown by sera from naïve mice. (**B**) Sera from rCSP-immunized mice reacted to Pb-CS(Pf) sporozoites demonstrated the expected pattern of surface fluorescence by IFA. (**C**) Sera from rCSP-immunized mice diluted 1:100 and assayed for the ability to block Pf sporozoite infection of hepatocytes in an ISI assay demonstrated the highest ISI activity for anti-CSP-1 protein sera. The percent inhibition is shown relative to the number of sporozoites after incubating with the sera of naïve mice. Percent inhibition of LS parasite development in rCSP-immunized mice challenged with Pb-CS(Pf) sporozoites normalized to naïve control mice is also shown. The reduction of LS parasites in livers of mice immunized with CSP-1 or CSP-3 proteins was statistically significant based on the mean parasite-specific 18 s rRNA copy number compared the level of LS parasites in livers of naïve and adjuvant-only administered mice (as noted by asterisks).

tions resulted in significant inhibition of LS parasite development (compared to the naïve and adjuvant-only control mice), achieving >80% inhibition when administered as a 5 µg dose of rCSP and >90% inhibition when administered as a 20 µg dose of rCSP with these adjuvants (Figure 5a). Likewise, an ISI assay showed that a 1:100 dilution of the sera obtained from mice immunized with rCSP in the context of either GLA formulation resulted in a comparable reduction in sporozoite invasion to that observed with sera from mice immunized with rCSP emulsified in incomplete Freund's adjuvant (Figure 5b). Since neither of the rCSP alum-based formulations showed effectiveness in the LS inhibition assay, the immune sera was not tested by ISI, and these formulations were not considered for further testing.

Sterile Protection Induced by Adjuvanted rCSP in Mice

While assessment of the amounts of parasite-specific rRNA in the liver by a real-time qRT-PCR translates to the percentage reduction of liver parasite load and provides a quantitative measure of protective efficacy, we also wanted to measure the level of sterile protection achieved upon vaccination with our lead candidate antigen-adjuvant combinations by monitoring the presence or absence of blood stage parasites following sporozoite challenge. Therefore, mice were immunized three times with 25 µg of rCSP formulated in either GLA-SE or GLA-LSQ and

challenged 2 weeks after the last immunization. The presence of parasites in the blood was monitored by daily blood smear; sterile protection was determined as lack of parasites in thin blood smears (Table 1). The GLA-SE formulation exhibited slightly higher sterile protection (50%) as compared to the GLA-LSQ formulation (40%). No protection was seen in the adjuvant control or naïve mouse groups.

Cell-Mediated Immunity Elicited by Adjuvanted rCSP

Although the humoral response to the full-length rCSP is encouraging, the long-term effectiveness of a malaria vaccine may also require activation of the cell-mediated immune response. The level of cell-mediated immunity induced by the vaccine candidates was measured by ELISpot. Briefly, liver and spleen from immunized BALB/c mice (MHC haplotype H-2^d) were harvested on study days 46 and 56, and lymphocytes were re-stimulated with peptides containing either an H-2A^d (I-A^d)-restricted CD4 epitope (EYLNKIQNSLSTEWSPCSVT), which is located in the C-terminal region of CSP, or an H-2K^d-restricted CD8 epitope (NYDNAGTNL), which is located in the N-terminal region of CSP. The relative number of epitope-specific, IFN-γ-secreting T-cells among lymphocytes of immunized mice were determined, as previously described [34]. Results showed that re-stimulation with the CD4 epitope-containing peptide produced a much stronger T-

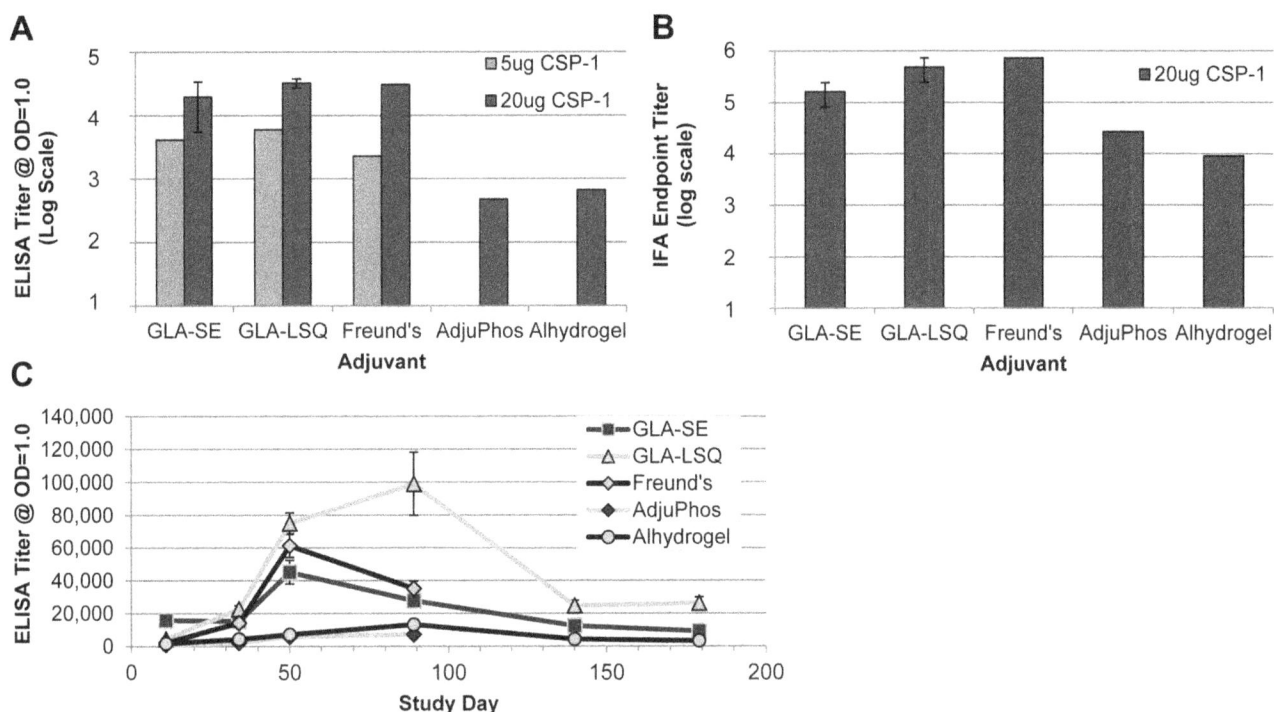

Figure 4. Comparison of Humoral Responses with rCSP Adjuvant Formulations in Mice. C57BL/6 mice were immunized 3x with 2.5 µg, 5 µg, or 20 µg of rCSP (CSP-1) formulated in different adjuvants. (**A**) Pooled sera (7 mice per cohort) collected 2 weeks after the final immunization from mice immunized with GLA-SE, GLA-LSQ and incomplete Freund's adjuvant showed similar ELISA titers to a $[NANP]_6$ peptide at OD = 1.0 based on 4-parameter logistic curve fits. The titers of the sera from mice injected with rCSP formulated in alum-based adjuvants were lower. The immunogenicity of GLA-SE and GLA-LSQ formulations with 20 µg rCSP was assessed in multiple independent studies (5–7 mice per cohort were used for each study); average titer is shown with error bars represented as standard error. (**B**) Pooled sera (7 mice per cohort) collected 2 weeks after the final immunization from mice immunized with GLA-SE, GLA-LSQ and incomplete Freund's adjuvant demonstrated the highest IFA endpoint titers based on the lowest serum dilution to give Pb-CS(Pf) sporozoite surface fluorescence above negative sera control level. The titers of the sera from mice immunized with rCSP formulated in alum-based adjuvants were lower. The immunogenicity of GLA-SE and GLA-LSQ formulations with 20 µg rCSP were assessed in multiple independent studies (5–7 mice per cohort were used for each study); average endpoint titer is shown with error bars represented as standard error. (**C**) Individual immune mouse sera (10 mice per cohort) evaluated in a long-range time course study demonstrated the highest peak ELISA titer to rCSP with the GLA-LSQ formulation ~8 weeks after the final immunization with a 2.5 µg of rCSP administered per dose for each formulation tested. Average titer at OD = 1 is shown with error bars represented as standard error. Note that a standard scale (rather than log scale) is shown for the y-axis.

cell response than that of the CD8 epitope-containing peptide. Nevertheless, lymphocytes harvested from mice immunized with rCSP formulated with GLA-LSQ demonstrated higher responses to re-stimulation compared to those from mice immunized with rCSP formulated in GLA-SE (Figure 6a). A statistically significant difference in splenocytes harvested from mice immunized with the two formulations was seen with re-stimulation using both the CD4 peptide (p = 0.007 for day 46 and p = 0.015 for day 56) and CD8 peptide (p = 0.008 for day 46). In a similar experiment, three separate pools of overlapping peptides (15-mers) comprising the N-terminal region, central repeat region, or the C-terminal region of CSP were also used for re-stimulation. Similar results, albeit with slightly lower levels of IFN-γ secreting T lymphocytes, were found with the N-terminal and C-terminal peptide pools as compared to the re-stimulation with CD8 and CD4 epitope-containing peptides, respectively. As anticipated, no response above baseline was seen when the lymphocytes were re-stimulated with the central repeat peptide pool (data not shown), which has a putative H-$2A^b$ (I-A^b)-restricted CD4 epitope but no putative H-2^d epitope.

An ELISA was performed with sera collected from the same mice used for ELISpot assays, with rCSP as an antigen. We confirmed that rCSP formulated in either of the GLA adjuvants induced a strong CSP-specific antibody response as compared to

either naïve animals (Figure 6b) or mice immunized with adjuvants only (data not shown). Consistent with our previous ELISA results, in which the antibody response was measured to either the CSP central repeat region or to the recombinant protein (see Figure 4), sera titers in mice immunized with the rCSP GLA-LSQ formulation trended higher than in mice immunized with the rCSP GLA-SE formulation with a statistically significant difference seen on day 46 (p = 0.089, p = 0.001, and p = 0.056 on study days 36, 46, and 56, respectively).

Heterologous Prime-Boost Strategies Using rCSP and Ad35CS

Prime-boost vaccination regimens are standard for many licensed products; however, the concept of priming and boosting with different vaccine delivery methods (i.e., viral vectored and recombinant protein) while utilizing the same antigen (heterologous prime-boost) is a more recent strategy that has shown promise for improving the breadth of immunogenicity and protection [18,35,36]. For CSP-based vaccines, preclinical studies have demonstrated the effectiveness of this strategy [37,38]. In particular, adenovirus has been shown to be an excellent delivery vehicle for induction of cellular-mediated immunity [39,40,41,42] and has been used in other heterologous prime-boost studies [43].

A

B

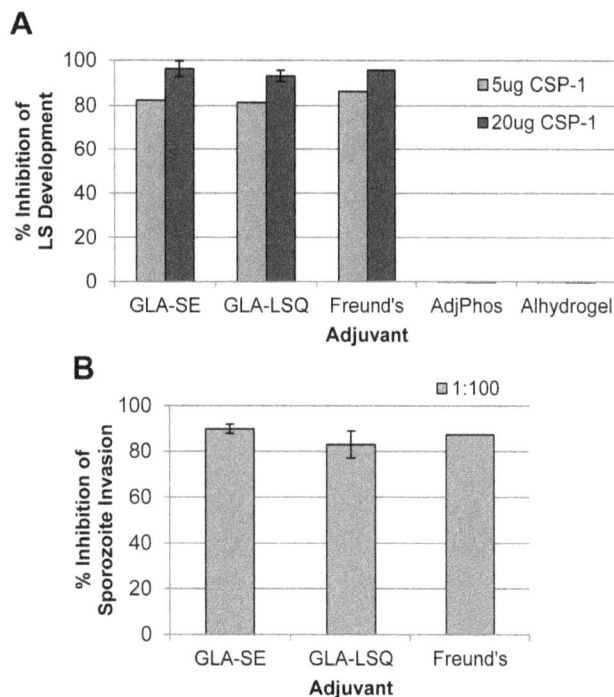

Figure 5. ISI and Transgenic Parasite Protection Comparisons of rCSP Adjuvant Formulations. Seven C57BL/6 mice per cohort were immunized 3x with 5 μg or 20 μg of rCSP (CSP-1) formulated in different adjuvants. Sera were collected 2 weeks after the last immunization and pooled for analysis. Mice were challenged with *Pb*-CS(*Pf*) parasites 3 weeks after the last immunization. (**A**) Percent inhibition of LS parasite development in rCSP-immunized mice versus naïve mice is shown upon *Pb*-CS(*Pf*) sporozoites challenge. A statistically significant reduction of LS parasites in livers of mice immunized with the GLA-SE or GLA-LSQ formulations was seen compared to the level of LS parasites in livers of naïve and adjuvant-only control mice. However, no reduction of LS parasites was seen with alum-based adjuvants. GLA-SE and GLA-LSQ formulations with 20 μg rCSP were assessed in multiple independent studies (5–7 mice per cohort for each study); average percent inhibition is shown with error bars represented as standard error. (**B**) Sera from mice immunized with rCSP in different GLA formulations were diluted 1:100 or 1:500 and assayed for the ability to block *Pf* sporozoite infection of hepatocytes in an ISI assay. Percent inhibition is shown relative to a naïve sera control. Sera from GLA-SE adjuvanted mice demonstrated higher ISI activity compared to sera from those adjuvanted with GLA-LSQ. Where multiple independent experiments were performed, average percent inhibition is shown with error bars represented as standard error.

Therefore, a heterologous prime-boost strategy was evaluated using rCSP formulated with GLA-SE and Ad35CS, a replication-deficient, recombinant human adenovirus serotype 35 vector expressing the full-length PfCSP, previously demonstrated as safe in Phase 1 clinical trials [44,45,46]. GLA-SE was selected as the adjuvant for rCSP in these studies since this adjuvant has previously been used in the clinic, whereas GLA-LSQ has not yet been tested in humans. While the reported data generally point to priming with a viral or DNA vectored construct, followed by a boost with the recombinant protein (i.e., boosting T-cell responses followed by B-cell responses) as a successful strategy [47], we tested all of the permutations using a 3-dose regimen so as not to bias the results (Table 2).

Efficacy of the different prime-boost regimens was evaluated via challenge studies in C57BL/6 mice (Figure 7A). The GLA-SE 3x, GLA-SE 1x/Ad35CS 2x, GLA-SE 2x/Ad35CS 1x, and Ad35CS 2x/GLA-SE 1x regimens all resulted in significant decrease in liver parasite load compared to naïve mice. However, contrary to the previously established consensus, the greatest inhibition (>85%) was seen in regimens where rCSP + GLA-SE was administered first (GLA-SE 3x; GLA-SE 1x/Ad35CS 2x; and GLA-SE 2x/Ad35CS 1x). No reduction in liver parasite load was found in mice immunized with 3 injections of Ad35CS as compared to the naïve control group.

Humoral response was measured in the sera of immunized mice via ELISA, using either a repeat region peptide, [NANP]$_6$, or rCSP as an antigen (Figure 7B). Sera titers (calculated at OD = 1) assessed with either antigen were similar across the different regimens except for animals immunized with the Ad35CS 3x regimen, for which titer was comparatively lower.

T-cell response was evaluated via ELISpot by harvesting lymphocytes (from both spleen and liver) 10 days post final immunization. Cells were re-stimulated using overlapping peptides (15-mers) comprising full-length CSP and divided into several pools encompassing specific regions of CSP (N-terminal, repeat, and C-terminal regions). Note that as C57BL/6 mice (H-2^b) were used for this study, the H-2A^d (I-A^d)- and H-2K^d-restricted CD4 and CD8 peptides used in the previous study with BALB/c (H-2^d) were not appropriate for re-stimulation, and, as expected, no activity was seen with either the N-terminal or C-terminal peptide pools (data not shown). However, positive responses were seen with lymphocytes harvested from both the spleen and liver re-stimulated with the repeat region peptide pool, likely resulting from the putative H-2A^d (I-A^d)-restricted CD4 epitope present within the repeat region (Figure 7C). Striking differences in the levels of T-cell response were seen with the different regimens. Interestingly, the levels of T-cell response induced by different immunization regimens appear to trend with the levels of reduction in the liver parasite load in a challenge study

Table 1. Sterile Protection with rCSP.

Immunization	No. Infected/No. Challenged Mice	Sterile Protection (%)
rCSP (25 μg) + GLA-SE (5 μg)	5/10	50
rCSP (25 μg) + GLA-LSQ (5 μg)	6/10	40
GLA-SE (5 μg)	10/10	0
None (naïve)	10/10	0

BALB/c mice were immunized 3x with 25 μg of rCSP formulated in GLA-SE or GLA-LSQ, followed by challenge with Pb-CS(Pf) parasites 2 weeks after the last immunization. Parasitemia was monitored by Giemsa-stained blood smears daily from 7 to 11 days post challenge. Animals with no detectable parasitemia on all days monitored were considered sterilely protected.

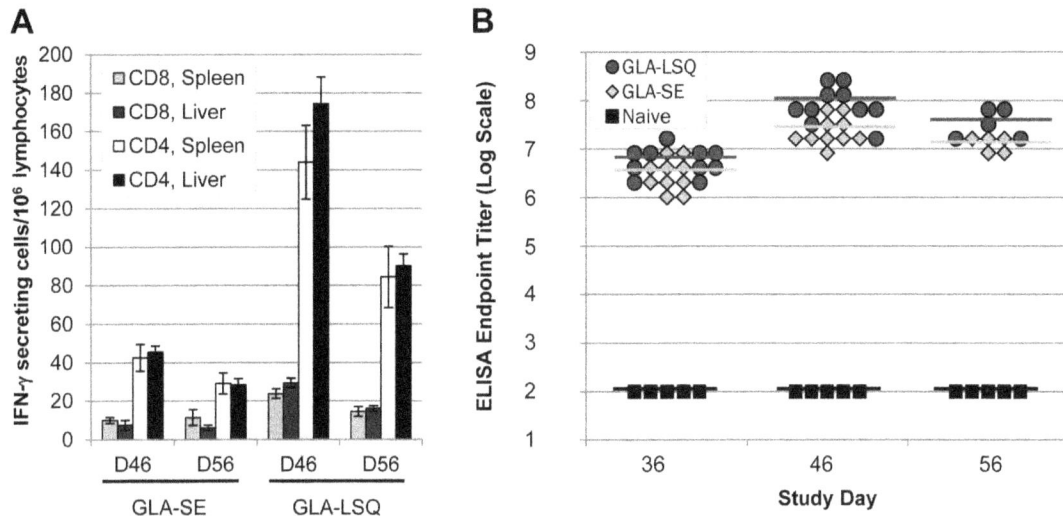

Figure 6. CSP-specific T-cell Response in Mice. Ten BALB/c mice (MHC haplotype H-2^d) per cohort were immunized 3x with 25 µg of rCSP formulated in GLA-SE or GLA-LSQ. Lymphocytes (harvested from spleens and livers) were collected 10 and 20 days after the final immunization (i.e., study days 46 and 56, respectively) from 5 mice per cohort at each time point. Sera were collected on the day of the final immunization (before injection) and at 10 and 20 days post final immunization (i.e., study days 36, 46, and 56, respectively). No more than 2 IFN-γ producing cells/10^6 lymphocytes were counted with cells harvested from mice injected adjuvant only (data not shown). (**A**) ELISpot assay was performed using either a H-2K^d-restricted CD8 epitope containing peptide located in the N-terminus of the CSP, or a I-A^d-restricted CD4 epitope containing peptide located in the C-terminus of the CSP for in vitro re-stimulation. Average numbers of IFN-γ producing cells/10^6 lymphocytes are shown with error bars representing standard error of the mean. (**B**) Endpoint ELISA titers of sera collected from the same mice used for T-cell response assessments demonstrated high level of humoral responses (using rCSP to coat microtiter plates) induced in mice immunized with rCSP in both the GLA-SE and GLA-LSQ formulations.

(Figure 7A). Specifically, cells from mice immunized with the GLA-SE 3x; GLA-SE 1x/Ad35CS 2x; GLA-SE 2x/Ad35CS 1x; and Ad35CS 2x/GLA-SE 1x regimens showed the highest cellular responses, and those from mice immunized with the Ad35CS 3x and Ad35CS 1x/GLA-SE 2x regimens showed lower cellular response, comparatively. A statistically significant increase in cellular response was seen in splenocytes harvested from mice immunized with the GLA-SE 2x/Ad35CS 1x regimen as compared to all other regimens tested (p≤0.0159).

Altogether, the results from our heterologous prime-boost experiments indicate that although the vaccination with Ad35CS construct alone (Ad35CS 3x) is not very effective in the mouse model, a single dose of Ad35CS vaccination, following two doses of rCSP-GLA-SE (GLA-SE 2x/AD35CS 1x), appears to modestly improve the T-cell response over the rCSP-GLA-SE alone (GLA-SE 3x). Surprisingly, reversing the order of the prime-boost

regimen - a single dose of Ad35CS followed by two rCSP-GLA-SE boosters (Ad35CS 1x/GLA-SE 2x) - markedly reduces the ability to elicit T-cell response and ultimately anti-malarial protection. These observations not only support the use of a heterologous prime-boost strategy as an effective vaccination approach but also highlight the importance in the order of delivery. Therefore, further studies are required to determine the optimal strategy to achieve the desired protection in the clinical setting.

Discussion

In this study, manufacture of a full-length recombinant CSP was accomplished using a *P. fluorescens* expression system. The advantages of this expression system include the obviation of the need for antibiotics as it exerted positive selective pressure on resulting strains, and the simplification of the downstream

Table 2. rCSP/Ad35CS Heterologous Prime-Boost Regimens.

Group	Injection Type and Day (D)			Regimen Abbreviation
	D = 0	D = 14	D = 36	
1	Ad35CS	Ad35CS	Ad35CS	Ad35CS 3x
2	GLA-SE+rCSP	GLA-SE+rCSP	GLA-SE+rCSP	GLA-SE 3x
3	GLA-SE+rCSP	Ad35CS	Ad35CS	GLA-SE 1x/Ad35CS 2x
4	GLA-SE+rCSP	GLA-SE+rCSP	Ad35CS	GLA-SE 2x/Ad35CS 1x
5	Ad35CS	GLA-SE+rCSP	GLA-SE+rCSP	Ad35CS 1x/GLA-SE 2x
6	Ad35CS	Ad35CS	GLA-SE+rCSP	Ad35CS 2x/GLA-SE 1x
7	Naïve			

C57BL/6 mice were immunized with rCSP formulated in GLA-SE or Ad35CS according to the regimens shown.

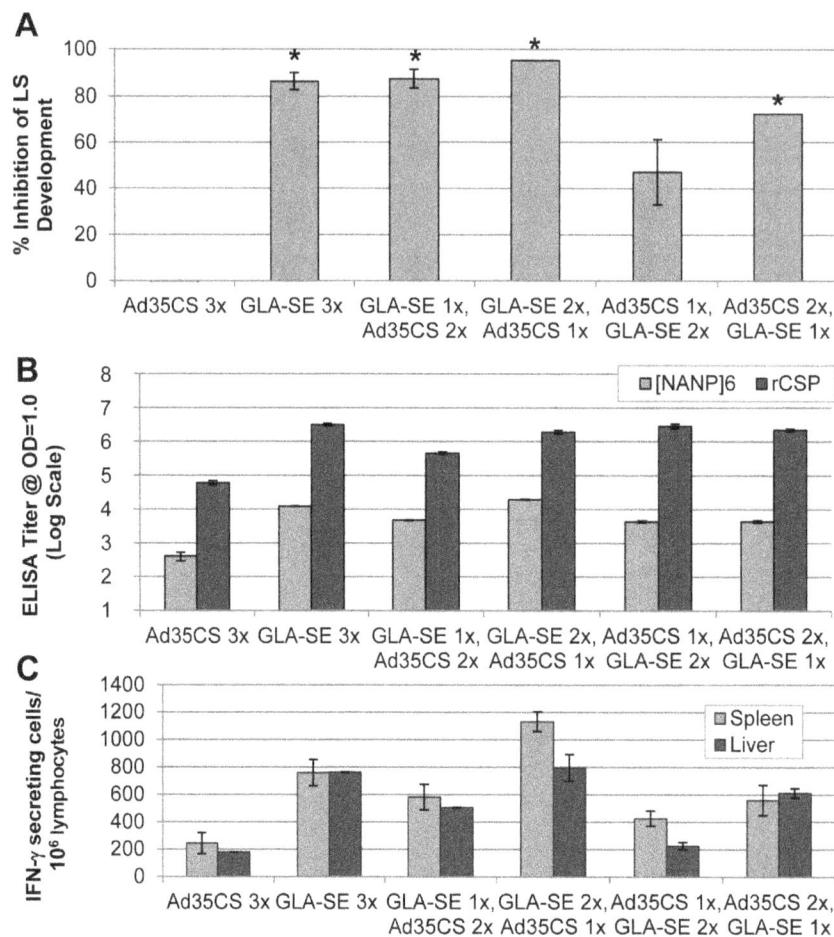

Figure 7. Activity of Prime-Boost Regimens in Mice. C57BL/6 mice (5–7 per cohort) were immunized 3x with 20 µg of rCSP formulated in GLA-SE, $1-2 \times 10^{10}$ v.p. of Ad35CS, or a combination of the two. Spleens and livers were collected 10 days after the last immunization; sera were collected 10 or 14 days after the last immunization. To assess reduction in parasite load in the liver or sterile protection, mice were challenged with Pb-CS(Pf) parasites 3 weeks after the last immunization. (**A**) Percent inhibition of LS parasite development in immunized mice challenged with Pb-CS(Pf) sporozoites normalized to naïve mice received the challenge, and percent sterile protection in immunized BALB/c mice upon challenge with a fewer number of Pb-CS(Pf) sporozoites are shown. A statistically significant reduction of LS parasites in the livers of mice immunized with all regimens, except Ad35CS 3x and Ad35CS (1x), GLA-SE (2x), was seen upon sporozoites challenge, compared to the level of LS parasites in the livers of negative control mice following sporozoites challenge, based on the mean parasite's 18s rRNA copy number (as noted by asterisks). Three regimens were assessed in two independent studies (GLA-SE 3x; GLA-SE 1x, Ad35CS 2x; and Ad35CS 1x, GLA-SE 2x); for these regimens, average percent inhibition is shown with error bars represented as standard error. Statistical significance with the GLA-SE 3x; GLA-SE 1x, Ad35CS 2x was seen in both studies. (**B**) ELISA was used to evaluate the level of humoral response in immunized mice to the repeat region of CSP with pooled mouse sera and to whole rCSP with individual mouse sera. Titer was assessed at OD = 1.0 based on 4-parameter logistic curve fits. Average titers are shown with error bars representing standard error of the mean. (**C**) The level of T-cell response was determined via ELISpot using lymphocytes isolated from the spleens and livers of immunized mice (5 mice per cohort were used for this study) and overlapping peptides (15-mers encompassing the CSP repeat region) for in vitro restimulation. Average numbers of IFN-γ producing cells/10^6 lymphocytes are shown with error bars representing standard error of the mean. No more than 5 IFN-γ producing cells/10^6 lymphocytes were counted with cells harvested from naïve mice (data not shown).

processing scheme due to rCSP being secreted into the periplasmic space of the down-selected *P. fluorescens* strains. Significant challenges for rCSP expression were overcome through selection of a *P. fluorescens* expression strain with a specific protease deletion that results in improved levels of full length rCSP (strain CSP-1), implementation of steps to rapidly separate rCSP from any residual endogenous proteases that may progressively truncated the N-terminus of the protein, and suppression of rCSP aggregation during the purification process (presumably due to the presence of an unpaired N-terminal cysteine residue) by freezing the lysate, which was found to be a preferred step prior to initiating the entrained chromatographic purification process, as well as by inclusion of a disaggregate (urea) and a reducing agent in all load and process buffers. Although a 2-column purification

process was sufficient to produce protein for analytical and biological activity screening, a 3-column purification process was ultimately implemented to reduce the level of rCSP dimer in the final product. Additionally, a suite of analytical assays were developed, including RP-HPLC, SEC-HPLC, SDS-PAGE, peptide mapping, and capillary iso-electric focusing, to biophysically monitor and characterize integrity of the molecule during in-process manufacturing, release, and long-term storage stability monitoring that will be used for clinical material development using the process described herein. The fermentation and purification processes described in this manuscript have been scaled up to accommodate manufacture at the 30 L scale. Therefore, to our knowledge, this is the first description of a process for generating clinical-grade full-length recombinant CSP that is suitable for economical large-scale production, clinical evaluation, and advanced product development.

To demonstrate that rCSP produced in the *P. fluorescens* expression system could be a potent immunogenic antigen against malaria infection in mice, we first measured immunogenic activity with a panel of biological assays upon rCSP vaccination. Humoral immune responses in mice were assessed via ELISA and IFA. An ISI assay was used to quantify the ability of functional antibodies to inhibit *Pf* sporozoite invasion of hepatocytes. Finally, protective efficacy of the rCSP-elicited immune response was determined by assessing the degree of liver stage parasite development upon challenge with *Pb*-CS(*Pf*) sporozoites in a mouse model. Upon vaccination, all three initial strains of rCSP elicited high levels of antibodies reactive against the repeat region of the CSP, but more importantly the antibodies raised by rCSP could recognize the native conformation of the protein present on the surface of the parasite. Among the 3 strains tested, one strain (CSP-1) induced a higher trend in ISI activity as compared to the other two strains. Intriguingly, while the titers of antibodies against the central repeat region of the CSP did not vary amongst the three strains, there was a marked decrease in biological activity by the CSP-2 strain derived protein. This is likely because the CSP-2 strain demonstrated the most N-terminally clipped protein (54%) and greatest number of amino acids clipped from the N-terminus (up to 17) as compared to the other strains. Not only did this negatively impact efficacy of the CSP-2 expression product but also may have changed the immune response to the protein such that a less efficacious response was generated. This may correlate with the fact that the central repeat region is the immunodominant B-cell domain and has been implicated as an immunogenic "decoy" in an immune evasion strategy of the parasite [48]. These observations may be of biological importance because the CSP N-terminus contains a proteolytic processing site that is cleaved prior to successful parasite invasion of the liver [10,49]. Indeed, we have identified an N-terminus mAb that binds near or on the cleavage site, and have shown through passive transfer studies that this mAb is able to block the invasion (>90%) of malaria parasites in vivo (data not shown). Therefore, our data further bolsters the rationale for the development of full-length CSP vaccine candidate that includes the N-terminus of the protein.

CSP-1 was selected as the strain to move forward for further development of GMP-grade rCSP, and we proceeded with the selection of an appropriate adjuvant to take forward to the clinic. We chose two adjuvants containing a TLR4 agonist (GLA-SE and GLA-LSQ) and two alum-based adjuvants (alhydrogel and adjuphos) to combine with rCSP for these studies. CSP-specific antibody responses were augmented in the presence of TLR4 agonist-based adjuvants, but not with the two alum-based products. Accordingly, the TLR4 agonist-based adjuvants, but not the alum-based adjuvants, showed good reduction of liver parasite load at a 5 μg dose of rCSP (80%–90%) with a slight improvement at the higher 20 μg dose of rCSP (90%–99%). Furthermore, sera from mice immunized with the GLA formulations were capable of inhibiting *Pf* sporozoite invasion into human hepatocytes in vitro, as measured by an ISI assay. It is important to note that the ISI assay measures the ability of anti-CSP antibodies to neutralize sporozoite invasion; however, it has not yet been clearly established with regards to the mechanism of the inhibitory action by the antibodies. Interestingly, the degree of protection (reduction in parasite liver load) seen with alum-formulated rCSP correlates with the low serum titers in immunized mice, which is consistent with field study reports showing a correlation between anti-CSP antibody titer and degree of protection from *P. falciparum* infection among people living in

endemic areas. However, the relationship between anti-CSP antibody titer and efficacy of RTS, S vaccine is more complicated in that a correlation was seen in naïve adults and young children of some endemic regions but not in other African populations [50,51]. Recent analysis of the RTS, S studies has revealed a synergistic effect between the antibody response and cellular response, particularly of CD4+ T cells, to CSP, leading to the modest protection now seen in their Phase 3 trials [52]. These studies underscore the need to elicit both arms of robust T-cell and B-cell responses in order to effectively protect from malaria infection.

Importantly, GLA-SE has been shown to elicit robust T-cell and B-cell responses induced by several malaria antigens including CelTOS and other forms of recombinant CSP [27,53]. This is the first reported study using GLA-LSQ, and we observed both an increased antibody response as well as an enhanced CSP-specific CD4+ T cell response rather than CD8+ T cell response, by the rCSP GLA-LSQ formulation as compared to the rCSP GLA-SE formulation. Note that the GLA-LSQ adjuvant is currently under development for use in clinical trials. Two completed clinical studies of pre-erythrocytic vaccines containing GLA-SE have demonstrated that GLA-SE is safe and well-tolerated [54,55], and there are multiple ongoing clinical studies using GLA-SE as adjuvant for malaria and other vaccines. In view of its pre-clinical performance with the rCSP and its safety profile, we plan to proceed with this adjuvant in combination with rCSP in the initial clinical trial.

Although its precise mechanism of action is still unknown, saponin clearly augments antibody and T-cell responses in synergy with TLR4 ligands. Nanoparticle adjuvant formulations similar to GLA-LSQ or GLA-SE are amenable to uptake by antigen presenting cells (APCs), which specialize in the phagocytosis of invading particulate pathogens and foreign particles. Additionally, delivery of antigens and adjuvants as complex particles can provide a sustained release mechanism, allowing for multimeric presentation to TLRs and APCs, and providing a delivery vehicle for otherwise insoluble components. Our data suggest different immunological profiles are elicited for GLA-SE and GLA-LSQ in combination with rCSP when administered to mice, which may also occur in humans and lead to different clinical outcomes for the two adjuvants. We are currently proceeding with the GMP process development of the GLA-LSQ so that we can measure its safety and immunogenicity in combination with rCSP in the clinic as well.

While our efforts to choose an adjuvant involve the enhancement of cell-mediated immunity targeted against the liver stage of parasite, we are well aware that immunization with recombinant proteins does not typically promote optimal T-cell response. Indeed in our initial study with BALB/c mice, we were able to enhance CSP-specific CD4+ T-cell response only modestly over baseline levels using rCSP formulated in GLA adjuvants. Therefore, we sought to test the effects of a heterologous prime-boost strategy in a series of studies using rCSP adjuvanted with GLA-SE in combination with an adenovirus-based CSP vaccine, since adenoviral vectors are known to induce a potent T-cell response [39,40,41]. Our results showed that the regimen where GLA-SE + rCSP was administered prior to Ad35CS demonstrated the most potent efficacy and resulted in the highest reduction in the development of liver stage parasites among the regimens tested. Additionally, this immunization regimen induced strong responses mediated by both B-cell and T-cell compartments, as assessed by measuring the humoral and cellular responses. There is precedent for heterologous prime-boost strategies to improve the magnitude, specificity or breadth of the host immune response

[43]; however, it is typically seen as the viral vectored or DNA as prime followed by a recombinant protein boost [18,47]. Interestingly, there have been studies with recombinant CSP and adenovirus-delivered CSP that follow this dogma [38,43,56]. Of note is a prime boost study conducted with adenovirus and GLA-SE adjuvanted recombinant protein vaccine candidates against *Mycobacterium tuberculosis* wherein long lived protection in mice was observed only when the adjuvanted protein vaccine candidate was administered prior to the adenovirus vaccine candidate [57]. Our results too suggest that administration of recombinant protein as prime followed by a recombinant adenovirus expressing the same protein as boost may improve efficacy of the vaccine, underlining the need to test all possible regimen combinations in order to achieve optimal performance. We are currently exploring the use of VLPs as a means to deliver critical epitopes of the CSP using several different technology platforms, which may be able to efficiently engage and induce potent "protective" anti-plasmodial immune responses mediated by both antibodies and T cells.

Materials and Methods

CSP Cloning, Strain Selection, Production, and Purification

The full-length CSP gene sequence, encoding *P. falciparum* 3D7 isolate (CAB38998) amino acids 21–382, was optimized for expression in *P. fluorescens* (DNA2.0; Menlo Park, California) and cloned into sixteen expression plasmids (Pfenex, Inc.; San Diego, California). Twenty *P. fluorescens* host strains were electroporated with each plasmid in a 96-well format, resulting in 320 expression strains [58]. Small-scale fermentation was performed with protein expression analysis by SDS-PAGE and western blotting to determine expression product levels. An anti-PfCSP conformational monoclonal antibody (mAb 4C2) used in screening was kindly provided by the Laboratory of Malaria Immunology and Vaccinology (NIAID) [28]. Bioreactor fermentations at a 1 L scale were performed for down-selected strains. Dissolved oxygen was maintained at a positive level in the liquid culture by regulating the sparging air flow, oxygen flow, and agitation rates after inoculation. The pH was controlled at the desired set-point through the addition of aqueous ammonia. The fed-batch high cell density fermentation process was divided into an initial growth phase and gene expression phase in which IPTG was added to initiate recombinant gene expression and the expression phase of the fermentation was allowed to proceed for 24 hours. Frozen cell pastes thawed at room temperature were suspended in a Tris-urea buffer and lysed by microfluidization at 15,000 psi. Lysates were clarified by continuous flow centrifugation, filtered (0.2 μm), and frozen at −70°C. Just prior to downstream purification, frozen lyates (thawed in a water bath at room temperature) were batch centrifuged and filtered (0.2 μm). 2 M urea was used in all load and process buffers to suppress protein aggregation. A two-step purification process was developed using Q-Sepharose (GE Healthcare) anion-exchange chromatography and hydrophobic interaction chromatography (Butyl 650S, Tosoh Bioscience LLC). The final eluate was exchanged into 1X PBS buffer by dialysis (regenerated cellulose, 10K MWCO; Thermo Scientific), followed by concentration with centrifugal ultrafiltration membrane filters (regenerated cellulose, 10K MWCO, Millipore).

CSP Analytical Characterization

Purity assessments of rCSP were made via SE chromatography and RP chromatography. SE chromatography was performed using a TSKgel G3000SWXL column (Tosoh Bioscience LLC) equipped to an Agilent 1100 HPLC system. PBS was used as the mobile phase with a 0.5 mL/minute flow rate and 50–100 μL injection volume; absorbance was monitored at 280 nm. RP-HPLC was performed on an Agilent 1100 HPLC system using a Jupiter C4 column (Phenomenex). Mobile phase A contained 0.1% trifluoreacetic acid (TFA) in water (v/v); solvent B contained 0.1% TFA in acetonitrile (v/v). Gradient conditions were 22%–32% solvent B with a 1 mL/minute flow rate and 30–60 μL injection volume (samples were diluted in PBS); absorbance was monitored at 214 nm and 280 nm. Integrity and identity of each rCSP were determined by assessment of intact mass and N-terminal sequencing. Intact mass was assessed with liquid chromatography coupled to mass spectrometry. Briefly, reduced samples were analyzed using an Agilent 1100 HPLC system coupled to a Q-T micro mass spectrometer (Waters) with an electrospray interface. A C_8 column (Zorbax 5 μm, Agilent) was used for separation and UV absorbance collected from 180–500 nm. UV chromatograms and MS total ion current chromatograms were generated and spectra deconvoluted using MaxEnt 1 (Waters) scanning for a molecular weight range of 30,000–50,000 at a resolution of 1 Da per channel. For N-terminal sequencing (American International Biotechnology, LLC; Richmond, Virginia), samples were loaded onto a PVDF sequencing membrane and subjected to 36 cycles of Edman degradation using a Procise Protein Sequencer (Applied Biosystems, Inc.). After each cycle, amino acids were separated and quantified using the integrated chromatographic system and the resultant sequence(s) compared to the expected *P. falciparum* 3D7 CS protein sequence.

Mice

Animal studies were conducted at The Rockefeller University (OLAW number A3081-01), Johns Hopkins University (OLAW number A3272-01), and Molecular Diagnostic Services, Inc. (OLAW number A4202-01). Six- to 8-week old female C57BL/6 mice were purchased from NCI (Frederick, Maryland) or Taconic (Germantown, New York). Six- to 8-week old female BALB/c mice were purchased from Taconic. All mice were maintained under standard conditions by the facility conducting the study, and all animal studies were approved by the respective Institutional Animal Care and Use Committee.

Immunizations

Mice were injected via the intramuscular (IM) route on study days 0, 14, and 36 with 2.5 μg, 5 μg, 20 μg, or 25 μg of rCSP formulated with an adjuvant. Several adjuvants were utilized including 50% Alhydrogel (Brenntag), 50% Adjuphos (Brenntag), 5 μg GLA-SE (Infectious Disease Research Institute) and 5 μg GLA-LSQ (Infectious Disease Research Institute) in combination with antigen. In addition, incomplete Freund's adjuvant (Sigma) was used as a positive control. Naïve mice and/or mice immunized with adjuvant alone served as negative controls. For some experiments, mice immunized via the IM route with a recombinant adenovirus expressing PfCSP were included – 2×10^{10} AdCS [34] or $1–2 \times 10^{10}$ Ad35CS (produced by Crucell) [40] viral particles (v.p.) per dose.

Humoral Response in Mice

Immunogenicity of C57BL/6 pooled mouse sera (collected on study day 50) to the PfCSP repeat region was evaluated via ELISA using a [NANP]$_6$ peptide coated on the ELISA plates as described in Kastenmuller et al., 2013 [27]. Immunogenicity of individual mouse serum to the full length rCSP was also evaluated by ELISA using a method similar to that used for the extended time course study detailed below. Titer at an optical density (OD) of 1.0 was determined based on a 4-parameter logistic regression model with the calibFit package [59] for R [60]. Statistical significance for assessments of individual mouse sera was calculated using Mann-Whitney U.

An immunogenicity study of sera collected from immunized C57BL/6 mice over an extended period of time was performed by Molecular Diagnostic Services, Inc. (San Diego, CA). Individual mouse sera collected on study days -2 (pre-bleed), 11, 34, 50, 89, 140, and 189 were tested by ELISA. Briefly, Maxisorp ELISA plates were coated with 100 or 200 ng/well of rCSP, incubated at 4°C for overnight, washed and blocked with PBS containing 1% BSA (PBS-1% BSA). Plates were washed again and incubated with either 2- or 3-fold serial dilutions of serum starting at a 1:2,000 dilution. After 1-hour incubation at room temperature (RT), plates were washed and incubated with an HRP-labeled goat anti-mouse secondary antibody, followed by addition of 3, 3′, 5, 5′ tetramethyl benzidine (TMB) substrate and stop buffer. Finally, the plates were read at 450 nm by spectrophotometer and titers calculated either based on a 4-parameter curve fit at an OD of 1.0 or determined by the maximal dilution that gave a value of OD = 0.1 (endpoint titer). Statistical analysis of the extended time course study data was done by repeated measures ANOVA and post hoc comparisons to compare titers achieved for each formulation across the study, and significant comparisons (Tukey) from a one-way ANOVA to compare titers at the defined collection time points. For the analysis, data were log+1 transformed to better meet assumptions of ANOVA and adjustments made using corrections based on the Mauchly's Test for sphericity.

Immunofluorescence Assay (IFA)

Reactivity of the sera collected from immunized mice to transgenic *P. berghei* sporozoites (*Pb*-CS(*Pf*) expressing the *P. falciparum* CSP repeat region and a portion of the N-terminal region including the cleavage site), kindly provided by Dr. Elizabeth Nardin [61], was tested by IFA. Briefly, IFA slides were first coated with the sporozoites and fixed with 0.05% glutaraldehyde and were incubated for 30 minutes with 3-fold serial dilutions of the sera. Slides were then washed with PBS-1% BSA, and a secondary antibody, Alexa Fluor 488-labeled F(ab')2 fragment of goat anti-mouse IgG (H+L), was added for 30 minutes. Finally, the slides were visualized with a fluorescence microscope (Nikon Eclipse 90i) and endpoint titers determined as the lowest serum dilution to give sporozoite surface fluorescence above negative sera control level.

Inhibition of Sporozoite Invasion (ISI)

HepG2-A16 cells (a human hepatocyte line) were seeded on 8-chamber slides and grown to monolayer. Diluted pool serum samples followed by 25,000 *P. falciparum* (NF54) sporozoites per well at a final dilution of 1:50, 1:100, or 1:500 were added to triplicate wells on the chamber slides. Cultures were incubated for 3 hours at 37°C, then washed 2X with PBS and fixed with 4% paraformaldehyde. The slides were then immuno-stained with mAb 2A10 (specific for CSP) to identify invaded sporozoites. The number of sporozoites, having invaded the hepatoma cells, was then determined by microscopic counting of the early-invaded forms using an epifluorescence microscope. Percent inhibition was calculated as mean sporozoites/well in negative control well minus

the mean sporozoites/well in the test sample wells divided by the mean sporozoites/well in negative control wells multiplied by 100.

T-cell Response in Mice

Liver and spleen from immunized BALB/c and C57BL/6 mice were harvested on study days 46 and 56, and lymphocytes were isolated from respective organ to determine the level of T-cell responses via an ELISpot assay. Individual spleens were analyzed; whereas 2–3 livers were pooled for the analysis. For the experiment performed using lymphocytes derived from BALB/c mice, peptides containing either an I-A^d-restricted CD4 epitope (EYLNKIQNSLSTEWSPCSVT) located in the C-terminal region of the CSP and an H-2K^d-restricted CD8 epitope (NYD-NAGTNL) located in the N-terminal region of the CSP were used for in vitro restimulation. For the experiment conducted using lymphocytes derived from C57BL/6 mice, overlapping peptides (15-mers) encompassing the N-terminal, repeat region, and C-terminal regions of CSP were used for the restimulation. The relative number of epitope-specific, IFN-γ-secreting T cells among lymphocytes of immunized mice were then determined, as previously described in Shiratsuchi et al., 2010 [34]. Statistical significance was calculated using Mann-Whitney U.

Sporozoite Challenge

For evaluation of the protective efficacy of the vaccine candidates, the amount of parasite load in the liver was determined upon challenging C57BL/6 mice by intravenous administration of 1×10^4 transgenic *Pb*-CS(*Pf*) sporozoites. Briefly, 42 hours post challenge, mice were euthanized and the liver was excised from each mouse. Estimation of the amount of parasite-specific 18S rRNA in each liver was determined by quantitative RT-PCR assay, and percent inhibition of the parasite development in the liver was calculated as previously described [62]. Statistical significance was calculated using Mann-Whitney U.

For evaluation of sterile protection, BALB/c mice were immunized as described above, except the time interval between the second and third immunizations was 1 month. Mice were challenged 2 weeks after the third immunization by intravenous administration of 5000 transgenic *Pb*-CS(*Pf*) sporozoites. The presence of parasitemia was determined via microscopic evaluation of Giemsa-stained blood smears daily on days 7–11 post challenge.

Acknowledgments

We thank Dr. Annie Mo, Program Officer for the Parasitology and International Programs Branch, Division of Microbiology and Infectious Diseases, National Institute of Allergy and Infectious Diseases, for helpful discussion regarding study design and manuscript review.

Author Contributions

Conceived and designed the experiments: ARN RH JRA BR SBW JS MT FZ GMG. Analyzed the data: ARN DE XL JGAC RF SG RA JS MT FZ GMG. Contributed reagents/materials/analysis tools: ARN DE XL JGAC RF HJ DMR RH JRA TSV CBF SGR JS MT FZ GMG. Wrote the paper: ARN DE SG RA MT FZ GMG.

References

1. WHO (2013) Malaria Fact Sheet. Available: http://www.who.int/mediacentre/factsheets/fs094/en/. Accessed 7 March 2014.
2. Greenwood BM, Bojang K, Whitty CJ, Targett GA (2005) Malaria. Lancet 365: 1487–1498.
3. Hoffman SL, Goh LM, Luke TC, Schneider I, Le TP, et al. (2002) Protection of humans against malaria by immunization with radiation-attenuated Plasmodium falciparum sporozoites. J Infect Dis 185: 1155–1164.

4. Roestenberg M, McCall M, Hopman J, Wiersma J, Luty AJ, et al. (2009) Protection against a malaria challenge by sporozoite inoculation. N Engl J Med 361: 468–477.

5. Kumar KA, Sano G, Boscardin S, Nussenzweig RS, Nussenzweig MC, et al. (2006) The circumsporozoite protein is an immunodominant protective antigen in irradiated sporozoites. Nature 444: 937–940.

6. Trieu A, Kayala MA, Burk C, Molina DM, Freilich DA, et al. (2011) Sterile protective immunity to malaria is associated with a panel of novel P. falciparum antigens. Mol Cell Proteomics 10: M111 007948.

7. Rathore D, Sacci JB, de la Vega P, McCutchan TF (2002) Binding and invasion of liver cells by Plasmodium falciparum sporozoites. Essential involvement of the amino terminus of circumsporozoite protein. J Biol Chem 277: 7092–7098.

8. Calvo-Calle JM, Hammer J, Sinigaglia F, Clavijo P, Moya-Castro ZR, et al. (1997) Binding of malaria T cell epitopes to DR and DQ molecules in vitro correlates with immunogenicity in vivo: identification of a universal T cell epitope in the Plasmodium falciparum circumsporozoite protein. J Immunol 159: 1362–1373.

9. Doolan DL, Hoffman SL, Southwood S, Wentworth PA, Sidney J, et al. (1997) Degenerate cytotoxic T cell epitopes from P. falciparum restricted by multiple HLA-A and HLA-B supertype alleles. Immunity 7: 97–112.

10. Coppi A, Pinzon-Ortiz C, Hutter C, Sinnis P (2005) The Plasmodium circumsporozoite protein is proteolytically processed during cell invasion. J Exp Med 201: 27–33.

11. Ballou WR (2009) The development of the RTS, S malaria vaccine candidate: challenges and lessons. Parasite Immunol 31: 492–500.

12. Tam JP, Clavijo P, Lu YA, Nussenzweig V, Nussenzweig R, et al. (1990) Incorporation of T and B epitopes of the circumsporozoite protein in a chemically defined synthetic vaccine against malaria. J Exp Med 171: 299–306.

13. Yoshida N, Nussenzweig RS, Potocnjak P, Nussenzweig V, Aikawa M (1980) Hybridoma produces protective antibodies directed against the sporozoite stage of malaria parasite. Science 207: 71–73.

14. Zavala F, Tam JP, Barr PJ, Romero PJ, Ley V, et al. (1987) Synthetic peptide vaccine confers protection against murine malaria. J Exp Med 166: 1591–1596.

15. Breman JG, Alilio MS, Mills A (2004) Conquering the intolerable burden of malaria: what's new, what's needed: a summary. Am J Trop Med Hyg 71: 1–15.

16. Clyde DF, Most H, McCarthy VC, Vanderberg JP (1973) Immunization of man against sporozite-induced falciparum malaria. Am J Med Sci 266: 169–177.

17. Gupta S, Snow RW, Donnelly CA, Marsh K, Newbold C (1999) Immunity to non-cerebral severe malaria is acquired after one or two infections. Nat Med 5: 340–343.

18. Lu S (2009) Heterologous prime-boost vaccination. Curr Opin Immunol 21: 346–351.

19. Nussenzweig RS, Vanderberg J, Most H, Orton C (1967) Protective immunity produced by the injection of x-irradiated sporozoites of plasmodium berghei. Nature 216: 160–162.

20. Renia L, Marussig MS, Grillot D, Pied S, Corradin G, et al. (1991) In vitro activity of CD4+ and CD8+ T lymphocytes from mice immunized with a synthetic malaria peptide. Proc Natl Acad Sci U S A 88: 7963–7967.

21. Rieckmann KH, Beaudoin RL, Cassells JS, Sell KW (1979) Use of attenuated sporozoites in the immunization of human volunteers against falciparum malaria. Bull World Health Organ 57 Suppl 1: 261–265.

22. Rodrigues M, Nussenzweig RS, Zavala F (1993) The relative contribution of antibodies, CD4+ and CD8+ T cells to sporozoite-induced protection against malaria. Immunology 80: 1–5.

23. Snow RW, Guerra CA, Noor AM, Myint HY, Hay SI (2005) The global distribution of clinical episodes of Plasmodium falciparum malaria. Nature 434: 214–217.

24. Agnandji ST, Lell B, Fernandes JF, Abossolo BP, Methogo BG, et al. (2012) A phase 3 trial of RTS, S/AS01 malaria vaccine in African infants. N Engl J Med 367: 2284–2295.

25. Bongfen SE, Ntsama PM, Offner S, Smith T, Felger I, et al. (2009) The N-terminal domain of Plasmodium falciparum circumsporozoite protein represents a target of protective immunity. Vaccine 27: 328–335.

26. Vergara U, Ruiz A, Ferreira A, Nussenzweig RS, Nussenzweig V (1985) Conserved group-specific epitopes of the circumsporozoite proteins revealed by antibodies to synthetic peptides. J Immunol 134: 3445–3448.

27. Kastenmuller K, Espinosa DA, Trager L, Stoyanov C, Salazar AM, et al. (2013) Full-length Plasmodium falciparum circumsporozoite protein administered with long-chain poly(I.C) or the Toll-like receptor 4 agonist glucopyranosyl lipid adjuvant-stable emulsion elicits potent antibody and CD4+ T cell immunity and protection in mice. Infect Immun 81: 789–800.

28. Plassmeyer ML, Reiter K, Shimp RL, Jr., Kotova S, Smith PD, et al. (2009) Structure of the Plasmodium falciparum circumsporozoite protein, a leading malaria vaccine candidate. J Biol Chem 284: 26951–26963.

29. Ophorst OJ, Radosevic K, Havenga MJ, Pau MG, Holterman L, et al. (2006) Immunogenicity and protection of a recombinant human adenovirus serotype 35-based malaria vaccine against Plasmodium yoelii in mice. Infect Immun 74: 313–320.

30. Anderson RC, Fox CB, Dutill TS, Shaverdian N, Evers TL, et al. (2010) Physicochemical characterization and biological activity of synthetic TLR4 agonist formulations. Colloids Surf B Biointerfaces 75: 123–132.

31. Coler RN, Baldwin SL, Shaverdian N, Bertholet S, Reed SJ, et al. (2010) A synthetic adjuvant to enhance and expand immune responses to influenza vaccines. PLoS One 5: e13677.

32. Gomes R, Teixeira C, Oliveira F, Lawyer PG, Elnaiem DE, et al. (2012) KSAC, a defined Leishmania antigen, plus adjuvant protects against the virulence of L. major transmitted by its natural vector Phlebotomus duboscqi. PLoS Negl Trop Dis 6: e1610.

33. Windish HP, Duthie MS, Misquith A, Ireton G, Lucas E, et al. (2011) Protection of mice from Mycobacterium tuberculosis by ID87/GLA-SE, a novel tuberculosis subunit vaccine candidate. Vaccine 29: 7842–7848.

34. Shiratsuchi T, Rai U, Krause A, Worgall S, Tsuji M (2010) Replacing adenoviral vector HVR1 with a malaria B cell epitope improves immunogenicity and circumvents preexisting immunity to adenovirus in mice. J Clin Invest 120: 3688–3701.

35. Hu SL, Abrams K, Barber GN, Moran P, Zarling JM, et al. (1992) Protection of macaques against SIV infection by subunit vaccines of SIV envelope glycoprotein gp160. Science 255: 456–459.

36. Lu S (2006) Combination DNA plus protein HIV vaccines. Springer Semin Immunopathol 28: 255–265.

37. Arama C, Assefaw-Redda Y, Rodriguez A, Fernandez C, Corradin G, et al. (2012) Heterologous prime-boost regimen adenovector 35-circumsporozoite protein vaccine/recombinant Bacillus Calmette-Guerin expressing the Plasmodium falciparum circumsporozoite induces enhanced long-term memory immunity in BALB/c mice. Vaccine 30: 4040–4045.

38. Li S, Rodrigues M, Rodriguez D, Rodriguez JR, Esteban M, et al. (1993) Priming with recombinant influenza virus followed by administration of recombinant vaccinia virus induces CD8+ T-cell-mediated protective immunity against malaria. Proc Natl Acad Sci U S A 90: 5214–5218.

39. Reyes-Sandoval A, Harty JT, Todryk SM (2007) Viral vector vaccines make memory T cells against malaria. Immunology 121: 158–165.

40. Shott JP, McGrath SM, Pau MG, Custers JH, Ophorst O, et al. (2008) Adenovirus 5 and 35 vectors expressing Plasmodium falciparum circumsporozoite surface protein elicit potent antigen-specific cellular IFN-gamma and antibody responses in mice. Vaccine 26: 2818–2823.

41. Yang TC, Millar JB, Grinshtein N, Bassett J, Finn J, et al. (2007) T-cell immunity generated by recombinant adenovirus vaccines. Expert Rev Vaccines 6: 347–356.

42. Rodrigues EG, Zavala F, Eichinger D, Wilson JM, Tsuji M (1997) Single immunizing dose of recombinant adenovirus efficiently induces CD8+ T cell-mediated protective immunity against malaria. J Immunol 158: 1268–1274.

43. Stewart VA, McGrath SM, Dubois PM, Pau MG, Mettens P, et al. (2007) Priming with an adenovirus 35-circumsporozoite protein (CS) vaccine followed by RTS, S/AS01B boosting significantly improves immunogenicity to Plasmodium falciparum CS compared to that with either malaria vaccine alone. Infect Immun 75: 2283–2290.

44. Creech CB, Dekker CL, Ho D, Phillips S, Mackey S, et al. (2013) Randomized, placebo-controlled trial to assess the safety and immunogenicity of an adenovirus type 35-based circumsporozoite malaria vaccine in healthy adults. Hum Vaccin Immunother 9: 2548–2557.

45. Keefer MC, Gilmour J, Hayes P, Gill D, Kopycinski J, et al. (2012) A phase I double blind, placebo-controlled, randomized study of a multigenic HIV-1 adenovirus subtype 35 vector vaccine in healthy uninfected adults. PLoS One 7: e41936.

46. Ouedraogo A, Tiono AB, Kargougou D, Yaro JB, Ouedraogo E, et al. (2013) A phase 1b randomized, controlled, double-blinded dosage-escalation trial to evaluate the safety, reactogenicity and immunogenicity of an adenovirus type 35 based circumsporozoite malaria vaccine in Burkinabe healthy adults 18 to 45 years of age. PLoS One 8: e78679.

47. Dunachie SJ, Hill AV (2003) Prime-boost strategies for malaria vaccine development. J Exp Biol 206: 3771–3779.

48. Schofield L (1990) The circumsporozoite protein of Plasmodium: a mechanism of immune evasion by the malaria parasite? Bull World Health Organ 68 Suppl: 66–73.

49. Coppi A, Tewari R, Bishop JR, Bennett BL, Lawrence R, et al. (2007) Heparan sulfate proteoglycans provide a signal to Plasmodium sporozoites to stop migrating and productively invade host cells. Cell Host Microbe 2: 316–327.

50. Alonso PL, Sacarlal J, Aponte JJ, Leach A, Macete E, et al. (2004) Efficacy of the RTS, S/AS02A vaccine against Plasmodium falciparum infection and disease in young African children: randomised controlled trial. Lancet 364: 1411–1420.

51. Asante KP, Abdulla S, Agnandji S, Lyimo J, Vekemans J, et al. (2011) Safety and efficacy of the RTS, S/AS01E candidate malaria vaccine given with expanded-programme-on-immunisation vaccines: 19 month follow-up of a randomised, open-label, phase 2 trial. Lancet Infect Dis 11: 741–749.

52. Ndungu FM, Mwacharo J, Kimani D, Kai O, Moris P, et al. (2012) A statistical interaction between circumsporozoite protein-specific T cell and antibody responses and risk of clinical malaria episodes following vaccination with RTS, S/AS01E. PLoS One 7: e52870.

53. Fox CB, Baldwin SL, Vedvick TS, Angov E, Reed SG (2012) Effects on immunogenicity by formulations of emulsion-based adjuvants for malaria vaccines. Clin Vaccine Immunol 19: 1633–1640.

54. Behzad H, Huckriede AL, Haynes L, Gentleman B, Coyle K, et al. (2012) GLA-SE, a synthetic toll-like receptor 4 agonist, enhances T-cell responses to influenza vaccine in older adults. J Infect Dis 205: 466–473.

55. Treanor JJ, Essink B, Hull S, Reed S, Izikson R, et al. (2013) Evaluation of safety and immunogenicity of recombinant influenza hemagglutinin (H5/Indonesia/05/2005) formulated with and without a stable oil-in-water emulsion containing glucopyranosyl-lipid A (SE+GLA) adjuvant. Vaccine 31: 5760–5765.

56. Radosevic K, Rodriguez A, Lemckert AA, van der Meer M, Gillissen G, et al. (2010) The Th1 immune response to Plasmodium falciparum circumsporozoite protein is boosted by adenovirus vectors 35 and 26 with a homologous insert. Clin Vaccine Immunol 17: 1687–1694.

57. Baldwin SL, Ching LK, Pine SO, Moutaftsi M, Lucas E, et al. (2013) Protection against tuberculosis with homologous or heterologous protein/vector vaccine approaches is not dependent on CD8+ T cells. J Immunol 191: 2514–2525.

58. Retallack DM, Jin H, Chew L (2012) Reliable protein production in a Pseudomonas fluorescens expression system. Protein Expr Purif 81: 157–165.

59. Haaland P, Samarov D, McVey E (2011) calibFit: Statistical models and tools for assay calibration (R package version 2.1.0). Available at: http://cran.r-project.org/web/packages/calibFit/index.html (Accessed on 15 December 2012).

60. R Core Team (2012) R: A language and environment for statistical computing. Vienna, Austria. Available: http://www.r-project.org/. Accessed 11 March 2013.

61. Persson C, Oliveira GA, Sultan AA, Bhanot P, Nussenzweig V, et al. (2002) Cutting edge: a new tool to evaluate human pre-erythrocytic malaria vaccines: rodent parasites bearing a hybrid Plasmodium falciparum circumsporozoite protein. J Immunol 169: 6681–6685.

62. Bruna-Romero O, Hafalla JC, Gonzalez-Aseguinolaza G, Sano G, Tsuji M, et al. (2001) Detection of malaria liver-stages in mice infected through the bite of a single Anopheles mosquito using a highly sensitive real-time PCR. Int J Parasitol 31: 1499–1502.

13

Great Spotted Cuckoo Fledglings Often Receive Feedings from Other Magpie Adults than Their Foster Parents: Which Magpies Accept to Feed Foreign Cuckoo Fledglings?

Manuel Soler[1,2]*, **Tomás Pérez-Contreras**[1,2], **Juan Diego Ibáñez-Álamo**[1], **Gianluca Roncalli**[1], **Elena Macías-Sánchez**[1], **Liesbeth de Neve**[1,3]

1 Departamento de Zoología, Facultad de Ciencias, Universidad de Granada, Granada, Spain, **2** Grupo Coevolución, Unidad Asociada al Consejo Superior de Investigaciones Científicas (CSIC), Universidad de Granada, Granada, Spain, **3** Department of Biology, Terrestrial Ecology Unit, Ghent University, Gent, Belgium

Abstract

Natural selection penalizes individuals that provide costly parental care to non-relatives. However, feedings to brood-parasitic fledglings by individuals other than their foster parents, although anecdotic, have been commonly observed, also in the great spotted cuckoo (*Clamator glandarius*) – magpie (*Pica pica*) system, but this behaviour has never been studied in depth. In a first experiment, we here show that great spotted cuckoo fledglings that were translocated to a distant territory managed to survive. This implies that obtaining food from foreign magpies is a frequent and efficient strategy used by great spotted cuckoo fledglings. A second experiment, in which we presented a stuffed-cuckoo fledgling in magpie territories, showed that adult magpies caring for magpie fledglings responded aggressively in most of the trials and never tried to feed the stuffed cuckoo, whereas magpies that were caring for cuckoo fledglings reacted rarely with aggressive behavior and were sometimes disposed to feed the stuffed cuckoo. In a third experiment we observed feedings to post-fledgling cuckoos by marked adult magpies belonging to four different possibilities with respect to breeding status (i.e. composition of the brood: only cuckoos, only magpies, mixed, or failed breeding attempt). All non-parental feeding events to cuckoos were provided by magpies that were caring only for cuckoo fledglings. These results strongly support the conclusion that cuckoo fledglings that abandon their foster parents get fed by other adult magpies that are currently caring for other cuckoo fledglings. These findings are crucial to understand the co-evolutionary arms race between brood parasites and their hosts because they show that the presence of the host's own nestlings for comparison is likely a key clue to favour the evolution of fledgling discrimination and provide new insights on several relevant points such as learning mechanisms and multiparasitism.

Editor: Alexandre Roulin, University of Lausanne, Switzerland

Funding: This work was supported by the Spanish Ministerio de Economía y Competitividad/FEDER (research project CGL2011-25634/BOS). The funders had no role in study design, data collection and analysis, decision to publish, or preparation of the manuscript.

Competing Interests: The authors have declared that no competing interests exist.

* Email: msoler@ugr.es

Introduction

Parental behaviours that enhance the fitness of offspring while provoking a cost to the parents are considered to be parental care [1]. Parental care is costly, not only in terms of time and energy, but also because parental investment in current reproduction implies a reduction in the number of future offspring [1,2]. Natural selection also penalizes individuals that provide costly parental care to non-relatives. Therefore, parental investment theory predicts that parents should have adaptations allowing them to both optimize the balance between investment in current and future reproduction and to recognize kin. For instance, in many taxa it is well known that parents usually favour offspring of higher reproductive value [3–5] and that parents become more insensitive to begging signals by their offspring when they are

ready to initiate a second breeding attempt [6]. Furthermore, offspring desertion occurs when fitness costs related to investment in the current brood exceed the expected fitness benefits [7–9], for example in response to partial egg predation [7,9,10–12]. In addition, males reduce parental investment in cases of reduced certainty of paternity [13–16].

Many parental care behaviours are susceptible to parasitism resulting in a huge variety of interactions in which various combinations of nest, food and offspring care are parasitized [17]. In fact, parental care parasitism by unrelated individuals is widely distributed within the animal kingdom (reviewed in [17]).

Alloparental care could be considered a type of parental-care parasitism in which adult animals feed unrelated juveniles [17]. This behaviour, as it occurs with nest switching [18], has been treated by the literature as adoption behaviour by foster parents. However, since alloparental care behaviours are promoted by the young themselves rather than by the foster parents, they should be considered cases of parental-care parasitism [17].

Alloparental care is especially frequent at the intraspecific level in cooperatively breeding species, in which the parasitized individual is related to the juveniles and receives benefits from inclusive fitness [17,19]. However, it is also very frequent in birds at the interspecific level in situations of brood parasitism. When brood parasitic fledglings leave the host nest, they continue being fed by their foster parents [20–23]. This is not surprising because host foster parents could learn the begging calls of a parasitic nestling at the end of the nestling period, and later, continue feeding it because they had learnt the vocal signature of that fledgling as one of their own nestlings [24–28].

However, reports of brood-parasitic fledglings being fed by individuals other than their foster parents, or even individuals from a different host species are also common [20,22,29–38], especially for the pallid cuckoo (*Cuculus pallidus*; [38]) and the great spotted cuckoo (*Clamator glandarius*; [36,37]). However, all these reports were of anecdotic nature and based on very few observations [38]. Still, observations of parasitic fledglings being fed by individuals other than their foster parents could be explained by the hypothesis that the parasitized parents were tricked into feeding because they were exposed to the proper stimulus (i.e. a begging fledgling), which is the most important prerequisite for expressing feeding behaviour [39].

The great spotted cuckoo is a non-evictor brood parasite that uses the magpie (*Pica pica*) as its primary host. The nestling period of great spotted cuckoo chicks is considerably shorter than that of magpie chicks (between17–22 days and 23–24 days, respectively for cuckoos and magpies [40]). Great spotted cuckoo fledglings have a post-fledging dependence period that is highly variable (between 25 and 59 days, [21]). Between one and three weeks after leaving the nest, usually join in groups that are communally fed by more magpies than those involved in rearing those cuckoos in the nest [37,41]. In a recent experimental study, we found that great spotted cuckoo fledglings that had been reared together with magpie nestlings in the same nest were disadvantaged (in terms of feeding patterns) by magpie adults compared to cuckoo fledglings that had been reared in only cuckoo broods [23]. They were fed less frequently than those reared in only cuckoo broods, and magpie adults approached less frequently to feed cuckoos from mixed broods than cuckoos from only cuckoo broods [23]. These, probably undernourished, great spotted cuckoo fledglings might abandon their less-efficient foster parents and join other cuckoo fledglings, to obtain a higher feeding rate in a communal fed group [37]. This ability to look for more efficient caregivers has frequently been reported in some cooperative species [42–44] but, as far as we know, it has never been found in fledglings of any other brood parasite species. The study of the relationships between brood parasites and their hosts during the post-fledgling period is very important because of two reasons. First, because hardly anything is known about the post-fledging period in brood parasites and hosts [45] in spite that this period of care is critical for juvenile survival [46]. Second, coevolutionary adaptations and counter-adaptations can evolve at any stage of the breeding cycle, including the fledgling stage [22,45,47,48] (host defences have also been found at the nestling stage in several brood parasite – host systems [49–51]; reviewed in [52,53]). The clearest example of an arms race at the fledgling stage comes from the bay-winged cowbird *Agelaioides badius* that feeds fledglings of its specialist parasitic screaming cowbird *Molothrus rufoaxillaris*, which visually and vocally mimic host fledglings, but it refuses to feed non-mimetic fledglings from a generalist brood parasite, the shiny cowbird *M. bonariensis* [22,47].

The main aim of this study is to study in depth the relationships between great spotted cuckoos and their magpie hosts during the post-fledgling stage. We were especially interested in the most surprising behaviour reported during this stage; i.e. that magpie adults sometimes feed cuckoo fledglings that have abandoned their foster parents. We performed three different experiments during four breeding seasons to answer the following crucial questions related to this behaviour: (1) is obtaining food from magpies other than their foster parents a frequently used strategy by cuckoo fledglings or are they only anecdotic cases as reported in brood parasitic fledglings of other species? (2) Which magpies accept to feed foreign cuckoo fledglings that beg for food?

Material and Methods

Ethics Statement

Research has been conducted according to relevant Spanish national (Real Decreto 1201/2005, de 10 de Octubre) and regional guidelines. All necessary permits were obtained from the Consejería de Medio Ambiente de la Junta de Andalucía, Spain. Approval for this study was not required according to Spanish law because it is not a laboratory study in which experimental animals have to be surgically manipulated and/or euthanatized. Our study area is not protected but privately owned, and the owners allow us to work in their properties. This study did not involve endangered or protected species. The great spotted cuckoo is included in both Spanish national (R. D. 139/2011, 4 February) and regional (D. 23/2012, 14 February) lists of species under special protection, but not in the catalog of endangered species of any of the two entities.

Plastic patagial wing tags used to mark adult magpies cause no damage to the individuals (e.g. [54]). After releasing the magpies, we made observations in each territory to ensure that the captured individuals flew correctly, continued in the territory and maintained their nests. None of the adult magpies showed problems to fly or abandoned the nest or their nestlings because of the capture and marking process. Cross-fostering manipulations were made by carefully transporting the nestlings in an artificial cotton nest lined with tissues, maintaining the temperature in the car between 25–30°C. Cross-fostering per se does not affect nestlings or host parents' behaviour [40,41,55]. In some cases in which we took one great spotted cuckoo chick that was alone in the nest we left another cuckoo chick of similar age from a multiparasitized nest to avoid nest abandonment. Transmitters were attached using the leg-loop harness method, which has been demonstrated to be effective without causing skin or plumage damage, or interfering with behaviour [56,57]. Attaching the transmitters several days before the fledglings leave the nest allows the nestlings to become accustomed to it and allow the harness to fit the nestling body.

Study species, study area and general field methods

The relationships between great spotted cuckoo fledglings and adult magpies have been extensively studied [21,23,27] and it has been observed that the feeding of one fledgling great spotted cuckoo by more than two magpies is frequent [37].

This study was carried out during the breeding seasons of 2007, 2010, 2011 and 2012 in a population of great spotted cuckoos located in the Hoya de Guadix in southern Spain (37° 10′ N, 3° 11′ W; 1000 m.a.s.l.). This area is a high-altitude plateau (approx. 1000 m a.s.l.) with extensive non-cultivated areas, cereal crops (especially barley), some areas with dispersed holm-oak trees (*Quercus rotundifolia*) and groves of almond trees (*Prunus dulcis*) and pines (*Pinus halepensis* and *P. pinaster*), in which magpies build their nests [58]. The land in the Hoya the Guadix is privately owned by many different landowners, but most fields are not fenced and so the magpie territories are freely accessible. Most proprietors allow us to follow up nests on their land during the breeding season.

Occurrence of brood parasitism was frequent in the study area with 56.8% of magpie nests parasitized by great spotted cuckoos for the period 2008–2012 [23].

Every year we searched for new magpie nests during the complete breeding season (mid March – early June) and recorded the main breeding data, i.e. laying date, clutch size, presence and number of great spotted cuckoo eggs and the number of great spotted cuckoos and magpies that successfully left each nest.

Experiment 1: translocation of fledglings

During June-July of 2007, we captured great spotted cuckoo fledglings of about 40 days old (i.e. about 20 days after fledging; with wings and tails completely developed) using mist nets and provided them with a radio-transmitter (Holohil PD-2, weight: approximately 4.5 g, back-pack harness included, range of 1000 m and a battery life of 14–26 weeks). Transmitters were attached using the leg-loop harness method [59]. Further details about radio-transmitters and attaching method can be found in [23]. These fledglings were randomly assigned to one of the two following treatments: (i) experimental treatment: fledglings were retained temporally within a cloth bag while we transported them by car to a different area with cuckoo fledglings (a mean ± SD of 7.00±3.51 km away from their original area) where they were released), and (ii) control treatment: captured fledglings were also retained temporally within a cloth bag, and after about 15 minutes they were released in the same area where they were captured. All captures and releases were done with good weather conditions and in the presence of no potential predators of cuckoo fledglings (i.e. raptors).

Every 3–5 days we went to search for the released fledglings with a reception antenna (Televilt (now Followit) O-5/8, receptor RX-98H). Once close enough to the signal (about 150 meters) we searched visually for the activity of the individual using binoculars. If we failed to detect the fledgling after 15 min, despite the close signal, we approached the place in order to look for the cuckoo fledgling more closely and determine its situation (if it was dead or alive). We followed the activity of all marked cuckoo fledglings for two weeks, a period of time long enough to allow us to conclude that the fledgling survived in the new situation (i.e. after the experimental translocation).

During our inspections of each fledgling we did not wait to make observations of feedings by magpies, however, we assume that survival of the translocated fledglings implied that magpies were feeding them. This assumption is based on four points. First, great spotted cuckoo fledglings have never been seen feeding themselves [21, 23, this study]. Second, magpies other than foster parents have frequently been reported feeding great spotted cuckoo fledglings [36,37]. Third, cuckoo fledglings could not suddenly change to feed themselves given that fledglings of altricial species need a long period to achieve foraging skills (e.g. [46]). And fourth, the assumption has been supported by Experiment 3 (see below).

The translocation experiment was done to test the hypothesis that cuckoo fledglings frequently use the strategy to obtain food from magpies other than their foster parents, instead of being only anecdotic cases of alloparental feedings. Thus, we predicted that well developed fledglings translocated to a different area from their rearing territory, should be able to survive equally well than those that remain in their natal area (Prediction 1).

Experiment 2: playback-stuffed-cuckoo presentation

We actively searched for adult magpies and great spotted cuckoo fledglings in the study area at the end of the breeding seasons of 2011 and 2012. Once adult magpies or cuckoo fledglings were detected in the field, we observed the location for a variable period of time from a distance of about 200 m using binoculars to determine the number, species (cuckoo or magpie) and age (adult or fledgling) of individuals. When this information was collected we presented a stuffed great spotted cuckoo fledgling (6 different stuffed-cuckoo fledglings mounted with the beak closed and in a non-begging display), with begging calls (4 different playbacks; 60 s of begging and 45 s of silence) for 30 min in the following two situations: in the presence of adult magpies that were together with cuckoo fledglings alone (experimental treatment) or in the presence of adult magpies accompanied by only magpie fledglings (i.e. family group; control treatment). We avoided testing the same individuals by doing the experiments in clearly separated locations (more than two km).

The experimental procedure consisted in the placement of a stuffed-cuckoo fledgling in a visible location (usually on the ground in an open area) close to the group of birds but far enough to avoid frightening them (between 100–200 m). We placed the playback device on the ground, as close as possible to the stuffed cuckoo, covered with a camouflage fabric. The playback device consisted in a MP3 device, an amplifier and two speakers powered by a battery. We observed the behaviour of adult magpies for 30 min from a hidden location about 200 m away from the experimental setup using binoculars and recorded the following variables: latency (the time elapsed between the start of observation and the first adult magpie approaching the stuffed cuckoo closer than 50 m), the number of different magpies approaching the stuffed cuckoo, the number of times that magpies approached, and the approach behaviour (negative, positive or neutral). An approach was classified as negative when the adult magpie showed an aggressive behaviour against the stuffed cuckoo usually involving scolding calls and/or flying over it repeatedly. Positive approaches involved carrying food to the stuffed cuckoo, while neutral approaches were attributed to those observations where the adult magpies ignored the playback and the stuffed cuckoo.

This playback-stuffed-cuckoo experiment was performed to test the hypothesis that cuckoo fledglings that abandon their foster parents are only fed by other magpies that are already caring for cuckoo fledglings. This hypothesis is based on previous findings showing that cuckoo fledglings reared together with magpie fledglings were fed less frequently than those reared in only cuckoo broods [23]. We predicted that the stuffed cuckoo will receive neutral or aggressive responses by magpies attending magpie fledglings, and more positive responses by magpies attending cuckoo fledglings (Prediction 2).

Experiment 3: non-parental-feeding observations

During the breeding season of 2012 we found a total of 133 magpie nests. We assigned each nest to one of the following experimental treatments depending on the composition of the brood: (i) only cuckoos (1–3 great spotted cuckoo nestlings), (ii) only magpies (2–5 magpie nestlings), and (iii) mixed broods (one cuckoo and one or two magpie nestlings). These experimental groups were created by cross-fostering 1 or 2 day old nestlings by carefully transporting them to the corresponding nest. For further information on the cross-fostering manipulation see [40,55,60]. We decided to carry out this cross-fostering manipulation because in naturally parasitized nests, the parasitic chicks usually outcompete their host nestmates with a series of adaptations [58,61,62]. Thus, this manipulation was necessary in order to ensure the survival of both parasite and host nestlings in the same nest until fledging.

We tried to capture as many breeding adult magpies as possible at experimental nests during the entire breeding period (from April to June). In order to do so, we used decoy traps (with a live magpie inside; [63]), which were located in a visible place near the nest. Once captured, adult magpies were marked with numbered metal rings (Ministerio de Agricultura - ICONA) and two patagial wing tags (4 cm length x 3 cm width) with a unique alphanumeric combination. The wing tags were made of PVC fabric of high resistance that causes no damage to the individuals (e.g. [54]) and allows their visual identification from a distance. We managed to capture a total of 66 adult magpies (34 females, 32 males) belonging to a total of 39 different territories: 27 complete pairs (male and female) and 12 cases in which we captured only one of the two adults (Table 1). We measured standard biometrical parameters (weight, wing, tail and tarsus lengths) to identify the sex of each individual [64]. We likewise checked for the presence of a brood patch in incubating females. Additionally, we used behavioral differences between sexes during the breeding season to confirm the sex of pair members [65,66].

We regularly (every 2–3 days) checked all nests and some days before the expected date when cuckoo nestlings leave the nest, we equipped them with a radio-transmitter (Holohil PD-2, see Experiment 1) and patagial wing tags, similar to those attached to adult magpies, but smaller (3 cm length x 2 cm width). Transmitters were attached using the same method described in Experiment 1. A total of 47 great spotted cuckoo nestlings from 33 different nests were equipped with both radio-transmitters and patagial wing tags (Table 1). Six of them were from mixed broods and 41 from "only cuckoo" broods (Table 1).

All nests were monitored in detail until all chicks had left the nest, so we could determine the breeding experience of each magpie pair during the current breeding season (hereafter referred to as "breeding status", Table 1): (i) only cuckoo: adults that have raised exclusively cuckoo chicks until fledging; (ii) mixed broods: adults that have raised at least one cuckoo and one magpie chick until fledging; (iii) only magpies: adults that only raised magpie chicks until fledging; (iv) magpies that failed to raise any chicks until fledging (due to natural causes; e.g. predation).

We carried out an intensive observation schedule of post-fledging feeding events to radio-tracked cuckoos (about 350 hours) following the methodology described in [23]. Basically it consisted in locating radio-tracked cuckoo fledglings using the radio-tracking method (3 element hand-held antennas O-5/8, receptor RX-98H (Televilt, now Followit)) and observing them carefully from the distance.

When a fledgling was detected, we noted its identity (i.e. the frequency of its radio-transmitter) and double-checked it through the observation of the alphanumeric combination of its wing tags. We continuously observed the focal fledgling until we lost sight of the individual. We obtained reliable observations from 23 fledglings, belonging to 15 only cuckoo territories and 4 mixed territories (Table 1), i.e. the fledgling was observed for more than two hours on each observation day (mean ± SE: 136.1±12.6 minutes: $N = 152$ observations). In each feeding event we carefully recorded the identity of the feeding adult magpie (i.e. the number of its wing tag). A total of 374 feeding events were observed in which 25 marked adult magpies were involved (Table 1). We made a strong effort to mark as many adult magpies as possible but not all individuals involved in the observed feeding events were marked birds. We considered a feeding interaction as "non-parental" only when both foster parents of the focal cuckoo fledgling were marked and the fledgling was fed by another adult magpie (either marked or not) than its foster parents, or when its foster parents fed another (marked or not) cuckoo fledgling that was not the one raised by them. In case that not both foster parents were marked, we only considered a feeding as non-parental when the focal fledgling was fed by another marked adult magpie that was not its foster parent. All the observations were made during the most active periods, i.e. from sunrise to 11 a.m. and from 6 p.m. until sunset.

With this experiment, we want to test two predictions. First, based on the hypothesis presented in experiment 2, we predict that non-parental feedings will be mainly provided by magpies caring for only cuckoo fledglings (Prediction 3a). Another non-mutually exclusive hypothesis can also be considered: magpies that failed to fledge any chick could also contribute to provide non-parental feedings to fledgling cuckoos. This hypothesis is based on two ideas: (1) only the proper stimulus (i.e. a fledgling cuckoo begging for food) is enough to provoke alloparental feedings [39], (2) caring for even unrelated fledglings could increase the probabilities of maintaining or acquiring breeding status [67]. This hypothesis predicts that we should find unsuccessful magpies (i.e. those that reared no nestlings until fledging) providing non-parental feedings to fledging cuckoos (Prediction 3b).

Statistical analyses

To analyze the data of our translocation study (Experiment 1), we used a generalized linear model (GLZ) to test if the probability of survival after two weeks (binomial error) was different between cuckoo fledglings from the two translocation groups (fixed factor).

To analyze data collected from the playback-stuffed-cuckoo experiment (Experiment 2), we carried out a generalized (GLZ) or general linear model (GLM), depending on the nature of the response variables. We considered the "approach behaviour" (multinomial distribution) and the number of different magpies approaching to the stuffed cuckoo (Poisson distribution) as the

Table 1. Sample sizes of the total number of marked and observed individuals (adult magpies and cuckoo nestlings) in relation to their breeding status.

Territory type	Marked		Observed	
	Adults	Fledglings	Adults	Fledglings
Only cuckoo	32 (18)	41 (27)	19 (11)	19 (15)
Only magpie	5 (3)	NA	0 (0)	NA
Mixed broods	11 (6)	6 (6)	6 (4)	4 (4)
Failed broods	18 (12)	NA	0(0)	NA

The number of nests to which the individuals belong are indicated between brackets. NA = Not applicable.

response variables for the generalized linear models. Time of latency and the total number of approaches fitted a normal distribution after transformation (log or sqrt) and were the response variables for the general linear models. Treatment (i.e. adult magpies in family groups either with magpie or cuckoo fledglings) and year were included as nominal independent variables, while the number of adult magpies present at the beginning of the experiment was considered as a covariable for all analyses.

The observations on non-parental feedings were analysed from both the adult's and the fledgling's point of view (see Results). We pooled information from different individuals from the same territory, either adults or cuckoo chicks, to avoid pseudoreplication (i.e. each nest is considered as an independent unit). We calculated the ratio of non-parental feedings out of all feeding observations, and analyzed with a General Linear Model (GLM) if the proportion of non-parental feedings differed between nests that only raised cuckoo chicks and those that raised mixed broods (i.e. the two types of territories in which both adults and fledglings have been observed; Table 1). In addition, we used a repeated measures ANOVA to investigate if the total number of non-parental and parental feeding events per hour differed between male and female adult magpies (sex and type of feeding as within factors). This final analysis was done using only adults of only cuckoo territories as they were the only adult magpies involved in non-parental feeding events (see Results).

All analyses were made with STATISTICA 7.0 (StatSoft Inc. 2001–2004).

Results

Experiment 1: translocation of fledglings

We managed to capture and radio-track 15 different great spotted cuckoo fledglings (7 control, 8 translocated; see Database S1). We only found one marked fledgling dead in each treatment group. Cuckoo fledglings survived equally well when released in their own area or when moved to another area (GLZ, $\chi^2_1 = 0.01$, $p = 0.91$). This result is in agreement with Prediction 1 (i.e. translocated fledglings should be able to survive equally well than those that remain in their natal area).

Experiment 2: playback-stuffed-cuckoo presentation

We managed to present the playback and stuffed cuckoo in 56 different magpie territories (11 territories with only magpie fledglings, 45 territories with only cuckoo fledglings) which is the sample size for all analyses except for time of latency (N = 42; see Database S2).

We found a significant difference in "approach behaviour" by adult magpies against the stuffed cuckoo between treatments (GLZ $\chi^2_2 = 12.60$, $p = 0.002$; N = 56). The frequency of negative behaviours against the stuffed cuckoo was much higher when adult magpies were in family groups with only magpie fledglings (aggressive response in 70% of the trials) than when they were observed together with cuckoo fledglings (only in 19% of the trials; Fig. 1). In addition, feeding trips to the stuffed-cuckoo model were only observed when adult magpies were together with great spotted cuckoo fledglings (23% of the trials; Fig. 1). These results strongly support Prediction 2 (i.e. the stuffed cuckoo will receive a neutral or aggressive response by magpies attending magpie fledglings, whereas magpies that were attending cuckoo fledglings will provide a more positive response).

We did not find any significant differences between treatments in the time of latency (GLM $F_{1, 37} = 0.04$, $p = 0.84$), the total number of approaches (GLM $F_{1, 51} = 0.26$, $p = 0.61$) or the number of different adult magpies approaching the stuffed cuckoo (GLZ $\chi^2_1 = 0.54$, $p = 0.46$). The number of adult magpies present in the area at the beginning of the experiment was positively related to the total number of approaches and the number of different magpies approaching (GLM, $F_{1, 51} = 5.60$, $p = 0.02$; and GLZ, $\chi^2_1 = 23.26$, $p < 0.0001$, respectively).

Experiment 3: non-parental-feeding observations

Data obtained in this experiment can be found in Database S3. From the adults' point of view (i.e. feedings provided to cuckoo fledglings by marked adult magpies), there were no significant differences in the ratio of non-parental feedings provided between adults that raised only cuckoo broods and those that raised mixed broods (GLM $F_{1, 13} = 2.16$, $p = 0.17$), although all (100%) non-parental feedings corresponded to adults that raised only cuckoo broods (Fig. 2) (which supports Prediction 3a). In fact, if we consider only those territories in which we observed non-parental feedings (N = 5), we found that these events involved between 22% and 100% of all the observed feedings (N = 201 feedings, mean 58.6±12.5% non-parental feedings). In contrast, we never observed adult magpies from families that raised only magpie fledglings (in agreement to Prediction 3a) or from families with failed breeding events providing non-parental feedings to any of the observed cuckoo fledglings (contrary to Prediction 3b).

From the fledglings' perspective (i.e. feeding interactions observed to marked cuckoo fledglings), we found that non-parental feeding events involved a marginally significantly higher proportion of the feedings to fledgling cuckoos from only cuckoo broods compared to those that were raised together with magpie fledglings (GLM $F_{1, 17} = 4.30$, $p = 0.05$; Fig. 2). Again, all (100%) observed non-parental events involved fledglings from only cuckoo territories. Furthermore, if we consider the territories in which we observed cuckoo fledglings receiving non-parental feedings (N = 9), we found that a mean of 59.1±10.3% (range 10.7% to 100%) of the feedings (N = 184) consisted of non-parental feedings.

Regarding sexual differences, we did not find significant differences in the total number of feedings provided by adult magpies to cuckoo fledglings between males (0.42±0.19 events/h) and females (0.17±0.07 events/h; RM-ANOVA $F_{1, 10} = 3.10$, $p = 0.11$). However we found a marginally non-significant effect for the interaction type of feeding and sex (RM-ANOVA $F_{1, 10} = 3.37$, $p = 0.09$) indicating that male, but not female magpies, tend to feed their own cuckoo fledglings more frequently than other unknown cuckoos (Tukey HSD posthoc: $p = 0.07$; Fig. 3). Furthermore, males significantly feed more actively their cuckoo fledglings than females (Tukey HSD posthoc: $p = 0.03$; Fig. 3).

Discussion

The translocation experiment (Experiment 1) has demonstrated that cuckoo fledglings survived well when moved far away from their natal territories, which indirectly suggest that they managed to be fed by other magpies than their foster parents (those that reared them in the nest). In fact, our third experiment showed that non-parental feeding interactions (i.e. feedings provided by magpies other than their foster parents) involve a high percentage (43%) of all observed feedings to cuckoo fledglings. These data support the hypothesis that obtaining food from magpies other than their foster parents can be a successful and frequently used strategy by some great spotted cuckoo fledglings.

It is well known, and considered to be adaptive, that adult birds frequently attack unrelated offspring that beg for food [68,69]. Then, why do brood parasitic fledglings manage to get fed by other adults different than their foster parents? The results of our

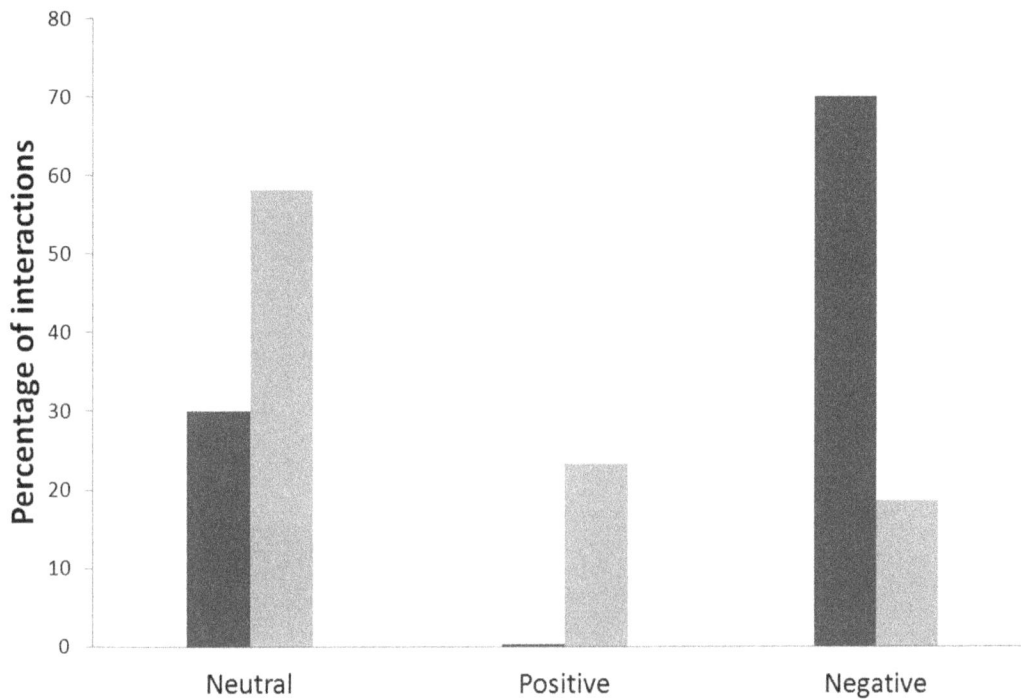

Figure 1. Adult magpies' "approach behaviour" (neutral, positive or negative) to the presentation of the stuffed cuckoo depending on the presence of other cuckoo fledglings (soft grey) or only magpie fledglings (dark grey). See Material and Methods section for a detailed explanation for each type of "approach behaviour".

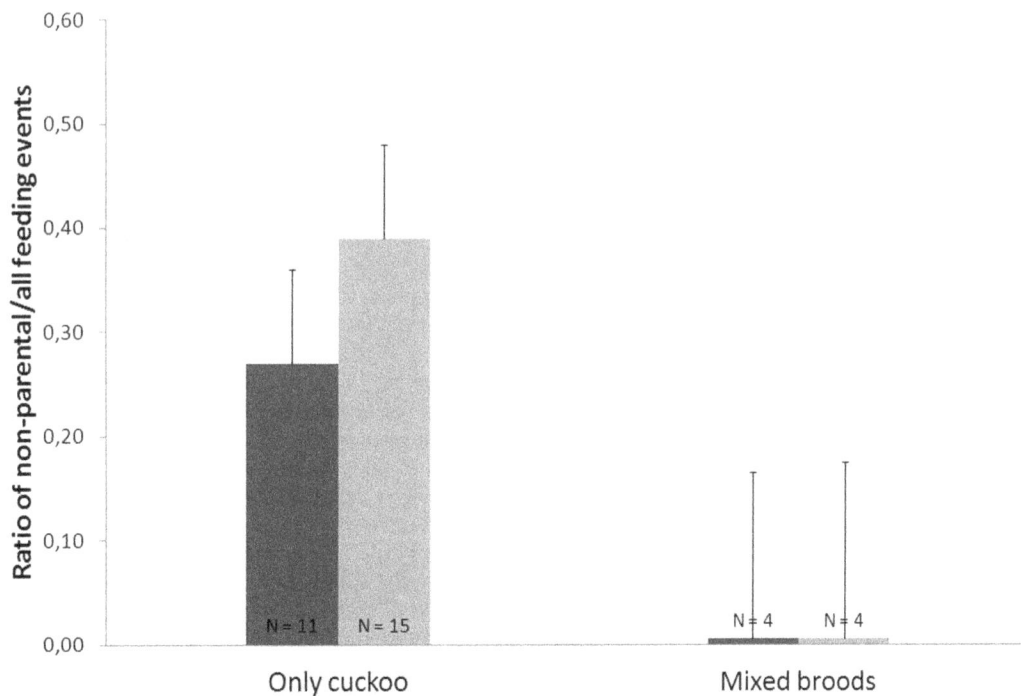

Figure 2. Ratio of non-parental feeding events out of all observed feedings provided by adult magpies (dark grey) or received by cuckoo fledglings (soft grey) for each territory type. Note that "only magpie" and "failed broods" territories are not represented as adults of these kinds of territories were never observed feeding great spotted cuckoo fledglings. Data represented are Least Square Means ± SE.

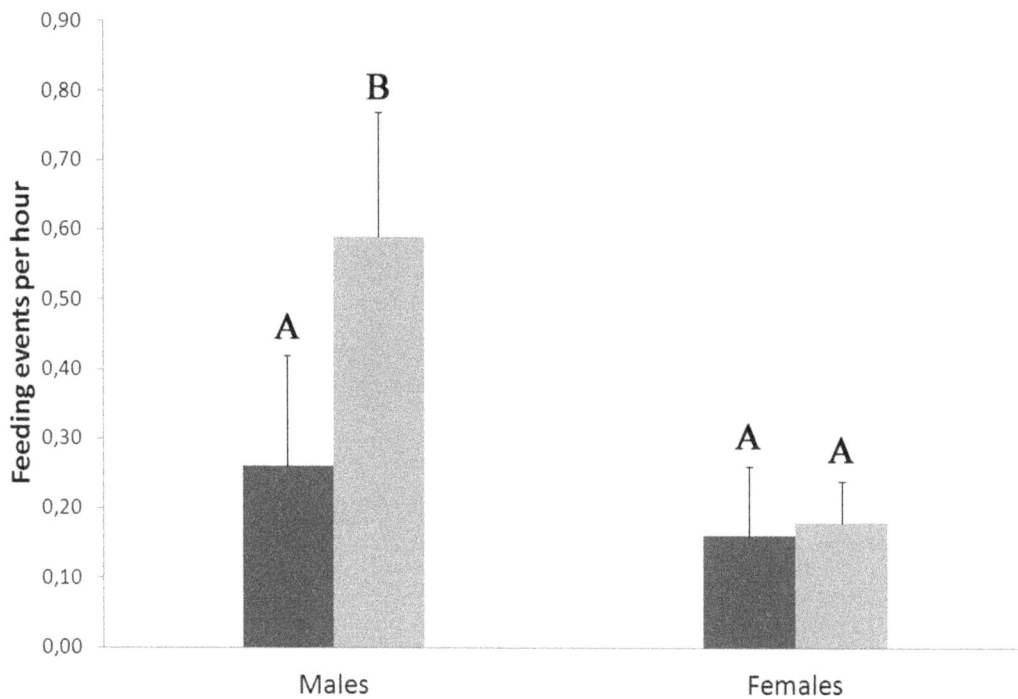

Figure 3. Sex differences for adult magpies in the number of non-parental (dark grey) and parental (soft grey) feedings per hour provided to great spotted cuckoo fledglings.

translocation experiment imply that cuckoo fledglings have adaptations that enable them deceive magpies into feeding them. The existence of such adaptations has also been suggested by Eastzer et al. [70] who found that barn swallows (*Hirundo rustica*) that had raised nestlings from three different species only continued to feed the brood parasitic brown-headed cowbird (*Molothrus ater*) after leaving the nest.

A key factor of this potential adaptation is begging behaviour, which is more exuberant in fledglings of brood parasites than in fledglings of host species [20,37,70,71] and could have an important role in attracting host attention. Another aspect related to begging is vocal mimicry, because host ability to discriminate parasitic fledglings would select for mimetic begging calls in brood parasite fledglings [22,47]. However, this does not seem to be the case in great spotted cuckoo fledglings because their begging calls are very different from those of magpie fledglings (own observations), and neither during the nestling stage vocal mimicry exists in this brood parasite species [72].

The fact that some magpies after hearing the begging calls of great spotted cuckoo fledglings approached the stuffed-cuckoo fledgling disposed to feed it indicates that the begging calls are the stimuli responsible for the fact that magpies other than their foster parents feed them. Although it could be possible that magpies cue on the begging calls to locate the fledgling and then decide whether to feed it or not based on visual cues too. Two pieces of evidence support the former statement. First, in a first try-out of performing Experiment 3, the same stuffed-cuckoo fledglings (which were mounted with the beak closed and in a non-begging display; see Methods) did not provoke any approaching by magpies, neither in absence of the begging calls ($N = 8$) nor when accompanied by bad quality recordings of fledgling begging calls (i.e. recordings taken at long distance without the appropriate recording material, $N = 18$) (Soler et al. unpublished). Second,

begging call is the most important component of communication between brood parasite fledglings and their foster parents [20,22,37].

However, interestingly, the response of magpies to the playback-stuffed cuckoo (Experiment 2) depended on the social situation: adult magpies showed some level of predisposition to feed the stuffed-cuckoo fledgling only when they were observed together with cuckoo fledglings (i.e. magpies that were attending a group of cuckoo fledglings). Adult magpies that were in family groups together with magpie fledglings never approached to feed the stuffed cuckoo and the response against it was frequently aggressive (Fig. 1). These results are in agreement with previous findings. Magpies rearing a cuckoo nestling accepted and fed other cuckoo nestlings experimentally introduced into their nests during the last phase of the nestling period, while magpies from non-parasitized nests frequently were reluctant to feed the experimental cuckoo nestling, and even, sometimes rejected it [27]. In addition, after leaving the nest, cuckoo fledglings that had been reared together with magpie fledglings were less intensely defended than their magpie nestmates, less frequently fed than cuckoo fledglings reared in "only cuckoo broods", and more importantly in relation to this study, magpie adults approached to feed cuckoos from mixed broods less frequently than cuckoos from "only cuckoo broods" [23]. These results suggest that the presence of host's own nestlings for comparison may be a crucial clue favouring the evolution of fledgling discrimination [22, 23, 47, this study]. An effect of opportunities for hosts to compare own and foreign chicks on nestling discrimination was also suggested [73,74], however, recent reviews have showed that such an effect is not important [52,75].

Our playback-stuffed-cuckoo presentation study (Experiment 2) has shown that only some magpies that are already caring for cuckoo fledglings are willing to feed the stuffed cuckoo (i.e.

unknown cuckoo fledglings that try to join their group of cuckoos). Which magpies are those that have the motivation to feed foreign cuckoo fledglings that beg for food? This is a very important question for the understanding of the evolution of the arms race between great spotted cuckoos and their magpie hosts at the fledgling stage. Our study on feeding interactions (Experiment 3) was conducted in a population of previously marked magpies to answer this question and we found, although with small sample sizes, that 100% of non-parental feeding events were provided by magpies that were caring only for cuckoo fledglings (Fig. 2), which support Prediction 3a. This result confirms previous findings obtained in Experiment 2. Thus, we can conclude that possibly undernourished great spotted cuckoo fledglings that abandon their foster parents and join a group of other fledglings [23] are fed by adult magpies that only have reared cuckoos. The alternative hypothesis that unsuccessful magpies that fail to rear any chick could also provide non-parental feedings (Prediction 3b) has not been supported by our study. None of the 21 marked magpies that did not successfully rear any chick were observed feeding cuckoo fledglings.

Magpies, in natural non-parasitized populations ignore begging by unrelated fledglings and at the end of the fledgling period become more reluctant to feed their own fledglings in spite of their intensive begging behaviour [76]. Then, why do some magpies in our parasitized population behave "maladaptively" by feeding unrelated parasitic fledglings? In those cases in which magpies are the foster parents that reared the cuckoo chicks in their nests, they continue feeding them after leaving the nest because they learnt the begging calls of those chicks at the end of the nestling period (see references above) and parents are also able to distinguish later the begging calls of their fledglings [77].

Experiments 2 and 3 have demonstrated that some magpies feed foreign cuckoo fledglings, and that alloparental feedings to brood parasitic fledglings are frequently observed (see above). In Experiment 3, when considering only those territories in which we observed non-parental feedings (the adults' point of view) or those in which we observed cuckoo fledglings receiving non-parental feedings (the fledglings' perspective), more than 50% of the feedings consisted of non-parental feedings (see Results, Fig. 2). Although the number of territories in which non-parental feedings were observed was low, probably causing only marginally significant results, these observations do suggest that adults that do provide non-parental feedings do this frequently and, from the fledglings' perspective, that non-parental feedings involve an important source of energy for these fledglings.

Why do some host individuals accept to feed brood parasitic fledglings that were not reared in their nests? The first answer to this question was suggested by Sealy and Lorenzana [38] who proposed that parasitic fledglings may display a supernormal stimulus [78] which would manipulate hosts into feeding them. The best candidate responsible for such a supernormal stimulus would be begging behaviour (see above) because begging is the most important component of avian adult-young communication [79] and begging calls of brood parasitic fledglings are louder, more persistent and more exaggerated than those of host fledglings [20,37,70,71]. However, Experiment 2 has shown that magpie response is highly variable: sometimes they are disposed to feed the stuffed-cuckoo fledgling, but in other cases they show an aggressive behaviour against the stuffed cuckoo, especially dependent on the status of their social group (see above). This result does not support the supernormal stimulus hypothesis because this hypothesis predicts that a supernormal stimulus should be able to manipulate magpie adults into feeding the fledglings regardless the social situation [78].

The second answer to the above-mentioned question is based on the fact that the learning mechanism between parents and offspring at the end of the nestling period usually involves the evolution of highly variable vocal signatures among nestlings that allow parents to recognize and differentiate them later from other fledglings [80–82]. We could speculate that great spotted cuckoos developed begging calls that are very attractive to magpies, but characterized by invariable vocal signatures, which would make it difficult for magpies to differentiate between cuckoo fledglings. This scenario would facilitate that any great spotted cuckoo fledgling begging at a high intensity could be fed by magpies that are already feeding other cuckoo fledglings. However, the variability of vocal signatures remains to be studied in fledglings of the great spotted cuckoo and of any other brood parasitic species.

Our results also provide new insights to understand the mechanisms underlying egg or chick discrimination. Magpies are long-lived birds [66]. Therefore, the fact that the current social situation (i.e. caring only for cuckoo fledglings) is a main factor determining the propensity to feed foreign cuckoo fledglings implies that recognition templates (i.e. internal representation of the appearance of parasitic chicks [83]) are not inherited or learned during the female's first breeding attempt as traditional theory assumed [52,74,75,84–88] but that they are acquired again at each new breeding attempt, as has been suggested in several more recent studies [89,90] and has recently experimentally been demonstrated [91].

Magpie nests are frequently multiparasitized (i.e. with more than one great spotted cuckoo egg per nest) either by different females or by the same female laying several eggs in the same nest [58,92,93]. The existence of more than one parasitic nestling per nest increases competition and can trigger the starvation of some of them [93]. Now we have two pieces of evidence showing that multiparasitism is selected because of benefits for the cuckoos provided at the fledging stage. Multiparasitism usually prevents survival of any of the host young and this benefits cuckoos during the post-fledgling stage: first, great spotted cuckoo fledglings reared together with magpie nestlings are disadvantaged by magpie foster parents [23], and second, magpies caring for only cuckoo broods are more prone to feed cuckoo fledglings [23], even those that were not reared in their nests (this study).

Interestingly, we have found that male magpies more actively feed their own cuckoo fledglings compared to their female partner. In addition, males, but not females, also tend to feed own cuckoo fledglings more frequently than other unknown cuckoos (Fig. 3). We did however not find any information about sexual differences in feeding frequency of magpie parents to their own fledglings [76], or about feeding frequency by males versus females in any host species of brood parasitic fledglings [20,22]. It has only been reported that both males and females baywings refused to feed shiny cowbird fledglings [22].

In conclusion, the results obtained in this study are very important for the understanding of the evolution of the arms race between great spotted cuckoos and their magpie hosts because they provide new insights on several relevant points. They indicate that (1) the presence of host's own nestlings for comparison may be a crucial clue favouring the evolution of fledgling discrimination, (2) the benefits provided at the fledging stage could select for multiparasitism, and (3) the results offer new evidence that recognition templates, the basis for the mechanisms underlying egg or chick discrimination, are not inherited or learned during the female's first breeding attempt but are acquired again at each new breeding attempt.

Supporting Information

Database S1 Data used for statistical analyses of experiment 1 (translocation of fledglings).

Database S2 Data used for statistical analyses of experiment 2 (playback-stuffed-cuckoo presentation).

Database S3 Data used for statistical analyses of experiment 3 (non-parental-feeding observations).

References

1. Clutton-Brock TH (1991) The evolution of parental care. Princeton: Princeton University Press.
2. Alonso-Alvarez C, Velando V (2012) Benefits and costs of parental care. In Royle NJ, Smiseth PT ,Kölliker M, editors. The evolution of parental care. Oxford: Oxford University Press. pp. 40–61.
3. Rytkönen S (2002) Nest defence in great tits Parus major: support for the parental investment theory. Behav Ecol Sociobiol 52: 379–384.
4. Bize P, Piault R, Moreau B (2006) A UV signal of offspring condition mediates context-dependent parental favoritism. Proc R Soc B 273: 2063–2068.
5. Griggio M, Morosinotto C, Pilastro A (2009) Nestlings' carotenoid feather ornament affects parental allocation strategy and reduces maternal survival. J Evol Biol 22: 2077–2085.
6. Thorogood R, Ewen JG, Kilner RM (2011) Sense and sensitivity: responsiveness to offspring signals varies with the parents' potential to breed again. Proc R Soc B 278: 2638–2645.
7. Winkler DW (1991) Parental investment decision rules in tree swallows: parental defense, abandonment, and the so-called Concorde Fallacy. Behav Ecol 2: 133–142.
8. Székely T, Webb JN, Houston AI, McNamara JM (1996) An evolutionary approach to offspring desertion in birds. Curr Ornithol 13: 271–330.
9. Ward RJS, Cotter SC, Kilner RM (2009) Current brood size and residual reproductive value predict offspring desertion in the burying beetle Nicrophorus vespilloides. Behav Ecol 20: 1274–1281.
10. Jennions MD, Polakow DA (2001) The effect of partial brood loss on male desertion in a cichlid fish: an experimental test. Behav Ecol 12: 84–92.
11. Ackerman JT, Eadie JM (2003) Current versus future reproduction: an experimental test of parental investment decisions using nest desertion by mallards (Anas platyrhynchos). Behav Ecol Sociobiol 54: 264–73.
12. Zink AG (2003) Quantifying the costs and benefits of parental care in female treehoppers. Behav Ecol 14: 687–693.
13. Zeh DW, Smith RL (1985) Paternal investment by terrestrial arthropods. Am Zool 25: 785–805?
14. Sheldon BC, Räsänen K, Dias PC (1997) Certainty of paternity and paternal effort in the collared flycatcher. Behav Ecol 8: 421–428.
15. Møller AP, Cuervo JJ (2000) The evolution of paternity and paternal care in birds. Behav Ecol 11: 472–485.
16. Neff BD (2003) Decisions about parental care in response to perceived paternity. Nature 422: 716–719.
17. Roldán M, Soler M (2011) Parental care parasitism: how unrelated offspring attain acceptance by foster parents? Behav Ecol 22: 679–691.
18. Riedman ML (1982) The evolution of alloparental care and adoption in mammals and birds. Q Rev Biol 57: 405–435.
19. Roulin A (2002) Why do lactating females nurse alien offspring? A review of hypotheses and empirical evidence. Anim Behav 63: 201–208.
20. Woodward PW (1983). Behavioral ecology of fledgling Brown-headed Cowbirds and their hosts. Condor 85, 151–163.
21. Soler M, Palomino JJ, Martínez JG, Soler JJ (1994) Activity, survival, independence and migration of fledgling great spotted cuckoos. Condor 96: 802–805.
22. De Mársico MC, Gantchoff MG, Reboreda JC (2012) Host–parasite coevolution beyond the nestling stage: activity of host fledglings by the specialist screaming cowbird. Proc R Soc B 279: 3401–3408.
23. Soler M, de Neve L, Roncalli G, Macías-Sánchez E, Ibáñez-Álamo JD, et al. (2013) Great spotted cuckoo fledglings are disadvantaged by magpie host parents when reared together with magpie nestlings. Behav Ecol Sociobiol 68: 333–342.
24. Beecher MD (1988) Kin recognition in birds. Behav Genet 18: 465–482.
25. Lessells CM, Coulthard ND, Hodgson PJ, Krebs JR (1991) Chick recognition in European bee-eaters: acoustic playback experiments. Anim Behav 42: 1031–1033.
26. Medvin M, Stoddard PK, Beecher MD (1993) Signals for parents-offspring recognition: a comparative analysis of the begging calls of cliff swallows and barn swallows. Anim Behav 45: 841–850.
27. Soler M, Soler JJ, Martínez JG (1995a) Chick recognition and acceptance: a weakness in magpies exploited by the parasitic great spotted cuckoo. Behav Ecol Sociobiol 37: 243–248.
28. Levréro F, Durand L, Vignal C, Blanc A, Mathevon N (2009) Begging calls support offspring individual identity and recognition by zebra finches. C R Biol 332: 579–589.
29. Friedmann H (1929) The cowbirds: a study in the biology of social parasitism. Springfield: Charles C. Thomas.
30. Jubb RA (1966) Red-billed hoopoe and a greater honey-guide. Bokmakierie 18: 66.
31. Lack D (1968) Ecological adaptations for breeding in birds. London: Methuen.
32. Klein NK, Rosenberg KV (1986) Feeding of Brown-headed cowbird (Molothrus ater) fledglings by more than one "host" species. Auk 103: 213–214.
33. Brooker MG, Brooker LC (1989) Cuckoo hosts in Australia. Aust Zool Rev 2: 1–67.
34. Hatton AIG (1989) A juvenile Fan-tailed Cuckoo fed by two different species of small birds. Aust Birds 22: 81.
35. Smith LH (1989) Feeding of young Pallid Cuckoo by four passerine species. Aust. Bird Watcher 13: 99–100.
36. Zuñiga JM, Redondo T (1992) Adoption of Great Spotted Cuckoo Clamator glandarius fledglings by magpies Pica pica. Bird Study 39: 200–202.
37. Soler M, Palomino JJ, Martínez JG, Soler JJ (1995b) Communal parental care by monogamous magpie hosts of fledgling great spotted cuckoos. Condor 97: 804–810.
38. Sealy SG, Lorenzana JC (1997) Feeding of nestling and fledgling brood parasites by individuals other than the foster parents: a review. Can J Zool 75: 1739–1752.
39. Eisner E (1960) The relationships of hormones to the reproductive behaviour of birds, referring specially to parental behaviour. A review. Anim Behav 8: 155–179.
40. Soler M, de Neve L (2013) Brood mate eviction or brood mate acceptance by brood parasitic nestlings? An experimental study with the non-evictor great spotted cuckoo and its magpie host. Behav Ecol Sociobiol 67: 601–607.
41. Soler M, Soler JJ (1999) Innate versus learned recognition of conspecifics in great spotted cuckoos Clamator glandarius. Anim Cogn 2: 97–102.
42. McGowan KJ, Woolfenden GE (1990) Contributions to fledgling feeding in the Florida scrub-jay. J Anim Ecol 59: 691–707.
43. Hodge SJ, Flower TP, Clutton-Brock TH (2007) Offspring competition and helper associations in cooperative meerkats. Anim Behav 74: 957–964.
44. Thompson AM, Ridley AR (2013) Do fledglings choose wisely? An experimental investigation into social foraging behaviour. Behav Ecol Sociobiol 67: 69–78.
45. Soler M (2013) Long-term coevolution between avian brood parasites and their hosts. Biol Rev doi: 10.1111/brv.12075.
46. Grüebler MU, Naef-Daenzer B (2010) Survival benefits of post-fledging care: Experimental approach to a critical part of avian reproductive strategies. J Anim Ecol 79: 334–341.
47. Fraga RM (1998) Interactions of the parasitic screaming and shiny cowbirds Molothrus rufoaxillaris and M. bonariensis) with a shared host, the bay-winged cowbird (M. badius). In: Rothstein SI, Robinson SK editors. Parasitic birds and their hosts: studies in coevolution. New York: Oxford University Press. pp. 173–193.
48. Davies NB (2011) Cuckoo adaptations: trickery and tuning. J Zool 284: 1–14.
49. Sato NJ, Tokue K, Noske RA, Mikami OK, Ueda K (2010) Evicting cuckoo nestlings from the nest: a new anti-parasitism behaviour. Biol Lett 6: 67–69.
50. Tokue K, Ueda K (2010) Mangrove gerygones Gerygone laevigaster eject little bronze-cuckoo Chalcites minutillus hatchlings from parasitized nests. Ibis 152: 835–839.
51. Delhey K, Carrizo M, Verniere L, Mahler B, Peters A (2011) Rejection of brood-parasitic shiny cowbird Molothrus bonariensis nestlings by the firewood gatherer Anumbius annumbi? J Avian Biol 42: 463–467.
52. Grim T (2006) The evolution of nestling discrimination by hosts of parasitic birds: why is rejection so rare? Evol Ecol Res 8: 785–802.
53. Soler M (2009) Co-evolutionary arms race between brood parasites and their hosts at the nestling stage. J Avian Biol 40: 237–240.
54. Canestrari D, Marcos JM, Baglione V (2007) Costs of chick provisioning in cooperatively breeding crows: an experimental study. Anim Behav 73: 349–357.
55. Soler M, de Neve L (2012) Great spotted cuckoo nestlings but not magpie nestlings starve in experimental age-matched broods. Ethology 118: 1036–1044.

Acknowledgments

We thank Francisco Espinosa, Francisco Ferri, Danail M. Ivanov, David Martín-Gálvez, Juan G. Martínez, Patricia Molina and Juan S. Sánchez for their help with field work. We also want to thank Adela González-Megías who provided useful information regarding statistical analyses.

Author Contributions

Conceived and designed the experiments: MS TPC LDN. Performed the experiments: MS TPC JDIA GR EMS LDN. Analyzed the data: MS TPC JDIA GR LDN. Contributed reagents/materials/analysis tools: MS TPC JDIA GR EMS LDN. Wrote the paper: MS TPC JDIA GR EMS LDN.

56. Hill IF, Cresswell BH, Kenward RE (1999) Field-testing the suitability of a new back-pack harness for radio-tagging passerines. J Avian Biol 30: 135–142.

57. Naef-Daenzer B, Widmer F, Nuber M (2001) Differential post-fledging survival of great and coal tits in relation to their condition and fledging date. J Anim Ecol 70: 730–738.

58. Soler M (1990) Relationship between the Great Spotted Cuckoo *Clamator glandarius* and its corvid hosts in a recently colonized area. Ornis Scand 21: 212–223.

59. Rappole JH, Tipton AR (1991) New harness design for attachment of radio transmitters to small passerines. J Field Ornithol 62: 335–337.

60. Soler M, de Neve L, Roldán M, Macías-Sánchez E, Martín-Gálvez D (2012) Do great spotted cuckoo nestlings beg dishonestly? Anim Behav 83: 163–169.

61. Soler M, Martínez JG, Soler JJ, Møller AP (1995c) Preferential allocation of food by magpie *Pica pica* to great spotted cuckoo *Clamator glandarius* chicks. Behav Ecol Sociobiol 37: 7–13.

62. Soler M, Martínez JG, Soler JJ (1996) Effects of brood parasitism by the great spotted cuckoo on the breeding success of the magpie host: An experimental study. Ardeola 43: 87–96.

63. Díaz-Ruiz F, García JT, Pérez-Rodríguez L, Ferreras F (2010) Experimental evaluation of live cage-traps for black-billed magpies *Pica pica* management in Spain. Eur J Wildl Res 56: 239–248.

64. Reese KP, Kadlec JA (1982) Determining the sex of Black-billed magpies by external measurements. J Field Orn 53: 417–418.

65. Erpino MJ (1969) Seasonal cycle of reproductive physiology in the black-billed magpie. Condor 71: 267–279.

66. Birkhead T (1991) The Magpies. The ecology and behaviour of Black-billed and Yellow-billed Magpies. London: Poyser.

67. Fitzpatrick JW, Woolfenden GE (1986) Demographic routes to cooperative breeding in some New World jays. In Nitecki MH, Kitchell JA, editors. Evolution of animal behaviour: paleontological and field approaches. New York: Oxford University Press. pp. 137–160.

68. Beecher MD, Beecher IM, Lumpkin S (1981) Parent-offspring recognition in Bank Swallows (*Riparia riparia*).1. Natural history. Anim Behav 29: 86–94.

69. Proffitt FM, McLean IG (1991) Recognition of parents calls by chicks of the snares crested penguin. Bird Behav 9: 103–113.

70. Eastzer D, Chu RR, King AP (1980) The young cowbird: average or optimal nestling? Condor 82: 417–425.

71. Dearborn DC (1998) Begging behavior and food acquisition by brown-headed cowbird nestlings. Behav Ecol Sociobiol 43: 259–270.

72. Roldán M, Soler M, Márquez R, Soler JJ (2013) The vocal begging display of great spotted cuckoo *Clamator glandarius* nestlings in nests of its two main host species: genetic differences or developmental plasticity? Ibis 155: 867–876.

73. Davies NB, Brooke M de L (1988) Cuckoos versus reed warblers: adaptations and counteradaptations. Anim Behav 36: 262–284.

74. Lotem A (1993) Learning to recognize nestlings is maladaptive for cuckoo *Cuculus canorus* hosts. Nature 362: 743–745.

75. Grim T (2011) Ejecting chick cheats: a changing paradigm? Front Zool 8: 14.

76. Husby M, Slgasvold T (1992) Post-fledging behaviour and survival in male and female magpies *Pica pica*. Ornis Scand 23: 483–490.

77. Draganoiu T, Nagle L, Musseau R, Kreutzer M (2006) In a songbird, the black redstart, parents use acoustic cues to discriminate between their different fledglings. Anim Behav 71: 1039–1046.

78. Dawkins R, Krebs JR (1979) Arms races between and within species. Proc R Soc B 205: 489–511.

79. Wright J, Leonard M (2002) The evolution of begging: competition, cooperation and communication. Dordrecht: Kluwer Academic Publishers.

80. Stoddard PK, Beecher MD (1983) Parental recognition of offspring in the cliff swallow. Auk 100: 795–799.

81. Tibbetts EA, Dale J (2007) Individual recognition: it is good to be different. Trends Ecol Evol 22: 529–537.

82. Reers H, Jacot A, Forstmeier W (2011) Do zebra finch parents fail to recognize their own offspring? PLoS ONE, 6, e18466.

83. Hauber ME, Sherman PW (2001) Self-referent phenotype matching: theoretical considerations and empirical evidence. Trends Neurosci 24: 609–616.

84. Rothstein SI (1974) Mechanisms of avian egg recognition: possible learned and innate factors. Auk 91: 796–807.

85. Rothstein SI (1975) Mechanisms of avian egg-recognition: do birds know their own eggs? Anim Behav 23: 268–278.

86. Rothstein SI (1978) Mechanisms of avian egg-recognition: additional evidence for learned components. Anim Behav 26: 671–677.

87. Lotem A, Nakamura H, Zahavi A (1992) Rejection of cuckoo eggs in relation to host age: a possible evolutionary equilibrium. Behav Ecol 3: 128–132.

88. Lotem A, Nakamura H, Zahavi A (1995) Constraints on egg discrimination and cuckoo-host co-evolution. Anim Behav 49: 1185–1209.

89. Moskát C, Hauber ME (2007) Conflict between egg recognition and rejection decisions in common cuckoo (*Cuculus canorus*) hosts. Anim Cogn 10: 377–386.

90. Langmore NE, Cockburn A, Russell AF, Kilner RM (2009) Flexible cuckoo chick rejection rules in the superb fairy-wren. Behav Ecol 20: 978–984.

91. Soler M, Ruiz-Castellano C, Carra LG, Ontanilla J, Martín-Galvez D (2013) Do first-time breeding females imprint on their own eggs? Proc. R. Soc. B 280, 20122518.

92. Martínez JG, Burke T, Dawson D, Soler JJ, Soler M, Møller AP (1998) Microsatellite typing reveals mating patterns in the brood parasitic great spotted cuckoo (*Clamator glandarius*). Mol Ecol 7: 289–297.

93. Soler M, Soler JJ, Martínez JG (1998) Duration of sympatry and coevolution between the great spotted cuckoo *Clamator glandarius* and its primary host, the magpie *Pica pica*. In: Rothstein SI, Robinson SK, editors. Parasitic birds and their hosts: studies in coevolution. New York: Oxford University Press. Pp 113–118.

14

Gene Expression Profiling in *Entamoeba histolytica* Identifies Key Components in Iron Uptake and Metabolism

Nora Adriana Hernández-Cuevas[1,2]*, **Christian Weber**[1,2], **Chung-Chau Hon**[1,2], **Nancy Guillen**[1,2]

1 Institut Pasteur, Unité Biologie Cellulaire du Parasitisme, Paris, France, **2** INSERM U786, Paris, France

Abstract

Entamoeba histolytica is an ameboid parasite that causes colonic dysentery and liver abscesses in humans. The parasite encounters dramatic changes in iron concentration during its invasion of the host, with relatively low levels in the intestinal lumen and then relatively high levels in the blood and liver. The liver notably contains sources of iron; therefore, the parasite's ability to use these sources might be relevant to its survival in the liver and thus the pathogenesis of liver abscesses. The objective of the present study was to identify factors involved in iron uptake, use and storage in *E. histolytica*. We compared the respective transcriptomes of *E. histolytica* trophozoites grown in normal medium (containing around 169 µM iron), low-iron medium (around 123 µM iron), iron-deficient medium (around 91 µM iron), and iron-deficient medium replenished with hemoglobin. The differentially expressed genes included those coding for the ATP-binding cassette transporters and major facilitator transporters (which share homology with bacterial siderophores and heme transporters) and genes involved in heme biosynthesis and degradation. Iron deficiency was associated with increased transcription of genes encoding a subset of cell signaling molecules, some of which have previously been linked to adaptation to the intestinal environment and virulence. The present study is the first to have assessed the transcriptome of *E. histolytica* grown under various iron concentrations. Our results provide insights into the pathways involved in iron uptake and metabolism in this parasite.

Editor: Pedro Lagerblad Oliveira, Universidade Federal do Rio de Janeiro, Brazil

Funding: The work was funded by the French National Research Agency through grants to NG (ANR-MIE-08 Intestinalamibe. The funders had no role in study design, data collection and analysis, decision to publish, or preparation of the manuscript.

Competing Interests: The authors have declared that no competing interests exist.

* Email: noraahc@gmail.com

Introduction

Entamoeba histolytica is a unicellular eukaryote that causes amebiasis in humans after the ingestion of contaminated water or food-containing cysts. Trophozoites (the vegetative form of *E. histolytica*) released into the intestinal lumen by excystation may then penetrate the intestinal mucosal layer and cause colitis and bloody diarrhea [1]. In some cases, virulent trophozoites may also invade the liver and cause liver abscesses. Given that the host environments of the intestinal lumen and the liver are dramatically different, the parasite's abilities to adapt to these environments are considered to be crucial for its parasitic lifestyle [2]. *Entamoeba histolytica*'s ability to efficiently use iron sources in the liver might be relevant to the parasite's survival and the pathogenesis of liver abscesses. In fact, amebic proteins such as ferredoxin, alcohol dehydrogenase 2 [3] and superoxide dismutase [4] require iron in order to function. Moreover, the fact that trophozoites grown under low-iron concentration *in vitro* show low cell adherence and cytotoxic activities suggests that iron metabolism has a role in the pathogenesis of amebiasis [3,5]. Despite iron's seemingly important roles in parasite survival and pathogenesis, the mechanisms underlying iron uptake, use and storage in *E. histolytica* have yet to be characterized.

Iron can enter eukaryotic cells by three major pathways: (i) uptake of ferrous iron (i.e., Fe^{2+}); (ii) endocytosis of iron-binding proteins; and (iii) acquisition of heme. In the first pathway, ferric iron (i.e., Fe^{3+}, the predominant form in the diet) is reduced to ferrous iron (Fe^{2+}) by intestinal ferric reductase (for a review, see [6,7]). Next, Fe^{2+} can be transported into enterocytes by the divalent metal transporter 1 (DMT1). In the second pathway, certain cell types acquire iron via transferrin receptor (TfR1)-mediated endocytosis of holotransferrin (transferrin (Tf) is the main iron carrier in plasma) [8]. Iron is transported into the cytoplasm by DMT1 after internalization of the transferrin-TfR1 complex. In the third pathway, iron is obtained from hemoglobin (Hb) or heme upon erythrocyte lysis. Heme and Hb are scavenged by hemopexin and haptoglobin, respectively. Hemoglobin's functions include oxygen transport from the lungs to the tissues, the removal of carbon dioxide and carbon monoxide from the body and the regulation of vascular tone through nitric oxide binding [7]. Heme is a prosthetic group in hemoproteins such as hemoglobin, myoglobin, cytochromes, catalases, and peroxidases. Enterocytes can internalize heme via the heme carrier protein 1 (HCP-1) transporter. Heme oxygenase-1 (HO-1) then cleaves the tetrapyrrole ring to yield α-biliverdin, carbon monoxide, and iron

[6,9]. Iron participates readily in redox reactions and produces reactive oxygen species (ROS) [10]. Since free iron is toxic to the cell, it is rapidly bound to ferritin and/or released into the circulation via the ferrous iron transporter ferroportin [11]. In the mitochondria, iron is used to synthesize heme and iron-sulfur (Fe-S) clusters. In microbial pathogens, heme acquisition and the endocytosis of iron-binding proteins are the main routes for iron uptake in host environments. Mechanisms of iron uptake have been best studied in bacteria and include (i) receptor-mediated binding of transferrin, lactoferrin and Hb, (ii) secretion of Fe^{3+}-chelating siderophores and (iii) hemophores (heme-chelating proteins that remove heme from diverse sources). The iron-siderophore and heme-hemophore complexes enter the cell via specialized receptors (for a review, see [12]).

It has been reported that *E. histolytica* is able to obtain iron from the bacterial flora and the host's iron-binding proteins (e.g., transferrin, hemoglobin, ferritin, and lactoferrin). These proteins are usually released when the amoeba lyses and phagocytes cells [13]. However, the fact that hemolytic activities have been detected in *E. histolytica* trophozoites [13] suggests that the parasites might be able to lyse red blood cells (RBCs) and liberate the host's heme via an non-phagocytic route. Furthermore, it has been suggested that two heme-binding proteins secreted by *E. histolytica* facilitate the scavenging of host heme [14]. Little is known about iron uptake, utilization and storage pathways in *E. histolytica*, with the exception of a few enzymes (i.e., NifS, NifU and rubrerythrin) involved in Fe-S cluster pathways [15,16].

The objective of the present study was to identify candidate factors involved in iron uptake, use and storage mechanisms in *E. histolytica*. We compared the transcriptome of *E. histolytica* trophozoites grown *in vitro* under four conditions: (i) normal iron concentrations (around 169 μM iron); (ii) low iron concentrations (around 123 μM); (iii) iron deficiency (91 μM) and (iv) iron deficiency with Hb replenishment. Considering the parasite's hemolytic activities and its phagocytosis of RBCs, Hb replenishment was used to mimic the transition from low iron levels in the intestine to high (Hb-derived) iron levels in the blood and liver. The differentially expressed candidate genes coded for proteins with a function in iron uptake (i.e., major facilitator transporters (MFTs) and ATP-binding cassette (ABC) transporters, which share homology with bacterial siderophores or heme transporters) or were homologous to important genes involved in heme biosynthesis and degradation in bacteria, plants, and humans (e.g., S-adenosylmethionine synthetase, glutamyl-tRNA synthetase, and monooxygenases). This study is the first to have assessed the transcriptome of *E. histolytica* grown with various iron concentrations and enabled us to identify candidates involved in iron uptake and metabolism in the parasite.

Methods

Entamoeba histolytica culture

The axenic, virulent *E. histolytica* strain HM1:IMSS was cultured in the TYI-S-33 medium at 37°C [17]. The TYI-S-33 medium contains 15% serum, 78 μM ammonium ferric citrate (AFC) and vitamins (Table S1 in File S1). The serum and peptone contribute 90.7 μM iron, meaning that the "normal iron" condition corresponds to ≈169 μM iron. In the "low iron" condition, 2×10^3 amoebae mL^{-1} grown in normal iron medium were grown for 6 days in modified TYI-S-33 medium (Table S1 in File S1) supplemented with just 32 μM AFC (yielding a final iron concentration of around 123 μM). The medium was changed at 72 and 96 hrs post-inoculation. The trophozoites were grown for 3 days in low iron and then were supplemented with 52 μM Hb

(Sigma, H7379, France) for three days to determine the expression level of modulated genes by qRT-PCR assay. To obtain trophozoites adapted to "iron deficiency", trophozoites were grown for one month in TYI-S-33 medium in which the AFC supplementation was reduced in a step-wise manner every three days (from 80 μM AFC to 40 μM, 30 μM, 20 μM, 10 μM and 0 μM AFC). Hence, the "iron deficiency" condition corresponds to TYI-S-33ΔFe medium that contained only the iron contributed by serum and peptone (90.7 μM) (Table S1 in File S1). Trophozoites were growth for an additional seven weeks in iron-deficient medium prior to the extraction of RNA of transcriptomic analysis. Lastly, the iron-deficiency–adapted strain grown in TYI-S-33ΔFe was then supplemented with 84 μM Hb for 2 hours at 37°C (i.e. the "iron deficiency + Hb" condition). The number of amoebae was counted every day for five days using a hemocytometer (Neubauer, Germany). Reported values correspond to the mean of experiments performed in triplicate.

Quantification of iron in the culture medium

The ferrozine method [18] was used to quantify iron in the TYI-S-33 and TYI-S-33ΔFe media. Briefly, 100 μl of TYI-S-33 or TYI-S-33ΔFe medium were mixed with 100 μl of 10 mM HCl (the solvent of the iron standard $FeCl_3$) and 100 μl of the iron-releasing reagent (a freshly prepared solution containing equal volumes of 1.4 M HCl and 4.5% (w/v) $KMnO_4$) and were incubated for 2 hrs at 60°C. Next, 30 μl of the iron-detection reagent (6.5 mM ferrozine, 6.5 mM neocuproine, 2.5 M ammonium acetate, and 1 M ascorbic acid) were added to the samples. After a 30 min incubation, 280 μl aliquots of the sample solutions were transferred into the wells of a 96-well plate. Absorbance at 550 nm was measured in a microplate reader. Each sample's iron content was calculated with respect to a standard curve established with 0 to 300 μM $FeCl_3$ in the same reaction mixture. Reported values correspond to the mean of experiments performed in triplicate.

RNA preparation and microarray experiments

RNA prepared from trophozoites grown in the reference condition (normal iron) and the query conditions (i.e., iron deficiency, iron deficiency + Hb and low iron) was used for microarray analyses. RNA was obtained from three independent cultures grown with at least a one-week interval (i.e. three biological replicates). The differentially expressed genes mentioned below refer to genes expressed at significantly different levels in one or more of the three query conditions, relative to the reference condition. RNA was purified from 4×10^6 trophozoites using Trizol reagent (Invitrogen by LifeTechnologies Corp., USA), according to the manufacturer's protocol. The quality and integrity of the purified RNA was checked with spectrophotometry, electrophoresis on 0.8% agarose gels, and capillary electrophoresis in a Bioanalyser 2100 (Agilent Technologies, Les Ulis, France). The RNA was reverse-transcribed using Superscript III Reverse Transcriptase (45-0039, Invitrogen by LifeTechnologies Corp.) [19], according to the manufacturer's protocol. Cy3- or Cy5-labeled cDNA obtained from amoeba grown under each of the query conditions was cohybridized with Cy3- or Cy5-labeled cDNA from amoebae grown in the reference condition on EH2008 oligomicroarrays [19,20]. Standard dye-swap hybridizations were carried out. The EH2008 microarray is a custom-designed microarray developed by our group. It contains oligonucleotides covering the open reading frames from *E. histolytica* genome [21]. After pooling data from technical and biological replicates, differential expression analysis was carried out using both a paired Student's *t*-test and the variance estimating method in the VarMixt software package [22]. The raw P-values were adjusted using the Benjamini and Yekutieli method [23], which monitors the false discovery rate. Differentially expressed genes were considered to be those with a Benjamini and Yekutieli P-value <0.05 and a ≥2-fold change in expression level. Complete experimental details and data sets are available in the "ArrayExpress" MIAME-based database (www.ebi.ac.uk/arrayexpress/) with the accession number E-MTAB-1158.

Quantitative real-time PCR

RNA was reverse-transcribed with Superscript III Reverse Transcriptase (45-0039, Invitrogen by LifeTechnologies Corp., USA), according to the manufacturer's protocol and using the qRT-PCR primers listed in Table S7 in File S1. qRT-PCRs were performed on the RNA samples used in the microarray assay. After reverse transcription, qRT-PCRs were carried out using an ABI Prism 7900HT system (Applied BioSystems by LifeTechnologies Corp.). Reactions were performed in a 15 µl volume containing 667 nM of each primer, 1× PCR SYBR Green Master Mix (4309155, Applied Biosystems by LifeTechnologies Corp., USA) and 1 µg of template cDNA. A control curve (with 10-fold serial dilutions of cDNA) was used to check the amplification efficiency. The α-tubulin transcript was used to normalize cycle threshold (Ct) values because its expression levels were stable throughout all our microarray experiments. For each biological triplicate, the mean relative concentration was divided by the mean of the values obtained for the normal iron condition. This ratio represents the change in the tested gene's mRNA abundance under the different iron concentrations, relative to the normal iron condition.

Identification of iron-responsive-element-like structures in *Entamoeba histolytica*

The Searching for Iron-Responsive Elements (SIREs) web server (http://ccbg.imppc.org/sires/index.html) was used to identify iron-responsive-element (IRE)-like sequences in the set of *E. histolytica* mRNAs. The server's predictions have high, medium, and low levels of stringency. High- and medium-stringency predictions are mainly based on IREs that have been well characterized *in vivo* and/or *in vitro*, whereas low-level stringency predictions are based on mRNAs that interacted with novel iron-regulatory proteins (IRPs) in a recent genome-wide study. Some of these predictions have been validated *in vitro* but not *in vivo* [24]. The SIREs program also considers other criteria, including the motif type, apical loop, the nucleotide at position 25, the number of UG pairs, and the free energy of the stem-loop structures. The putative 5'- and 3'-UTRs (defined as the 100 nucleotides on each side of the coding sequences) and coding sequences (n = 8306, based on AmoebaDB version 1.3) were submitted for analysis using default parameters. The mouse ferredoxin sequence (NM_010239, which contains an IRE sequence) was included as a positive control in each submitted group.

Results

The growth conditions and their effects on trophozoites

The ferrozine method was used to quantify the iron levels in the TYI-S-33 and TYI-S-33ΔFe media (Table S1 in File S1). We measured the iron concentration in TYI-S-33 medium to be 168.9 µM±5.5 with 78 µM AFC supplementation and 90.7 µM±6.1 in the absence of supplementation (Table S1 in File S1). Serum and peptone contribute ≈55 µM and ≈40 µM, respectively (Table S1 in File S1). The iron concentration in low-iron medium is approximately 123 µM (Table S1 in File S1). Under low-iron conditions, the initial mortality rate was ≈30%. However, four days later, the surviving trophozoites grew at much the same rate as in normal TYI-S-33 medium. The iron concentration in the TYI-S-33ΔFe medium (90.7 µM) was supplemented with 84 µM Hb (Table S1 in File S1). In the iron deficiency condition, trophozoites showed a longer lag growth phase but achieved a normal growth rate after one day (Figure S1 in File S1).

To identify changes in gene expression in trophozoites grown with different iron concentrations, total RNA was purified and analyzed with microarrays. We analyzed the gene expression profiles of trophozoites grown with different iron concentrations in order to mimic the scenarios probably faced by *E. histolytica* during infection: adaptation to the low iron concentration in the intestine (mimicked by TYI-S-33 medium supplemented with only 32 µM AFC). We selected 32 µM AFC (Table S1 in File S1) because this was the threshold concentration for a subnormal growth rate (since trophozoites grew as well in TYI-S-33 medium supplemented with 40 µM AFC as they do in TYI-S-33 medium supplemented with around 78 µM AFC). Trophozoites initially grow in normal TYI-S-33 medium were able to adapt to and grow in iron-deficient medium (via stepwise reductions in the magnitude of AFC supplementation, see the Methods). These adapted trophozoites were then supplemented with Hb as an iron source for 2 hours, in order to mimic the higher iron concentrations encountered in the blood and liver blood. *E. histolytica* is known to use Hb as an iron source *in vitro* [13].

Genes differentially expressed under different iron conditions

The transcriptomes of trophozoites exposed to different iron concentrations were evaluated using our previously designed genome-wide microarray [21]. Differentially expressed genes refer to genes expressed at significantly different levels under low-iron, iron deficiency and/or iron deficiency + Hb conditions when compared with normal iron conditions. The gene expression profile of trophozoites in the iron deficiency condition revealed transcripts modulated upon adaptation to iron deficiency, whereas the gene expression of trophozoites in the iron deficiency + Hb condition revealed transcripts modulated upon uptake and further degradation of Hb (which is expected to rapidly increase intracellular iron levels). A comparison of the gene expression profiles under low-iron, iron deficiency and iron deficiency + Hb conditions with the profile in the normal iron condition revealed a total of 224 transcripts with significantly modulated ($P<0.05$ and at least a two-fold change) expression levels. The distribution and overlap of these transcripts under the various conditions is represented as a Venn diagram (Figure S2 in File S1). In iron deficiency, the transcriptome did not differ greatly from that seen with normal iron levels. Only a few genes were modulated (9 upregulated transcripts and 11 downregulated transcripts; Table S2 and Figure S2 in File S1), suggesting that the trophozoites were able to adapt to iron deficiency. Addition of Hb revealed 107 upregulated transcripts and 50 downregulated transcripts (Table S3 and Figure S2 in File S1). In cells grown under low-iron conditions, 34 transcripts were upregulated and 46 were downregulated (Table S4 and Figure S2 in File S1). Of the 42 genes extracted from AmoebaDB using the keyword "iron", 2 (5%) were downregulated in iron deficiency and 10 (24%) were upregulated however after Hb supplementation.

The microarray results were validated by quantitative real-time PCR assays (Table 1). We selected 6 transcripts that were downregulated under low-iron conditions and 5 that were

Table 1. Fold-changes for genes differentially expressed in low iron as detected by microarray and quantitative real-time PCR.

Gene description	AmoebaDB ID	Genbank ID	Microarray	qRT-PCR
Cell division control protein 42	EHI_154270	XM_645351	4.5	3.9±1.1
Glutamyl-tRNA synthetase	EHI_155570	XM_650693	3.1	3.3±1.4
Regulator of nonsense transcripts	EHI_110840	XM_649191	3.0	36±19.4
S-adenosylmethionine synthetase	EHI_195110	XM_001913755	2.9	3.3±2.0
Hypothetical protein	EHI_023330	XM_650547	2.6	5.1±1.5
Fe-hydrogenase	EHI_005060	XM_647747	−2.9	−3.1±0.1
Grainin 2	EHI_167310	XM_645265	−3.1	−1.7±0.6
Actobindin	EHI_039020	XM_651745	−3.3	−1.7±0.1
Grainin 2	EHI_111720	XM_001913814	−3.3	−1.8±0.1
Actobindin	EHI_158570	XM_644616	−4.0	−3.2±0.1
Monooxygenase	EHI_009840	XM_652013	−4.4	−2.6±0.3

AmoebaDB ID and Genbank ID refers to the accession number of the gene in AmoebaDB and NCBI GenBank, respectively; Microarray and qRT-PCR refers to fold-change in the low iron condition as compared with the normal iron condition as detected using microarray.

upregulated. The qRT-PCR data agreed with the microarray results in all cases. Putative functions for the proteins encoded by the identified genes were annotated and browsed using tools that we had previously implemented for *E. histolytica* gene discovery [21]. These data are summarized in the following sections.

Genes that are differentially expressed in iron deficiency

Genes that are differentially expressed in iron deficiency might be important for the parasite's adaptation to long-term iron deficiency (Table 2 and Table S2 in File S1). Upregulated transcripts included genes coding for acyl-CoA synthetase, ComEC competence proteins, androgen-inducible gene 1 (AIG1) and NADPH-dependent oxidoreductase (EhNO$_2$). Acyl-CoA synthetase (coded for by EHI_153060) is essential enzyme for *de novo* lipid synthesis, fatty acid catabolism, vesicular trafficking, membrane remodeling. It is also involved in the post-translational modification of proteins and the regulation of gene expression [25]. The ComEC competence proteins are putative channels for DNA uptake in bacteria [26]. They contain seven transmembrane domains, a competence (COM) domain and a metallo-β-lactamase domain. The EHI_169340 and EHI_156240 proteins (annotated as "amebic ComEC") are smaller than the bacterial homologs and do not appear to contain a COM domain (Figure S3 in File S1). Functional activity has not yet been described for amebic ComEC proteins. The ComEC EHI_169340 and EHI_156240 transcripts were upregulated in iron deficiency. After Hb supplementation, the ComEC EHI_169340 transcript declined to normal levels but the ComEC EHI_156240 was still upregulated (Table 2). AIG1 is a member of the GTPase immunity-associated protein family [27,28]. Three AIG1 transcripts (EHI_115160, EHI_022500, and EHI_195260) were upregulated in iron deficiency (Table 2). These three genes are also upregulated in an *E. histolytica* cell line that produce large liver abscesses in a gerbil model but are not modified in an *E. histolytica* cell line that does not produce abscesses. Hence, Biller et al coworkers have suggested that the AIG1 gene is a pathogenicity factor in *E. histolytica* [29]. The EHI_195260 transcript is downregulated in trifluoromethionine-resistant trophozoites [30]. Other members of the AIG gene family (such as EHI_144280 and EHI_144390) are known to be overexpressed in trophozoites during mouse intestinal colonization [31]. EhNO$_2$ is involved in

redox homeostasis through L-cysteine and iron reduction [32] and was also upregulated in iron deficiency (Table 2).

We observed that a number of transcripts were modulated upon Hb supplementation after iron deficiency (Table 2). However, most of these encoded proteins of unknown function. The genes for alcohol dehydrogenase 3 (EHI_160670), acyl-CoA synthetase (EH_131880), kinase (EHI_140330), serine/threonine-protein kinase RIO1 (EHI_170330) and serine/threonine rich protein (STIRP, EHI_004340) were also upregulated (Table 2). The kinase RIO1 is involved in ribosome biogenesis, whereas STIRP is associated with pathogenicity.

Genes that are potentially responsible for iron uptake and heme/iron metabolism

Comparison of gene expression in the three different conditions enabled us to highlight factors that are likely to be important for iron or heme uptake and trafficking (Table 2). Six putative transmembrane transporters were upregulated in the iron deficiency + Hb condition; (i) three P-glycoprotein-5 transporters (PgP5, EHI_175450, EHI_125030, EHI_075410) and (ii) three ABC transporters (EHI_095820, EHI_178050, EHI_178580). Furthermore, an MFT family member (EHI_173950) (Table 2) also showed increased transcript levels in the iron deficiency + Hb condition. Transcript levels for EHI_178580, EHI_175450 and EHI_125030 were also upregulated under low-iron conditions (Table 2). The ABC transporter (EHI_095820) had already been identified in a proteomic analysis of mitosomes [33]. P-glycoprotein-5 (EHI_175450 and EHI_125030) shows 45% homology with the ATP-binding cassette LABCG5 protein that is involved in intracellular heme trafficking in the parasite *Leishmania* [34]. Furthermore, P-glycoprotein-5 (EHI_175450) and ABC transporters (EHI_095820, EHI_178580, and EHI_178050) show similarities with the domain architecture of the iroC siderophore exporter in *Salmonella typhi* (Figure S3 in File S1) [35]. P-glycoprotein-5 (EHI_125030) shows homology with the PvdE siderophore pyoverdin exporter in *Pseudomonas aeruginosa* [36] (Figure S3 in File S1). Moreover, MFT (EHI_173950) shows similarities with both MFS1 (the azotochelin siderophore exporter in *Azotobacter vineldii* [37]) and FLVCR1 (a human cytoplasmic heme exporter) [38,39]. Furthermore, MFT is overexpressed in response to L-cysteine deprivation, which suggests that it is involved in metabolite intake or efflux in *E. histolytica* [40].

Table 2. Differentially expressed genes relevant to the iron uptake, utilization and storage.

Gene description	AmoebaDB ID	Genebank ID	Iron deficiency	Iron def + Hb	Low iron
Genes that are preferentially upregulated in iron deficiency					
Acyl-CoA synthetase	EHI_153060	XM_651318	5.7	5.4	NM
AIG1 family protein	EHI_195260	XM_643194	2.6	NM	NM
AIG1 family protein	EHI_022500	XM_642923	2.3	NM	NM
AIG1 family protein	EHI_115160	XM_644114	2.2	NM	NM
Competence protein ComEC	EHI_156240	XM_647820	6.5	4.1	NM
Competence protein ComEC	EHI_169340	XM_650745	2.4	NM	NM
EhNO2	EHI_045340	XM_648481	2.2	NM	NM
Genes that are preferentially upregulated after Hb supplementation					
Acyl-CoA synthetase	EHI_131880	XM_647524	NM	2.2	NM
Alcohol dehydrogenase 3	EHI_160670	XM_001914165	NM	3.7	NM
Hypothetical protein	EHI_058920	XM_646161	2.4	4.2	NM
Hypothetical protein	EHI_009990	XM_651985	NM	4.1	NM
Hypothetical protein	EHI_187790	XM_643995	NM	3.7	NM
Hypothetical protein	EHI_174580	XM_648227	NM	3.7	NM
Hypothetical protein	EHI_075990	XM_649526	NM	3.7	NM
Hypothetical protein	EHI_159670	XM_645153	NM	3.2	NM
Hypothetical protein	EHI_148870	XM_652487	NM	2.6	NM
Hypothetical protein	EHI_112830	XM_651012	NM	2.6	NM
Hypothetical protein	EHI_031640	XM_648447	NM	2.1	NM
Hypothetical protein	EHI_151930	XM_652295	NM	4.5	NM
Hypothetical protein	EHI_169830	XM_650917	NM	3.6	NM
Hypothetical protein	EHI_087110	XM_651036	NM	2.5	NM
Hypothetical protein	EHI_050590	XM_651508	NM	2.4	NM
Protein kinase	EHI_140330	XM_646643	NM	3.7	NM
RIO1 family protein	EHI_170330	XM_645949	NM	4.5	NM
STIRP	EHI_004340	XM_001913561	NM	2.2	NM
ABC and Major Facilitator Transporters					
ATP binding cassette	EHI_095820	XM_649804	NM	2.9	NM
ATP binding cassette	EHI_178050	XM_646404	NM	2.0	NM
ATP binding cassette	EHI_178580	XM_001913406	NM	2.2	2.2
P-glycoprotein 5	EHI_175450	XM_644247	NM	3.2	2.4
P-glycoprotein 5	EHI_125030	XM_644884	NM	4.8	2.1
P-glycoprotein 5	EHI_075410	XM_001914252	NM	2.1	NM
Transporter major facilitator	EHI_173950	XM_647419	NM	2.5	NM
GluRS, SAMS and Monooxygenase					
Glutamyl-tRNA synthetase	EHI_155570	XM_650693	NM	3.4	3.1
Monooxygenase	EHI_009840	XM_652013	−3.7	−20.6	−4.4
S-adenosylmethionine synthetase	EHI_004920	XM_001913609	−2.3	NM	2.9
S-adenosylmethionine synthetase	EHI_174250	XM_647762	−3.5	NM	2.7
S-adenosylmethionine synthetase	EHI_195110	XM_001913755	−2.9	NM	2.5

AmoebaDB ID and Genbank ID refers to the accession number of the gene in AmoebaDB (http://amoebadb.org/amoeba/) and NCBI GenBank (http://www.ncbi.nlm.nih.gov/genbank/), respectively; iron deficiency refers to the fold changes of expression level in the iron deficiency condition as compared with the normal condition; Iron def + Hb refers to the fold-change of the expression level in iron deficiency following by hemoglobin supplementation as compared with the normal condition; low iron refers to the fold-change of expression level in low iron as compared with the normal condition; Please see methods for definition of iron deficiency, low iron, and normal iron condition. NM: non-modulated.

Further experiments are necessary to clarify the role of ABC transporters and MFTs in siderophore/hemophore transport in *E. histolytica*.

It is noteworthy that genes relevant for heme metabolism in bacteria were modulated in trophozoites grown under both iron deficiency and low-iron conditions. These include genes coding for three S-adenosylmethionine synthetases (SAMS; EHI_174250,

EHI_004920, and EHI_195110), a monooxygenase (EHI_009840) and glutamyl-tRNA synthetase (GluRS, EHI_155570) (Table 2). The SAMS transcripts were downregulated in iron deficiency but normal levels were restored after Hb supplementation (Table 2). S-adenosylmethionine synthetase is responsible for the biosynthesis of S-adenosylmethione (SAM); the latter is necessary in many biosynthetic pathways, including methionine-cysteine conversion [40] and heme biosynthesis [41,42]. S-adenosylmethionine synthetase is also involved in iron acquisition in plants and it is necessary for the synthesis of mugineic acid (MA) phytosiderophores through a well described pathway [43].

The monooxygenase-encoding transcript was downregulated moderately under low-iron and iron deficiency conditions and dramatically (20-fold) in the iron deficiency + Hb condition (Table 2). The monooxygenase (EHI_009840) presents significant homology (Figure in File S1) with bacterial monooxygenases (such as the heme-degrading monooxygenase (isdG) from *Staphylococcus aureus*). The *E. histolytica* gene coding for glutamyl-tRNA synthetase (EHI_155570) was upregulated under iron deficiency + Hb and low-iron conditions (Table 2). Expression of the amebic GluRS encoding gene did not change during iron starvation but was upregulated by a factor of 3.4 in the iron deficiency + Hb condition. Glutamyl-tRNA synthetase is involved in heme/ tetrapyrrole biosynthesis in bacteria and plants.

We used qRT-PCR to further examine expression levels of SAMS, monooxygenase and GluRS transcripts by (Tables 3, S5 and S6 in File S1). The SAMS transcript (EHI_195110) was upregulated under low-iron conditions and was downregulated in iron deficiency. However, Hb supplementation was associated with a return to normal expression levels of SAMS (Tables 3). Glutamyl-tRNA synthetase gene expression was not modulated in the iron-deficient environment but was upregulated under both low-iron and iron deficiency + Hb environments. Moreover, we confirmed that SAMS and GluRS transcripts were modulated only in trophozoites having been exposed to below-normal iron concentrations, since the transcript levels did not vary significantly in trophozoites grown in normal iron medium supplemented with Hb for 2 hrs (qRT-PCR, Table S5 in File S1). Levels of monooxygenase transcript increased by a factor of 1.8 in normal iron medium supplemented with Hb for 2 hours (Table S5 in File S1) and decreased by a factor of 20 in the iron deficiency + Hb condition (Table 2), as observed with transcriptome analyses.

To confirm that SAMS, GluRS, and monooxygenase transcript levels were regulated by iron deficiency and that iron from Hb and/or AFC supplementation reestablished their expression, transcript amounts were determined in trophozoites grown in iron-deficient medium for only 24 hrs. In this condition, expression was similar to that observed under low-iron conditions; SAMS and GluRS expression levels were upregulated 2.9- and 2.5-fold, respectively (Table 2 and Table S6 in File S1). However, levels of the monooxygenase transcript were 6.3-fold lower. Transcript expression returned to baseline levels when the iron-

deficient medium was replaced by normal medium for 24 hours (Table S6 in File S1). Overall, we confirmed the modulation of SAMS, GluRS, and monooxygenase transcripts by iron deficiency and thus validated the microarray data.

The differential regulation of the genes coding for GluRS and SAMS in *E. histolytica* grown in various iron concentrations may indicate a potential role in heme biosynthesis, given the enzymes' reported role in bacteria, plants, and humans. However, a heme biosynthesis pathway has not yet been described in *E. histolytica*. In a BLAST search for orthologous genes encoding enzymes responsible for the production of components within the heme biosynthesis pathway (from aminolevulinic acid to heme production), we observed that only two genes in *E. histolytica* (the hypothetical proteins EHI_138420 and EHI_095090) present homology with intermediate enzymes. The hypothetical protein EHI_138420 shows homology with uroporphyrin-III C-methyltransferase from *Actinobacillus pleuropneumoniae* (32% similarity), whereas the hypothetical protein EHI_095090 presents homology with ferrochelatase from *Nitrosomonas sp* (27.4% similarity) (Figure S3 in File S1). Ferrochelatase is an enzyme that catalyzes the terminal step in heme biosynthesis (conversion of protoporphyrin IX (PIX) into heme). Thus, further experiments are necessary to characterize the role of the hypothetical protein EHI_138420 and the ferrochelatase-like protein (EHI_095090) in *E. histolytica*.

Molecular functions that are potentially regulated by changes in iron levels

When considering the functional category, we found that differentially expressed genes belonged to groups related to oxidoreductase activity, sulfur-containing amino acid metabolism, general stress responses, DNA repair, RNA synthesis, cysteine proteinases (CP), and actin cytoskeleton rearrangements. The genes regulated within each of these categories are described below.

Oxidoreductase activity. Nineteen genes encoding proteins with oxidoreductase activity were upregulated in the iron deficiency + Hb condition (Table 4). *Entamoeba histolytica* lacks the components for MA siderophore biosynthesis. However, we found that *E. histolytica* aldose reductase shows homology with the deoxymugineic acid synthases from *Zea mays* (maize) and *Oryza sativa* (rice). Furthermore, an aldo-keto reductase from the green algae *Chlorella vulgaris* with aldehyde reductase activity is capable of functioning as a ferric reductase and driving the Fenton reaction. In the presence of Fe^{2+} as an electron donor, hydrogen peroxide is univalently reduced to produce the hydroxyl radical, which can then produce toxic ROS [44]. Three aldose reductase transcripts (EHI_157010, EHI_039190, and EHI_107560) were upregulated 5.0- to 5.6-fold under the iron deficiency + Hb condition only (Table 4). Hence, further work will have to clarify the function of aldose reductase in the siderophores biosynthesis and/or ROS formation in *E. histolytica*.

Table 3. Fold-changes for genes differentially expressed in low iron condition detected by quantitative real-time PCR.

Gene description	AmoebaDB ID	Genbank ID	Low iron	Low iron + Hb
S-adenosylmethionine synthetase	EHI_195110	XM_001913755	3.3±2.0	1.1±0.1
Glutamyl-tRNA synthetase	EHI_155570	XM_650693	3.3±1.4	1.2±0.5

AmoebaDB ID and Genbank ID refers to the accession number of the gene in AmoebaDB and NCBI GenBank, respectively; low iron refers to the to fold-change in low iron condition compared with normal iron condition detected by quantitative real-time PCR; Low iron + Hb refers to the to fold-change in low iron condition with hemoglobin supplementation compared with normal iron condition detected by quantitative real-time PCR.

Table 4. Differentially expressed genes relevant to oxidoreductase activity and sulfur-containing amino acid metabolism.

Gene description	AmoebaDB ID	Genbank ID	Iron deficiency	Iron def + Hb	Low iron
Oxidation-reduction					
Alcohol dehydrogenase	EHI_125950	XM_645327	NM	−3.9	−3.6
Alcohol dehydrogenase ADH3	EHI_160670	XM_001914165	NM	3.7	NM
Alcohol dehydrogenase ADH2	EHI_150490	XM_647208	NM	2.5	NM
Alcohol dehydrogenase ADH3	EHI_088020	XM_643978	NM	4.2	2.8
Aldehyde-alcohol dehydrogenase ADH2	EHI_160940	XM_650725	NM	2.2	NM
Aldehyde-alcohol dehydrogenase ADH2	EHI_024240	XM_001913618	NM	2.3	NM
Aldose reductase NADPH-dependent oxidoreductase	EHI_157010	XM_001914421	NM	5.6	NM
Aldose reductase NADPH-dependent oxidoreductase	EHI_039190	XM_648674	NM	5.1	NM
Aldose reductase NADPH-dependent oxidoreductase	EHI_107560	XM_001914234	NM	5.0	NM
Fe-hydrogenase	EHI_005060	XM_647747	NM	2.0	−2.9
Fe-S cluster assembly NifU	EHI_049620	XM_650796	NM	2.7	NM
Hydroxylamine reductase	EHI_004600	XM_644914	NM	2.5	NM
Iron sulfur flavoprotein	EHI_022600	XM_643169	−2.9	2.2	NM
Iron sulfur flavoprotein	EHI_067720	XM_643101	−2.6	2.0	NM
Iron sulfur flavoprotein	EHI_022270	XM_644774	−2.3	2.0	NM
Iron sulfur flavoprotein	EHI_103260	XM_001913434	−2.3	NM	NM
Iron sulfur flavoprotein, FprB2	EHI_138480	XM_650038	NM	3.3	2.8
Iron-containing superoxide dismutase (Fe-SOD)	EHI_159160	XM_643735	NM	2.5	NM
Malate dehydrogenase	EHI_014410	XM_001913511	NM	2.4	NM
Malate dehydrogenase	EHI_165350	XM_644664	NM	2.1	NM
NADP-dependent alcohol dehydrogenase	EHI_023110	XM_648415	NM	−2.4	−2.7
Sulfotransferase	EHI_197340	XM_646583	NM	2.0	2.8
Sulfur-containing amino acid metabolism					
Cysteine desulfurase NifS	EHI_136380	XM_650165	NM	3.2	NM
Cysteine synthase CS1	EHI_171750	XM_648014	NM	3.8	NM
Cysteine synthase CS2	EHI_160930	XM_643199	NM	3.9	2.5
D 3 phosphoglycerate dehydrogenase PGDH	EHI_060860	XM_647048	−2.0	−2.1	NM
Methionine gamma lyase	EHI_057550	XM_001913898	NM	−2.6	NM
Methionine gamma lyase MGL1	EHI_144610	XM_647004	NM	−2.1	NM
Phosphoglycerate dehydrogenase PSAT	EHI_026360	XM_650291	−2.0	−2.2	NM

For definition of columns please refer to footnote of Table 2.

An iron-containing superoxide dismutase transcript was also identified. Levels of this transcript were not altered in trophozoites grown under iron deficiency or low-iron conditions. Iron-containing superoxide dismutase has been identified as a dimeric enzyme responsible for superoxide radical ($O_2\bullet^-$) detoxification in *E. histolytica* [45]. The downregulated genes in the iron deficiency condition included four iron sulfur flavoproteins (ISFs) (EHI_022600, EHI_022270, EHI_067720, and EHI_103260; Tables 4 and S2 in File S1). Expression of the ISF EHI_138480 gene expression was not altered by iron deficiency but was upregulated upon Hb supplementation and under low-iron conditions. The ISFs constitute a novel family of proteins that are broadly distributed across distantly related anaerobes. *Trichomonas vaginalis* and *E. histolytica* are the only eukaryotes to possess ISFs [40,46]. The ISF EHI_067720, EHI_103260, and EHI_138480 transcripts were upregulated during L-cysteine deprivation [40], and the ISF EHI_067720 was downregulated in a mouse model of intestinal amebiasis [31].

Alcohol dehydrogenase transcripts were upregulated in the iron deficiency + Hb condition; they included aldehyde-alcohol dehydrogenase 2 (EHI_024240; a 2.3-fold increase), which known to be involved in the internalization of human transferrin [47] and is regulated by iron [3]. Alcohol dehydrogenase (EHI_125950) and NADP-dependent alcohol dehydrogenase (EHI_023110) transcripts were downregulated in trophozoites grown under low iron concentrations and upon Hb supplementation. The Fe-hydrogenase (EHI_005060) transcript was downregulated 2.9-fold under low iron conditions (Table 4) and upregulated 2-fold after supplementation with Hb. Transcripts such as alcohol dehydrogenase (EHI_088020) and sulfotransferase (EHI_197340) were also differentially expressed in trophozoites cultured under low iron conditions (Table 4).

Sulfur-containing amino acid metabolism. Cysteine synthase CS1 (EHI_171750) and CS2 (EHI_190630) transcripts and cysteine desulfurase NifS (EHI_136380) transcripts were upregulated in the iron deficiency + Hb condition (Table 4). Together with the changes observed for SAMS (Table 2), this upregulation

Table 5. Differentially expressed genes relevant to heat shock stress, DNA repair, and RNA synthesis.

Gene description	AmoebaDB ID	Genbank ID	Iron deficiency	Iron def + Hb	Low iron
Stress response					
Chaperone clpB	EHI_090840	XM_001914528	NM	2.9	3.9
Chaperone clpB	EHI_155060	XM_001914269	NM	2.1	3.5
Chaperone clpB	EHI_094680	XM_001914511	NM	NM	2.7
Heat shock protein 101	EHI_076480	XM_001914516	NM	NM	3.3
Heat shock protein 101	EHI_013550	XM_001914488	NM	NM	2.8
Heat shock protein 101	EHI_156560	XM_001914239	NM	2.2	4.1
Heat shock protein 101	EHI_183680	XM_001914603	NM	2.1	3.9
Heat shock protein 101	EHI_178230	XM_001914365	NM	NM	3.8
Heat shock protein 101	EHI_094470	XM_001914472	NM	NM	2.5
Heat shock protein 90	EHI_196940	XM_648040	NM	2.0	NM
Heat shock protein Hsp20	EHI_125830	XM_651403	NM	−2.1	NM
Heat shock protein70 (hsp70A2)	EHI_015390	XM_001913629	NM	2.1	NM
DNA repair					
DNA directed RNA polymerase II subunit	EHI_056690	XM_643999	NM	2.1	NM
DNA directed RNA polymerase III subunit	EHI_050830	XM_651537	NM	2.0	NM
Double strand break repair protein MRE11	EHI_125910	XM_651393	NM	3.9	NM
RNA synthesis					
DEAD box ATP dependent RNA helicase 42	EHI_197990	XM_001914267	NM	2.0	NM
DEAD/DEAH box helicase	EHI_131080	XM_650428	NM	3.7	NM
Regulator of nonsense transcripts	EHI_070810	XM_649317	NM	3.7	3.8
Regulator of nonsense transcripts	EHI_110840	XM_649191	NM	2.7	3.0
Regulator of nonsense transcripts	EHI_035550	XM_651038	NM	2.0	NM
Regulator of nonsense transcripts	EHI_193520	XM_646961	NM	NM	2.2
RNA binding protein	EHI_151990	XM_652289	NM	2.2	NM

For definition of columns please refer to footnote of Table 2.

suggests that cysteine metabolism is activated and forms a key part of the metabolism of sulfur-containing amino acids in *E. histolytica* [48,49]. Upregulation of the NifS transcript also suggests that Fe-S cluster synthesis is activated in the presence of Hb as an iron source (Table 4). In contrast, serine metabolism appears to be downregulated, since transcripts of genes coding for D-3-phosphoglycerate dehydrogenase (PGDH, EHI_060860), methionine γ-lyase 1 (MGL1, EHI_144610), and phosphoserine aminotransferase (PSAT, EHI_026360) were downregulated in iron deficiency. The PGDH and PSAT transcripts were still downregulated after Hb supplementation.

Stress responses, DNA repair and RNA synthesis. A number of transcripts involved in stress responses were upregulated upon Hb replenishment, including heat shock proteins (HSPs) such as Hsp101 and ClpB (Table 5) and the chaperone Hsp90. Hsp101 and ClpB transcripts were also upregulated in the low-iron condition but not in the iron deficiency condition. Upon Hb replenishment, Hsp20 was downregulated and Hsp70A2 was upregulated. Thus, in an iron-deficient environment, the sudden supply of heme might induce a stress response in *E. histolytica*. Gene transcripts involved in DNA/RNA synthesis and DNA repair (such as the double strand break repair protein MRE11, DEAD/DEAH box helicase and the regulators of nonsense transcripts) were modulated by Hb supplementation (Table 5). Expression of three regulators of nonsense transcripts was observed in amoebae grown in the low-iron condition (Table 5)

and upon supplementation with Hb. The regulators of nonsense transcripts present homology with the product of the human *Upf1* gene (regulator of nonsense transcripts 1, a RNA helicase that detects mRNA containing premature stop codons).

Cysteine proteinases and actin-related cytoskeleton. It has already been suggested that iron-limited conditions regulate cysteine proteinase levels in *E. histolytica* [50]. In the present analysis, we identified three annotated cysteine proteinases (CP-A4, CP-A5, and CP-A7) and the putative cysteine proteinase EHI_010850 (XP_001914429). Upon Hb supplementation, expression of CP-A5, CP-A7, and EHI_010850 was upregulated 2.4-, 4.5-, and 4.5-fold, respectively. Furthermore, CP-A4 was upregulated 3.0-fold in iron-starved trophozoites (Table 6). Transcripts encoding proteins involved in actin cytoskeleton organization (such as actobindin (EHI_158570, EHI_039020), cofilin (EHI_186840), and ARP 2/3 complex subunits (EHI_199690, EHI_045000)) (Table 6) were downregulated only in low-iron conditions. Reorganization of the actin cytoskeleton may thus be correlated with the previously reported loss of parasite adherence under low-iron conditions [5].

Overall, the results of our microarray analysis provide a new vision of iron-related functions in *E. histolytica*. There are clear similarities with bacterial iron-related pathways. Furthermore, our work is the first to highlight candidates for iron uptake and utilization studies *in E. histolytica*.

Table 6. Differentially expressed genes relevant to other notable molecular functions.

Gene description	AmoebaDB ID	Genbank ID	Iron deficiency	Iron def + Hb	Low iron
Cysteine Proteinases					
Cysteine proteinase	EHI_010850	XM_001914394	NM	4.5	NM
Cysteine proteinase (CP-A7)	EHI_039610	XM_643904	NM	4.5	NM
Cysteine proteinase (CP-A4)	EHI_050570	XM_651510	3.0	NM	NM
Cysteine proteinase (CP-A5)	EHI_168240	XM_645845	NM	2.4	NM
Actin cytoskeleton organization					
Actobindin	EHI_158570	XM_644616	NM	NM	−4.0
Actin binding protein cofilin/tropomyosin	EHI_186840	XM_651345	NM	NM	−2.4
Actin related protein 2/3 complex subunit 1A	EHI_045000	XM_652168	NM	NM	−2.0
Actobindin	EHI_039020	XM_651745	NM	NM	−3.3
ARP2/3 complex 34 kDa subunit	EHI_199690	XM_645477	NM	NM	−2.0
Other amoebic transcripts					
Acetyltransferase	EHI_039180	XM_001913968	NM	2.0	NM
Acetyltransferase	EHI_167080	XM_649690	NM	−2.1	NM
Aspartate ammonia lyase	EHI_082270	XM_001913840	NM	−2.3	−2.3
Aspartate ammonia lyase	EHI_150390	XM_650734	NM	−2.5	−2.0
Calmodulin	EHI_010020	XM_651988	NM	−2.0	−3.0
Grainin 1	EHI_167300	XM_645280	NM	NM	−2.8
Grainin 2	EHI_111720	XM_001913814	NM	NM	−3.3
Grainin 2	EHI_167310	XM_645265	NM	NM	−3.1
Molybdenum cofactor sulfurase	EHI_194600	XM_649637	NM	−2.1	NM
N system amino acid transporter 1	EHI_050900	XM_649086	NM	2.4	NM
WD domain containing protein	EHI_126260	XM_647361	NM	−2.0	NM

For definition of columns please refer to footnote of Table 2.

Identification of IRE-like sequences

By taking advantage of existing bioinformatics approaches, we screened *E. histolytica*'s entire genome for post-transcriptional regulatory elements controlling cellular iron homeostasis. In eukaryotes, key proteins involved in iron transport and storage (e.g., TfR1 and ferroportin) are regulated post-transcriptionally by *cis* elements present in their mRNA. These IREs are stem-loop structures containing the canonical sequences required to bind IRPs. Iron-responsive elements have been found in both the 5′ and 3′ untranslated regions (UTRs) of mRNAs and also in regions coding for proteins involved in iron transport and storage [51–53]. We used the SIREs web server to identify potential IRE structures in the *E. histolytica* transcripts identified by the microarray analysis [24]. We analyzed the 5′-UTR, 3′-UTR and coding mRNA sequences extracted from the *E. histolytica* genome. The results were then categorized into SIREs' three stringency levels and the presence of putative IRE-like stem-loop structures was determined by inspection of the SIREs output files. In the whole-genome analysis, we identified 550 transcripts containing IRE structures. Of these transcripts, 173 had high- or medium-stringency IREs, which were more present in coding regions than in 5′- and 3′-UTRs. Eighteen transcripts with putative IREs appeared to be significantly regulated in our microarray analyses (Table 7). Five of the latter had IRE-like structures with high and medium stringency levels. The proportion of differentially expressed genes with IRE-containing transcripts is similar to that in the genome as a whole. Furthermore, stem-loop IRE-like structures were identified in transcripts that were upregulated in

iron-starved parasites or upon Hb replenishment (e.g., MFT, DEAD/DEAH helicase, AIP family members, and acyl-CoA synthetase) (Table 7). Stem-loop folds were also identified in transcripts that were not modulated (according to the microarray analysis) by iron starvation or Hb replenishment (e.g. thioredoxin and CP-A8). Thus, the presence of IRE-like structures in these mRNAs may control translation during changes in intracellular iron levels.

Discussion

Entamoeba histolytica must adapt its metabolism as a function of the changing iron concentrations encountered during host infection. It also needs to acquire iron from host sources. In the present study, we searched for genes whose *in vitro* expression level was modulated by differing iron conditions (i.e. cues for iron uptake, utilization and storage in *E. histolytica*). *Entamoeba histolytica* has hemophore-like proteins [14] and receptors for transferrin and lactoferrin [13]. The mechanisms of heme-hemophore and iron-siderophore uptake and the subsequent iron utilization and storage have not been described. In our transcriptome analysis, we identified several genes that encode factors with transporter functions (such as ABC proteins, P-glycoprotein 5 transporters and MFTs). Since homologs of these transporter families are involved in siderophore export/import in bacteria and heme export in humans [35–38], they are likely to have a role in siderophore or hemophore export in *E. histolytica* (although this remains to be explored).

Table 7. IRE-like sequences in differentially expressed genes.

Gene description	AmoebaDB ID	Genbank ID	Iron deficiency	Iron def + Hb	Low iron	RNA	IRE quality	Stem-loop
DEAD/DEAH box helicase	EHI_131080	XM_650428	NM	3.7	NM	CD	H	-
STIR	EHI_025700	XM_644280	NM	2.3	NM	CD	H	-
Hypothetical protein	EHI_092110	XM_649471	NM	NM	-2.4	5' UTR	H	-
Transporter major facilitator	EHI_173950	XM_647419	NM	2.5	1.6	CD	M	YES
Calmodulin	EHI_023500	XM_650529	NM	-1.4	-2.0	5' UTR	M	-
Proliferating cell nuclear antigen	EHI_128450	XM_646418	NM	4.2	NM	CD	L	YES
Leucine rich repeat / protein phosphatase 2C	EHI_137760	XM_648955	NM	3.9	NM	CD	L	-
Cysteine synthase A	EHI_160930	XM_643199	NM	3.9	2.5	CD	L	YES
Hypothetical protein	EHI_197520	XM_646601	NM	2.8	NM	CD	L	YES
DEAD/DEAH box helicase	EHI_119920	XM_644837	NM	2.0	NM	CD	L	YES
AIG1 family protein	EHI_022500	XM_642923	2.5	NM	NM	5' UTR	L	YES
AIG1 family protein	EHI_195260	XM_643194	2.3	NM	NM	5' UTR	L	YES
AIG1 family protein	EHI_115160	XM_644114	2.2	NM	NM	5' UTR	L	YES
Acyl-CoA synthetase	EHI_131880	XM_647524	NM	2.2	NM	5' UTR	L	YES
ARP2/3 complex 20 kDa subunit	EHI_152660	XM_643475	NM	-1.5	-1.8	CD	L	-
Threonine dehydratase	EHI_049910	XM_652079	NM	-1.7	-1.9	CD	L	YES
Hypothetical protein	EHI_053140	XM_001913728	NM	NM	-2.2	CD	L	-
Amino acid transporter	EHI_072120	XM_648362	NM	-2.0	NM	CD	L	-

NM: non-modulated; CD: coding region; 5'UTR: 5' untranslated region; H: high stringency; M: medium stringency, and L: low stringency.

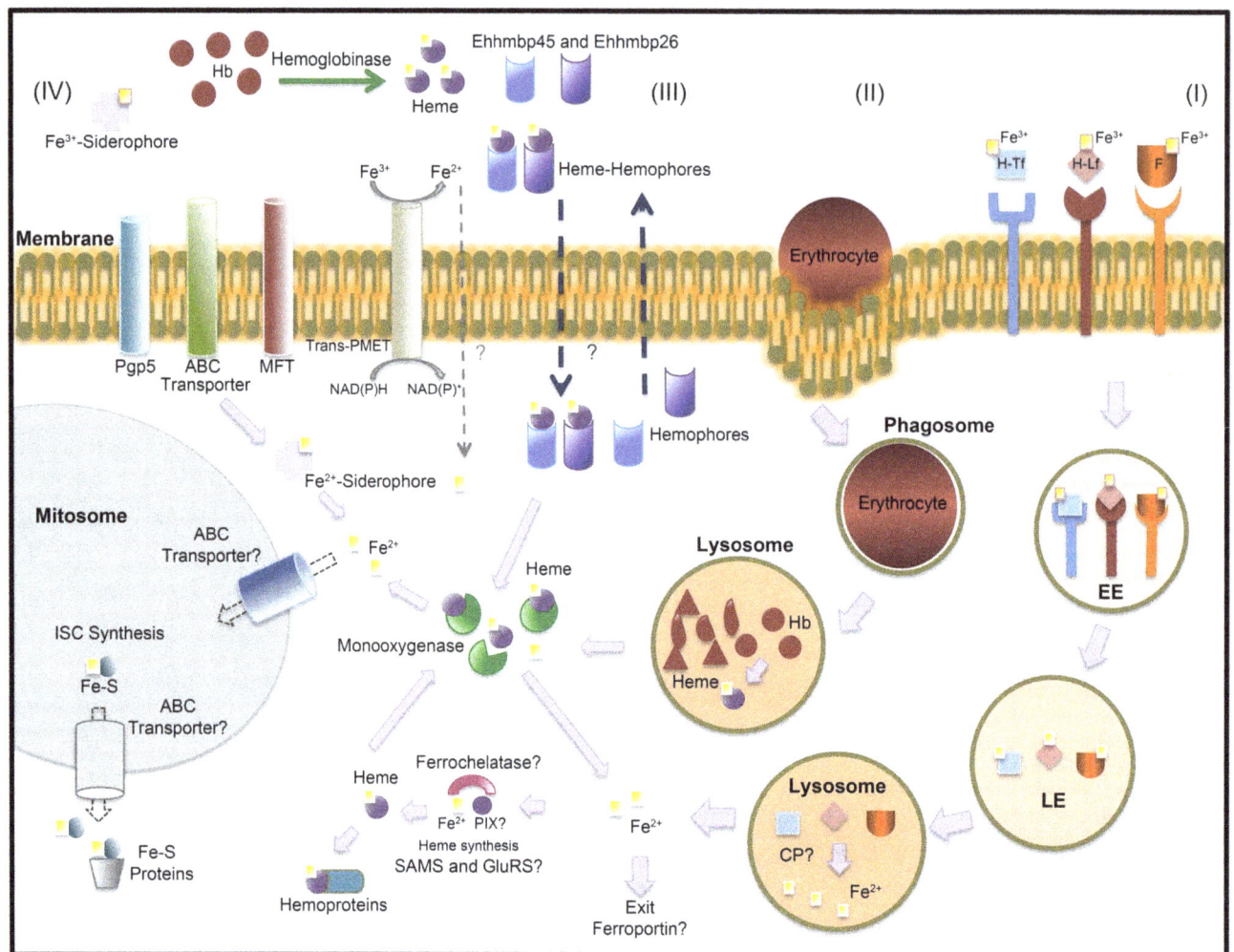

Figure 1. A model summarizing our hypothesis for iron uptake and metabolism in *Entamoeba histolytica*. The pathways described in (I), (II), (III), and (IV) refer to potential routes for iron entry into cells. (I) Trophozoites are able to obtain iron from host proteins such as holotransferrin (H-Tf), hololactoferrin (H-Lf), ferritin (F), and Hb that bind to cell surface proteins. The complexes are internalized in clathrin-coated vesicles (for ferritin and holotransferrin) or caveolin-like coated vesicles (for hololactoferrin). The ligand-receptor complexes dissociate in the endosomes (early endosome; EE and late endosome; LE) and the iron bound to proteins is released in the lysosomes. It is not clear whether iron transporter proteins are degraded by cysteine proteinases (as has been suggested by López-Soto et al [13]) or whether these transporters and cognate receptors are shuttled to the plasma membrane for reuse. (II) *E. histolytica* also acquires iron from Hb via heme internalization during the phagocytosis of human erythrocytes, which are degraded by hemolysin and phospholipases. It has been proposed that hemoglobinases and cysteine proteinases degrade Hb [13]. The monooxygenase heme oxygenase degrades heme and releases iron; we suggest that a monooxygenase (EHI_009840) could be responsible for heme degradation in *E. histolytica*. (III) Heme can also be scavenged by secreted hemophores such as hemoglobin-binding proteins 45 and 26 (Ehhmbp 45 and Ehhmbp 26), which are able to bind Hb and heme [14]. However, the mechanisms of hemophore export and heme-hemophore complex uptake are unknown (blue dashed arrows). An MFT (EHI_173950) and an ABC transporter (EHI_178050) may be involved in heme-hemophore uptake in *E. histolytica*. Once heme has entered the cell, it can be degraded by monooxygenase. (IV) Iron may be taken up by siderophores, since the ABC transporters, P-glycoprotein 5 transporters and the MFT family share homology with bacteria siderophore export systems. Lastly, a trans-plasma membrane electron transport (trans-PMET) system capable of transferring electrons from cytosolic electron donors to non-permeable electron acceptors may have an important role in iron reduction and acquisition (gray dashed arrow) [61]. The above-listed mechanisms deliver iron to cells. The iron could then be used (for example) in iron-sulfur cluster (ISC) biosynthesis in the mitosomes [16]. Heme synthesis has not been described in *E. histolytica*. Both GluRS and SAMS may be key sensors of iron status in this pathway. Ferroportin has not yet been identified in *E. histolytica*.

In eukaryotes, iron can be recovered from heme via the action of the endoplasmic reticulum heme oxygenase-1 (HO-1) [54]. A gene encoding a monooxygenase was strongly downregulated (20-fold) in iron-deficient trophozoites supplemented with Hb. In bacteria and human cells, monooxygenase is responsible for heme-degradation. Hence, the repression of monooxygenase in the iron deficiency + Hb condition may be required to avoid the production of ROS and Fe^{2+} via heme catabolism [55]. The

identification of a monooxygenase protein in *E. histolytica* is noteworthy because the parasite may be able to recover iron from heme that is directly internalized or degraded after erythrophagocytosis.

We found that GluRS and SAMS transcripts were modulated by changes in the iron concentration. These enzymes are involved in several metabolic pathways, including heme biosynthesis [41,42]. The first step in heme biosynthesis is the production of

5-aminolevulinic acid (ALA) via two unrelated pathways: the C_5-pathway and the Shemin pathway. In the C_5-pathway, glutamate is metabolized to glutamyl-t-RNA, L-glutamate-1-semialdehyde and then ALA through the action of GluRS, glutamyl t-RNA reductase and glutamate-1-semialdehyde aminotransferase, respectively. In the Shemin pathway, ALA is produced from the condensation of glycine and succinyl co-enzyme A [56,57]. The condensation of eight molecules of ALA is necessary to form uroporphyrinogen III, which is then converted to protoporphyrin IX. In a reaction catalyzed by ferrochelatase, protoporphyrin IX can incorporate Fe^{2+} to produce heme [57]. S-adenosylmethionine synthetases are involved in this complex pathway by producing SAM (an essential cosubstrate for many enzymatic reactions, including the final steps of heme synthesis) [42]. For instance, a mutation in SAMS in *Rhodobacter sphaeroides* leads to the accumulation of the intermediate coproporphyrin III [42]. In parasites such as *Leishmania* and *Crithidia*, some (but not all) of the enzymatic machinery required for *de novo* heme synthesis is present [58]; genes in the genomes of symbiotic bacteria complement the absence of full heme synthesis in these parasites. Although the heme biosynthesis pathways in *E. histolytica* have not yet been characterized, the presence of SAMS, GluRS, and a hypothetical (EHI_095090) non-conserved ferrochelatase homolog suggests that iron-related biosynthetic pathways merit further exploration. In higher eukaryotes, heme and Fe-S cluster biosynthesis occurs in the mitochondria. However, heme biosynthesis also occurs in bacteria, which lack mitochondria. *E. histolytica* lacks mitochondria but does have mitosomes. Nevertheless, there is currently no evidence to suggest that amebic mitosomes are involved in heme biosynthesis.

On the basis of our current knowledge of iron uptake and metabolism in *E. histolytica*, we propose the following hypotheses: (i) *E. histolytica* is able to obtain iron and heme via phagocytosis and directly from host proteins through import mediated by hemophores [14] and siderophores; (ii) transporter and trafficking proteins (e.g., P-glycoprotein 5 proteins, ABC transporters and MFTs) are involved in the internalization and salvaging of iron/heme, as observed for *Leishmania* [34,58]; (iii) within the cell, iron ions may be released from these complexes through the action of monooxygenase and cysteine proteinases; and (iv) *E. histolytica* maintains a balance between free and protein-bound iron ions using Fe-S clusters. Experimental validation of these hypotheses is now required to confirm the roles of these factors in iron uptake and metabolism in *E. histolytica*. Our working hypothesis is summarized in Figure 1.

Putative IREs are present in prokaryotes and protozoans (e.g., *Plasmodium falciparum* [59] and *Trichomonas vaginalis*). For example, IRE-like structures are present in the mRNA for TVCP4 cysteine proteinase in *Trichomonas vaginalis* [53]. Furthermore, the α-actinin 3 protein (TvACTN3) is able to bind to this IRE-like structure and may be involved in the post-translational mechanism for iron regulation [60]. The amebic cysteine proteinases CP-A7, CP-A5, and CP-A4 were modulated in iron-starved amoebae supplemented with Hb. However, we did not identify IRE-like structures in these mRNAs in a SIREs screen. In contrast, we identified IRE-like structures in amebic STIRP, helicase, MFT, AIG1 and acyl-CoA synthetase mRNAs. Interesting, it has been suggested that the AIG1 genes are *E. histolytica* pathogenicity factors [29]. Iron-regulatory protein homologs that work with IRE elements have yet to be identified in *E. histolytica*. Thus, active IRE-like structures and α-actinin's role in iron regulation in *E. histolytica* have yet to be functionally confirmed.

In summary, we identified candidates for subsequent functional studies of iron/heme uptake, trafficking, and utilization in *E.*

histolytica. These included ABC transporters, MFTs, GluRS, SAMS, a monooxygenase, and a ferrochelatase homolog. The regulation of these proteins in presence of ferritin (which is highly abundant in the liver) remains to be investigated.

Supporting Information

File S1 Supporting information. Figure S1, Growth of *Entamoeba histolytica* trophozoites under normal iron conditions (▲) and in iron-deficient medium (■). Trophozoites were counted in a Neubauer chamber every 24 hrs for 5 days. The cell count corresponds to the mean of three observations. Figure S2, A Venn diagram of genes differentially expressed in three conditions. Upper panel: upregulated genes. Lower panel: downregulated genes. A total of 224 transcripts were significantly modulated in one or more of the three conditions. In iron deficiency, 9 transcripts were upregulated and 11 were downregulated. In iron deficiency + Hb, 107 transcripts were upregulated and 50 were downregulated. Under low-iron conditions, 34 transcripts were upregulated and 46 were downregulated. Figure S3, Alignments of selected *E. histolytica* genes. A) amoebic ComEC orthologs (EHI_169340 and EHI_156240) and ComEC from *Bacillus licheniformis* (WP_003183688.1, 20% similarity), B) amebic P-glycoprotein-5 (EHI_125030) and PvdE from *Pseudomonas aeruginosa* (YP_002440490.1, 33% similarity), C) amoebic MFT (EHI_173950) and MFS1 exporter from *Azotobacter vinelndii* (YP_002797373.1, 33% similarity), D) amoebic MFT (EHI_173950) and human FLVCR1 (NP_054772.1), E) amoebic hypothetical protein (EHI_009840) and heme-degrading monooxygenase (isdG) from *Staphylococcus aureus* (NP_645835.1, 38% similarity), F) amoebic hypothetical protein (EHI_095090) and ferrochelatase from *Nitrosomonas sp.* (YP_004294842.1, 27% similarity), and G) amoebic hypothetical protein (EHI_138420) and uroporphyrin-III C-methyltransferase from *Actinobacillus pleuropneumoniae* (WP_005597783.1, 32% similarity). Residues with 100% and 80% homology are highlight in gray and black, respectively. The comparisons were performed using the CLUSTALW alignment tool from the WebExPASy Molecular Biology Server (http://ca.expasy.org). Table S1, Quantification of iron in the TYI-S-33 and TYI-S-33ΔFe medium. *Footnote*: The ferrozine method described in the Material and Methods section was used to quantify iron in the TYI-S-33 medium, incomplete TYI-S-33ΔFe medium (no supplementation with AFC, vitamins and serum) and TYI-S-33ΔFe complete (no supplementation with AFC but supplemented with vitamins and serum) medium. Serum accounts for 55.5 μM iron, i.e. the difference between complete and incomplete media. Peptone accounts for 39.7 μM iron, i.e. the level determined in the incomplete TYI-S-33ΔFe medium. AFC accounts for 78 μM iron, i.e. the difference between the complete TYI-S-33 and complete TYI-S-33ΔFe media. AFC: ammonium ferric citrate; Hb: hemoglobin. Values correspond to the mean of experiments performed in triplicate. Table S2, Differentially expressed genes in iron deficiency. *Footnote*: FC: fold-change; BY: the false discovery rate according to Benjamini and Yekutieli multiple testing; rawp: the unadjusted *P*-value. Table S3, Differentially expressed genes in iron deficiency with Hb supplementation. *Footnote*: FC: fold-change; BY: the false discovery rate according to Benjamini and Yekutieli multiple testing; rawp: the unadjusted *P*-value. Table S4, Differentially expressed genes under low-iron conditions. *Footnote*: FC: fold-change; BY: the false discovery rate according to Benjamini and Yekutieli multiple testing; rawp: the unadjusted *P*-value. Table S5, Fold-changes for genes differentially expressed in normal medium + Hb for 2 hours. *Footnote*: The AmoebaDB ID and GenBank ID numbers refer to the gene's accession number in AmoebaDB and NCBI GenBank, respectively; "Normal iron + Hb for 2 h" refers to the fold-change in expression in TYI-S-33 medium supplemented with Hb for 2 hours, compared with the normal iron condition (as detected by quantitative real-time PCRs). Table S6, Fold-changes for genes differentially expressed in iron deficiency for 24 hours. *Footnote*: The AmoebaDB ID and GenBank ID numbers refer to the gene's accession number in AmoebaDB and NCBI GenBank, respectively; "Normal iron to iron deficiency for 24 h" refers to the fold-change in expression in TYI-S-33 medium without AFC supplementation for 24 hours, when compared with the normal iron condition (as detected by quantitative real-time PCRs). The trophozoites incubated in iron-deficient medium for 24 hours were recovered and incubated for an additional 24 hours in normal iron medium ("iron deficiency 24 h to normal iron 24 h"). Gene expression was detected using quantitative real-time PCRs. Table S7, List of primers used for real time-PCRs. *Footnote*: Position: the relative nucleotide position of the primer's 5' end, where 0 refers to the first nucleotide of the start codon; Sequence: sequence of the primer from 5' to 3'. Note that the reverse primer's sequence is reversed and complemented.

Acknowledgments

The authors thank Jean-Yves Coppée, Marie-Agnes Dillies, and Caroline Proux for advice on the statistical analysis and for technical assistance during the microarray experiments. We acknowledge Sherri Smith and Daniela Faust for critical reading and correction of the manuscript.

Author Contributions

Conceived and designed the experiments: NAHC NG CW. Performed the experiments: NAHC CCH. Analyzed the data: NAHC NG CCH. Contributed reagents/materials/analysis tools: NAHC NG CW CCH. Contributed to the writing of the manuscript: NAHC NG CCH.

References

1. Ralston KS, Petri Jr WA (2011) Tissue destruction and invasion by *Entamoeba histolytica*. Trends in Parasitology 27: 254–263.
2. Olivos-García A, Saavedra E, Ramos-Martínez E, Nequiz M, Pérez-Tamayo R (2009) Molecular nature of virulence in *Entamoeba histolytica*. Infection, Genetics and Evolution 9: 1033–1037.
3. Espinosa A, Perdrizet G, Paz-y-Mino CG, Lanfranchi R, Phay M (2009) Effects of iron depletion on *Entamoeba histolytica* alcohol dehydrogenase 2 (EhADH2) and trophozoite growth: implications for antiamoebic therapy. J Antimicrob Chemother 63: 675–678.
4. Tannich E, Bruchhaus I, Walter RD, Horstmann RD (1991) Pathogenic and nonpathogenic *Entamoeba histolytica*: identification and molecular cloning of an iron-containing superoxide dismutase. Molecular and Biochemical Parasitology 49: 61–71.
5. Lee J, Park S-J, Yong T-S (2008) Effect of Iron on Adherence and Cytotoxicity of *Entamoeba histolytica* to CHO Cell Monolayers. Korean J Parasitol 46: 37–40.
6. Evstatiev R, Gasche C (2012) Iron sensing and signalling. Gut 61: 933–952.
7. Cassat JE, Skaar EP (2013) Iron in Infection and Immunity. Cell Host & Microbe 13: 509–519.
8. Dunn LL, Rahmanto YS, Richardson DR (2007) Iron uptake and metabolism in the new millennium. Trends in Cell Biology 17: 93–100.
9. Wilks A, Schmitt MP (1998) Expression and Characterization of a Heme Oxygenase (Hmu O) from *Corynebacterium diphtheriae*. Journal of Biological Chemistry 273: 837–841.
10. Puntarulo S (2005) Iron, oxidative stress and human health. Molecular Aspects of Medicine 26: 299–312.
11. Richardson D, Huang M, Whitnall M, Becker E, Ponka P, et al. (2010) The ins and outs of mitochondrial iron-loading: the metabolic defect in Friedreich's ataxia. Journal of Molecular Medicine 88: 323–329.
12. Wandersman Cc, Delepelaire P (2004) BACTERIAL IRON SOURCES: From Siderophores to Hemophores. Annual Review of Microbiology 58: 611–647.
13. López-Soto F, León-Sicairos N, Reyes-López M, Serrano-Luna J, Ordaz-Pichardo C, et al. (2009) Use and endocytosis of iron-containing proteins by *Entamoeba histolytica* trophozoites. Infection, Genetics and Evolution 9: 1038–1050.
14. Cruz-Castañeda A, López-Casamichana M, Olivares-Trejo JJ (2011) *Entamoeba histolytica* secretes two haem-binding proteins to scavenge haem. Biochemical Journal 434: 105–111.
15. Ali V, Shigeta Y, Tokumoto U, Takahashi Y, Nozaki T (2004) An Intestinal Parasitic Protist, Entamoeba histolytica, Possesses a Non-redundant Nitrogen Fixation-like System for Iron-Sulfur Cluster Assembly under Anaerobic Conditions. Journal of Biological Chemistry 279: 16863–16874.
16. Maralikova B, Ali V, Nakada-Tsukui K, Nozaki T, Van Der Giezen M, et al. (2010) Bacterial-type oxygen detoxification and iron-sulfur cluster assembly in amoebal relict mitochondria. Cellular Microbiology 12: 331–342.
17. Diamond LS (1961) Axenic cultivation of *Entamoeba hitolytica*. Science 4: 336–337.
18. Riemer J, Hoepken HH, Czerwinska H, Robinson SR, Dringen R (2004) Colorimetric ferrozine-based assay for the quantitation of iron in cultured cells. Analytical Biochemistry 331: 370–375.
19. Weber C, Guigon G, Bouchier C, Frangeul L, Moreira S, et al. (2006) Stress by Heat Shock Induces Massive Down Regulation of Genes and Allows Differential Allelic Expression of the Gal/GalNAc Lectin in *Entamoeba histolytica*. Eukaryotic Cell 5: 871–875.
20. Santi-Rocca J, Weber C, Guigon G, Sismeiro O, Coppée J-Y, et al. (2008) The lysine- and glutamic acid-rich protein KERP1 plays a role in *Entamoeba histolytica* liver abscess pathogenesis. Cellular Microbiology 10: 202–217.
21. Santi-Rocca J, Smith S, Weber C, Pineda E, Hon C-C, et al. (2012) Endoplasmic Reticulum Stress-Sensing Mechanism Is Activated in *Entamoeba histolytica* upon Treatment with Nitric Oxide. PLoS ONE 7: e31777.
22. Delmar P, Robin Sp, Daudin JJ (2005) VarMixt: efficient variance modelling for the differential analysis of replicated gene expression data. Bioinformatics 21: 502–508.
23. Reiner A, Yekutieli D, Benjamini Y (2003) Identifying differentially expressed genes using false discovery rate controlling procedures. Bioinformatics 19: 368–375.
24. Campillos M, Cases I, Hentze MW, Sanchez M (2010) SIREs: searching for iron-responsive elements. Nucleic Acids Research 38: W360–W367.
25. Black PN, DiRusso CC (2007) Yeast acyl-CoA synthetases at the crossroads of fatty acid metabolism and regulation. Biochimica et Biophysica Acta (BBA) - Molecular and Cell Biology of Lipids 1771: 286–298.
26. Draskovic I, Dubnau D (2005) Biogenesis of a putative channel protein, ComEC, required for DNA uptake: membrane topology, oligomerization and formation of disulphide bonds. Molecular Microbiology 55: 881–896.
27. Wang Z, Li X (2009) IAN/GIMAPs are conserved and novel regulators in vertebrates and angiosperm plants. Plant Signaling & Behavior 4: 165–167.
28. Krücken J, Schroetel RMU, Müller IU, Saïdani N, Marinovski P, et al. (2004) Comparative analysis of the human gimap gene cluster encoding a novel GTPase family. Gene 341: 291–304.
29. Biller L, Davis P, Tillack M, Matthiesen J, Lotter H, et al. (2010) Differences in the transcriptome signatures of two genetically related *Entamoeba histolytica* cell lines derived from the same isolate with different pathogenic properties. BMC Genomics 11: 63.
30. Penuliar GM, Furukawa A, Nakada-Tsukui K, Husain A, Sato D, et al. (2012) Transcriptional and functional analysis of trifluoromethionine resistance in Entamoeba histolytica. Journal of Antimicrobial Chemotherapy 67: 375–386.
31. Gilchrist CA, Houpt E, Trapaidze N, Fei Z, Crasta O, et al. (2006) Impact of intestinal colonization and invasion on the *Entamoeba histolytica* transcriptome. Molecular and Biochemical Parasitology 147: 163–176.
32. Jeelani G, Husain A, Sato D, Ali V, Suematsu M, et al. (2010) Two Atypical l-Cysteine-regulated NADPH-dependent Oxidoreductases Involved in Redox Maintenance, l-Cystine and Iron Reduction, and Metronidazole Activation in the Enteric Protozoan *Entamoeba histolytica*. Journal of Biological Chemistry 285: 26889–26899.
33. Mi-ichi F, Yousuf MA, Nakada-Tsukui K, Nozaki T (2009) Mitosomes in *Entamoeba histolytica* contain a sulfate activation pathway. Proceedings of the National Academy of Sciences 106: 21731–21736.
34. Flannery AR, Renberg RL, Andrews NW (2013) Pathways of iron acquisition and utilization in Leishmania. Current Opinion in Microbiology 16: 716–721.
35. Crouch M-LV, Castor M, Karlinsey JE, Kalhorn T, Fang FC (2008) Biosynthesis and IroC-dependent export of the siderophore salmochelin are essential for virulence of *Salmonella* enterica serovar *Typhimurium*. Molecular Microbiology 67: 971–983.
36. Reimmann C, Patel HM, Serino L, Barone M, Walsh CT, et al. (2001) Essential PchG-Dependent Reduction in Pyochelin Biosynthesis of *Pseudomonas aeruginosa*. J Bacteriol 183: 813–820.
37. Page W, Kwon E, Cornish A, Tindale A (2003) The csbX gene of *Azotobacter vinelandii* encodes an MFS efflux pump required for catecholate siderophore export. FEMS Microbiology Letters 228: 211–216.
38. Keel SB, Doty RT, Yang Z, Quigley JG, Chen J, et al. (2008) A Heme Export Protein Is Required for Red Blood Cell Differentiation and Iron Homeostasis. Science 319: 825–828.

39. Quigley JG, Yang Z, Worthington MT, Phillips JD, Sabo KM, et al. (2004) Identification of a Human Heme Exporter that Is Essential for Erythropoiesis. Cell 118: 757–766.

40. Husain A, Jeelani G, Sato D, Nozaki T (2011) Global analysis of gene expression in response to L-Cysteine deprivation in the anaerobic protozoan parasite Entamoeba histolytica. BMC Genomics 12: 275.

41. Levicán G, Katz A, de Armas M, Núñez H, Orellana O (2007) Regulation of a glutamyl-tRNA synthetase by the heme status. Proceedings of the National Academy of Sciences 104: 3135–3140.

42. Sabaty M, Adryanczyk G, Roustan C, Cuine S, Lamouroux C, et al. (2010) Coproporphyrin Excretion and Low Thiol Levels Caused by Point Mutation in the *Rhodobacter sphaeroides* S-Adenosylmethionine Synthetase Gene. J Bacteriol 192: 1238–1248.

43. Kobayashi T, Nishizawa NK (2012) Iron Uptake, Translocation, and Regulation in Higher Plants. Annual Review of Plant Biology 63: 131–152.

44. Sato J, Takeda K, Nishiyama R, Fusayama K, Arai T, et al. (2010) *Chlorella vulgaris* Aldehyde Reductase Is Capable of Functioning as Ferric Reductase and of Driving the Fenton Reaction in the Presence of Free Flavin. Bioscience, Biotechnology, and Biochemistry 74: 854–857.

45. Tannich E, Bruchhaus I, Walter RD, Horstmann RD (1991) Pathogenic and nonpathogenicEntamoeba histolytica: identification and molecular cloning of an iron-containing superoxide dismutase. Molecular and Biochemical Parasitology 49: 61–71.

46. Zhao T, Cruz F, Ferry JG (2001) Iron-Sulfur Flavoprotein (Isf) from *Methanosarcina thermophila* Is the Prototype of a Widely Distributed Family. Journal of Bacteriology 183: 6225–6233.

47. Reyes-López M, Bermúdez-Cruz RM, Avila EE, De la Garza M (2011) Acetaldehyde/alcohol dehydrogenase-2 (EhADH2) and clathrin are involved in internalization of human transferrin by *Entamoeba histolytica*. Microbiology 157: 209–219.

48. Ali V, Hashimoto T, Shigeta Y, Nozaki T (2004) Molecular and biochemical characterization of d-phosphoglycerate dehydrogenase from Entamoeba histolytica. European Journal of Biochemistry 271: 2670–2681.

49. Mishra V, Kumar A, Ali V, Nozaki T, Zhang KYJ, et al. (2012) Glu-108 is essential for subunit assembly and dimer stability of d-phosphoglycerate dehydrogenase from *Entamoeba histolytica*. Molecular and Biochemical Parasitology 181: 117–124.

50. Park S-J, Lee S-M, Lee J, Yong T-S (2001) Differential gene expression by iron-limitation in *Entamoeba histolytica*. Molecular and Biochemical Parasitology 114: 257–260.

51. Hentze MW, Muckenthaler MU, Galy B, Camaschella C (2010) Two to Tango: Regulation of Mammalian Iron Metabolism. Cell 142: 24–38.

52. Leipuviene R, Theil E (2007) The family of iron responsive RNA structures regulated by changes in cellular iron and oxygen. Cellular and Molecular Life Sciences 64: 2945–2955.

53. Solano-González E, Burrola-Barraza E, León-Sicairos C, Avila-González L, Gutiérrez-Escolano L, et al. (2007) The trichomonad cysteine proteinase TVCP4 transcript contains an iron-responsive element. FEBS Letters 581: 2919–2928.

54. Kim HP, Pae H-O, Back SH, Chung SW, Woo JM, et al. (2011) Heme oxygenase-1 comes back to endoplasmic reticulum. Biochemical and Biophysical Research Communications 404: 1–5.

55. Tronel C, Rochefort GY, Arlicot N, Bodard S, Chalon S, et al. (2013) Oxidative Stress Is Related to the Deleterious Effects of Heme Oxygenase-1 in an In Vivo Neuroinflammatory Rat Model. Oxidative medicine and cellular longevity 2013: 1–10.

56. Frankenberg N, Moser J, Jahn D (2003) Bacterial heme biosynthesis and its biotechnological application. Applied Microbiology and Biotechnology 63: 115–127.

57. Heinemann IU, Jahn M, Jahn D (2008) The biochemistry of heme biosynthesis. Archives of Biochemistry and Biophysics 474: 238–251.

58. Saurin W, Hofnung M, Dassa E (1999) Getting In or Out: Early Segregation Between Importers and Exporters in the Evolution of ATP-Binding Cassette (ABC) Transporters. Journal of Molecular Evolution 48: 22–41.

59. Loyevsky M, Mompoint F, Yikilmaz E, Altschul SF, Madden T, et al. (2003) Expression of a recombinant IRP-like *Plasmodium falciparum* protein that specifically binds putative plasmodial IREs. Molecular and Biochemical Parasitology 126: 231–238.

60. Calla-Choque JS, Figueroa-Angulo EE, Ávila-González L, Arroyo R (2014) α-Actinin TvACTN3 of Trichomonas vaginalis Is an RNA-Binding Protein That Could Participate in Its Posttranscriptional Iron Regulatory Mechanism. BioMed Research International 2014: 20.

61. Bera T, Nandi N, Sudhahar D, Akbar M, Sen A, et al. (2006) Preliminary evidence on existence of transplasma membrane electron transport in Entamoeba histolytica trophozoites: a key mechanism for maintaining optimal redox balance. Journal of Bioenergetics and Biomembranes 38: 299–308.

15

Identification and Characterization of the RouenBd1987 *Babesia divergens* Rhopty-Associated Protein 1

Marilis Rodriguez[1], Andy Alhassan[1], Rosalynn L. Ord[1], Jeny R. Cursino-Santos[1], Manpreet Singh[1], Jeremy Gray[2], Cheryl A. Lobo[1]*

1 Department of Blood-Borne Parasites, New York Blood Center, New York, New York, United States of America, **2** University College Dublin School of Biology and Environmental Science, Dublin, Republic of Ireland

Abstract

Human babesiosis is caused by one of several babesial species transmitted by ixodid ticks that have distinct geographical distributions based on the presence of competent animal hosts. The pathology of babesiosis, like malaria, is a consequence of the parasitaemia which develops through the cyclical replication of *Babesia* parasites in a patient's red blood cells, though symptoms typically are nonspecific. We have identified the gene encoding Rhoptry-Associated Protein −1 (RAP-1) from a human isolate of *B. divergens*, Rouen1987 and characterized its protein product at the molecular and cellular level. Consistent with other *Babesia* RAP-1 homologues, BdRAP-1 is expressed as a 46 kDa protein in the parasite rhoptries, suggesting a possible role in red cell invasion. Native BdRAP-1 binds to an unidentified red cell receptor(s) that appears to be non-sialylated and non-proteinacious in nature, but we do not find significant reduction in growth with anti-rRAP1 antibodies *in vitro*, highlighting the possibility the *B. divergens* is able to use alternative pathways for invasion, or there is an alternative, complementary, role for BdRAP-1 during the invasion process. As it is the parasite's ability to recognize and then invade host cells which is central to clinical disease, characterising and understanding the role of *Babesia*-derived proteins involved in these steps are of great interest for the development of an effective prophylaxis.

Editor: Tobias Spielmann, Bernhard Nocht Institute for Tropical Medicine, Germany

Funding: Funding for this study was provided by a grant from the National Institutes of Health (NIH) to CAL: RO1-HL105694. The funders had no role in study design, data collection and analysis, decision to publish, or preparation of the manuscript.

Competing Interests: The authors have declared that no competing interests exist.

* Email: clobo@nybloodcenter.org

Introduction

The phylum Apicomplexa consists of protozoan parasites, among which are a number of etiological agents of medical and veterinary importance, including *Plasmodium, Toxoplasma, Babesia, Neospora, Cryptosporidium, Eimeria* and *Theileria*. They are defined by a common set of apically located secretory organelles that are required for host cell invasion [1–3]. Although the host cell range of Apicomplexan parasites is large and varied, the mechanism of host cell entry is thought to be conserved [4].

The genus *Babesia* comprises many species of parasites that invade RBCs of many different vertebrate hosts [5]. They are transmitted by their tick vectors during the taking of a blood meal from the vertebrate host [5,6]. In the last 50 years babesiosis has emerged as a major public health concern due to the expansion of the geographic range of the vectors, ixodid ticks [7,8]. Additionally, fatality rates average 30% to 45% in susceptible hosts, which include older subjects, new-born infants, and people that are immune-compromised [9]. As a consequence, since 2011, babesiosis is now a nationally notifiable disease in 18 states in the United States [10]. Human babesiosis is caused by one of several babesial species that have distinct geographical distributions based on the presence of competent animal hosts [11]. In Europe, babesiosis in man is caused by the bovine pathogen *B.*

divergens [12,13]; in North America, it is caused predominantly by *B. microti* [14], a rodent borne parasite.

The pathology of babesiosis, like malaria, is a consequence of the parasitaemia which develops through the cyclical replication of *Babesia* parasites in a patient's RBC's, though symptoms typically are nonspecific (fever, headache, and myalgia) [15]. The parasite's ability to first recognize and then invade host RBCs is central to the disease process, and thus *Babesia*-derived proteins involved in these recognition and invasion steps are of great interest for the development of an effective prophylaxis.

Rhoptries are club-shaped organelles located at the apical end of the invading merozoite. They are divided into two distinct intra-organelle compartments, the posterior bulb and the more anterior duct (neck) through which rhoptry proteins are secreted [16,17]. Rhoptry contents include both protein and lipid components, which assemble to form membrane-like structures. Proteins derived from the rhoptries are crucial for the invasion and survival of these parasites. Among these proteins in malaria, two rhoptry-associated protein complexes are considered attractive anti-parasitic targets: the high molecular weight (HMW or RhopH complex) and the low molecular weight (LMW) complexes; the first one containing the RhopH-CLAG proteins and the second one consisting of the rhoptry-associated proteins (RAP)-1, RAP-2 and RAP-3 [16,18]. Homologs of these two sets of rhoptry proteins have not been identified in any of the Babesia species. However a

```
Bd-KJ699102   MKGFSRIVVALCVVATKSTFALRHAAQNGAMVALGEPDSVAEAIETVDSVMAAIQESVKT  60
Bd-Z49818     MKGFSRIVVALCVVATKSTFALRHAAQNGAMVALGEPDSVAEAIETVDSVMAAIQESVKT  60
Bb-M38218     MRIISGVVGCLFLVFSHHVSAFRHNQRVGSLAPAEVVGDLTSTLETADTIMTLRDHMHNI  60
Bc-M91168     MRLVKQAIGTLVLAFTTSAFAYKVPGFGGILSKSDGAEKTLSTLLNVDASTRAALEGYPM  60
              *:..:  :  *:.:  .  .*:  .  *  :  . .  .::..*:. .

Bd-KJ699102   MDNMIAHLSTRSEGFVDDVCQGVNEESQCQESVRTYVSRCMQHDCFKIDNERYPAEKEYQ  120
Bd-Z49818     MDNMIAHLSTRSEGFVDDVCQGVNEESQCQESVRTYVSRCIQRGCLKIDHERYAAEKGNQ  120
Bb-M38218     TKIMKHVLSNGREQIVNDVCSNAPEDSNCREVVNNYADRCEMYGCFTIDNVKYPLYQEYQ  120
Bc-M91168     NAAMANFSNGRREEEEEAVCGNIAEETECQKSVAEYVESCVRYDCFSIENQKYPQEKEYQ  120
              *  .  :  *  :** *:::*!:: * *..* *:.*:.:* *

Bd-KJ699102   PLALPNPYQLDAAFQLFKECDSNATKEPIDEFLMRFKHGGRYGAYQSFIMNVVTSSYTQL  180
Bd-Z49818     PLAFPNPYQVGAAFQLFQECDSDPTQEPNGNFGMRFKHGGRYGAYQSFIMNVVTSSYTQL  180
Bb-M38218     PLSLPNPYQLDAAFRLFKESASNPAKNSVKREWLRFRNGANHGDYHYFVTGLLNNNVVHE  180
Bc-M91168     PLTLPNPYQLEAAFYVFRNSESNPIKNPTEAFWMRFHGGRYGAYHNFLVNILYKNLSDS  180
              **:;*****; *** :*::. *: ::   :**:.*..:* *: *: :.

Bd-KJ699102   SDNSEVEDLVNRYLYMATMYYKTYLILDTTKAHLINKIDFAHHIFGKSIKHMLEKIIRNH  240
Bd-Z49818     SDNTEVEDLVNRYLYMATMYYKTYLILDTTKVHLINKIDFGHHISGKSIKHMVEKIIRNH  240
Bb-M38218     EGTTDVEYLVNKVLYMATMNYKTYLTVNSMNAKFFNRFSFTTKIFSRRIRQTLSDIIRWN  240
Bc-M91168     MVDDNLEGFVRKYAYMATMYYKTYTALDVVNARIINKIAFSPHLFGRQIRNALTNIIRSN  240
              ::* :*.: ***** **** :: :.::::*:: *  .:: *:. .**.

Bd-KJ699102   LPKDFGIYSIERLSHISAGYGDYILKQVPHMPAFSTRFNSMVVDTLHKIIGKYQKMPWYK  300
Bd-Z49818     LPRDFGIYSIERLSHISAGYGDYILKQVPYMPAFSTRFNRMVVDTLHKIIGKYQKMPWYK  300
Bb-M38218     VPEDFEERSIERITQLTSSYEDYMLTQIPTLSKFARRYADMVKKVLLGSLTSYVEAPWYK  300
Bc-M91168     IPEDFGKYNVDRLRHVMGGYEEYMMKQVPSLPNFAKKYAGMVVKSLIKNVGAYQKQPWFK  300
              :*.**  .::*:. ::  ..* :*::.*:*: :  **  .*  * *  * ::.

Bd-KJ699102   KWFNSVADFFKNSIGKPIKNLFSKRAP-----------SSSTEGPWHKKVTHSVKKMLNE  349
Bd-Z49818     KWFNSVADFFKNSIGKPIKNLFSKRAP-----------SSSTEGAWHKKVSHSVKKMLNE  349
Bb-M38218     RWIKKFRDFFSKNVTQPTKKFIEDTNEVTKNYLKANVAEPTK--KFMQDTHEKTKGYLKE  358
Bc-M91168     KLNNQIRNFFVNKIHEPTKEFFVNKIH-----------EPTKE-FFVNKIHEPTKEFFVN  348
              :  :..,:**  .: :* *:::   .  :  :  :  . .* : :

Bd-KJ699102   KI-PVVKNFF-----------KDGIRKSS---L--------------GDLLHGPKASSMKEDE  383
Bd-Z49818     KI-PVVKNFF-----------KDGIRKSS---L--------------GDLLHGPKASSMKEDE  383
Bb-M38218     NVAEPTKTFFKEAPQVTKHFFDENIGQPTKEFFREAPQATKHFLDENIGQPTKEFFREAP  418
Bc-M91168     KIHEPTKEFF-----------VNKLHEPTKEFF---------VNKLHEPTKEFFS-NM  385
              ::  .* **   :  :  :  :   :  : *.. .:

Bd-KJ699102   VPLTEAMETNESSGENLRGTG----------------------------------- 404
Bd-Z49818     VPLTEAMETNESSGENLRGYR----------------------------------- 404
Bb-M38218     Q-----------ATKHFLGENIAQPTKEFFKDVPQVTKKVITENIAQPTKEFRREVPHAT 467
Bc-M91168     VPG-AFQKISEKAGRHLRSSKTVVPEDE-----PS--------------SSLENEA---- 421
                 : ...:.

Bd-KJ699102   MD-----SA----------------------------------- 408
Bd-Z49818     HG-----LR----------------------------------- 408
Bb-M38218     MKVLNENIAQPAKEIIHEFGTGAKNFISAAHEGTKQFLNETVGQPTKEFLNGALETTKDA 527
Bc-M91168     VE-----DGQLTMGDVTDFEMRTPTYE----QGSQESLNE----------------- 452

Bd-KJ699102   ----------------------------------- 408
Bd-Z49818     ----------------------------------- 408
Bb-M38218     LHHLGKSSEEANLYDATENTTQANDSTTSNGEDTAGYL 565
Bc-M91168     ---VGNE----------------------------- 456
```

Figure 1. Multiple sequence alignment of BdRAP-1 sequences shows homology to other babesial RAP-1 proteins is within the N-terminal portion. Clustal analysis of RAP-1 sequences of *B. divergens* (human host source) (accession number KJ699102) identified in this manuscript, *B. divergens* (bovine host origin) (accession number Z49818), *B. bovis* (accession number M38218) and *B. canis* (accession number M91168) shows that homology is restricted to the N-terminus region, which corresponds to the location of the RAP-1 superfamily structure (indication above the sequence by a sold line from BdRAP-1 Ala[41] to Gly[291]) as identified by BLAST analysis. The sites of non-synonymous differences

between the BdRAP-1 proteins are shown in grey shading. Positions of identity are indicated by asterisk and similarity by dots, below the sequences. Four cysteine residues, at sites BdRAP-1 80, 89, 100, and 105 (bold underlined), are conserved across all species, and are also limited to the N-terminus region, further suggesting this portion of RAP-1 may contain the RBC binding site.

number of rhoptry molecules classified as RAP-1 and RAP-2 have been characterized in some of the veterinary parasite species.

There is a brief report on the gene sequence of a RAP-1 in *B. divergens* [19] isolated from a bovine source, although there are no further details on the characterization of this antigen. As *B. divergens* is an important human zoonosis, it is prudent to fully assess genes responsible for RBC invasion from parasites that result in human infection to aid the most effective interventions. We have identified and characterised a Rhoptry Associated Protein −1 (RAP-1) homolog of BdRouen1987 isolated from, a human infection, and provide additional support for this antigen role in invasion.

Materials and Methods

Animal Protocol Work and Ethics Statement

Cattle and gerbil sera were produced in 1995 under license in the Republic of Ireland (License number B100/702, Department of Health, Cruelty to Animal Act, 1876 (European Directive 86/609/EC) as part of studies on vaccination against bovine babesiosis. The license provided permitted the use of infesting with ticks, infecting with *Babesia divergens* by intraperitoneal injection, taking of blood spots from the tail for blood smears and bleeding while under halothane anaesthesia without recovery (euthanasia). As no formal ethics (IACUC) committees for animal experimentation existed in Ireland at the time (1995), the conditions of the licence were approved at the University College Dublin by both the Director of the Biomedical Facility (the departmental animal facility) and the Dean of the Faculty of Veterinary Sciences.

Briefly, yearling cattle housed on the University College Dublin farm were inoculated subcutaneously behind the left shoulder with 1×10^7 gerbil (*Meriones unguiculatus*) RBCs infected with the TL strain of *B. divergens* and suspended in 60% RPMI 1640 in HEPES with added L-glutamine and 40% fetal calf serum. Four weeks later approximately 100 mL of venous blood were taken by syringe from the jugular vein, and antibody levels of 1:256 titre in the resulting sera determined by IFA. The sera were then stored at −70°C.

Gerbils were housed at the Biomedical Facility, University College Dublin and sera were prepared by intraperitoneal inoculation of a group of 8-week old gerbils with PBS-suspended 1×10^2 erythrocytes infected with *B. divergens* (DR strain). Three weeks later the gerbils in which transient parasitaemia had been observed were challenged with 1×10^3 infected erythrocytes subdermally and then after a further three weeks with 1×10^7 infected erythrocytes. After a further 4 weeks, all gerbils were bled by cardiac puncture under halothane anaesthesia and sacrificed after bleeding by intraperitoneal injection of sodium pentobarbitone formulated for euthanasia, and the resulting sera stored at −70°C.

Babesia divergens strain BdRouen1987 was first isolated from a French patient in 1986 [20] and propagated in culture, and has been used by many laboratories as the reference strain for *B. divergens* studies [21–25].

Parasite propagation

Asexual erythrocytic cultures of *B. divergens* (BdRouen1987 isolated from a French patient) [20] were maintained *in vitro* in human A$^+$ blood using RPMI 1640 medium (Life Technologies Corporation, Carlsbad, CA) supplemented with 10% human serum and sodium bicarbonate solution 7.5% (w/v) (Life Technologies Corporation, Carlsbad, CA). Cells were cultured at 37°C in 90% C02, 5% nitrogen and 5% oxygen.

Identification and confirmation of the cDNA and gDNA Bdrap-1 sequences

The Bdrap-1 sequence was obtained by initially screening a *B. divergens* cDNA library [22] by PCR with the universal primers, T3 and T7, in various combinations with degenerate primers based upon published rap-1 genes of alternative *Babesia* spp. Amplified PCR products representing partial Bdrap-1 sequences were separated on 1% agarose gels, cleaned using QIAquick gel extraction kit (Qiagen), cloned into TOPO TA vector (Life Technologies Corporation, Carlsbad, CA), and transformed into TOP10 *Echerichia coli* strain (Life Technologies Corporation, Carlsbad, CA) prior to sequencing. Partial Bdrap-1 sequences obtained were aligned using ClustalW (http://www.ebi.ac.uk/Tools/msa/clustalw2/) to design specific Bdrap-1 primers. Using various combinations of the T3 and T7 universal and Bdrap-1 specific primers, the full length sequence of the cDNA Bdrap-1 gene was obtained using the same amplification, and cloning strategy as above. The genomic sequence of the Bdrap-1 gene was obtained by amplifying 100 ng genomic *B. divergens* DNA (prepared as described in [25]) using specific primers Bdrap-1F1 (5′-GCTTTCGAAGCTTATTGACTCT-3′) and Bdrap-1R1 (5′ GTTAAAACCTTAGGGATAAATTGGTA 3′). The PCR product was separated on 1% agarose gel and cleaned using QIAquick gel extraction kit (Qiagen, Hilden, Germany), cloned into TOPO TA vector (Life Technologies Corporation, Carlsbad, CA), and transformed into TOP10 *Echerichia coli* strain (Life Technologies Corporation, Carlsbad, CA) prior to sequencing.

Protein Expression and purification and rabbit antiserum production

Genomic *B. divergens* DNA encoding residues Val^{32} to Met^{379} of the BdRAP-1 protein was amplified using the modified primers Bdrap-1F2 (5′ CGCGGATCCGTGGCTCTTGGCGAG 3′), containing *BamH*I restriction site, and Bdrap-1R2 (5′ CCGCTCGAGTCAGAGTCCATGCCTGTA 3′), containing *Xho*I restriction site and stop codon, using SuperMix kit (Life Technologies Corporation, Carlsbad, CA). Amplified products were cloned into the expression vector, pET28(a) vector (Life Technologies Corporation, Carlsbad, CA) according to the manufacturer's instructions. *E. coli* strain BL21 was transformed with pET28a-Bdrap1 in 250 mL of SOB medium containing 100 μg/mL Ampicillin (Sigma-Aldrich, St Louis, MO) and inoculated with 1 mL of fresh overnight culture and grown at 37°C to A$_{600}$ = 0.6 prior to induction with 0.2 isopropyl-β-D-thiogalactopyranoside. After 4 h of induction, recombinant BdRAP-1 protein was purified from bacterial cells using Ni-NTA Buffer kit as per manufactures instructions (EMD Millipore Corporation, Billerica, MA) supplemented with a protease inhibitor cocktail (Sigma-Aldrich, St Louis, MO). The insoluble material was removed by centrifugation and the soluble fraction was used to purify the rBdRAP-1 protein. Protein concentration was determined by the Bradfords Dye (Bio-Rad Laboratories, Hercules, CA). Polyclonal

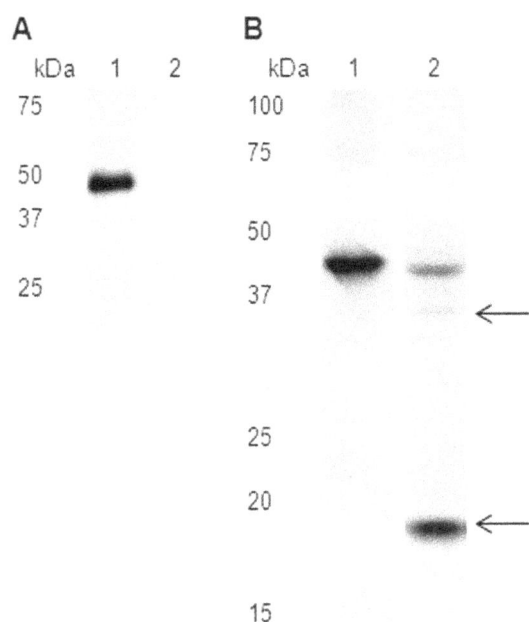

Figure 2. Anti-rBdRAP-1 antibodies identify a specific ~46 kDa protein in parasite lysate. A, Immunoblotting of proteins from in vitro cultured *B. divergens* lysate showed anti-rBdRAP-1 antibodies (lane 1) are able to detect native BdRAP-1, as shown by the presence of a single band at ~46 kDa. Pre-immune rabbit serum (lane 2) did not react with any native antigens. B, Immunoprecipitation with lysate from [^{35}S] methionine/cysteine-labelled *B. divergens* cultures showed native BdRAP-1 is present in the culture pellet (lane 1), as observed by the presence of a single band at ~46 kDa. BdRAP-1 is secreted into the culture supernatant as a processed product, as shown by the presence of an abundant ~17 kDa product, and a much less abundant product of ~34 kDa, in the culture supernatant (lane 2; indicated by arrows).

serum was then raised against the rBdRAP-1 in rabbits (Strategic Diagnostics Inc., Newark, DE).

Immunoprecipitation

Freshly cultured parasites were washed and resuspended in methionine-free RPMI 1640 (MP Biomedicals Inc., Aurora, OH). 200 μCi/mL of [^{35}S] methionine/cysteine (PerkinElmer, Boston, MA) was added and parasites were incubated at 37°C for 2 h. Parasites were lysed in NETT buffer [10 mM Tris pH 7.5, 150 mM NaCl, 0.5 mM EDTA, 0.1% Triton X-100] using a protease inhibitor cocktail (Sigma-Aldrich, St Louis, MO, USA)) and centrifuged to collect the supernatant. Lysates were pre-cleared with Protein G-Sepharose beads (GE Healthcare, Waukesha, WI) before antibody addition. Protein G-Sepharose beads were added and washed extensively with NETTS (10 mM Tris, pH 7.5, 500 mM, NaCl, 5 mM EDTA and 0.1% Triton X-100) and NETT buffers. Protein was eluted from the beads using elution buffer [10 mM Tris pH 7.5, 150 mM NaCl, 5 mM EDTA], boiled and run on SDS-PAGE gel. Gels were fixed with fixing solution [25% Isopropanol alcohol; 10% Acetic Acid] for 30 min and enhanced with Amplify^{TM} fluorographic solution (GE Healthcare, Waukesha, WI) for 45 min, dried under vacuum, and exposed to X-ray film for autoradiography.

Immunofluorescence assay (IFA)

Cultured parasites with a high parasitaemia (~70%) or purified free merozoites were prepared for immunofluorescence (IFA). The parasites were smeared on glass slides and fixed in cold 90%

acetone/10% methanol. The slides were incubated for 30 min with purified anti-rBdRAP-1 antibodies diluted 1:200 in 1× PBS/1%BSA. The slides were washed three times in 1× PBS and incubated for 30 min with a polyclonal anti-rabbit immunoglobulin (Dako, Produktionsvej Denmark) diluted 1:200 in 1× PBS/1% BSA. The slides were washed again three times in 1× PBS, rinsed once in distilled water, and mounted using Vectashield with DAPI solution for microscopy (Vector laboratories Inc., Burlingame, CA).

Electron microscopy

Cultured parasites and free merozoites were fixed with 1% paraformaldehyde and 0.1% glutaraldehyde in 0.1 M cacodylate buffer, for 1 h at 4°C, washed in 0.1 M buffer, pH 7.4 and treated with 50 mM ammonium chloride to quench the remaining aldehydes. The fixed infected RBCs and the free merozoites were than dehydrated and embedded in LR-White Resin (Electron Microscopy Sciences, Hatfield, PA). Thin sections of embedded parasites were mounted on parlodion covered nickel grids, blocked in 2% BSA and probed with purified anti-rBdRAP-1 antibodies overnight at 4°C, washed in buffer containing 1× BSA and Tween-20 and incubated with goat-anti-Rabbit IgG coupled with 6 nm gold particles (Electron Microscopy Sciences, Hatfield, PA) or Goat-anti-Mouse IgG coupled with 5 nm gold particles (GE Healthcare, Waukesha, WI). After staining with uranyl acetate, sections were observed using a Philips-410 electron microscope.

RBC Binding assays

RBCs. Erythrocytes collected in 10% citrate-phosphate-dextrose were washed 3 times in 1× PBS and treated with 1.0 mg/mL of neuraminidase and chymotripson, and 0.1 and 1.0 mg/mL of trypsin as described [25] or treated with Proteinase K (Life Technologies Corporation, Carlsbad, CA) and papain (Sigma-Aldrich, St Louis, MO) as follows: 500 uL (packed cell volume) of washed erythrocytes were incubated with 0.5 mg/mL Proteinase K for 30 in at 37°C in a final volume of 4 mL with 1×PBS. 500 uL (pvc) of washed erythrocytes were incubated with 0.5 mg/mL Papain at r.t for 30 min in a total volume of 2 mL.

Parasite Supernatant. Freshly cultured parasites were washed and re-suspended in methionine-free RPMI 1640 (MP Biomedicals Inc., Aurora, OH). 200 μCi/mL of [^{35}S] methionine/cysteine (PerkinElmer, Boston, MA) was added, and parasites were incubated at 37°C overnight. The supernatant was collected by centrifugation at 14,000 rpm for 15 min and a protease inhibitor mixture (Sigma-Aldrich, St Louis, MO) was added. The parasite supernatant was pre-cleared with protein G (GE Healthcare, Waukesha, WI) before to use.

Binding assay. An aliquot of 500 μL of labelled parasite supernatant, confirmed as having Bd-RAP-1 by immunoprecipitation, was mixed with 100 μL target erythrocytes (wild type or enzyme treated) for 30 min at r.t. The mixture was then spun at 6000 *g* for 1 min through sodium phthalate oil to remove unbound material. Oil and supernatant was aspirated, leaving the RBC/protein complexes. Cells were lysed in NETT [0.5% Triton-X100, 150 mM NaCl, 10 mM EDTA 50 mM Tris, pH 7.4] for 30 minutes at 4°C and spun at 14, 000 rpm for 5 min. The soluble extract containing the BdRAP1-receptor complex was immunoprecipitated with anti-rBdRAP1 antibodies. All experiments were repeated at least 3 times with identical pattern of binding observed.

ELISAs

Microtiter ELISA plates (Costar, Corning Life Sciences, Corning, NY) were coated with 1 μg/mL of recombinant BdRAP-1 protein

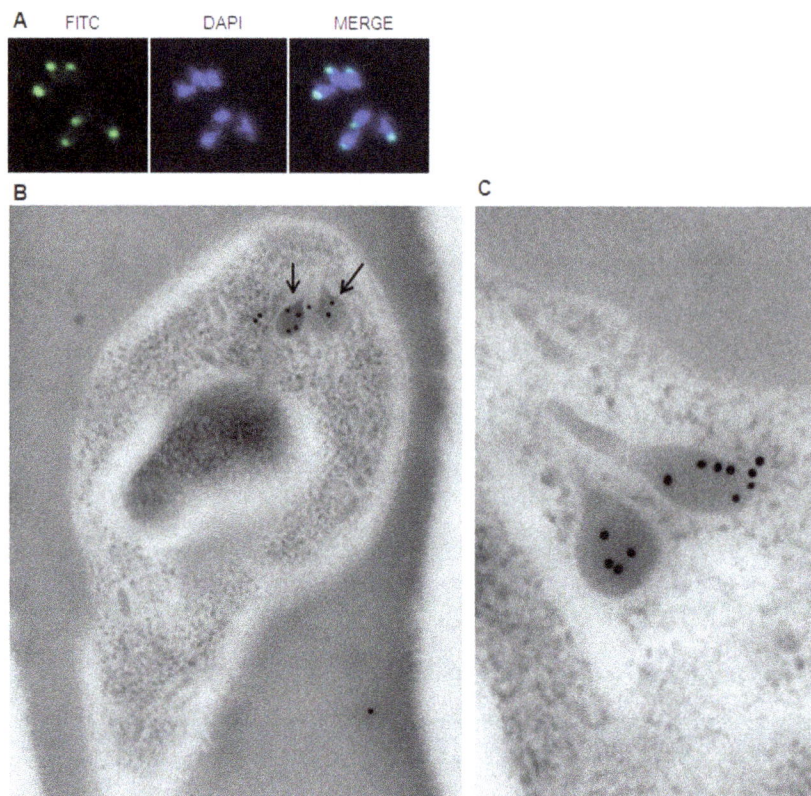

Figure 3. Native BdRAP-1 localised to apical complex. A, Immunofluorescence staining with anti-rBdRAP-1 antibodies (FITC) on fixed cells infected with *B. divergens* parasites (DAPI) clearly shows that BdRAP-1 is localised to the apical ends of all 4 parasites in this Maltese Cross form of the intracellular parasite (see Merged panel). B, Electron microscopy of a singly-infected RBC (direct magnification of 30,000×, print magnification of ~50,000×) shows the location of the rhoptry organelles and the presence of BdRAP-1. C, An enlarged section (direct magnification 49,000×, print magnification of ~140,000×) clearly shows localisation of native BdRAP-1 is within the bulb of the rhoptry organelle.

diluted in 0.05 M carbonate buffer, pH 9.6. After incubation overnight at 4°C, the plates were washed 5 times with 1× phosphate-buffered saline (PBS) with 0.05% Tween-20 (1× PBS-T) and blocked with blocking buffer (3% BSA in 1× PBS-T) for 1 h at 37°C. Serum samples, diluted in blocking buffer, were reacted with the bound antigens by incubating for 1 h at 37°C, in triplicate wells. Serum from the cows was diluted from 1:200 to 1:12,800, and serum from the gerbils was diluted 1:200 to 1:25,600. The bound antibodies in the cow sera were detected after incubation for 1 h at 37°C with goat anti-cow conjugated HRP at 1:1000 (Abcam, Cambridge, MA) diluted in blocking buffer; antibodies in the gerbil sera were detected after incubation for 1 h at 37°C with goat anti-rat conjugated HRP (Pierce Antibody Products, Thermo Fisher Scientific Inc, Rockford, IL) diluted 1:10,000 in blocking buffer. Tetramethylbenzidine (Sigma, St Louis, MO) was used as the substrate for all ELISAs for 30 min. Sulphuric acid (2M) was used to stop the substrate reaction, and the optical density (OD) was read at 450 nm immediately on a SpectraMax 190 ELISA Reader (Molecular Devices, Sunnyvale, CA).

In vitro growth inhibition assays (GIA)

Total preimmune rabbit (PI) and BdRAP-1 IgGs were purified using protein-G Sepharose (GE Healthcare, Waukesha, WI) with IgG binding and elution buffers (Thermo Fisher Scientific, Rockford, IL) according to the manufacturer's recommendation and dialyzed with DPBS. Purified pre-immune IgG (Sigma-Aldrich, St Louis, MO) and rBdRAP1 were added to pre-warmed medium and 5% haematocrit of RBCs at ~4% parasitaemia. The

final concentration of antibodies in the GIA was 2 mg/mL in a 500 uL final volume. Each antibody was tested in triplicate. The culture medium was removed at 24 h and replaced with fresh medium, 10% human serum, and purified IgG to a final concentration of 2 mg/mL. Samples were incubated at 37°C with 90% C02, 5% nitrogen, 5% oxygen. Smears were made at 0 h, 12 h, 24 h, 36 h. The slides were fixed in 100% methanol and stained with Giemsa (Sigma-Aldrich, St Louis, MO). Parasitaemia was determined after counting the total number of intracellular parasites present in 1×10^4 RBC at ×100 magnification using a Nikon Eclipse E 600 microscope, and the invasion inhibition with respect to the controls was determined for each antibody tested.

Results and Discussion

Cloning of a RAP-1 homolog (BdRAP-1) of *B. divergensRouen1987*

The *B. divergens* genome has not yet been sequenced. Although a *B. divergens* RAP-1 sequence from a bovine infection has been identified [19], no BdRAP-1 homolog from a human infection has been identified, and it is possible that sequence polymorphisms may exist between strains that have alternative host preferences to enable specific host cell interactions. To clone the gene encoding BdRAP-1, a *B. divergens* cDNA library, created from material obtained from a strain isolated from a human *B. divergens* infection and maintained in culture [20,22], was screened by PCR with the universal primers, T3/T7, in conjunction with degenerate primers based upon homology within published rap-1 genes of

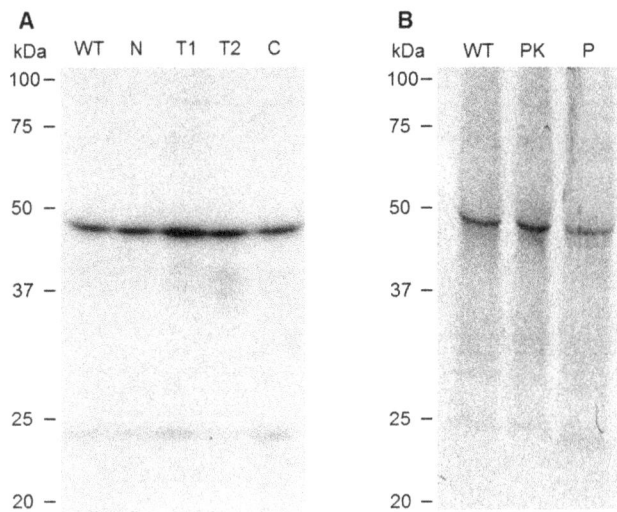

Figure 4. A, BdRap-1 binds to a neuraminidase, trypsin and chymotrypsin-resistant RBC receptor. ^{35}S-labelled proteins derived from parasite culture supernatants were mixed with untreated RBCs treated (WT lane), or with RBCs treated with 0.1 U/mL of Neuraminidase (N), 0.1 mg/mL trypsin (T1), 1 mg/mL trypsin (T2), and 1 mg/mL. Separated cells were then lysed and the soluble fraction was immuno-precipitated with anti-rBdRAP-1 antibodies and separated on SDS-PAGE gel. BdRAP-1 appears to be an adhesin that participates in invasion by binding to the RBC surface, as indicated by the presence of a band at the expected size of ~46 kDa. However, none of the enzyme treatments inhibited or decreased BdRAP-1 binding, as shown the presence of the ~46 kDa band in all lanes representing treated cells, thus, the binding profile of native BdRAP-1 to the RBC suggests the participation of a novel red cell receptor in merozoite invasion. B, BdRAP-1 binds to a non-proteinacious receptor. Binding profile of BdRAP-1 to untreated RBCs treated (WT lane), or with RBCs treated with 0.5 mg/mL of Proteinase K (PK) or 0.5 mg/mL Papain (P).

alternative *Babesia* spp. A band of ~1 Kb was obtained and sequenced. BLAST analysis identified this sequence as encoding a putative conserved domain corresponding to the RAP-1 super-family, as well as high homology to the rap-1 gene from other *Babesia* spp. This sequence was used to design specific Bdrap-1 primers, which were used with the universal primers, to continue screening the cDNA library. Amplified fragments were cloned into TOPO vector and sequenced multiple times. Nucleotide sequences were aligned using ClustalW2 (http://www.ebi.ac.uk/Tools/msa/clustalw2/) and assembled into a contig of ~1,300 bp with at least 3× sequence coverage across all nucleotides. This contig contained 1,255 bp of the full ORF encoding BdRAP-1, which has been submitted into GenBank (accession number KJ699101). Multiple clones containing the ORF were isolated and sequenced and all yielded identical sequence. PCR amplification of the Bdrap-1 gene from genomic DNA with specific primers derived from the full-length cDNA Bdrap-1 contig produced a single fragment of 1,225 bp (GenBank accession KJ699102). Cloning and sequencing showed the genomic Bdrap-1 gene to be 1,225 bp, the same length and nucleotide sequence as the cDNA Bdrap-1 contig, suggesting the absence of introns in the gene encoding BdRAP-1.

Comparison of BdRAP-1 sequence with other Apicomplexan sequences

Conceptual translation of the 1,225 bp open reading frame resulted in prediction of a ~46 kDa protein. A BLAST search

with the full translated sequence identifies the region encompassing Ala^{41} to Gly^{291} as a member of the RAP-1 superfamily, as shown in Figure 1. The BdRAP-1 antigen we identified shows a 93% homology to a putative *B. divergens* RAP-1 protein isolated from a bovine infection [19], 47% identity to *B. canis* [26], and 33% identity to *B. bovis* [27], although there is very little identity to RAP-1 antigens of either *Plasmodium* or *Toxoplasma* spp (not shown). Sequence alignment with these RAP-1 proteins from other babesial species shows the N-terminal region that encompasses the predicted location of the RAP-1 superfamily in *B. divergens* is relatively conserved (~40% homology) in both *B. canis* and *B. bovis*, and includes 4 conserved cysteine residues across the species, indicated by bold underline in Figure 1. There are 19 non-synonymous polymorphisms between the BdRAP-1 amino acid sequence identified here from a human infection and the BdRAP-1 identified from a bovine infection (highlighted in grey in Figure 1). Although these polymorphisms appear spread throughout the amino acid sequence, it is interesting to note that all but one of these occur down-stream of the 4 conserved cysteine residues, suggesting the N-terminus region is the location of the RBC receptor binding site for all babesial RAP-1 antigens. It remains to be determined if the polymorphisms have a role in determining host cell preference for parasite invasion.

Expression of BdRAP-1 as a hexa-His fusion protein and generation of antibodies

The sequence encoding Val^{32} to Met^{379} of BdRAP-1 was cloned into the expression vector pET28(a) and expressed in *E. coli* as a HIS fusion protein. The predicted mass of the recombinant product was ~42 kDa, including the hexa-His-tag. The recombinant RAP-1 protein was purified from the soluble fraction and Coomassie staining shows the purified BdRAP-1 recombinant product is observed at ~42 kDa (data not shown), as predicted from the gene sequence, and was used to immunize rabbits.

BdRAP-1 is expressed as a 46 kDA protein in the erythrocytic parasite

Immune sera from rabbits immunized with the purified rBdRAP-1 protein were used to characterize the native protein in the parasite. Immuno-blotting with native parasite lysate from infected cells showed a distinct single band of 46 kDa (Figure 2A, lane 1), but no products were detected with pre-immune sera (Figure 2A, lane 2). Immunoprecipitation (IP) with lysate from infected cells detected the same single band of 46 kDa, confirming the size of the native BdRAP-1 protein (Figure 2B, lane 1). Many rhoptry proteins undergo proteolytic processing during invasion. To investigate if BdRAP-1 was similarly processed, immunoprecipitations were carried out on spent culture supernatants from in vitro cultures of the parasite. These experiments revealed the 46 kDa native BdRAP-1 (Figure 2B, lane 2) along with a secondary, lower molecular-weight band of ~20 kDa (Figure 2B, lane 2). These analyses suggest that the native BdRAP-1 undergoes proteolytic processing, similar to RAP-1 of *P. falciparum* [28,29].

BdRAP-1 localizes to the rhoptry organelles in the parasite

All the RAP-1 proteins that have been described so far have been localized to the apical, invasive end of the merozoite, and should be true of BdRAP-1 if it also plays a role in parasite entry. The subcellular localization of BdRAP-1 was determined by immunofluorescence assay (IFA) using anti-rBdRAP-1 rabbit antibodies. Figure 3A shows the staining pattern obtained on smears of asexual parasites demonstrating that BdRAP-1 is indeed

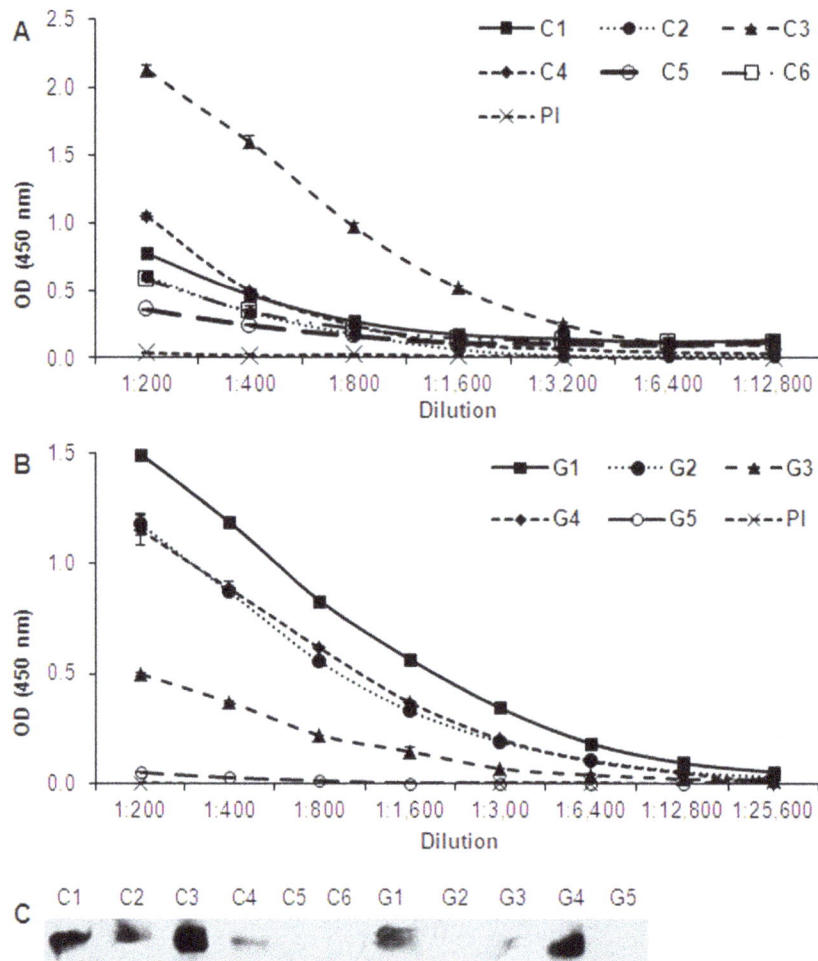

Figure 5. Antigenicity of recombinant BdRAP-1 against sera from *B. divergens* –infected cows and hamsters. A, The reactivity of rBdRAP-1 in ELISA with pre-immune (PI) sera and experimentally *B. divergens*-infected sera from cows (n =6). All infected bovine sera showed higher OD values than non-infected sera in a dilution-dependant manner at all dilutions. B, The reactivity of rBdRAP-1 in ELISA with pre-immune (PI) sera and experimentally *B. divergens*-infected sera from gerbils (n =5). Four of the five gerbil sera showed s higher OD values than non-infected sera in a dilution-dependant manner at dilutions 1:200 to 1:6,400. One gerbil serum (G5) showed no reactivity. At dilutions 1:12,800 and 1:26,600, there is no difference in the reactivity of the gerbil sera and the pre-immune sera. C, Specificity for the animal serum against recombinant BdRAP-1 was confirmed independently by immunoblotting. Four of the six bovine sera and two of the three gerbil sera reacted strongly against rBdRAP-1. One gerbil sera (G3) reacted weakly. Sera which showed the greatest reactivity in ELISA (C3, G1 and G4) also showed greatest reactivity on the blots. Sera G2 and G3 showed moderate reactivity in ELISA but surprisingly very low reactivity in the Western analysis. Sera which did not show any significant reactivity in the ELISA (C5,C6 and G5) also did not react against rBdRAP-1 by immune-blotting.

located at the apical end of merozoites. The punctate staining in this region, wherein a single dot occasionally resolves into double foci, is a hallmark of rhoptry localization. We confirmed this specific localization of BdRAP-1 in rhoptries by immuno-electronmicroscopic analysis (IEM). The IEM was carried out on sections of parasite-infected RBC with anti-rBdRAP-1 antibodies. As can be seen in Figure 3B, discrete antibody reactivity was observed in these electron-dense organelles with the morphological characteristics of rhoptries (indicated by arrows in Figure 3B). Increased magnification showed that BdRAP-1 is localized to the bulb of the rhoptries (Figure 3C), and thus may have an active role in host cell invasion.

BdRAP-1 binds to erythrocytes via a trypsin-, chymotrypsin- and neuraminidase-resistant RBC receptor

Parasite proteins that mediate interaction with host receptors during invasion are commonly localized in the apical organelles.

Having established its location at the invasive apical end of the merozoite and homology with other known *Babesia* RAP-1 proteins, we performed assays to determine whether native BdRAP-1 binds to RBCs. Members of the RAP-1 protein family in *Babesia* are involved during the complex process of invasion of host RBCs through direct binding with receptor(s) on the host-cell surface [1,30]. Identifying both the parasite ligands and their complementary RBC receptor in each part of the invasion process is critical to understanding the invasion biology of *B. divergens* in the human host, and provides continuing information for the development of effective intervention and prevention strategies. Studies have shown that extracellular merozoites release parasite proteins into the culture, and such culture supernatants can be a source of parasite ligands that bind erythrocytes. Thus, [35S]me-thionine/cysteine-labeled spent merozoite supernatants were used as the source of RBC-binding proteins, and when the eluate was immunoprecipitated with anti-rBdRAP-1 antiserum, a dominant band at ~46 kDa was seen (Figure 4, Panel A, lane WT). Thus,

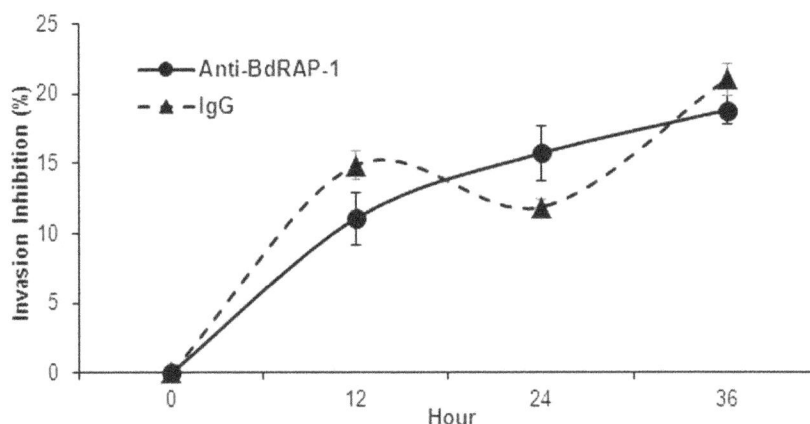

Figure 6. Antibodies against BdRAP-1 do not significantly inhibit parasite invasion *in vitro*. B. divergens cultures in human cells were maintained for 36 h in the presence of 2 mg/mL of the purified IgG fraction of anti-rBdRAP-1 antibodies from rabbit serum or in the presence of the equivalent concentration of purified rabbit pre-immune IgG. The level of growth inhibition compared to a no-serum control was determined every 12 h. The presence of BdRAP-1 (indicated by a solid line) was able to inhibit growth of B. divergens by 11% and 16% at 12 h and 24 h, respectively, to a maximum inhibition of 19% at 36 h. however, this is not significantly different from the inhibition due to the presence of pre-immune IgG (indicated by a dashed line), which inhibited growth by 20% at 36 h, and the lack of inhibition specifically due to anti-BdRAP-1 antibodies suggests multiple alternative pathways of invasion are available to the parasite, or the role of this ligand is not restricted to invasion only.

BdRAP-1 appears to be an adhesin that participates in invasion by binding to the RBC surface. Treatment of erythrocytes with enzymes that selectively cleave moieties of surface proteins, followed by an analysis of the resulting effects on BdRAP-1 binding, afforded a first indication of RBC molecules that could serve as receptor(s) for BdRAP-1 during invasion. RBCs were treated with neuraminidase, trypsin and chymotrypsin, and used for binding assays with radio-labeled *B. divergens* parasite supernatant in several independent experiments to detect the receptor recognized by native BdRAP-1. Figure 4 shows the RBC receptor profile for BdRAP-1. None of the enzyme treatments inhibited or decreased BdRAP-1 binding. Thus, the binding profile of native BdRAP-1 to the RBC initially suggests the participation of a novel red cell receptor in merozoite invasion.

BdRAP-1 binds to erythrocytes via non-proteinacious receptor

To further investigate if the receptor was composed of protein binding assays were carried out on RBCs that were treated with proteinase K and papain, treatments which have been shown to denude the red cell of almost all surface protein. Surprisingly, BdRAP-1 was found to bind these treated RBCs indicating that the RBC moieties involved in binding are not protein and may well be composed of lipid. Future work may elucidate the precise nature of this interaction.

BdRAP-1 appears to be an immuno-dominant antigen recognized by infected bovine and gerbil sera

Sera from six cows and five gerbils infected with *B. divergens* was assayed for reactivity against the rBdRAP-1 using an enzyme-linked immunosorbent assay (ELISA). Positive reactivity was defined as a reading of greater than twofold the values obtained for the negative control samples (pre-immune from the same animals). The highest bovine serum responder, C3, reacted at 37-fold higher, and the lowest responder, C5, reacted at 14-fold higher than the negative control, respectively, at 1:200 dilution. Further, the samples that showed reactivity remained significantly higher than the negative control reactivity at dilutions of 1:1,600–1:12,800 (Figure 5A). A similar pattern of reactivity was observed

when the infected gerbil sera was used to assess antibodies to BdRAP-1, with the highest gerbil serum, G4, showing >700-fold reactivity against the negative control at 1:200 dilution (Figure 5B). One of the gerbil sera (G5) did not react at any dilution, and the reactivity observed for the remaining four gerbil sera were not significantly higher than the negative control at dilutions 1:6,400 to 1:25,600. To confirm that these sera are specifically recognizing BdRAP-1, a confirmatory Western Blot was run using purified BdRAP-1. As can be seen from Figure 5C, the animals that gave the highest reactivity on ELISA assays (C3, G1 and G4), also recognized rBdRAP-1 on the blots. On the whole, sera which did not show any significant reactivity in the ELISA (C5,C6 and G5) also did not react against rBdRAP-1 by immune-blotting. However, sera from gerbils G2 and G3, which showed reactivity in the ELISA did not show reactivity during immunoblotting. This discrepancy may be due to the denaturation of immuno-reactive epitopes during SDS-PAGE. The high level of reactivity of anti-BdRAP-1 antibodies within these sera suggest that the BdRAP-1 protein is highly immunogenic and suggests it could form the basis of a diagnostic tool. This would be particularly useful in transfusion medicine where there is currently no licensed screening assay for the detection of *Babesia* parasites. Recipients of blood products are at greater risk of developing severe babesiosis due to factors such as extremes in age, lack of a spleen, hemoglobinopathies, cancers, HIV, and use of immunosuppressive therapy [31], yet the donor population largely consists of healthy, asymptomatic adults. The high reactivity also suggests that BdRAP-1 might be a potential vaccine candidate.

Anti-BdRAP-1 antibodies do not significantly inhibit invasion *in vitro*

Purified IgG from the rabbit anti-rBdRAP-1 serum was used to evaluate the role of native BdRAP-1 during the invasion process. Cultures were established with a starting parasitaemia of ~4% and purified anti-rBdRAP-1 IgG was added to a final concentration of 2 mg/mL to the cultures at 0 h and refreshed at 24 h to the same concentration. Smears were made and read at 0 h, 12 h, 24 h, and 36 h. Purified rabbit IgG from pre-immune serum was used at the same concentration as a control. The growth inhibition observed in the presence of anti-BdRAP-1 antibodies after 12 h was 11% (Figure 6), and the maximum inhibition observed was 19% after 36 h. However, there was no significant difference in the growth inhibition observed for the pre-immune serum, which showed 14% and 21% inhibition at 12 h and 36 h, respectively. The limited growth inhibition observed suggests the binding of BdRAP-1 to the RBC receptor during invasion is not the primary pathway this species uses to invade host cells, or,, there is significant redundancy in the invasion mechanisms and multiple invasion pathways are available to the parasite. Additionally, the full repertoire of RAP-1 and other RAP-1 related molecules in the parasite genome has not been accounted for, as it has in other *Babesia* species [32–35], and thus, homologs of this antigen may serve in its role when BdRAP-1 is neutralized by antibodies. It seems increasingly likely that RAP-1 in malaria may not directly play a role in the invasive process, but like other proteins housed in the rhoptry bulb, may participate in events downstream of active entry, like enhancing parasite survival in the host cell [36–38]. Thus, binding of BdRAP-1 to RBCs may not reflect an *in vivo* role for the ligand but relate to potential interactions that would occur between RAP-1 and RBC proteins transferred to the parasitophorous vacuole membrane (PVM) during invasion, or maybe the result of non-specific binding of the ligand to lipid membranes, including both the RBC surface and the PVM. Future experiments directed to the dissection of this interaction may help elucidate the true role of these rhoptry proteins.

Conclusions

Although other *Babesia* RAP-1 proteins have been identified [19,26,27,39,40], we present the gene encoding a RAP-1 homologue of a human babesia parasite BdRouen1987, the ~46 kDa *Babesia divergens* RAP-1 protein and have characterized this important antigen. BdRAP-1 is located in the bulb of the rhoptries and binds to an RBC receptor(s) that is resistant to sialidases and various proteinases. We do not find significant reduction in growth with anti-rRAP1 *in vitro*, suggesting the presence of alternative pathways of invasion or alternative roles for this ligand. Identification and assessment of the antigens used for RBC invasion and remodelling of RBC after invasion are essential for understanding the invasion biology of *B. divergens* to determine the most effective interventions.

Author Contributions

Conceived and designed the experiments: MR AA RO JCS JG CAL. Performed the experiments: MR AA RO JCS MS. Analyzed the data: MR AA RO JCS MS JG CAL. Contributed reagents/materials/analysis tools: MR AA RO JCS MS JG CAL. Contributed to the writing of the manuscript: RO JG CAL.

References

1. Yokoyama N, Okamura M, Igarashi I (2006) Erythrocyte invasion by Babesia parasites: current advances in the elucidation of the molecular interactions between the protozoan ligands and host receptors in the invasion stage. Vet Parasitol 138: 22–32.

2. Sibley LD (2004) Intracellular parasite invasion strategies. Science 304: 248–253.

3. Sam-Yellowe TY (1996) Rhoptry organelles of the apicomplexa: Their role in host cell invasion and intracellular survival. Parasitol Today 12: 308–316.

4. Besteiro S, Dubremetz JF, Lebrun M (2011) The moving junction of apicomplexan parasites: a key structure for invasion. Cell Microbiol 13: 797–805.

5. Spielman A, Wilson ML, Levine JF, Piesman J (1985) Ecology of Ixodes dammini-borne human babesiosis and Lyme disease. Annu Rev Entomol 30: 439–460.

6. Lantos PM, Krause PJ (2002) Babesiosis: similar to malaria but different. Pediatr Ann 31: 192–197.

7. Kjemtrup AM, Conrad PA (2000) Human babesiosis: an emerging tick-borne disease. Int J Parasitol 30: 1323–1337.

8. Vannier E, Krause PJ (2009) Update on babesiosis. Interdiscip Perspect Infect Dis 2009: 984568.

9. Gray J, Zintl A, Hildebrandt A, Hunfeld KP, Weiss L (2010) Zoonotic babesiosis: overview of the disease and novel aspects of pathogen identity. Ticks Tick Borne Dis 1: 3–10.

10. Lobo CA, Cursino-Santos JR, Alhassan A, Rodrigues M (2013) Babesia: an emerging infectious threat in transfusion medicine. PLoS Pathog 9: e1003387.

11. Homer MJ, Aguilar-Delfin I, Telford SR, 3rd, Krause PJ, Persing DH (2000) Babesiosis. Clin Microbiol Rev 13: 451–469.

12. Garnham PC (1980) Human babesiosis: European aspects. Trans R Soc Trop Med Hyg 74: 153–155.

13. Zintl A, Mulcahy G, Skerrett HE, Taylor SM, Gray JS (2003) Babesia divergens, a bovine blood parasite of veterinary and zoonotic importance. Clin Microbiol Rev 16: 622–636.

14. Dammin GJ, Spielman A, Benach JL, Piesman J (1981) The rising incidence of clinical Babesia microti infection. Hum Pathol 12: 398–400.

15. Ruebush TK, 2nd, Juranek DD, Chisholm ES, Snow PC, Healy GR, et al. (1977) Human babesiosis on Nantucket Island. Evidence for self-limited and subclinical infections. N Engl J Med 297: 825–827.

16. Kats LM, Black CG, Proellocks NI, Coppel RL (2006) Plasmodium rhoptries: how things went pear-shaped. Trends Parasitol 22: 269–276.

17. Preiser P, Kaviratne M, Khan S, Bannister L, Jarra W (2000) The apical organelles of malaria merozoites: host cell selection, invasion, host immunity and immune evasion. Microbes Infect 2: 1461–1477.

18. Kaneko O (2007) Erythrocyte invasion: vocabulary and grammar of the Plasmodium rhoptry. Parasitol Int 56: 255–262.

19. Skuce PJ, Mallon TR, Taylor SM (1996) Molecular cloning of a putative rhoptry associated protein homologue from Babesia divergens. Mol Biochem Parasitol 77: 99–102.

20. Gorenflot A, Brasseur P, Precigout E, L'Hostis M, Marchand A, et al. (1991) Cytological and immunological responses to Babesia divergens in different hosts: ox, gerbil, man. Parasitol Res 77: 3–12.

21. Lobo CA (2005) Babesia divergens and Plasmodium falciparum use common receptors, glycophorins A and B, to invade the human red blood cell. Infect Immun 73: 649–651.

22. Florin-Christensen M, Suarez CE, Hines SA, Palmer GH, Brown WC, et al. (2002) The Babesia bovis merozoite surface antigen 2 locus contains four tandemly arranged and expressed genes encoding immunologically distinct proteins. Infect Immun 70: 3566–3575.

23. Suarez CE, Florin-Christensen M, Hines SA, Palmer GH, Brown WC, et al. (2000) Characterization of allelic variation in the Babesia bovis merozoite surface antigen 1 (MSA-1) locus and identification of a cross-reactive inhibition-sensitive MSA-1 epitope. Infect Immun 68: 6865–6870.

24. Kania SA, Allred DR, Barbet AF (1995) Babesia bigemina: host factors affecting the invasion of erythrocytes. Exp Parasitol 80: 76–84.

25. Wittner M, Rowin KS, Tanowitz HB, Hobbs JF, Saltzman S, et al. (1982) Successful chemotherapy of transfusion babesiosis. Ann Intern Med 96: 601–604.

26. Dalrymple BP, Casu RE, Peters JM, Dimmock CM, Gale KR, et al. (1993) Characterisation of a family of multi-copy genes encoding rhoptry protein homologues in Babesia bovis, Babesia ovis and Babesia canis. Mol Biochem Parasitol 57: 181–192.

27. Suarez CE, Palmer GH, Jasmer DP, Hines SA, Perryman LE, et al. (1991) Characterization of the gene encoding a 60-kilodalton Babesia bovis merozoite protein with conserved and surface exposed epitopes. Mol Biochem Parasitol 46: 45–52.

28. Howard RF, Reese RT (1990) Plasmodium falciparum: hetero-oligomeric complexes of rhoptry polypeptides. Exp Parasitol 71: 330–342.

29. Howard RF, Narum DL, Blackman M, Thurman J (1998) Analysis of the processing of Plasmodium falciparum rhoptry-associated protein 1 and localization of Pr86 to schizont rhoptries and p67 to free merozoites. Mol Biochem Parasitol 92: 111–122.

30. Yokoyama N, Suthisak B, Hirata H, Matsuo T, Inoue N, et al. (2002) Cellular localization of Babesia bovis merozoite rhoptry-associated protein 1 and its erythrocyte-binding activity. Infect Immun 70: 5822–5826.

31. Gubernot DM, Lucey CT, Lee KC, Conley GB, Holness LG, et al. (2009) Babesia infection through blood transfusions: reports received by the US Food and Drug Administration, 1997–2007. Clin Infect Dis 48: 25–30.

32. Suarez CE, Palmer GH, Florin-Christensen M, Hines SA, Hotzel I, et al. (2003) Organization, transcription, and expression of rhoptry associated protein genes in the Babesia bigemina rap-1 locus. Mol Biochem Parasitol 127: 101–112.

33. Hotzel I, Suarez CE, McElwain TF, Palmer GH (1997) Genetic variation in the dimorphic regions of RAP-1 genes and rap-1 loci of Babesia bigemina. Mol Biochem Parasitol 90: 479–489.

34. Mishra VS, McElwain TF, Dame JB, Stephens EB (1992) Isolation, sequence and differential expression of the p58 gene family of Babesia bigemina. Mol Biochem Parasitol 53: 149–158.

35. Suarez CE, Laughery JM, Bastos RG, Johnson WC, Norimine J, et al. (2011) A novel neutralization sensitive and subdominant RAP-1-related antigen (RRA) is expressed by Babesia bovis merozoites. Parasitology 138: 809–818.

36. Ling IT, Florens L, Dluzewski AR, Kaneko O, Grainger M, et al. (2004) The Plasmodium falciparum clag9 gene encodes a rhoptry protein that is transferred to the host erythrocyte upon invasion. Mol Microbiol 52: 107–118.

37. Nguitragool W, Bokhari AA, Pillai AD, Rayavara K, Sharma P, et al. (2011) Malaria parasite clag3 genes determine channel-mediated nutrient uptake by infected red blood cells. Cell 145: 665–677.

38. Riglar DT, Richard D, Wilson DW, Boyle MJ, Dekiwadia C, et al. (2011) Super-resolution dissection of coordinated events during malaria parasite invasion of the human erythrocyte. Cell Host Microbe 9: 9–20.

39. Machado RZ, McElwain TF, Suarez CE, Hines SA, Palmer GH (1993) Babesia bigemina: isolation and characterization of merozoite rhoptries. Exp Parasitol 77: 315–325.

40. Ikadai H, Xuan X, Igarashi I, Tanaka S, Kanemaru T, et al. (1999) Cloning and expression of a 48-kilodalton Babesia caballi merozoite rhoptry protein and potential use of the recombinant antigen in an enzyme-linked immunosorbent assay. J Clin Microbiol 37: 3475–3480.

16

Evaluation of a gp63–PCR Based Assay as a Molecular Diagnosis Tool in Canine Leishmaniasis in Tunisia

Souheila Guerbouj[1,2], Fattouma Djilani[2], Jihene Bettaieb[3], Bronwen Lambson[4¤a],
Mohamed Fethi Diouani[2¤b], Afif Ben Salah[3], Riadh Ben Ismail[2¤c], Ikram Guizani[1,2]*

1 Laboratory of Molecular Epidemiology and Experimental Pathology, Pasteur Institute of Tunis, Université de Tunis el Manar, Tunis, Tunisia, 2 Laboratory of Epidemiology and Ecology of Parasitic Diseases, Pasteur Institute of Tunis, Tunis, Tunisia, 3 Laboratory of Medical Epidemiology, Pasteur Institute of Tunis, Tunis, Tunisia, 4 Molteno Institute for Parasitology, Department of Pathology, University of Cambridge, Cambridge, United Kingdom

Abstract

A gp63PCR method was evaluated for the detection and characterization of *Leishmania (Leishmania)* (L.) parasites in canine lymph node aspirates. This tool was tested and compared to other PCRs based on the amplification of 18S ribosomal genes, a *L. infantum* specific repetitive sequence and kinetoplastic DNA minicircles, and to classical parasitological (smear examination and/or culture) or serological (IFAT) techniques on a sample of 40 dogs, originating from different *L. infantum* endemic regions in Tunisia. Sensitivity and specificity of all the PCR assays were evaluated on parasitologically confirmed dogs within this sample (N = 18) and control dogs (N = 45) originating from non–endemic countries in northern Europe and Australia. The gp63 PCR had 83.5% sensitivity and 100% specificity, a performance comparable to the kinetoplast PCR assay and better than the other assays. These assays had comparable results when the gels were southern transferred and hybridized with a radioactive probe. As different infection rates were found according to the technique, concordance of the results was estimated by (κ) test. Best concordance values were between the gp63PCR and parasitological methods (74.6%, 95% confidence intervals CI: 58.8–95.4%) or serology IFAT technique (47.4%, 95% CI: 23.5–71.3%). However, taken together Gp63 and Rib assays covered most of the samples found positive making of them a good alternative for determination of infection rates. Potential of the gp63PCR-RFLP assay for analysis of parasite genetic diversity within samples was also evaluated using 5 restriction enzymes. RFLP analysis confirmed assignment of the parasites infecting the dogs to *L. infantum* species and illustrated occurrence of multiple variants in the different endemic foci. Gp63 PCR assay thus constitutes a useful tool in molecular diagnosis of *L. infantum* infections in dogs in Tunisia.

Editor: Jason Mulvenna, Queensland Institute of Medical Research, Australia

Funding: This work received financial support from the EU-DGXII STD3 (CT930253) and INCO-DC (CT970256) programs and from the Ministry of Higher Education and Research in Tunisia (LR00SP04 & LR11IPT04). The funders had no role in study design, data collection and analysis, decision to publish, or preparation of the manuscript.

Competing Interests: The authors have declared that no competing interests exist.

* Email: ikram.guizani@pasteur.rns.tn

¤a Current address: Centre for HIV and STI, National Institute for Communicable Diseases of the National Health Laboratory Service, Johannesburg, South Africa
¤b Current address: Laboratory of Veterinary Epidemiology and Microbiology, Pasteur Institute of Tunis, Tunis, Tunisia
¤c Current address: World Health Organization – Eastern Mediterranean Regional Office (WHO – EMRO), Cairo, Egypt

Introduction

Visceral leishmaniasis due to *Leishmania infantum* is endemic in Mediterranean basin countries, Middle East, Latin America and Asia. Canines are the major reservoir of the infection [1]. Infected dogs present either a range of clinical manifestations of a viscero-cutaneous form or an asymptomatic status. These latter are considered as carriers since they are for sand flies as infectious as the symptomatic ones [2,3]. Canine leishmaniasis (CanL) is a major veterinary and public health problem not only in old endemic foci but also in non endemic areas where outbreaks are occasionally reported, such as in the United States and Canada [4] and in northern Europe [5]. For epidemiological purposes, there is a need for a precise estimation of the number of infected dogs to evaluate the real extent of infection and better elaborate control programs. Several diagnostic techniques are available for detection of canine infection or diagnosis of the disease. These can be achieved either by demonstrating the parasites microscopically in

stained smears [6] or after *in vitro* cultivation [7]. Indirect methods use mainly serological means, like the enzyme-linked immunosorbent assay (ELISA) [8,9], the indirect immunofluorescence assay (IFAT) [10,11] and the direct agglutination test (DAT) [12,13]. However, these diagnostic methods present limitations essentially due to their sensitivity and specificity: parasitological techniques are characteristically insensitive and serological tests are limited by their inability to distinguish between past and present infections and the possibility of cross-reactions with other infectious agents [8,14,15]. With the advent of DNA-based methods and the polymerase chain reaction particularly, more sensitive and rapid detection of parasites has become possible. Although several groups have tested PCR assays in different types of biological samples (fresh, frozen, formalin-fixed or paraffin-embedded biopsies) for the detection of *Leishmania* [9,16,17,18], their values in diagnosis of canine leishmaniasis were partially evaluated.

Here we evaluate a gp63PCR-based technique [19] for molecular diagnosis of *Leishmania* infection in dogs collected from *L. infantum* stable transmission areas in Tunisia, comparing it to classical parasitological and serological techniques and to other molecular assays based on PCR amplification of 18S ribosomal genes, an *L. infantum* specific repetitive sequence and kinetoplast DNA minicircles.

Materials and Methods

Ethics statement

All canine sampling was conducted during routine veterinary care in primary practices. Sampling of Tunisian dogs was conducted during a previous survey study within leishmaniasis endemic regions in Tunisia [20], where (with the exception of 7 dogs that were received at the clinic of the veterinary school for diagnosis) stray dogs were collected during campaigns performed jointly by the Ministries of health and of the Interior, integrating analysis of *Leishmania* prevalence to anti-rabies control programs (stray dog culling). At the time of the study as the veterinarians of our institute took care of these dogs under humane conditions, we did not request ethical consent from the recently installed ethics committee at our institution, to take and use samples of stray dogs that were caught for elimination, or dog samples taken to confirm diagnosis of clinically patent dogs. We used in this last case, remains of the aspirate taken for culture to extract DNAs. Nevertheless, collection of dogs was performed in compliance with the directive 86/609/EEC of the European parliament and the council on the protection of animals used for scientific purposes, in agreement with the guidelines of International Guiding Principles for Biomedical Research Involving Animals.

The second group is composed of 45 control dogs living in regions free from leishmaniasis in northern Europe and Australia (Table S2). Extracted DNAs from samples of these dogs were kindly provided by Dr. David Sargan (University of Cambridge, UK). Sampling of these control dogs was in a range of clinical tests. In all cases the owners gave informed consent that any excess of samples taken during clinical testing could be used in research so long as that excess formed a minority of the sample. This was in accordance with UK Home Office Guidelines. Dogs were not anesthetized. The samples represent, in the case of the Cardigan Welsh Corgis and the Irish setter, DNA from excess blood (<1 ml) after a DNA based test for the rcd3 and rcd1 PRA mutations, respectively. The blood (2–5 ml) was collected as clinical samples into EDTA tubes without anesthesia. This was done by veterinary surgeons in primary practices. In the case of the Irish wolfhound, surplus blood (<1 ml) from clinical collection (usually 2 ml) was taken by a veterinary surgeon for blood ammonia and other tests as part of work up in surveillance for portosystemic shunt. Collections took place at a number of referral clinics. In the cases of the Cocker spaniel and Labrador retriever these were excess (< 1 ml) from cases where EDTA blood, usually 2–5 ml, was collected for routine hematological work up in the clinics of the Queen's Veterinary Hospital, University of Cambridge.

Dogs and samples

Two groups of dogs were studied. The first group corresponded to a total of 40 dogs, from endemic areas of leishmaniasis in Tunisia (Table S1). Lymph node aspirates (∼100 μl) were taken by veterinary surgeons in primary practices and stored at −80°C before DNA extraction. Dogs were not anesthetized and sampling did not make any suffering. This dog group has been previously characterized in our laboratory, using parasitological (Giemsa stained smear examination and in vitro culture), serological (IFAT)

and molecular (PCR) tests (Table S1). Only seven dogs (J1–J7) from this group have acute leishmaniasis that presented with clinical and biological features of the disease to the clinic of Sidi Thabet Veterinary school (Tunisia). The remaining 33 dogs did not have patent leishmaniasis; some of them were however oligosymptomatic. The 18 dogs that were positive either by smear examination or culture inoculation (parasitologically confirmed dogs) constituted the positive control group.

The second group is composed of 45 control dogs living in regions free from leishmaniasis in northern Europe and Australia (Table S2). Extracted DNAs from samples of these dogs were kindly provided by Dr. David Sargan (University of Cambridge, UK). Details about the samples are provided in the ethics statement section.

DNA extraction

Frozen lymph node aspirates were suspended in a lysis solution (50 mM NaCl, 10 mM EDTA, 50 mM Tris–HCl pH 7.4) and incubated overnight at 55°C with 100 μg/ml Proteinase K and 0.05% SDS. Total DNA was phenol/chloroform purified and ethanol precipitated as previously described [21].

PCR amplification

Different PCR tests were applied to dog DNA samples. The first PCR targets the coding region of gp63 genes of *Leishmania*, using specific primers SG1 and SG2 as previously described [19]. The second PCR amplifies a central region of a ribosomal gene encoding for the 18S subunit (RIB PCR) present in all *Leishmania* parasites [22]. The third PCR used in this study targets a repetitive genomic sequence found in *L. infantum* species (INF PCR, Genebank Accession No. L42486.1) [23]. KIN PCR used primers KINF and KINR to amplify minicircles of the kinetoplastic DNA of *Leishmania* [24]. Table 1 summarizes the primers sequences used for the different PCRs and the amplified fragments sizes expected. Reaction and cycling conditions are also presented on Table 1.

Another PCR used in this study amplifies a dog gene (acidic ribosomal phosphoprotein fragment, PO) in order to assess for possible sample degradation prior to analysis or inhibition of amplification. PO primers (Table 1), designed from human, rat and mouse PO gene sequences, cross-react with dog DNA and allow amplification of a 470 bp fragment [25]. Products from the different PCRs were analyzed by agarose gel electrophoresis.

In all PCR reactions, multiple negative controls (no DNA) were included in order to monitor for possible contamination. Furthermore, to avoid contamination of samples during carryover and processing, separated laboratory spaces were used for PCR reaction preparation and for analysis of amplified products (gel preparation and migration). Filter-filled tips were also used to set up the PCR reactions. All results were confirmed by hybridization to specific radio-labeled probes.

Probes and hybridization

All PCR gels were transferred onto Hybond N+ membranes according to the Southern method [21] and hybridized to specific probes at 65°C. RIB, INF and KIN PCRs unique fragments (650 bp, 100 bp and 800 bp, respectively) amplified from a positive control corresponding to an *L. infantum* DNA (MHOM/TN/96/Drep15), were gel-extracted, purified using the Qiaquick gel extraction kit (Qiagen, Paris, France) and used as probes. Whereas, a 2 kb fragment corresponding to the coding region of the *L. infantum* gp63 gene was used for gp63PCRs, as previously described [19]. Probes were labeled with α^{32}P dCTP using the random primer labeling kit (Amersham–HVD, Athens,

Table 1. PCR primers sequences, reaction and cycling conditions used.

PCR code	Target	Forward primer Sequence (5' – 3')	Reverse primer sequence (5' – 3')	Amplified fragment size (bp)	Annealing temperature (°C)	Agarose gel percentage	Reaction conditions (25 µl final volume)	Cycling conditions
Gp63	Gp63 gene family	SG1: GTCTCCACCGAG GACCTCACCGA	SG2: TGATGTAGCC GCCCTCCTCGAAG	1300	65	1,2%	22,5 pmol each primer; 0,2 mM dNTPs; 1 mM MgCl$_2$; 0,5 units Taq DNA polymerase (PerkinElmer, France); 50 ng template DNA; 5% DMSO	94°C: 5 min; 35 cycles of 94°C: 30 sec; annealing: 30 sec; 72°C: 1 min; 72°C: 10 min
RIB	18S ribosomal gene	RIBF: GGTTCCTTTCC TGATTTACG	RIBR: GGCCGGTAA AGGCCGAATAG	650	60	1,6%	22,5 pmol each primer; 0,2 mM dNTPs; 1 mM MgCl$_2$; 0,5 units Taq DNA polymerase (PerkinElmer, France); 50 ng template DNA	94°C: 5 min; 35 cycles of 94°C: 30 sec; annealing: 30 sec; 72°C: 1 min; 72°C: 10 min
INF	Repetitive genomic sequence	INFF: ACGAGGTCAGC TCCACTCC	INFR: CTGCAACG CCTGTGTCTAC	100	59	2%	22,5 pmol each primer; 0,2 mM dNTPs; 1 mM MgCl$_2$; 0,5 units Taq DNA polymerase (PerkinElmer, France); 50 ng template DNA	94°C: 5 min; 35 cycles of 94°C: 30 sec; annealing: 30 sec; 72°C: 1 min; 72°C: 10 min
KIN	Minicircles of the kinetoplastic DNA	KINF: GGGGTTGGTGTAAA ATAGGGCCGG	KINR: CCAGTTT CCCGCCCCGGAG	800	67	1,5%	22,5 pmol each primer; 0,2 mM dNTPs; 1 mM MgCl$_2$; 0,5 units Taq DNA polymerase (PerkinElmer, France); 50 ng template DNA	94°C: 5 min; 35 cycles of 94°C: 30 sec; annealing: 30 sec; 72°C: 1 min; 72°C: 10 min
PO	Acidic ribosomal phosphoprotein	POF: TCATTGTGGGA GCAGACA	POR: GGAGAAG GGGGAGATGTT	470	51	1,5%	20 pmol each primer; 0,2 mM dNTPs; 1,5 mM MgCl$_2$; 1,25 units Taq DNA polymerase (PerkinElmer, France); 20 ng template DNA	94°C: 5 min; 35 cycles of 94°C: 30 sec; annealing: 30 sec; 72°C: 1 min; 72°C: 10 min

Greece) and used to hybridize blots of all the gels. After high stringency washes, labeled hybrid DNA was visualized on X–ray sensitive auto-radiographic films.

Gp63 PCR-RFLP and cluster analysis

Amplified gp63 fragments were digested with *BsiE*I, *Msc*I, *Hinc*II, *BsmB*I and *Sal*I restriction enzymes (Amersham–HVD, Athens, Greece) as previously described [19]. PCR-RFLP profiles were analyzed after overnight electrophoresis in 3% agarose gels and subsequently hybridized to the ^{32}P-labelled gp63 probe. Restriction bands obtained with all the restriction enzymes were scored 1 or 0 for presence or absence of bands, respectively. Genetic distances according to the Nei-modified method were calculated from RFLP data. This data served to construct a dendrogram according to the Kitch method [26], using the PHYLIP package (version 3.69). The Kitch program constructs a tree by successive (agglomerative) clustering, using an average-linkage method of clustering, similar to that used in the UPGMA method. However, this method was chosen as it assumes a molecular clock, allowing the total length of branches from the root to any species to be the same. In addition, this program has options that allow after the tree is constructed to remove and re-add each group, and to try alternative topologies, thus improving the result [26].

Concordance test

The concordance between results of the parasitological, serological or molecular tests was estimated by determining the kappa coefficient (95% CI) using the kappa (k) test of concordance [27]. The Kappa coefficient is interpreted in accordance with Landis and Koch [28] as almost perfect (1.00–0.81), substantial (0.80–0.61), moderate (0.60–0.41), fair (0.40–0.21) and slight (0.20–0.0). The statistical analysis was carried out using the SPSS software package (Version 13.0).

Results

Assessment of DNA quality by PO PCR

The PO primers expected to amplify a 470 bp fragment of a mitochondrial phosphoprotein gene present in mammals [25] were used to evaluate occurrence of inhibition during amplification in dog samples DNAs. All the 40 Tunisian dogs'DNAs generated the expected fragment size of 470 bp (Table 2 and Table S1). Among the control group (N = 45), 3 DNAs were negative (Table 2 and Table S2), indicating PCR inhibition or DNA degradation.

Leishmania DNA PCR amplification from dog samples

The different PCR tests applied to the 85 dog DNA samples of the study showed that 26, 24, 20 and 17 dogs, all from the Tunisian group, were positive using RIB, INF, KIN and gp63 PCRs, respectively (Table 2) with fragments at the expected 650 bp, 100 bp, 800 bp and 1300 bp size, respectively. All control dogs, originating from non–endemic regions for leishmaniasis, did not present any amplification with the different PCR tests (Table 2).

In order to verify the specificity of the PCR products obtained and to assess the possibility of false negative results, all electrophoresis gels were Southern transferred onto a Nylon membrane and amplified products were hybridized to corresponding ^{32}P labeled probes. Results obtained after Ethidium bromide (EtBr) staining and UV observation (before hybridization) and after probe hybridization were compared (Table 2 and Table S1). All bands observed on the gels were confirmed by the probe

Table 2. Results of parasitology, serology and PCR investigations on Tunisian and control dogs.

		Parasitology	Serology	PO PCR[a]	RIB PCR[b]		INF PCR[b]		KIN PCR[b]		gp63 PCR[b]	
					EtBr	^{32}p	EtBr	^{32}p	EtBr	^{32}p	EtBr	^{32}p
Tunisian dogs (N = 40)	Positive results	18/40	26/40	40/40	26/40	35/40	24/40	39/40	20/40	33/40	17/40	17/40
	Sensitivity[c] (%)				55.6 (10/18)	83.5 (15/18)	83.5 (15/18)	100 (18/18)	61.1 (11/18)	83.5 (15/18)	83.5 (15/18)	83.5 (15/18)
	Infection rate[d] (%)	45.0 (18/40)	65.0 (26/40)			87.5 (35/40)		97.5 (39/40)		82.5 (33/40)		42.5 (17/40)
Control dogs (N = 45)	Positive results			42/45	0/45	3/45	0/45	5/45	0/45	0/45	0/45	1/45
	Specificity[e] (%)				100	93.3 (42/45)	100	88.9 (40/45)	100	100	100	97.8 (44/45)

[a]PO PCR targets a mammalian mitochondrial phosphoprotein gene.
[b]RIB, INF, KIN and gp63 PCRs target a central region of 18S ribosomal gene, a repetitive genomic sequence, minicircles of the kinetoplastic DNA and gp63 family coding sequences, respectively in *Leishmania*.
[c]sensitivity of the different PCR assays corresponds to the proportion of positive dogs among the 18 parasitologically confirmed ones.
[d]infection rates are the proportion of positive dogs among the total dog number.
[e]specificity of the different PCR assays corresponds to the number of negative dogs among the control dog group.
Abbreviations: EtBr, Ethidium bromide staining and reading under UV light; ^{32}P, autoradiographic reading after hybridization with a ^{32}P labeled probe.

hybridizations. This step also increased the number of positive Tunisian dogs to 35, 39 and 33 with RIB, INF and KIN PCRs, respectively while the number of gp63 PCR positive dogs did not change after hybridization (Table 2).

Sensitivity and Specificity of PCR tests

Sensitivity of the different PCR tests was measured as the proportion of positive dogs among the positive control group of 18 parasitologically confirmed dogs (Table 2). RIB, INF, KIN and gp63 PCRs had a sensitivity of 55.6% (10/18), 83.5% (15/18), 61.1% (11/18) and 83.5% (15/18), respectively. After ^{32}P labeled probe hybridization, sensitivity changed to 83.5% (15/18), 100% and 83.5% (15/18) for RIB, INF, and KIN PCRs, respectively but it remained the same for gp63PCR (Table 2).

Specificity of the PCR tests was estimated as the proportion of the negative control dogs (originating from non–endemic regions for leishmaniasis) that were negative in the assays. 100% of specificity was achieved by all PCR tests after analysis of EtBr stained gels, while 93.3% (42/45), 88.9% (40/45) and 97.8% (44/45) were found for RIB, INF and gp63 PCRs, respectively, after autoradiography analysis (Table 2 and Table S2). This decrease was due to the presence of positive signals after hybridization in the case of several dogs (9/45) (Table S2). No change was observed for KIN PCR, after hybridization (Table 2).

Comparative evaluation of parasitological, serological and molecular tools for diagnosis of canine leishmaniasis

Infection rate within our sample was estimated using the parasitological, serological and molecular techniques. Parasitological tests using direct examination of amastigotes within biopsies (stained smears) and in vitro isolation in culture media of promastigotes indicated a 45% (18/40) infection rate (Table 2). Using IFAT, the infection rate was 65% (26/40) (Table 2). With the PCR assays, considering the dogs were infected when positive signals were observed before or after hybridization, the infection rates reached 87.5% (35/40), 97.5% (39/40) and 82.5% (33/40) for RIB, INF and KIN PCRs, respectively while with the gp63PCR it was 42.5% (17/40) (Table 2). Given the differences in measures of infection rates, concordance of the results was investigated computing the proportion of identical results found by different tools and comparing them in a pair wise way (Table 3). The best concordance kappa (κ) values were found between the gp63PCR test and parasitological (74.6%, 95% confidence interval (CI): 0.588, 0.954) or serological IFAT (47.4%, 95%CI: 0.235, 0.713) methods (Table 3), with a substantial and moderate agreement between these tools, respectively. Concordance between the RIB, INF and KIN PCRs and parasitological or serological methods showed negative or close to zero (−0.304 to 0.041) kappa values, indicating a disagreement (Table 3). In addition, when the PCR tools were pair-wise compared, fair concordance kappa values were found between KIN and RIB (22%, 95%CI: −0.158, 0.598) and between KIN and INF (21.6%, 95%CI: −0.143, 0.575) PCRs (Table 3).

Species identification and analysis of intra-specific parasite polymorphism by gp63 PCR-RFLP

The gp63PCR products obtained for 15 parasitological positive dogs and representative strains of L. infantum, L. donovani, L. archibaldi, L. major, L. tropica and L. aethiopica species were purified from the gels, digested with BsiEI, SalI, MscI, BsmBI and HincII restriction enzymes and analyzed for restriction length polymorphisms by electrophoresis and southern blot analysis using a ^{32}P labeled gp63 probe, as previously described [19]. Restriction

Table 3. Pair wise concordance values calculated using the kappa coefficient for parasitology, serology and PCR investigations.

		Parasitology		Serology (IFAT)		PCRs					
						RIB		INF		KIN	
		Kappa	95% CI*	Kappa	95% CI*	Kappa	95% CI*	Kappa	95% CI*	Kappa	95% CI*
Serology (IFAT)		0.515	0.274, 0.756								
PCRs	RIB	−0.070	−0.264, 0.124	−0.226	−0.387, −0.065						
	INF	0.041	−0.039, 0.121	−0.049	−0.141, 0.043	−0.043	−0.116, 0.029				
	KIN	0.014	−0.205, 0.233	−0.304	−0.473, −0.135	0.220	−0.158, 0.598	0.216	−0.143, 0.575		
	gp63	0.746	0.588, 0.954	0.474	0.235, 0.713	0.011	−0.169, 0.191	0.037	−0.035, 0.109	0.089	−0.117, 0.295

*CI, Confidence interval.
RIB, INF, KIN and gp63 PCRs target a central region of 18S ribosomal gene, a repetitive genomic sequence, minicercles of the kinetoplastic DNA and gp63 family coding sequences, respectively in Leishmania.

A

B

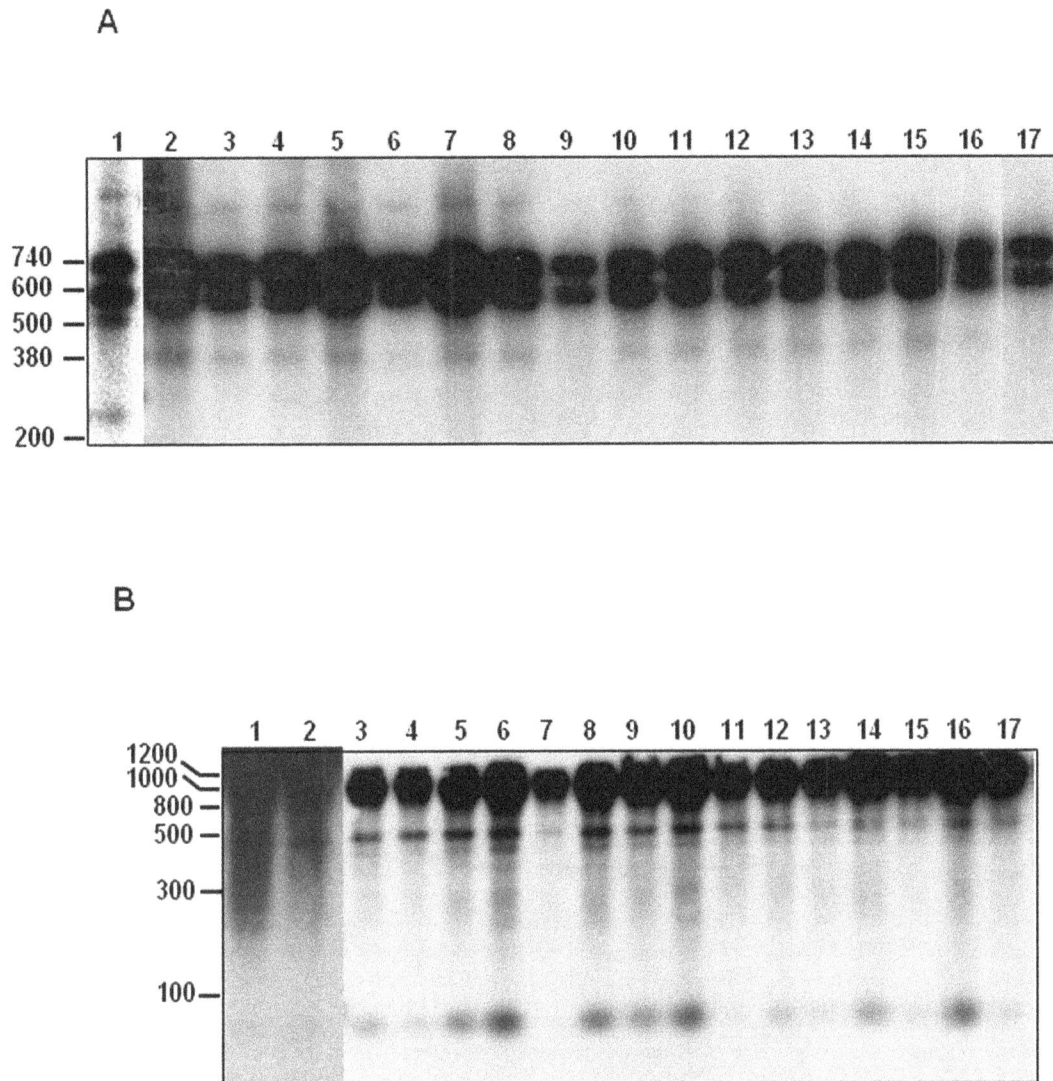

Figure 1. Gp63PCR-RFLP patterns of *Leishmania* **parasites obtained from dog biopsies.** A: digestion with *Msc*I restriction enzyme; B: digestion with *Sal*I restriction enzyme. 1: *L. donovani*, 2: *L. infantum*, 3: LN112, 4: LN129, 5: LN26, 6: LN11, 7: LN80, 8: LN2, 9: LN39, 10: LN77, 11: LN102, 12: LN110, 13: J1, 14: J3, 15: J5, 16: J6, 17: J7. All sizes are indicated in bp.

profiles were polymorphic but species- specific fragments like the presence of an *L. infantum* specific 380 bp *Msc*I fragment and the absence of a 500 bp and a 220 bp *L. donovani* specific *Msc*I fragments [19], allowed identification of the dog *Leishmania* parasites as members of the *L. donovani* complex, more precisely belonging to the *L. infantum* species (Figure 1). In order to better illustrate diversity and phenetic relationships of the amastigotes infecting the studied dogs, the restriction profiles were used to calculate Nei-modified distances and the generated data matrix was then used to construct a dendrogram (Kitch-Margoliash, Phylip package). All dog parasites clustered together with the *L. infantum* reference strain, distinctly from *L. donovani* and *L. archibaldi* representative strains (Figure 2). Whereas, strains representing other Old World species, *L. major*, *L. tropica* and *L. aethiopica* were individualized on separate branches (Figure 2). However, within the dogs' clade, small clusters were observed that were not correlated to epidemiological features like geographical origin, sex or age of dogs. This however highlights occurrence of multiple parasite variants (variability index $= 0.47$ $(7/15)$)

characterized by gp63 genes coding for surface antigens having variable, either exposed or buried residues. Three of these variants were shared by 11 parasites (Figure 2). Of relevance to molecular tracking of parasites, different variants were observed in the same transmission area while a same variant was observed in different endemic regions. Moreover, the study here brings information on 2 parasites (LN36 and LN100) that could not be isolated and maintained by *in vitro* culture (Table S1), which illustrates an additional value of the gp63 PCR based assays for molecular diagnosis of canine leishmaniasis.

Discussion

Variable clinical and biological manifestations characterize *Leishmania* canine infection and leishmaniasis. Its diagnosis still constitutes a major epidemiological problem. Within Tunisian endemic foci, for instance, 50% to 90% of infected dogs are asymptomatic [29,30], which further show the necessity to use sensitive diagnostic techniques. In spite of their limits, parasito-

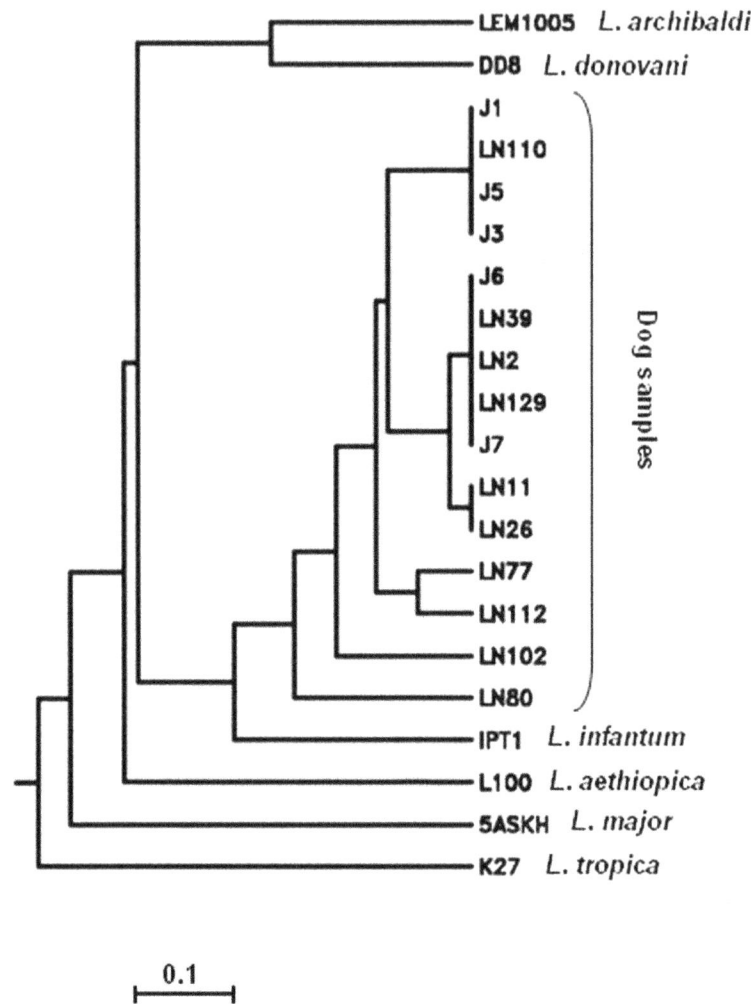

Figure 2. Kitch dendrogram constructed using Nei-modified distances calculated from gp63PCR-RFLP results obtained with lymph node biopsies of dogs from Tunisia. Branches corresponding to Old World representative *Leishmania* strains are indicated.

logical and serological techniques are still the most common methods used to diagnose canine leishmaniasis and are considered as gold standards [31,32]. With the advent of molecular tools like PCR, a more sensitive and rapid detection of parasites has become possible. The potential of a gp63 amplification–based tool in molecular diagnosis of canine leishmaniasis was here evaluated. Sensitivity and specificity of this gp63PCR were estimated and compared to parasitological (direct examination and *in vitro* culture), and serological (IFAT) techniques, as well as to other PCR assays. DNAs purified from lymph node (LN) biopsies, taken from 40 Tunisian dogs were PCR amplified in 5 assays targeting, the intra-genic regions of gp63 genes (gp63 PCR), a central region of 18S encoding ribosomal gene (RIB PCR), an *L. infantum* specific repetitive sequence (INF PCR), minicircles of kinetoplast DNA (KIN PCR) and a mammalian phosphorotein gene (PO PCR). This latter PCR was used to verify the absence of inhibitors within the DNA preparations. All PCR products were hybridized to their respective ^{32}P labeled probe, which allowed (i) to validate specificity of the obtained signal and (ii) to check for absence of false negative results. An 83.5% sensitivity was calculated for the gp63PCR tool while in controlled laboratory conditions using purified DNA the test could detect 0.01 pg of DNA, which correspond to 0.1 promastigote [19]. This could be explained by

differences at the level of amplification efficacy, in relation with parasite burden, presence of host material (including inhibitors) or reaction conditions. Sensitivity of PCR tests was shown to vary outstandingly when applied either on cultured parasites or on peripheral blood samples of infected dogs [18,33]. Specificity of gp63PCR was estimated on a group of control dogs, constituted by non-infected dogs originating from countries where leishmaniasis is not endemic. Consequently, 100% of specificity was found after EtBr reading. This percentage decreased to a value of 97.8% after hybridization with a ^{32}P labeled gp63 probe. Using the same dog sample, 100% of specificity was achieved with EtBr reading for PCRs amplifying genomic sequences (RIB and INF PCRs) while a decrease of specificity (93% for RIB PCR and 89% for INF PCR) was noticed with ^{32}P reading. Previous studies that used the INF PCR showed a specificity of 97% when this tool was applied to detect *Leishmania* DNA from patients presenting with visceral leishmaniasis [23]. Here, no certified explanation for the positive (and very faint) signals observed could be provided, travel history of the dogs could be a reason; non–specific amplification or cross-reactivity with other microbes could be other reasons. Nevertheless, it is important to notice that only a few studies have used hybridization with ^{32}P labeled probes consecutively after EtBr detection of PCR signals.

Parasitological, serological and molecular techniques, in addition to gp63PCR tool, are methods that showed different rates of infection when tested on our study sample. In other studies that amplified the conserved region of *Leishmania* kDNA minicircles, a prevalence of 67% was found in Spain [16], 24.7% is detected in dogs from an urban area in Brazil [34] while 51.88% was found in western China [35] and 29% in Greece [36]. PCRs using ribosomal genes detected 79.8% of dog infection in southern France [37] and 11.5% in the West Bank, Palestine [38]. However, 58.1% was found in Greece with INF PCR targeting a specific fragment of a repetitive sequence of *L. infantum* DNA [36]. Disparity of the measured rates of infection could be explained by (i) epidemiological differences, (ii) biases introduced when selecting negative dogs or (iii) by efficiency in amplification of different size fragments. Consequently, measure of concordance between results of different tools would allow a better evaluation. Thus, κ concordance values between the different tools showed a substantial and moderate agreement between gp63PCR and parasitological methods (74.6%) and serology IFAT technique (47.4%), respectively, which was higher than that achieved with the other PCR tools. This underscores need for using several methods to diagnose canine infection. Previous studies have in fact shown that a canine infection could be under-estimated when only one technique is used [31,32]. However, if we consider the appropriate situations where these different techniques could be used, several comments could be advanced. Indeed, although its weak sensitivity, parasitological diagnosis remains a method of choice in individual cases like veterinary consultation, where dogs having patent disease are specifically recruited and for which this kind of examination is sensitive [39]. *In vitro* culture, in these cases, constitutes a limitation, since up to several weeks could be necessary before a result could be advanced. Besides, it is a laborious diagnostic technique mostly performed in research laboratories. It is precisely in these cases that PCR tools are the most useful, since they allow the detection of *Leishmania* DNA with a high sensitivity and specificity, thus providing a rapid diagnosis [32]. Concerning serological methods, these are based on the presence of IgG antibodies within dogs' sera. However, a positive result may indicate an exposure to the parasite but not necessarily an active infection [2,3]. Moreover, these techniques are not able to reveal the real prevalence, nor the transmission intensity of the infection, since it is difficult to differentiate between an active and a non-active state of the infection [3,7,32]. Another challenge in comparative evaluation of tools is the selection of negative control dogs in countries endemic for leishmaniasis. Thus, using simultaneously several diagnostic methods, including PCR, seems to be necessary for canine leishmaniasis diagnosis. In this context, use of gp63PCR (EtBr reading) associated with another PCR tool like RIB PCR, which was most discordant, will allow maximizing possibilities to find positive responses. Gp63

PCR in addition to RIB PCR could constitute a good diagnostic procedure in canine leishmaniasis that would complement parasitological and IFAT methods or constitute alternatives to detect infection. Moreover, a gp63 PCR positive signal was found in the case of 2 dogs, from which *Leishmania* parasites could not be isolated in culture. This further emphasized the interesting potential of this gp63PCR as a molecular tool, able of detecting and studying parasites, bypassing *in vitro* culture steps. On the other hand, the gp63 assay did not detect parasites of cases that were positive by the classical techniques. Presence of specific inhibitors to this assay or parasites having polymorphic priming sites may be causes affecting PCR efficiency. PO primers amplified all the Tunisian dog samples inferring that other causes than PCR inhibition may explain this sensitivity default.

Intra-specific polymorphism of parasites within infected dogs' lymph nodes was evaluated by RFLP analysis of the amplified gp63PCR fragments by several restriction enzymes [19] followed by phenetic analysis using Nei-modified distances. The dendrogram confirmed the identity of the studied dog parasites as belonging to the *L. donovani* complex and more precisely to the *L. infantum* species. Polymorphic PCR-RFLP profiles highlighted their genetic variability constituting in some cases small groups that clustered together on the dendrogram. However this variability did not correlate with epidemiological features like geographical origin, sex or age of dogs. Gp63 PCR-RFLP highlighted geographical structuring of *L. infantum* and *L. donovani* parasites and importance of host selection pressures were hypothesized [19]. Thus, it appears important to develop further studies comparing parasites having diverse host origins. Nevertheless, this gp63PCR constitutes an innovative approach that allows study of *L. infantum* variability within the reservoir to assess occurrence of parasite variants in a concomitant way to their detection.

Supporting Information

Table S1 Panel of dogs collected from leishmaniasis endemic regions in Tunisia and results of parasitology, serology and PCR investigations.

Table S2 Panel of control dogs collected from non-endemic countries for leishmaniasis and PCR results.

Author Contributions

Conceived and designed the experiments: SG BL IG RBI. Performed the experiments: SG FD. Analyzed the data: SG JB IG FD RBI. Contributed reagents/materials/analysis tools: SG IG FD JB BL MFD ABS RBI. Wrote the paper: SG IG.

References

1. jeux P (2001) The increase in risk factors for leishmaniasis worldwide. Trans R Soc Trop Med Hyg 95: 239–243.
2. ina R, Amela C, Nieto J, San Andres M, Gonzalez F, et al. (1994) Infectivity of dogs naturally infected with *Leishmania infantum* to colonized *Phlebotomus perniciosus*. Trans Roy Soc Trop Med Hyg 88: 491–493.
3. halsky EM, Rocha MF, Da Rocha Lima AC, França-Silva JC, Pires MQ, et al. (2007) Infectivity of seropositive dogs, showing different clinical forms of leishmaniasis, to *Lutzomyia longipalpis* phlebotomine sand flies. Vet Parasitol 147: 67–76.
4. Schantz PM, Steurer FJ, Duprey ZH, Kurpel KP, Barr SC, et al. (2005) Autochthonous visceral leishmaniasis in dogs in North America. J Am Vet Med Assoc 226: 1316–1322.
5. Gramiccia M, Gradoni L (2007) The leishmaniases of Southern Europe. In: Takken W, Knols W, Bart GJ, editors.Emerging pests and vector-borne diseases

in Europe. Ecology and control of vector-borne diseases, vol. 1. Wageningen Academic Publishers, Wageningen. pp. 75–95.
6. Giudice, Passantino A (2011) Detection of eishmania mastigotes in peripheral blood from four dogs. Acta Vet Hung 59: 205–213.
7. Gramiccia M (2011) Recent advances in leishmaniasis in pet animals: Epidemiology, diagnostics and anti-vectorial prophylaxis. Vet Parasitol 181: 23–30.
8. Porrozzi R, Santos da Costa MV, Teva A, Falqueto A, Ferreira AL, et al. (2007) Comparative evaluation of enzyme-linked immunosorbent assays based on crude and recombinant leishmanial antigens for serodiagnosis of symptomatic and asymptomatic *Leishmania infantum* visceral infections in dogs. Clin Vaccine Immunol 14: 544–548.
9. Gomes Y, Paiva M, Cavalcanti M, Lira RA, Abath FG, et al. (2008) Diagnosis of canine visceral leishmaniasis: biotechnological advances. Vet J 175: 45–52.

10. Ferroglio E, Centaro E, Mignone W, Trisciuoglio A (2007) Evaluation of an ELISA rapid device for the serological diagnosis of *Leishmania infantum* infection in dog as compared with immunofluorescence assay and Western blot. Vet Parasitol 144: 162–166.

11. Maia C, Campino L (2008) Methods for diagnosis of canine leishmaniasis and immune response to infection. Vet Parasitol 158: 274–287.

12. Oskam L, Slappendel RJ, Beijer EG, Kroon NC, van Ingen CW, et al (1996) Dog–DAT: a direct agglutination test using stabilized, freeze–dried antigen for the serodiagnosis of canine visceral leishmaniasis. FEMS Immunol Med Microbiol 16: 235–239.

13. Ferreira Ede C, de Lana M, Carneiro M, Reis AB, Paes DV, et al. (2007) Comparison of serological assays for the diagnosis of canine visceral leishmaniasis in animals presenting different clinical manifestations. Vet Parasitol 146: 235–241.

14. Alvar J, Cañavate C, Molina R, Moreno J, Nieto J (2004) Canine leishmaniasis. Adv Parasitol 57: 1–88.

15. Srividya G, Kulshrestha A, Singh R, Salotra P (2012) Diagnosis of visceral leishmaniasis: developments over the last decade. Parasitol Res 110: 1065–1078.

16. Solano-Gallego L, Morell P, Arboix M, Alberola J, Ferrer L (2001) Prevalence of *Leishmania infantum* infection in dogs living in an area of canine leishmaniasis endemicity using PCR on several tissues and serology. J Clin Microbiol 39: 560–563.

17. Schönian G, Nasereddin A, Dinse N, Schweynoch C, Schallig HD, et al. (2003) PCR diagnosis and characterization of *Leishmania* in local and imported clinical samples. Diagn Microbiol Infect Dis 47: 349–358.

18. Solcà Mda S, Guedes CE, Nascimento EG, Oliveira GG, dos Santos WL, et al. (2012) Qualitative and quantitative polymerase chain reaction (PCR) for detection of *Leishmania* in spleen samples from naturally infected dogs. Vet Parasitol 184: 133–140.

19. Guerbouj S, Victoir K, Guizani I, Seridi N, Nuwayri-Salti N, et al. (2001) Gp63 gene polymorphism and population structure of *Leishmania donovani* complex: influence of the host selection pressure? Parasitology 122: 25–35.

20. Boelaert M, Aoun K, Liinev J, Goetghebeur E, Van der Stuyft P (1999) The potential of latent class analysis in diagnostic test validation for canine *Leishmania infantum* infection. Epidemiol Infect 123: 499–506.

21. Guizani I, Van Eys GJJM, Ben Ismail R, Dellagi K (1994) Use of recombinant DNA probes for species identification of Old World *Leishmania* isolates. Am J Trop Med Hyg 50: 632–640.

22. Van Eys GJJM, Schoone GJ, Kroon NCM, Ebeling SB (1992) Sequence analysis of small subunit ribosomal RNA genes and its use for detection and identification of *Leishmania* parasites. Mol Biochem Parasitol 51: 133–142.

23. Piarroux R, Azaiez R, Lossi AM, Reynier P, Muscatelli F, et al. (1993) Isolation and characterization of a repetitive DNA sequence from *Leishmania infantum*: development of a visceral leishmaniasis polymerase chain reaction. Am J Trop Med Hyg 49: 364–369.

24. Smyth AJ, Ghosh A, Hassan MQ, Basu D, De Bruijn MHL, et al. (1992) Rapid and sensitive detection of *Leishmania* kinetoplast DNA from spleen and blood samples of kala-azar patients. Parasitology 105: 183–192.

25. Ashford DA, Bozza M, Freire M, Miranda JC, Sherlock I, et al. (1995) Comparison of the polymerase chain reaction and serology for the detection of canine visceral leishmaniasis. Am J Trop Med Hyg 53: 251–255.

26. Felsenstein J (1984) Distance methods for inferring phylogenies: a justification. Evolution 38: 16–24.

27. Sim JJ, Wright CC (2005) The Kappa Statistic in Reliability Studies: Use, Interpretation, and Sample Size Requirements. Phys Therapy 85: 257–268.

28. Landis JR, Koch GG (1977) The measurement of observer agreement for categorical data. Biometrics 33: 159–174.

29. Chargui N, aouas N, orcii M, ahmar S, uesmi M, et al (2009) Use of PCR, IFAT and in vitro culture in the detection of Leishmania infantum infection in ogs nd evaluation of the prevalence of anine leishmaniasis n a low endemic area in unisia. Parasite 16: 65–69.

30. Diouani MF, en Alaya Bouafif N, ettaib J, ouzir H, edidi S, et al. (2008) Dogs. infantum infection from an endemic region of the north of unisia: a prospective study. Arch Inst Pasteur Tunis 85: 55–61.

31. Miró G, Cardoso L, Pennisi MG, Oliva G, Baneth G (2008) Canine leishmaniosis – new concepts and insights on an expanding zoonosis: part two. Trends Parasitol 24: 371–377.

32. Solano-Gallego L, Miró G, Koutinas A, Cardoso L, Pennisi MG, et al. (2011) LeishVet guidelines for the practical management of canine leishmaniosis. Parasites & Vectors 4: 86–101.

33. Lachaud L, Marchergui-Hammami S, Chabbert E, Dereure J, Dedet JP, et al. (2002) Comparison of six PCR methods using peripheral blood for detection of canine visceral leishmaniasis. J Clin Microbiol 40: 210–215.

34. Coura-Vital W, Marques MJ, Veloso VM, Roatt BM, Aguiar-Soares RD, et al. (2011) Prevalence and factors associated with *Leishmania infantum* infection of dogs from an urban area of Brazil as identified by molecular methods. PLoS Negl Trop Dis 5: e1291.

35. Wang JY, Ha Y, Gao CH, ang, Yang YT, et al. (2011) The prevalence of canine *eishmania nfantum* infection in western China detected by PCR and serological tests. Parasites & Vectors. 4: 69–76.

36. Andreadou, Liandris E, Kasampalidis IN, Taka S, Antoniou M, et al. (2012) Evaluation of the performance of selected in-house and commercially available PCR and real-time PCR assays for the detection of *Leishmania* DNA in canine clinical samples. Exp Parasitol 131: 419–424.

37. Lachaud L, habbert E, ubessay P, ereure J, amothe J, t al. (2002) Value of two PCR methods for the diagnosis of canine visceral leishmaniasis and the detection of asymptomatic carriers. Parasitology 125: 197–207.

38. Hamarsheh, Nasereddin A, Damaj S, Sawalha S, Al-Jawabreh H, et al. (2012) Serological and molecular survey of *Leishmania* parasites in apparently healthy dogs in the West Bank, Palestine. Parasites & Vectors 5: 183–190.

39. Oliva G, Scalone A, Foglia Manzillo V, Gramiccia M, Pagano A, et al. (2006) Incidence and time course of *eishmania nfantum* infections examined by parasitological, serologic, and nested–PCR techniques in a cohort of naive dogs exposed to three consecutive transmission seasons. J Clin Microbiol 44: 1318–1322.

17

P. berghei Telomerase Subunit TERT is Essential for Parasite Survival

Agnieszka A. Religa[1], Jai Ramesar[2], Chris J. Janse[2], Artur Scherf[3], Andrew P. Waters[1]*

1 Wellcome Trust Centre for Molecular Parasitology, Institute of Infection, Immunity and Inflammation, University of Glasgow, Glasgow, United Kingdom, **2** Leiden Malaria Research Group, Parasitology, Leiden University Medical Centre, Leiden, the Netherlands, **3** Biology of Host-Parasite Interactions Unit, Institut Pasteur, Paris, France

Abstract

Telomeres define the ends of chromosomes protecting eukaryotic cells from chromosome instability and eventual cell death. The complex regulation of telomeres involves various proteins including telomerase, which is a specialized ribonucleoprotein responsible for telomere maintenance. Telomeres of chromosomes of malaria parasites are kept at a constant length during blood stage proliferation. The 7-bp telomere repeat sequence is universal across different *Plasmodium* species (GGGTTT/CA), though the average telomere length varies. The catalytic subunit of telomerase, telomerase reverse transcriptase (TERT), is present in all sequenced *Plasmodium* species and is approximately three times larger than other eukaryotic TERTs. The *Plasmodium* RNA component of TERT has recently been identified *in silico*. A strategy to delete the gene encoding TERT via double cross-over (DXO) homologous recombination was undertaken to study the telomerase function in *P. berghei*. Expression of both TERT and the RNA component (TR) in *P. berghei* blood stages was analysed by Western blotting and Northern analysis. Average telomere length was measured in several *Plasmodium* species using Telomere Restriction Fragment (TRF) analysis. TERT and TR were detected in blood stages and an average telomere length of ~950 bp established. Deletion of the *tert* gene was performed using standard transfection methodologies and we show the presence of *tert⁻* mutants in the transfected parasite populations. Cloning of *tert*-mutants has been attempted multiple times without success. Thorough analysis of the transfected parasite populations and the parasite obtained from extensive parasite cloning from these populations provide evidence for a so called delayed death phenotype as observed in different organisms lacking TERT. The findings indicate that TERT is essential for *P. berghei* cell survival. The study extends our current knowledge on telomere biology in malaria parasites and validates further investigations to identify telomerase inhibitors to induce parasite cell death.

Editor: Georges Snounou, Université Pierre et Marie Curie, France

Funding: This work was also supported by Evimalar (Ref: 242095) a grant of the European Commission under FP7. APW is a Wellcome Trust Principal Research Fellow (REF: 083811/Z/07/Z); AAR was a MalParTraining PhD fellowship holder within FP6 Marie Curie Action under contract No. MEST-CT-2005-020492. The funders had no role in study design, data collection and analysis, decision to publish, or preparation of the manuscript.

Competing Interests: The authors have declared that no competing interests exist.

* Email: Andy.Waters@glasgow.ac.uk

Background

Telomerase is an RNA-dependent DNA polymerase complex functioning in extension and/or maintenance of telomeres [1] which are composed of a specialised conserved G-rich short (5-6 bp) tandem repeats [2,3]. Telomeres are essential for maintenance of eukaryote chromosome integrity and stability since the telomeric DNA repeats and associated proteins prevent chromosome end-to-end fusions and exonuclease degradation (for a review see [4]). The length of telomeres is a critical determinant of a cell's replicative life span [5,6] and telomere shortening has been linked to cell senescence, disease and ageing [5,7–10]. Telomeres shorten with each cell division due to the so called end-replication problem [11], resulting in loss of telomeric repeats. The loss of telomeric repeats is compensated for by synthesis of new repeats. The enzyme responsible for synthesis of new repeats is the telomerase holoenzyme, which is an RNA-dependent DNA polymerase complex. This enzyme synthesises new tandem telomeric repeats *de novo* at the 3' chromosome strand end [12–

17]. Telomerase-negative cells (e.g. human somatic cells) experience telomere shortening and lose on average between 30 and 200 bp of telomeric sequence per cell division, a loss which is ultimately lethal [18–20].

Telomerase consists of several subunits. The core subunit consists of Telomerase Reverse Transcriptase (TERT) together with its conserved RNA component (Telomerase RNA; TR), which acts as the template for the synthesis of telomeric repeats [6]. Telomere structure is dependent on multiple telomere-associated proteins (TAPs) and these proteins together with telomeric DNA form the so-called telosome. TAPs are part of the telomerase-mediated telomere maintenance and regulation mechanisms, including telomere loop formation [21]. The gene encoding TERT has been deleted or mutated in a number of organisms which led to cell senescence and eventual cell death ("delayed death") (e.g. [5,6,22]).

The genome of malaria parasites is arranged into 14 linear chromosomes which contain telomeres consisting of 7-bp telomeric repeat sequences (GGGTTT/CA) [23,24]. The average telomere length per species varies, ranging from 850 bp (estimated ~120 repeat copies) to 6700 bp (estimated ~955 repeat copies) in the human parasites *P. falciparum* and *P. vivax*, respectively [25]. *Plasmodium* telomere length appears to be kept constant during the erythrocytic cycle [26] The genomes of different *Plasmodium* parasites contain a single copy *tert* gene. For TERT of *P. falciparum* (gene ID PF3D7_1314200) it has been demonstrated that it is capable of *de novo* synthesis of telomeric repeats both to the 3′ telomeric overhang and to non-telomeric 3′ ends, thus contributing not only to telomere maintenance but also to adding new telomeric sequences to broken chromosomes [27,28]. Telomerase activity in *P. falciparum* is detectable both in late stage trophozoites and schizonts, stages where DNA replication occurs [27,29]. The RNA component of telomerase (encoded by PF3D7_0918500 in *P. falciparum*; 2148 nt) has been identified *in silico* in several *Plasmodium* species based on structural comparisons of conserved domains in TR domains from other organisms [30]. The telomerase RNA is detectable in all erythrocytic stages and the ookinete stage of *P. falciparum* according to the PlasmoDB expression data (www.PlasmoDB.org).

In order to analyse the importance of telomere metabolism/dynamics for asexual blood stage multiplication of *Plasmodium* parasites we have attempted to generate mutants that lack expression of TERT. We have used the rodent malaria parasite *P. berghei* because of the high efficiency of transfection and rapid selection of gene-deletion mutants, which might be essential when TERT-deficient parasites show a delayed death phenotype as has been shown for other organisms. We found that we can target the *P. berghei tert* gene for gene deletion. However, our results also indicate that TERT is an essential enzyme for survival of *P. berghei* blood stages since we were unable to clone and propagate TERT-deficient parasites.

Results

Expression of PbTERT and PbTR in blood stages

The *P. falciparum tert* gene sequence (PF3D7_1314200; 7557 bp) [29] was used to search the available *P. berghei* genome sequence. Two adjacent incomplete gene models have been identified in the genome of *P. berghei* (www.GeneDB.org) that encode *tert* genes: PBANKA_141260 (5691 bp) and PBANKA_141270 (1305 bp), that both share homology with the single copy *tert* genes of *P. falciparum* and *P. chabaudi* (PCHAS_141450; 6753 bp) (Fig. 1A). The *P. chabaudi tert* gene was subsequently used as a reference for assembling the complete *Pbtert* gene. Amplification of the predicted gap between the two *Pbtert* gene models using primers specific to the adjacent extremities of both gene models revealed a sequence duplication of 57 nt, which may explain the failure of correct assembly of *Pbtert* which has a total size of 6939 bp. The complete *Pbtert* gene shared 77%, 83% and 43% identity with the predicted *tert* genes of *P. yoelii* (17X), *P. chabaudi* and *P. falciparum*, respectively (table in Fig. 1A).

Expression of PbTERT in mixed blood stage parasites was assessed by Western blot analysis using anti-TERT antibodies raised against *P. falciparum* TERT [29] (Fig. 1B). In protein extracts of mixed blood stages two protein fragments hybridized to the antibodies; one with the expected size of PbTERT (240 kDa; 2312 aa) and a slightly smaller band of ~220 kDa. The smaller band suggests that PbTERT is processed. Attempts to visualise PbTERT in blood stages by immunofluorescence microscopy using the same antibodies were unsuccessful.

The presence of TR transcripts (PBANKA_081945) was analysed by Northern analysis of RNA of different blood stages

using a probe recognizing the complete transcript (Fig. 1C). Highest levels were detected in late trophozoites and schizonts consistent with the known expression pattern of TERT in *P. falciparum* blood stages [29] and the PlasmoDB expression data (www.PlasmoDB.org). TR was also readily detected in RNA collected from highly purified mature *P. berghei* gametocytes (data not shown). Our steady-state RNA sequencing data (Religa et al., in prep) shows a good correlation of TERT mRNA levels (rings, trophs, schizonts) with TR transcript levels obtained from Northern analysis (data not shown).

Telomere length and localisation in blood stages

The average telomere length of chromosomes in *P. berghei* and *P. yoelii* asexual blood stages was determined by telomere restriction fragment analysis (TRF) [31] using a previously described *Plasmodium* telomere-specific probe [32] that recognizes all chromosomes of the four rodent species, *P. berghei*, *P. chabaudi*, *P. yoelii* and *P. vinckei* (Fig. 2A, left panel). In *P. berghei* several chromosomes show a relatively strong hybridization signal (chromosomes 6, 7 and 13/14) as a result of additional 'internal' telomeric repeats in the subtelomeric 2.3 kb repeat element [33,34]. The TRF analysis generated a characteristic 'hybridisation smear' in both *P. berghei* and *P. yoelii* (Fig. 2A, middle panel and the graph) which has also been observed in other *Plasmodium* species [35] and results from the variation in telomere length of chromosomes within a population of parasites. To determine the average telomere length, the signal intensity in each lane was plotted on the graph, with each lane coloured accordingly to the sample depicted (Fig. 2A). The predicted average telomere length was determined at around 950 bp for telomeres of *P. berghei*. This value is in agreement with previously determined telomere length of *P. berghei* chromosomes [36,37] and is very similar to the telomere length of 960 bp reported for *P. chabaudi* [35]. The *P. yoelii* average telomere length is much higher and was determined to be approximately 2.5 kb, again similar to the size reported previously (approx. 2 kb) [35]. Fluorescence *in situ* hybridization (FISH) using a fluorescein-labelled telomere sequence-specific probe [35] revealed a clear nuclear foci signal in the nuclei of late blood stages of *P. berghei* (Fig. 2B). However, the analysis of trophozoite nuclei also suggests the presence of clusters of telomeres within a single nucleus as has been shown for blood stages of *P. falciparum* (Fig. 2B) [38].

Deletion of the tert gene and selection of tert-deletion mutants

In order to delete the *tert* gene of *P. berghei* we designed a standard double cross-over plasmid (pL1324) based on plasmid pL0001 plasmid (www.mr4.org: plasmid MRA-770, www.pberghei.eu). Plasmid pL1324 contains the *Tgdhfr-ts* selection cassette flanked by two genomic sequences of ~750 and 950 bp of the *tert* gene. The schematic representation of the construct and the expected integration event that leads to deletion of *tert* is shown in Fig. 3A. Wild type (WT) parasites were transfected using standard methods of transfection [39]. In five independent transfection experiments (exp. 1065, 1078, 1138, 1207, 1217), parasites were selected with pyrimethamine and parasite populations were collected between day 10–14 after infection with transfected parasites (at a parasitaemia of 2–5%). These parasite populations were genotyped by Southern analysis of separated chromosomes and diagnostic PCR analysis. Diagnostic PCR analysis clearly showed the presence of a correctly disrupted *tert* gene in all five parasite populations (Fig. 3B). However, these populations also consisted of parasites with a WT *tert* gene as shown for all experiments (Fig. 3B, lane wt). Southern analysis of

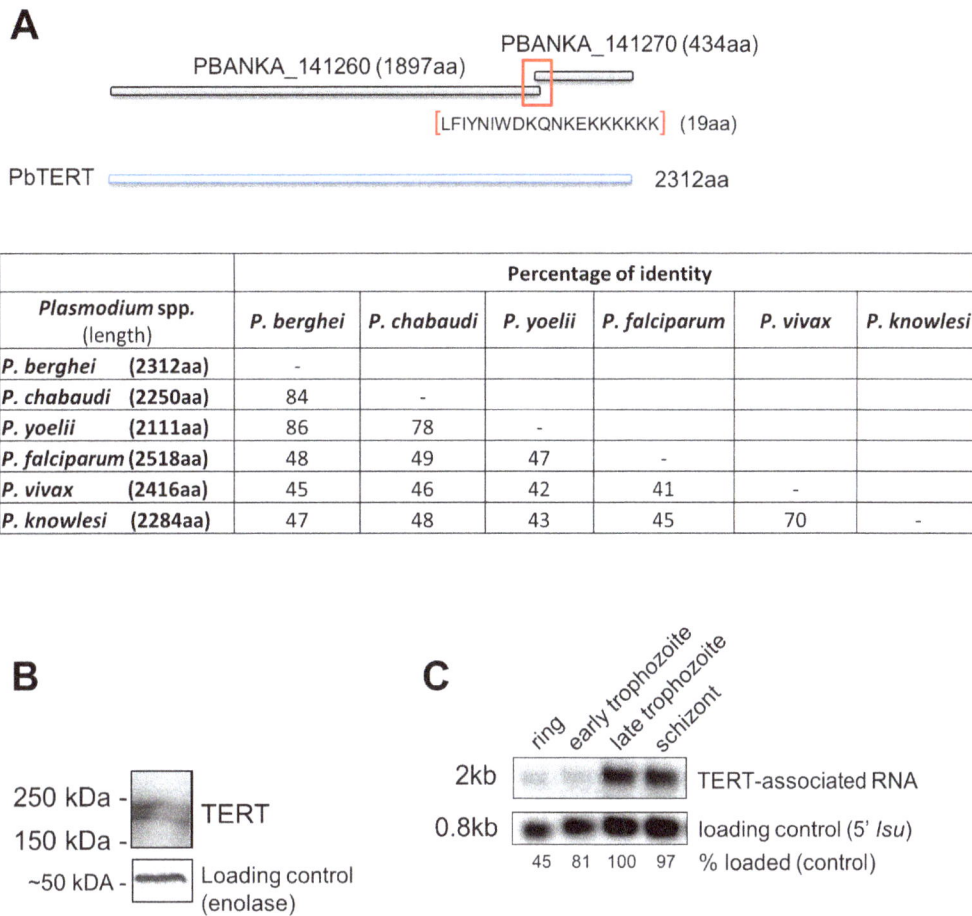

Figure 1. Pbtert gene structure (A) and PbTERT (B) and PbTR (C) expression. (A) The *tert* gene of *P. berghei* and homology (percentage identity) of TERT proteins in different *Plasmodium* species. Sequencing of the gap between two adjacent *tert* gene models available in PlasmoDB revealed a sequence duplication of 57 nt (19aa). The complete Pb*tert* gene encodes a protein of 2312aa, which is comparable to the size of other *Plasmodium tert* genes. (B) Western analysis of PbTERT protein in mixed blood stages. Two bands with a size between 150 and 250 kDa were detected (expected size of the TERT protein is ~240 kDa). (C) Northern analysis of Telomerase-associated RNA (TR) in different blood stages of *P. berghei*. RNA was hybridized with a probe recognizing TR (upper panel) (the expected size of TR is 2 kb) and as a loading control with a probe recognizing *large subunit ribosomal RNA* (expected size 0.8 kb). The "% loading" refers to the quantity of the loading control signal detected for each stage relative to the "late trophozoite" lane which is set as 100%.

separated chromosomes using a construct specific probe (*3'-UTR Pbdhfr-ts*) revealed in 3 experiments no or very weak hybridisation signals at chromosome 14 where the *tert* locus is located (Fig. 3C), indicating that the ratio of *tert*-deletion parasites and parasites with a WT-copy of *tert* is very low. In three experiments (1078, 1138, 1207) the intensity of the hybridization signal (compared to the signal of chromosome 7 on which the endogenous *Pbdhfr-ts* gene is located) and 'smeary' appearance of the signal is indicative of the presence of episomes containing the selectable marker (SM) cassette. Such episome-containing parasites are mainly selected when genes are targeted that are either essential for blood stage growth or strongly affect growth, resulting in selecting those rare parasites containing multiple copies of the episomes. However, a clear hybridisation with chromosome 14 was observed in experiment 1065 and 1217 (Fig. 3C). We therefore undertook large cloning experiments (20 mice per cloning experiment) to collect pure populations of *tert*-deletion mutants. However the attempts to clone *tert*-deficient parasites were unsuccessful and the only clones derived were those that contained an intact *tert* gene as shown by diagnostic PCR analysis (1065 clones 1–9, Fig. S1A).

Interestingly, Southern analysis of chromosomes of the clones of 1065 line still exhibited evidence for integration of the construct in chromosome 14 suggesting an incorrect integration event of the construct occurred that introduced the selectable marker *Tgdhfr-ts* on chromosome 14 but left the *tert*-gene intact (Fig. S1B). Integration of the *Tgdhfr-ts* gene in chromosome 7 has been excluded by both Southern and PCR analysis of the experiment 1065 clones (Fig. S1B and C, respectively). Such integration events have been observed before in selected parasite populations after transfection with constructs that target genes that result in lethal or affected-growth phenotypes (unpublished observations CJJ). Possible retention of the *tert* deletion plasmid was excluded (Fig. S2). The failure to clone the *tert*-deletion mutants might be explained by a low percentage of *tert*-deletion parasites in the populations, thereby reducing the chance of infecting a mouse with a single mutant. However, it is also possible that some mice have been injected with a single *tert*-deletion parasite but these mice became negative after the growth period of 8 days because the mutant parasites died as a result of telomere shortening during each multiplication cycle.

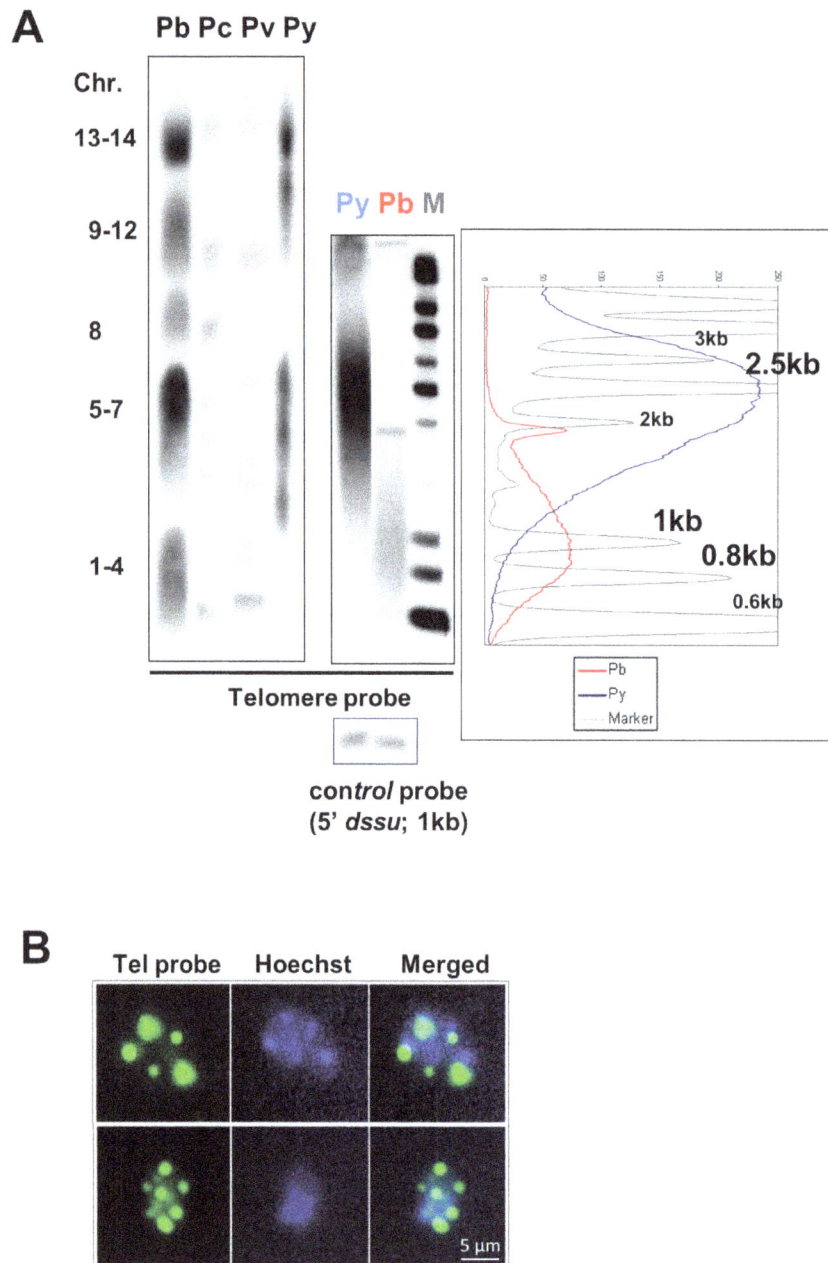

Figure 2. *P. berghei* **telomere characterisation.** (A) Determination of telomere length by Telomere Restriction Fragment (TRF) analysis. Left Panel: Southern analysis of separated chromosomes of *P. berghei* (Pb), *P. chabaudi* (Pc), *P. vinckei* (Pv) and *P. yoelii* (Py) showing hybridization of all chromosomes to a telomere-specific probe. The same probe was used for TRF analysis (middle, right panels). Middle panel: Southern analysis of digested *P. yoelii* (size control) and *P. berghei* gDNA probed with the telomeric probe showing the characteristic "smeared" hybridisation pattern in TRF analysis. Right panel: The average telomere length was measured as the highest peak of the signal intensity along the smear. Using the molecular marker ("M", grey line) as a size reference (relevant marker bands sizes are noted on the graph), the mean telomere length was estimated to be ~2500 bp and ~950 bp for *P. yoelii* (blue line) and *P. berghei* (red line), respectively. Complete digestion of gDNA was confirmed by hybridisation with a 5' *d-type small unit ribosomal RNA* probe. (B) Fluorescence *in situ* hybridisation with a telomere-specific probe. Fixed late blood stages of *P. berghei*. The telomeric probe (1.5 kb) was labelled with fluorescein (green). Hoechst (blue) was used for nuclear staining. The size bar is 5 μm.

In order to check if *tert*-mutants were disappearing from populations that undergo asexual multiplication, we injected 10^5 parasites of the uncloned populations of 1065, 1217 and 1207 in two mice and analysed the presence of *tert*-deletion parasites by diagnostic PCR during a period of 1–2 weeks of asexual multiplication. In all three populations the *tert*-deletion parasites could not be detected at the end of 1–2 week period (Fig. 3D). These results indicate that *tert*-deletion mutants have a growth disadvantage compared to parasites with a WT *tert* copy, possibly due to a delayed cell death phenotype of *tert*-deletion mutants.

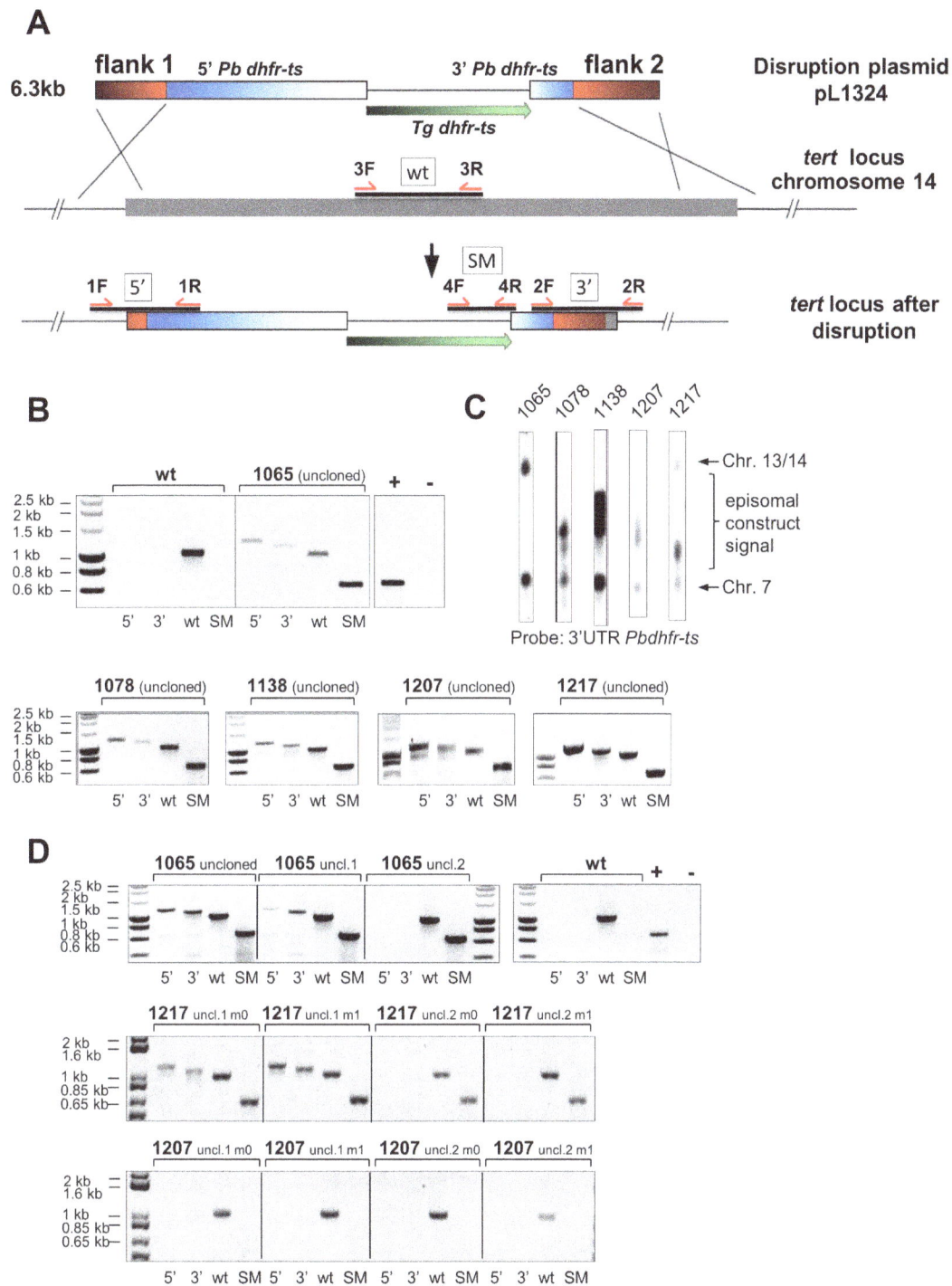

Figure 3. Pbtert deletion and selection of tert- mutants. (A) Schematic representation of the construct used to delete the *tert* gene. The construct, containing the *Tgdhfr-ts* selectable marker (SM) cassette, targets the *tert* gene at the flanking regions (red) by double cross-over integration. The red arrows indicate primers used for diagnostic PCR to confirm correct disruption of *tert*. Boxes correspond to lanes on the PCR gels in (B), (D) and Fig. S1A. (B) Diagnostic PCR of uncloned parasites transfected with a DNA construct to delete the *tert* gene. Parasites were collected and analysed directly after transfection and selection with pyrimethamine (parent populations). Diagnostic PCR shows the presence of parasites with correct disruption of the *tert* gene. In all experiments (1065, 1078, 1138, 1207, 1217) the 5′ and 3′ integration fragments (lanes 5′, 3′), as well as the *Tgdhfr-ts* fragment (lane SM) were amplified. However, all populations contained parasites with a wild type *tert* gene as shown by amplification of the wild type *tert* fragment (lane wt). The primer pairs used are shown in (A) and expected fragment sizes in Table S2. *pbs21*-specific primers were used as a positive control for all the PCR reactions ("+"). The water control is marked as "-". (C) Southern analysis of separated chromosomes using the 3′UTR *Pbdhfr-ts* probe shows only in experiment 1065 and 1217 hybridisation with chromosome 14 on which the *tert* gene is located. This probe recognizes the endogenous *Pbdhfr-ts* gene on chromosome 7 in all populations and additional chromosomes in experiments 1078, 1138, 1207 (possible episomal construct signal). (D) Diagnostic PCR of uncloned and propagated parasites transfected with a DNA construct to delete the *tert* gene. The

parent parasite populations of experiment 1065, 1207 and 1217 [see (B)] were propagated in mice (m0 = mouse 0, m1 = mouse 1) for another 1–2 weeks. Parasite populations collected were analysed by diagnostic PCR for the presence of parasites with correct disruption of the *tert* gene [primers same as in (B)]. In all populations no parasites with a disrupted *tert* gene could be detected by diagnostic PCR after 1 week (1207 all populations) or after two weeks of propagation (1065 uncl.2, 1217 uncl.2 m0 and 1217 uncl.2 m1).

Discussion

This study reports an analysis of telomeres and telomerase of the rodent malaria parasite *P. berghei* and the unsuccessful attempts to isolate gene-deletion mutants lacking the gene encoding telomerase reverse transcriptase (TERT). Although we are able to demonstrate the presence of *tert*-deficient parasites in transfected populations we were unable to clone these parasites indicating that TERT is essential for *P. berghei* cell survival in keeping with all studies reported to date on the essential role of TERT for survival of dividing cells of other eukaryotes (e.g. [40–42] and references therein).

Our inability to clone *tert*-deficient parasites suggests that the absence of TERT strongly affects the viability of blood stages and the rapid disappearance of *tert*-deficient parasites from dividing cell populations suggests a 'delayed cell-death' phenotype. The parasite lines obtained following parasite cloning all exhibited a possible non-specific integration event of the *tert* KO transfection construct (Fig. S1B, C), whereas *tert* KO plasmid presence was excluded (Fig. S2). Based on human cell studies where in the absence of TERT shortening of telomeres occurs at a rate of 50–100 bp per cell division [43,44] and assuming lethality in *tert*-deficient parasites occurs only after the complete loss of telomeres, the approximate survival time of the *P. berghei tert*-deficient parasites can be calculated. The number of nuclear/cell divisions in *P. berghei* is 3–4 per 24 hours. With a (mean) length of 950 bp and a loss of 50–100 bp per nuclear division, telomeres will be lost after 10-19 nuclear divisions which equals 2.5–6 days of multiplication in mice. This rate of telomere loss might explain the failure to obtain cloned parasite lines which takes a period of 14–18 days after transfection. However there exists the possibility that telomere shortening in *Plasmodium* has different kinetics than in other dividing cells. Generation of conditional KO parasites where *tert* can be synchronously disrupted in a homogenous population of transfected parasites might help to better define telomere turn-over in *Plasmodium*. The mean telomere length is not the same in different *Plasmodium* species and can differ significantly. For example *P. vivax* telomeres have an estimated length of 6.7 kb and *P. yoelii* 2–2.5 kb, which is significantly longer than the 950 bp of *P. berghei* telomeres (this paper; [35]). If telomere shortening per nuclear division is similar in *P. berghei* and *P. yoelii* it might therefore be possible to successfully clone *P. yoelli tert*-deficient parasites. However, TERT is possibly not only involved in telomere synthesis. TERT has been shown to interact with a range of other proteins in other organisms [45] and hence the absence of TERT might affect other pathways that are not necessarily associated with telomere shortening and chromosome instability. The function of TERT in telomere synthesis is dependent on the conserved RNA component of the TERT complex (TR). *Plasmodium* TR has first been identified in *P. falciparum*. The *P. berghei* orthologous molecule (PBANKA_081945) has been annotated based on strand specific transcriptome data and synteny to *P. falciparum* (U. Boehme, pers. comm.; www.GeneDB.org). Abolishing TR might be an alternative for the inactivation of the function of the TERT complex and might have a less strong lethal effect on blood stages compared with deletion of TERT. However, in other organisms cell dysfunctions resulting from the absence of TERT and TR loss overlap (e.g. [46]). Other approaches, such as the generation of parasite lines that carry mutated forms of TERT/TR or conditional mutagenesis [5,6,47–50] to inactivate TERT, might provide more insight into the role of TERT in telomere addition and help to define the minimal length of *Plasmodium* telomeres for chromosome stability, cell survival and delayed cell death. The availability of parasites that have a delayed cell death phenotype and that would result in self-resolving infections might also be useful tools for studies on the use of attenuated blood stage parasites to induce protective immune responses [51]. Such parasites would allow for analysing immune responses after (repeated) exposure to low numbers of replicating parasites without the need for clearing infections with antimalarials [52].

The studies reported here reveal TERT to be an essential protein for maintenance of *Plasmodium* blood stage parasite viability and can therefore be a legitimate target for the development of antimalarial drugs. Non-nucleoside reverse transcriptase analogues (NNRTIs) have been shown to kill *P. falciparum in vitro* after 3–5 blood stage cycles (A. Scherf, unpublished data), and efficiently inhibit *P. yoelii* liver stages *in vitro* [53] showing/indicating that *Plasmodium* TERT complex/activity can be targeted.

Materials and Methods

Experimental animals and parasite reference lines

All animal experiments performed in the Leiden malaria Research Group were approved by the Animal Experiments Committee of the Leiden University Medical Center (DEC 10099). All surgery was performed under isofluorane anaesthesia, and all efforts were made to minimise suffering. For all experiments the reference line cl15cy1 of the ANKA isolate of *P. berghei* was used [39].

Completion of the Pbtert gene sequence

The Pbtert sequence gap was amplified with primers GU2046 and GU2045 (see Table S1 for sequences), specific to the available *P. berghei* partial genes PBANKA_141260 and PBANKA_141270, respectively. The obtained PCR product was subcloned into pCR 2.1 TOPO vector (Invitrogen) and the inserts sequenced. Sequence assembly was performed in CLC Genomics Workbench (CLCbio).

Pbtert KO vector generation

The two homology arms of the TERT double cross-over (DXO) plasmid were amplified from *P. berghei* gDNA by PCR using primer pairs 3298/3299 [homology arm(HA) 1] and 3300/3301 [homology arm (HA) 2] (Table S1 primer sequences). PCR products were digested with Asp718I/HindIII (HA 1) and EcoRI/XbaI (HA 2) and subsequently cloned into the standard pL001 plasmid which contains the Toxoplasma gondii dihydrofolate reductase – thymidylate synthase gene (Tgdhfr-ts) conferring resistance to pyrimethamine [54]. The final TERT knockout vector (pL1324) was linearised with Asp718I and XbaI restriction enzymes, the 6.3 kb tert KO DNA fragment gel-purified and approximately 5 µg were used for transfection in order to obtain double cross-over tert knockout parasites.

P. berghei transfection, construct integration validation

Transfection of *P. berghei* parasites was performed as described [39]. Briefly, purified schizonts were transfected by electroporation and injected i.v. into a mouse tail (day 0). Pyrimethamine drug treatment was implemented at day 1. Parasites were collected at day 10–15 (parasitaemia of 2–5%) by cardiac puncture under anaesthesia. Erythrocytes were lysed in a cold 1× erythrocyte lysis buffer (10× stock: 1.5M of NH_4Cl, 0.1M $KHCO_3$, 0.1M EDTA in H_2O). Genomic DNA was extracted from blood stage parasites and analysed for construct integration by diagnostic PCR using the following primer pairs 1F/1R and 2F/2R (5′ and 3′ integration, respectively), 3F/3R (Pbtert gene fragment), 4F/4R (Tgdhfr-ts) (see Table S2 for expected products and Table S1 for primer sequences).

Telomere Restriction Fragment (TRF) analysis

Telomere Restriction Fragment analysis was performed as previously described [35] with modifications. 700 ng of gDNA was digested overnight at 37°C with AluI, MboII, RsaI and Sau3AI (5 units each), run on 1% agarose gel and blotted onto a positively charged nitrocellulose membrane (Hybond-N$^+$ membrane, GE Healthcare). The membrane was washed shortly in 2× SSC, pre-hybridised for at least 1 hour at 65°C, and probed with α-^{32}P dATP labelled double stranded telomere-specific probe from pTB4.1 plasmid [32] for 6 hours at 60°C. The blot was washed 3 times with 3× SSC/0.5% (v/v) SDS and once with 1× SSC/0.5% (v/v) SDS at 60°C (15 min. each wash), exposed to a Biorad phosphoimaging screen-K and results scanned using Typhoon 9410 from GE Healthcare. Image analysis was performed using ImageQuant TL (Amersham Biosciences).

Northern blot analysis

Northern analysis of total RNA obtained from synchronised *P. berghei* blood stages was performed as described previously [55,56]. Approximately 5 µg of RNA was hybridized to the P^{32} labelled DNA probe for telomerase-associated-RNA that was PCR amplified from wild type *P. berghei* gDNA using the primer set 3306/3307 (see Table S1). As a loading control primer L644R was used, which hybridizes to the blood stage large subunit ribosomal RNA [57]. The blots were exposed to a Biorad phosphoimaging screen-K and scanned using Typhoon 9410 (GE Healthcare). Quantification of the loading control signal intensities was done using ImageJ.

Pulse-field gel (PFG) electrophoresis

Separation of chromosomes by PFG electrophoresis was performed as previously described [39]. The blots were exposed to a Biorad phosphoimaging screen-K and scanned using Typhoon 9410 (GE Healthcare).

Western analysis

For Western analysis mixed *P. berghei* blood stages (2×10^7 parasites) were resuspended in 2× SDS gel-loading buffer (100 mM Tris-HCl pH 6.8; 200 mM dithiothreitol; 4% SDS; 0.2% bromophenol blue; 20% glycerol) with prior addition of 1 µl of PMSF ($C_i = 100$ mM, $C_f = 1$ mM). The samples were run on a 12% SDS-PAGE gel and transferred to a nitrocellulose membrane, which was subsequently blocked with 5% milk/0.1%

Tween in PBS, probed with the PfTERT primary antibody [29] in 5% milk/0.1% Tween in PBS (rabbit, 1:500, courtesy of A. Scherf), washed 3× with 0.1% Tween/PBS and probed with an HRP-labelled secondary goat anti-rabbit antibody (Dako) in 5% milk/0.1% Tween in PBS (1:5000). Same procedure was used for the control antibody against enolase (PBANKA_121430; rabbit, 1:1000, Biogenes). The complex was visualised using the ECL Plus^{TM} Western blotting detection reagents (Amersham Biosciences), and X-ray film (Kodak).

Fluorescent in situ hybridization (FISH)$^{DNA-DNA}$

FISH analysis used for telomeric DNA detection in *P. berghei* blood stage parasites was performed as described [32,38], omitting the step of saponin lysis. Fluorescence emission was analysed using a Leica DMRA Fluorescence microscope and ColourProc software.

Supporting Information

Figure S1 (A) Diagnostic PCR of 1065 clones 1–9. The 5′ and the 3′ integration fragments (lanes 5′ and 3′) were not amplified. The wild type *tert* (lane wt) and *Tgdhfr-ts* (lane SM) fragments were obtained for all clones and the uncloned population. Primers used and amplified products are same as in Fig. 3B and D. Schematic representation of the modified *tert* locus indicating the primers and products is shown in Fig. 3A. For primers and expected sizes see Table S2. **(B) Southern analysis of separated chromosomes in experiment 1065 clones 1, 2 and 4–9** using the 3′UTR *Pbdhfr-ts* probe shows hybridisation with chromosome 7 and 14. The *Tgdhfr-ts* probe shows the signal only in chromosome 14. **(C) Diagnostic PCR for *Tgdhfr-ts* presence in chromosome 7 in experiment 1065 clones 1–3.** 5′ and 3′ integration fragments (lanes 1, 2) were not amplified. Unspecific PCR product (lane 1) is caused by the GC-rich L301 primer as shown in PCR with both primers, and with each primer separately (lane 1, 3580 primer – lane 1a, L301 primer – lane 1b). The wt *tert* and *Tgdhfr-ts* fragments (lanes wt, SM) were amplified. For primers and expected sizes see Table S2. The "+" and "-" control primers same as in Fig. 3B.

Figure S2 PCR for *tert* KO plasmid presence in experiment 1065 clones 1–3. *tert* KO plasmid control fragments are not amplified in either of 1065 samples (lanes A, B). The *tert* and *Tgdhfr-ts* gene fragments (lanes wt, SM) are amplified in all 1065 samples and not in the *tert* KO plasmid. Schematic representation of the *tert* construct is shown indicating primers used (see Table S2 for expected products sizes).

Table S1 A list of primers used in this study.

Table S2 Primer combinations and expected product sizes in PCR analyses performed in this study.

Author Contributions

Conceived and designed the experiments: AAR CJJ AS APW. Performed the experiments: AAR JR CJJ. Analyzed the data: AAR CJJ APW. Contributed reagents/materials/analysis tools: AAR JR CJJ AS APW. Wrote the paper: AAR CJJ APW AS.

References

1. Greider CW, Blackburn EH (1985) Identification of a specific telomere terminal transferase activity in Tetrahymena extracts. Cell 43: 405–413. 0092-8674(85)90170-9 [pii].

2. Blackburn EH, Gall JG (1978) A tandemly repeated sequence at the termini of the extrachromosomal ribosomal RNA genes in Tetrahymena. J Mol Biol 120: 33–53. 0022-2836(78)90294-2 [pii].

3. Blackburn EH, Budarf ML, Challoner PB, Cherry JM, Howard EA, et al. (1983) DNA termini in ciliate macronuclei. Cold Spring Harb Symp Quant Biol 47 Pt 2: 1195–1207.

4. Frias C, Pampalona J, Genesca A, Tusell L (2012) Telomere dysfunction and genome instability. Front Biosci 17: 2181–2196. 4044 [pii].

5. Lundblad V, Szostak JW (1989) A mutant with a defect in telomere elongation leads to senescence in yeast. Cell 57: 633–643. 0092-8674(89)90132-3 [pii].

6. Yu GL, Bradley JD, Attardi LD, Blackburn EH (1990) In vivo alteration of telomere sequences and senescence caused by mutated Tetrahymena telomerase RNAs. Nature 344: 126–132. 10.1038/344126a0 [doi].

7. Dokal I (2001) Dyskeratosis congenita. A disease of premature ageing. Lancet 358 Suppl: S27. S0140673601070404 [pii].

8. Nelson ND, Bertuch AA (2012) Dyskeratosis congenita as a disorder of telomere maintenance. Mutat Res 730: 43–51. S0027-5107(11)00150-3 [pii];10.1016/j.mrfmmm.2011.06.008 [doi].

9. Vulliamy TJ, Knight SW, Mason PJ, Dokal I (2001) Very short telomeres in the peripheral blood of patients with X-linked and autosomal dyskeratosis congenita. Blood Cells Mol Dis 27: 353–357. 10.1006/bcmd.2001.0389 [doi];S1079-9796(01)90389-4 [pii].

10. Kirwan M, Dokal I (2009) Dyskeratosis congenita, stem cells and telomeres. Biochim Biophys Acta 1792: 371–379. S0925-4439(09)00025-8 [pii];10.1016/j.bbadis.2009.01.010 [doi].

11. Watson JD (1972) Origin of concatemeric T7 DNA. Nat New Biol 239: 197–201.

12. Baran N, Haviv Y, Paul B, Manor H (2002) Studies on the minimal lengths required for DNA primers to be extended by the Tetrahymena telomerase: implications for primer positioning by the enzyme. Nucleic Acids Res 30: 5570–5578.

13. Collins K, Greider CW (1993) Tetrahymena telomerase catalyzes nucleolytic cleavage and nonprocessive elongation. Genes Dev 7: 1364–1376.

14. Greider CW (1991) Telomerase is processive. Mol Cell Biol 11: 4572–4580.

15. Lue NF, Peng Y (1997) Identification and characterization of a telomerase activity from Schizosaccharomyces pombe. Nucleic Acids Res 25: 4331–4337. gka711 [pii].

16. Melek M, Davis BT, Shippen DE (1994) Oligonucleotides complementary to the Oxytricha nova telomerase RNA delineate the template domain and uncover a novel mode of primer utilization. Mol Cell Biol 14: 7827–7838.

17. Morin GB (1989) The human telomere terminal transferase enzyme is a ribonucleoprotein that synthesizes TTAGGG repeats. Cell 59: 521–529. 0092-8674(89)90035-4 [pii].

18. Pace T, Ponzi M, Dore E, Janse C, Mons B, et al. (1990) Long insertions within telomeres contribute to chromosome size polymorphism in Plasmodium berghei. Mol Cell Biol 10: 6759–6764.

19. Hayflick L (1965) The limited *in vitro* lifetime of human diploid cell strains. Exp Cell Res 37: 614–636.

20. Allsopp RC, Chang E, Kashefi-Aazam M, Rogaev EI, Piatyszek MA, et al. (1995) Telomere shortening is associated with cell division in vitro and in vivo. Exp Cell Res 220: 194–200. S0014-4827(85)71306-7 [pii];10.1006/excr.1995.1306 [doi].

21. Griffith JD, Comeau L, Rosenfield S, Stansel RM, Bianchi A, et al. (1999) Mammalian telomeres end in a large duplex loop. Cell 97: 503–514. S0092-8674(00)80760-6 [pii].

22. Chiang YJ, Hemann MT, Hathcock KS, Tessarollo L, Feigenbaum L, et al. (2004) Expression of telomerase RNA template, but not telomerase reverse transcriptase, is limiting for telomere length maintenance in vivo. Mol Cell Biol 24: 7024–7031. 10.1128/MCB.24.16.7024-7031.2004 [doi];24/16/7024 [pii].

23. Dore E, Pace T, Ponzi M, Scotti R, Frontali C (1986) Homologous telomeric sequences are present in different species of the genus Plasmodium. Mol Biochem Parasitol 21: 121–127. 0166-6851(86)90015-0 [pii].

24. Ponzi M, Pace T, Dore E, Frontali C (1985) Identification of a telomeric DNA sequence in Plasmodium berghei. EMBO J 4: 2991–2995.

25. Figueiredo LM, Freitas-Junior LH, Bottius E, Olivo-Marin JC, Scherf A (2002) A central role for Plasmodium falciparum subtelomeric regions in spatial positioning and telomere length regulation. EMBO J 21: 815–824. 10.1093/emboj/21.4.815 [doi].

26. Figueiredo LM, Pirrit LA, Scherf A (2000) Genomic organisation and chromatin structure of Plasmodium falciparum chromosome ends. Mol Biochem Parasitol 106: 169–174. S0166685199001991 [pii].

27. Bottius E, Bakhsis N, Scherf A (1998) Plasmodium falciparum telomerase: de novo telomere addition to telomeric and nontelomeric sequences and role in chromosome healing. Mol Cell Biol 18: 919–925.

28. Mattei D, Scherf A (1994) Subtelomeric chromosome instability in Plasmodium falciparum: short telomere-like sequence motifs found frequently at healed chromosome breakpoints. Mutat Res 324: 115–120.

29. Figueiredo LM, Rocha EP, Mancio-Silva L, Prevost C, Hernandez-Verdun D, et al. (2005) The unusually large Plasmodium telomerase reverse-transcriptase localizes in a discrete compartment associated with the nucleolus. Nucleic Acids Res 33: 1111–1122. 33/3/1111 [pii];10.1093/nar/gki260 [doi].

30. Chakrabarti K, Pearson M, Grate L, Sterne-Weiler T, Deans J, et al. (2007) Structural RNAs of known and unknown function identified in malaria parasites by comparative genomics and RNA analysis. RNA 13: 1923–1939. rna.751807 [pii];10.1261/rna.751807 [doi].

31. de Lange T, Shiue L, Myers RM, Cox DR, Naylor SL, et al. (1990) Structure and variability of human chromosome ends. Mol Cell Biol 10: 518–527.

32. Ponzi M, Janse CJ, Dore E, Scotti R, Pace T, et al. (1990) Generation of chromosome size polymorphism during in vivo mitotic multiplication of Plasmodium berghei involves both loss and addition of subtelomeric repeat sequences. Mol Biochem Parasitol 41: 73–82. 0166-6851(90)90098-7 [pii].

33. Dore E, Pace T, Ponzi M, Picci L, Frontali C (1990) Organization of subtelomeric repeats in Plasmodium berghei. Mol Cell Biol 10: 2423–2427.

34. Pace T, Ponzi M, Dore E, Frontali C (1987) Telomeric motifs are present in a highly repetitive element in the Plasmodium berghei genome. Mol Biochem Parasitol 24: 193–202. 0166-6851(87)90106-X [pii].

35. Figueiredo LM, Freitas-Junior LH, Bottius E, Olivo-Marin JC, Scherf A (2002) A central role for Plasmodium falciparum subtelomeric regions in spatial positioning and telomere length regulation. EMBO J 21: 815–824. 10.1093/emboj/21.4.815 [doi].

36. Ponzi M, Pace T, Dore E, Picci L, Pizzi E, et al. (1992) Extensive turnover of telomeric DNA at a Plasmodium berghei chromosomal extremity marked by a rare recombinational event. Nucleic Acids Res 20: 4491–4497.

37. Dore E, Pace T, Picci L, Pizzi E, Ponzi M, et al. (1994) Dynamics of telomere turnover in Plasmodium berghei. Mol Biol Rep 20: 27–33.

38. Freitas-Junior LH, Bottius E, Pirrit LA, Deitsch KW, Scheidig C, et al. (2000) Frequent ectopic recombination of virulence factor genes in telomeric chromosome clusters of P. falciparum. Nature 407: 1018–1022. 10.1038/35039531 [doi].

39. Janse CJ, Ramesar J, Waters AP (2006) High-efficiency transfection and drug selection of genetically transformed blood stages of the rodent malaria parasite Plasmodium berghei. Nat Protoc 1: 346–356. nprot.2006.53 [pii];10.1038/nprot.2006.53 [doi].

40. Amiard S, Da IO, Gallego ME, White CI (2014) Responses to telomere erosion in plants. PLoS One 9: e86220. 10.1371/journal.pone.0086220 [doi];PONE-D-13-35581 [pii].

41. Lee HW, Blasco MA, Gottlieb GJ, Horner JW, Greider CW, et al. (1998) Essential role of mouse telomerase in highly proliferative organs. Nature 392: 569–574. 10.1038/33345 [doi].

42. Shay JW, Wright WE (2010) Telomeres and telomerase in normal and cancer stem cells. FEBS Lett 584: 3819–3825. S0014-5793(10)00419-9 [pii];10.1016/j.febslet.2010.05.026 [doi].

43. Martin-Ruiz C, Saretzki G, Petrie J, Ladhoff J, Jeyapalan J, et al. (2004) Stochastic variation in telomere shortening rate causes heterogeneity of human fibroblast replicative life span. J Biol Chem 279: 17826–17833. 10.1074/jbc.M311980200 [doi];M311980200 [pii].

44. Stoppler H, Hartmann DP, Sherman L, Schlegel R (1997) The human papillomavirus type 16 E6 and E7 oncoproteins dissociate cellular telomerase activity from the maintenance of telomere length. J Biol Chem 272: 13332–13337.

45. Martinez P, Blasco MA (2011) Telomeric and extra-telomeric roles for telomerase and the telomere-binding proteins. Nat Rev Cancer 11: 161–176. nrc3025 [pii];10.1038/nrc3025 [doi].

46. Blackburn, E., de Lange, T., and Lundblad, V. (15-12-2005) Telomeres. Cold Spring Harbor Laboratory Press.

47. Blasco MA, Lee HW, Hande MP, Samper E, Lansdorp PM, et al. (1997) Telomere shortening and tumor formation by mouse cells lacking telomerase RNA. Cell 91: 25–34. S0092-8674(01)80006-4 [pii].

48. Herrera E, Samper E, Martin-Caballero J, Flores JM, Lee HW, et al. (1999) Disease states associated with telomerase deficiency appear earlier in mice with short telomeres. EMBO J 18: 2950–2960. 10.1093/emboj/18.11.2950 [doi].

49. Ujike-Asai A, Okada A, Du Y, Maruyama M, Yuan X, et al. (2007) Large defects of type I allergic response in telomerase reverse transcriptase knockout mice. J Leukoc Biol 82: 429–435. jlb.1006638 [pii];10.1189/jlb.1006638 [doi].

50. Yuan X, Ishibashi S, Hatakeyama S, Saito M, Nakayama J, et al. (1999) Presence of telomeric G-strand tails in the telomerase catalytic subunit TERT knockout mice. Genes Cells 4: 563–572. gtc284 [pii].

51. McCarthy JS, Good MF (2010) Whole parasite blood stage malaria vaccines: a convergence of evidence. Hum Vaccin 6: 114–123. 10394 [pii].

52. Stanisic DI, Barry AE, Good MF (2013) Escaping the immune system: How the malaria parasite makes vaccine development a challenge. Trends Parasitol 29: 612–622. S1471-4922(13)00162-1 [pii];10.1016/j.pt.2013.10.001 [doi].

53. Mahmoudi N, Garcia-Domenech R, Galvez J, Farhati K, Franetich JF, et al. (2008) New active drugs against liver stages of Plasmodium predicted by molecular topology. Antimicrob Agents Chemother 52: 1215–1220. AAC.01043-07 [pii];10.1128/AAC.01043-07 [doi].

54. van Dijk MR, Waters AP, Janse CJ (1995) Stable transfection of malaria parasite blood stages. Science 268: 1358–1362.

55. Kyes S, Pinches R, Newbold C (2000) A simple RNA analysis method shows var and rif multigene family expression patterns in Plasmodium falciparum. Mol Biochem Parasitol 105: 311–315. S0166-6851(99)00193-0 [pii].

56. Mair GR, Braks JA, Garver LS, Wiegant JC, Hall N, et al. (2006) Regulation of sexual development of Plasmodium by translational repression. Science 313: 667–669. 313/5787/667 [pii];10.1126/science.1125129 [doi].

57. van Spaendonk RM, Ramesar J, van WA, Eling W, Beetsma AL, et al. (2001) Functional equivalence of structurally distinct ribosomes in the malaria parasite, Plasmodium berghei. J Biol Chem 276: 22638–22647. 10.1074/jbc.M101234200 [doi];M101234200 [pii].

18

Giardia duodenalis Infection Reduces Granulocyte Infiltration in an *In Vivo* Model of Bacterial Toxin-Induced Colitis and Attenuates Inflammation in Human Intestinal Tissue

James A. Cotton[1,2,3], Jean-Paul Motta[1,2,3], L. Patrick Schenck[4,5], Simon A. Hirota[6,7], Paul L. Beck[5], Andre G. Buret[1,2,3]*

1 Department of Biological Sciences, University of Calgary, Calgary, Alberta, Canada, **2** Inflammation Research Network, University of Calgary, Calgary, Alberta, Canada, **3** Host-Parasite Interactions, University of Calgary, Calgary, Alberta, Canada, **4** Department of Biochemistry and Molecular Biology, University of Calgary, Calgary, Alberta, Canada, **5** Department of Medicine, University of Calgary, Calgary, Alberta, Canada, **6** Department of Physiology and Pharmacology, University of Calgary, Calgary, Alberta, Canada, **7** Department of Immunology, Microbiology and Infectious Diseases, University of Calgary, Calgary, Alberta, Canada

Abstract

Giardia duodenalis (syn. *G. intestinalis*, *G. lamblia*) is a predominant cause of waterborne diarrheal disease that may lead to post-infectious functional gastrointestinal disorders. Although *Giardia*-infected individuals could carry as much as 10^6 trophozoites per centimetre of gut, their intestinal mucosa is devoid of overt signs of inflammation. Recent studies have shown that in endemic countries where bacterial infectious diseases are common, *Giardia* infections can protect against the development of diarrheal disease and fever. Conversely, separate observations have indicated *Giardia* infections may enhance the severity of diarrheal disease from a co-infecting pathogen. Polymorphonuclear leukocytes or neutrophils (PMNs) are granulocytic, innate immune cells characteristic of acute intestinal inflammatory responses against bacterial pathogens that contribute to the development of diarrheal disease following recruitment into intestinal tissues. *Giardia* cathepsin B cysteine proteases have been shown to attenuate PMN chemotaxis towards IL-8/CXCL8, suggesting *Giardia* targets PMN accumulation. However, the ability of *Giardia* infections to attenuate PMN accumulation *in vivo* and how in turn this effect may alter the host inflammatory response in the intestine has yet to be demonstrated. Herein, we report that *Giardia* infection attenuates granulocyte tissue infiltration induced by intra-rectal instillation of *Clostridium difficile* toxin A and B in an isolate-dependent manner. This attenuation of granulocyte infiltration into colonic tissues paralled decreased expression of several cytokines associated with the recruitment of PMNs. *Giardia* trophozoite isolates that attenuated granulocyte infiltration *in vivo* also decreased protein expression of cytokines released from inflamed mucosal biopsy tissues collected from patients with active Crohn's disease, including several cytokines associated with PMN recruitment. These results demonstrate for the first time that certain *Giardia* infections may attenuate PMN accumulation by decreasing the expression of the mediators responsible for their recruitment.

Editor: John Wallace, University of Calgary, Canada

Funding: JA Cotton is a recipient of NSERC Alexander Graham Bell Scholarship graduate student scholarship and a joint IBD studentship from Alberta Innovates Health Solutions (AIHS) and the Crohn's and colitis foundation of Canada (CCFC). Research in AG Buret's lab is funded by grants from the Natural Sciences and Engineering Research Council of Canada (NSERC) RT 690446, and the CCFC 10000008. The funders had no role in study design, data collection and analysis, decision to publish, or preparation of the manuscript.

Competing Interests: The authors have declared that no competing interests exist.

* Email: aburet@ucalgary.ca

Introduction

Giardia duodenalis (syn. *G. intestinalis*, *G. lamblia*) is a non-invasive, small intestinal protozoan parasite infecting a variety of animal species, including humans. *Giardia duodenalis* is currently subdivided into eight distinct genetic assemblages, with only assemblages A and B being infective to humans [1,2]. Some reports suggest that these assemblages may be unique *Giardia* species, but much controversy remains on this topic [3,4]. This parasite induces a response within its host that may cause malabsorptive diarrheal disease ([5]). Recent evidence also indicates that giardiasis can lead to the development of post-infectious disorders, in the intestine as well as extra-intestinally, via mechanisms that remain obscure ([6,7]). During the acute stage of the infection, parasite loads can exceed 10^6 trophozoites/cm of gut but the intestinal mucosa of most *Giardia*-infected patients is devoid of overt signs of inflammation [8]. These observations are counter-intuitive, considering that *Giardia* infections increase small intestinal permeability via several mechanisms and, therefore, may indeed facilitate the translocation of luminal antigens into underlying host tissues [9–11]. However, microscopic duodenal inflammation has been observed in some patients with chronic assemblage B *Giardia* infections following metronidazole treatment [12]. *Giardia* assemblage B infections *in vivo* have been shown to induce small intestinal inflammation [13]. Therefore, it remains to be determined whether *Giardia* infections modulate pro-inflammatory responses within the intestinal mucosa and if these events are assemblage or isolate specific.

Transmission of *Giardia* infections occurs via ingestion of infectious cysts in contaminated food or water, or directly via the fecal-oral route (reviewed in [14]). This route of infection is adopted by many gastrointestinal pathogens, and as a result, *Giardia* co-infections may occur. Indeed, *Giardia* infections have been reported to occur concurrently with *Ascaris sp.* [15], *Cryptosporidium sp.* [16], *Helicobacter pylori* [16,17], *Vibrio cholerae* [18], *Salmonella* sp. [19] and rotavirus [18,20,21]. Many of these pathogens are known to promote inflammatory responses and simultaneously induce diarrheal disease within their hosts. A recent report suggested that Tanzanian children infected with *Giardia* have reduced incidence rates of diarrheal disease and fever, while also displaying lower serum C-reactive protein (CRP) levels [22]. *Giardia* infections have also been shown to attenuate symptoms of diarrheal disease during rotavirus infection [20]. The mechanisms remain unclear, it has been suggested that *Giardia* infections may create an advantageous environment for other co-infecting gastrointestinal pathogens [18]. Contradictorily, a separate study suggests that *Giardia* infections may instead enhance symptoms of diarrheal disease during rotavirus infection [16].

Polymorphonuclear leukocytes (PMNs) are heavily involved in inflammatory responses, and essential to host defence against many bacterial and fungal pathogens. Indeed, genetic mutations or drug therapy that result in defective PMN function make individuals highly susceptible to life-threatening infections [23–25]. These cells have been implicated in a variety of inflammatory disorders, in the gut and beyond [26,27]. PMN recruitment into intestinal tissues is a multistep process involving egression from bone marrow and circulation, migration through host tissues, and transepithelial migration [28–30]. Moreover, multiple mediators are known to promote tissue accumulation of PMNs. During acute inflammatory responses, granulocyte colony stimulating factor (G-CSF) promotes PMN bone marrow egression and increases the number of circulating PMNs [31,32]. Production of interleukin-

17A (IL-17A) by tissue-resident cells increases circulating G-CSF levels and promotes subsequent PMN bone marrow egression [32–34]. PMN chemokines containing a consecutive glutamate-leucine-arginine sequence (ELR+ chemokines) in their N-terminal region, such as CXCL1, CXCL2, and IL-8/CXCL8, are produced by a variety of cells within the intestinal mucosa. ELR+ chemokines are involved in multiple processes of PMN tissue recruitment, including bone marrow egression [31], exit from the blood stream [30,35,36], and migration through host tissues [37,38].

Previous research has suggested that *Giardia* excretory/secretory products induce IL-8/CXCL8 expression in intestinal epithelial cells [39]. However, we and others have demonstrated that *Giardia* trophozoites attenuate IL-8/CXCL8 secretion from *in vitro* epithelial monolayers [40,41]. Our research further demonstrated that *Giardia* cathepsin B family proteases contributed to the degradation of IL-8/CXCL8 and attenuated PMN chemotaxis [41]. As PMN accumulation contributes to the development of diarrheal disease via several different mechanisms [28,42–44], we hypothesized that during a *Giardia* infection, the attenuation of diarrheal disease or the creation of a favourable environment for other co-infecting gastrointestinal pathogens may stem, at least partially, from *Giardia*'s ability to attenuate PMN recruitment. These observations support the notion that *Giardia* parasites may reduce host inflammatory responses. Such effects and their mechanisms have yet to be established *in vivo*. The present report demonstrates that *Giardia* infections attenuate granulocyte infiltration during acute experimental colitis induced by *Clostridium difficile* toxins TcdA and TcdB, and markedly modulate the expression of several pro-inflammatory mediators, in an isolate-dependent manner. This study also shows that *Giardia* trophozoites decrease expression of a variety of inflammatory mediators when parasites are co-incubated with inflamed biopsy tissues from patients with Crohn's disease. Collectively, these results indicate for the first time that certain *Giardia* isolates are capable of attenuating intestinal inflammatory responses in inflammatory settings *in vivo and ex vivo*.

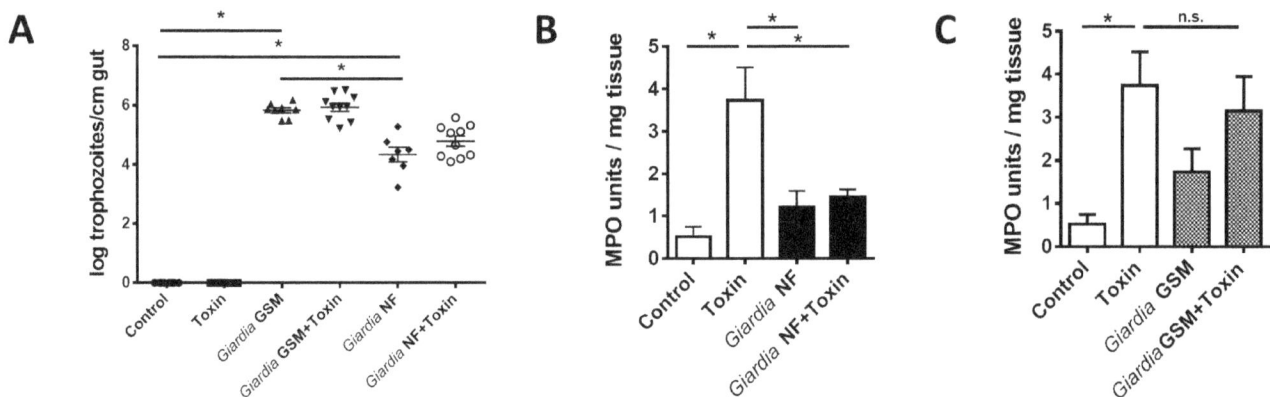

Figure 1. *Giardia* NF infections attenuate granulocyte recruitment in a model of *C. difficile* TcdAB-induced colitis. Male C57BL/6 mice aged 4 to 6 weeks were infected with *Giardia* NF or GS/M trophozoites. 7 days post-infection, animals were given 100 µg of TcdAB or PBS i.r. and, after 3 hours, animals were euthanized. (A) Trophozoites counts within the upper 3 cm of the jejunum were determined. (B) Colonic myeloperoxidase (MPO) activity was determined between uninfected controls and *Giardia* NF-infected animals given 100 µg TcdAB or PBS. (C) Colonic MPO activity was determined between uninfected controls and *Giardia* GS/M-infected animals given 100 µg TcdAB or PBS. (D) Values were determined via bead-based cytokine assay. All data are representative of two independent experiments (n = 5–11/group) and represented as mean ± SEM. * p<0.05.

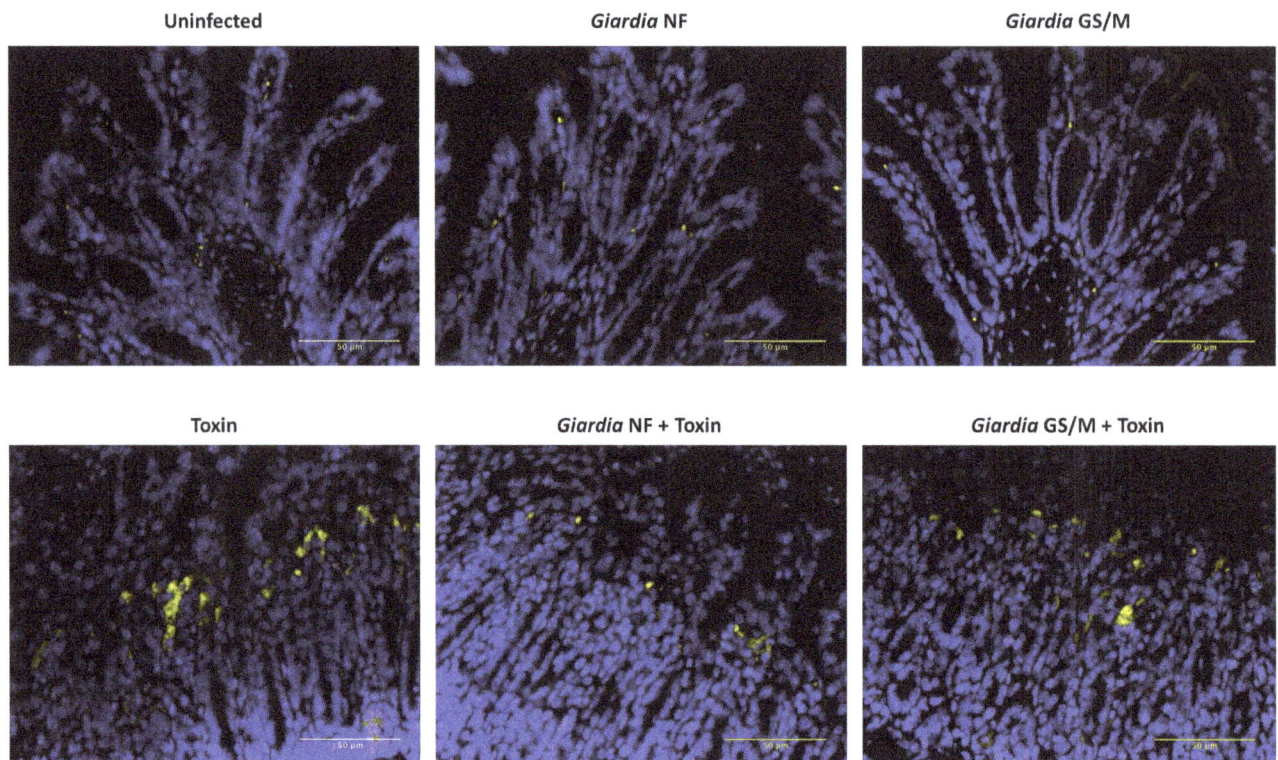

Figure 2. *Giardia* **NF-infected animals exhibit reduced MPO-positive cells following administration of** *C. difficile* **TcdAB.** Male C57BL/6 mice aged 4 to 6 weeks were infected with *Giardia* NF or GS/M trophozoites. 7 days post-infection, animals were given 100 µg of TcdAB or PBS i.r. and, after 3 hours, animals were euthanized. Representative immunohistochemical images of MPO-positive cells (yellow) counterstained with DAPI (blue) in colonic tissues of uninfected animals or animals infected with *Giardia* NF or GS/M trophozoites following i.r. instillation of 100 µg TcdAB or PBS. All data are representative of two independent experiments (n = 5–11/group) and taken at the same magnification (400×). Bars equal 50 µm for all micrographs.

Materials and Methods

Ethics statement

All studies involving human colonic mucosal biopsy tissues were approved by the Conjoint Health Research Ethics Board (CHREB) at the University of Calgary and the Calgary Health Region. In accordance with CHREB guidelines, adult subjects used in this study provided informed, written consent and a parent or guardian of any child participant provided informed, written consent on their behalf. Animal experiments were approved by the Life and Environmental Sciences Animal Care Committee of the University of Calgary, conducted in compliance with that approval, and followed guidelines established by the Canadian Council of Animal Care. All isolates used in this study have been used for over 15 years in the laboratory. Therefore, their isolation did not require their approval from an ethical review board at their time of collection.

Human biopsy tissues

Adapted from previous protocols [41,45], intestinal biopsy tissues were collected from the descending colon of patients with active Crohn's disease (CD). Upon collection, samples were washed once in Dulbecco's PBS (Sigma-Aldrich) containing 0.016% 1,4-Diothioerythritol (Sigma-Aldrich) to remove loosely adherent mucous and bacteria followed by three washes with phosphate-buffered saline (PBS). Following this, biopsy tissues were placed in 96-well plates and incubated in 300 µL of OptiMEM (Life Technologies) at 37°C, 5% CO_2, and 96%

humidity. All studies were performed in duplicate, whereby multiple biopsy tissues were collected from each patient.

Parasites

Giardia NF trophozoites were originally obtained from a water sample during an outbreak of giardiasis in Newfoundland, Canada [46], while *Giardia* GS/M clone H7 trophozoites were obtained from ATCC (50581) as previously described [47]. Trophozoites were grown axenically in 15 mL polystyrene tubes (Becton-Dickinson Falcon) in Keister's modified TYI-S-33 medium [48,49] supplemented with piperacillin (Sigma-Aldrich) and used at peak density culture.

Giardia trophozoite isolation

Confluent tubes of *Giardia* NF or GS/M trophozoites were harvested by cold shock on ice for 30 minutes, pooled into 50 mL polypropylene tubes (Falcon), and centrifuged at 500×g for 10 minutes. The resulting pellets were re-suspended in a total of 10 mL of ice cold PBS (Sigma-Aldrich) and centrifuged for 10 minutes at 500×g. Pellets were then re-suspended in 3 mL of fresh PBS, trophozoites were enumerated with a hemocytometer, and adjusted to the required concentration. For *ex vivo* human biopsy experiments, *Giardia* trophozoites were adjusted to a concentration of 5.0×10^6 trophozoites/well, while trophozoites were adjusted to a concentration of 1.0×10^8 trophozoites/mL for *in vivo* experiments. For *ex vivo* experiments, cells were co-incubated with *Giardia* trophozoites at 37°C, 5% CO_2, and 96% humidity for 6 hours.

Giardia NF

Giardia GS/M

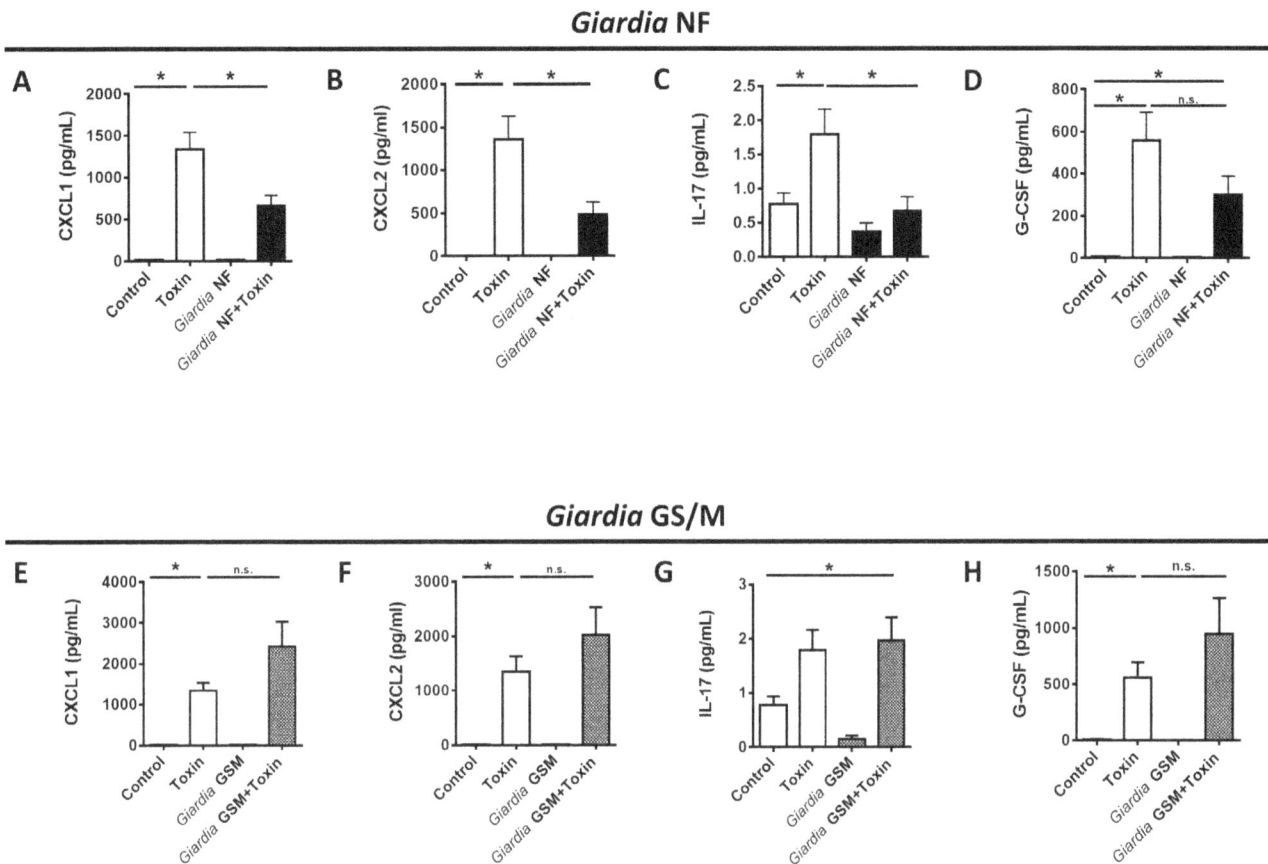

Figure 3. *Giardia* **NF infections attenuate colonic expression levels of several PMN-associated mediators following administration of** *C. difficile* **TcdAB.** Male C57BL/6 mice, aged 4 to 6 weeks, were infected with *Giardia* NF or GS/M trophozoites. 7 days post-infection, animals were given 100 µg of TcdAB or PBS i.r. and, after 3 hours, animals were euthanized. Levels of (A) CXCL1, (B) CXCL2, (C) IL-17, and (D) G-CSF were compared between *Giardia* NF-infected animals and uninfected animals given i.r. 100 µg of TcdAB or PBS. Levels of (E) CXCL1, (F) CXCL2, (G) IL-17, and (H) G-CSF were compared between *Giardia* GS/M-infected animals and uninfected animals given i.r. 100 µg of TcdAB or PBS. Values were determined via bead-based cytokine assay. All data are representative of two independent experiments (n = 5–11/group) and represented as mean ± SEM. n.s. = not significant * p<0.05.

Giardia in vivo infection

Using a previously described *Giardia* infection model [50,51], male C57BL/6 mice aged 4 to 6 weeks (Charles River Laboratories, Sherbrooke, QC) were acclimatized for 1 week prior to infection with *Giardia* NF or GS/M trophozoites. Forty-eight hours prior to oral gavage, all mice were administered broad-spectrum antibiotics to their drinking water ad libitum (1.4 mg/mL neomycin (Alfa Aesar), 1.0 mg/mL ampicillin (Alfa Aesar) and 1.0 mg/mL vancomycin (Alfa Aesar)). This regimen was maintained for the entire duration of the study. After 48 hours, mice were administered 10^7 *Giardia* NF or GS/M trophozoites in 0.1 mL of TYI-S-33 medium or 0.1 mL TYI-S-33 medium by oral gavage. After 7 days, small intestinal parasite loads were quantified. Mice were euthanized and the first 3 cm of the small intestine distal to the ligament of Treitz were opened longitudinally, placed in 1.5 mL Eppendorf tubes containing 1 mL of PBS, and kept on ice for 15 minutes. After 15-minute incubation, tubes were vortexed, and trophozoite numbers were visually enumerated under light microscope, and total concentration was extrapolated using a hemocytometer.

Clostridium difficile TcdAB-induced colitis

Following a previously described model of *C. difficile* toxin-induced colitis [52], mice infected with *Giardia* NF or GS/M

trophozoites for 7 days were intra-rectally (i.r.) administered a 100 µg solution of *C. difficile* A/B toxin (TcdA/TcdB) for 3 hours. Briefly, a 5F infant feeding tube catheter containing side ports (Mallinckrodt Inc.) was lubricated with a water-soluble personal lubricant and inserted 2.5 cm into the colon. With pressure applied to the anal area to prevent leakage, 100 µL of a 1.0 µg/µL TcdA/TcdB solution diluted in PBS was slowly administered over a period of 30 seconds. Following this, the tube was slowly removed and pressure was maintained for another 30 seconds. Vehicle control-treated control groups were administered PBS.

Tissue collection

Mice were euthanized via cervical dislocation 3 hours post intracolonic instillation of TcdAB or PBS. The colon was removed and samples were collected for bead-based cytokine analysis, myeloperoxidase (MPO) assays, or fixed in a fresh 4% paraformaldehyde solution for immunofluorescence, as described below. Colonic tissue samples collected for histology and immunohistochemistry were fixed in 4% paraformaldehyde solution overnight at 4°C, and, subsequently, washed and transferred to PBS at 4°C.

Giardia NF

Giardia GS/M

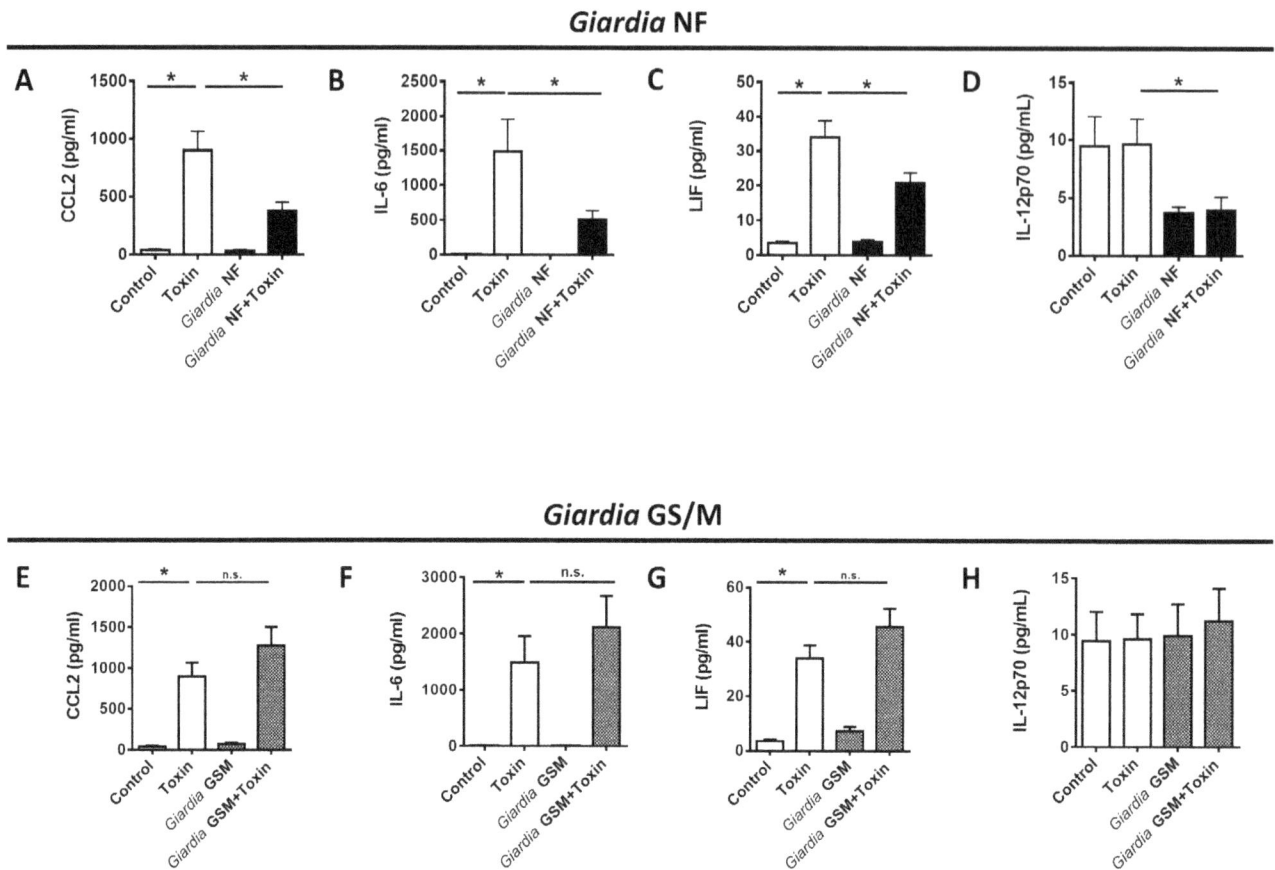

Figure 4. *Giardia* **NF infections attenuate colonic expression levels of other pro-inflammatory mediators following administration of** *C. difficile* **TcdAB.** Male C57BL/6 mice, aged 4 to 6 weeks, were infected with *Giardia* NF or GS/M trophozoites. 7 days post-infection, animals were given 100 µg of TcdAB or PBS i.r. and, after 3 hours, animals were euthanized. Levels of (A) CCL2, (B) IL-6, (C) LIF, and (D) IL-12p70 were compared between *Giardia* NF-infected animals and uninfected animals given i.r. 100 µg of TcdAB or PBS. Levels of (E) CCL2, (F) IL-6, (G) LIF, and (H) IL-12p70 were compared between *Giardia* GS/M-infected animals and uninfected animals given i.r. 100 µg of TcdAB or PBS. Values were determined via bead-based cytokine assay. All data are representative of two independent experiments (n = 5–11/group) and represented as mean ± SEM. n.s. = not significant * p < 0.05.

Cytokine analysis

Supernatants from *ex vivo* human biopsy tissue experiments were collected and centrifuged at 10,000 g for 15 minutes at 4°C. Resulting supernatants were decanted and stored at −70°C. Colonic tissue samples and biopsy tissues were weighed and collected in 2 mL Fast-Prep tubes (MP Biomedicals) containing a mixture of 0.9–2.0 mm stainless steel beads (NextAdvance). Tissue samples were suspended in lysis buffer [20 mM Tris-HCl (pH 7.5), 150 mM NaCl, 0.5% Tween-20 and a Complete Minitab protease inhibitor cocktail (Roche)] at a ratio of 50 mg tissue per 1 mL lysis buffer and homogenized using a Fast-Prep24 device (MP Biochemicals) at speed 6.0 for 40 seconds. The resulting homogenate solution was collected into pyrogen-free 1.5 mL Eppendorf tubes and centrifuged at 10,000 g for 15 minutes at 4°C. Supernatants and tissue homogenate cytokine levels were assessed using a Luminex XMap assay according to manufacturer's instructions (Luminex Corp.).

Tissue myeloperoxidase assay

Myeloperoxidase (MPO) activity was used as a marker of granulocyte infiltration [53]. Colonic tissue samples were weighed and collected in 2 mL Fast-Prep tubes (MP Biomedicals) containing a mixture of 0.9–2.0 mm stainless steel beads

(NextAdvance). Tissue samples were suspended in 50 mM potassium phosphate buffer containing 5 mg/mL hexadecyltri-methylammonium bromide (Sigma-Aldrich) at a ratio of 50 mg tissue per 1 mL lysis buffer. Samples were then homogenized using a Fast-Prep24 device (MP Biochemicals) at speed 6.0 for 40 seconds. The resulting homogenate solution was collected into pyrogen-free 1.5 mL Eppendorf tubes and centrifuged at 10,000×g for 15 minutes at 4°C. Seven µL of the resulting supernatant was added to a standard 96-well plate along with 200 µL of the reaction mixture (comprised of 0.005 g O-dianisidine (Sigma-Aldrich), 30 mL of distilled H_2O, 3.33 mL of potassium phosphate buffer, and 17 µL of 1% H_2O_2). Using a microplate scanner (SpectraMax M2^e, Molecular Devices, Sunnyvale, CA), three absorbance readings at 450 nm were recorded every 30 seconds. MPO activity was measured as units of activity per milligram of tissue, with 1 unit of MPO being defined as the amount required to degrade 1 µmol of H_2O_2 per minute at room temperature.

Immunohistochemistry

Samples were dehydrated and embedded in paraffin wax, cut on a cryostat into 8 µm sections, and mounted on poly-D-lysine coated slides. Slides were deparaffinised and incubated in antigen

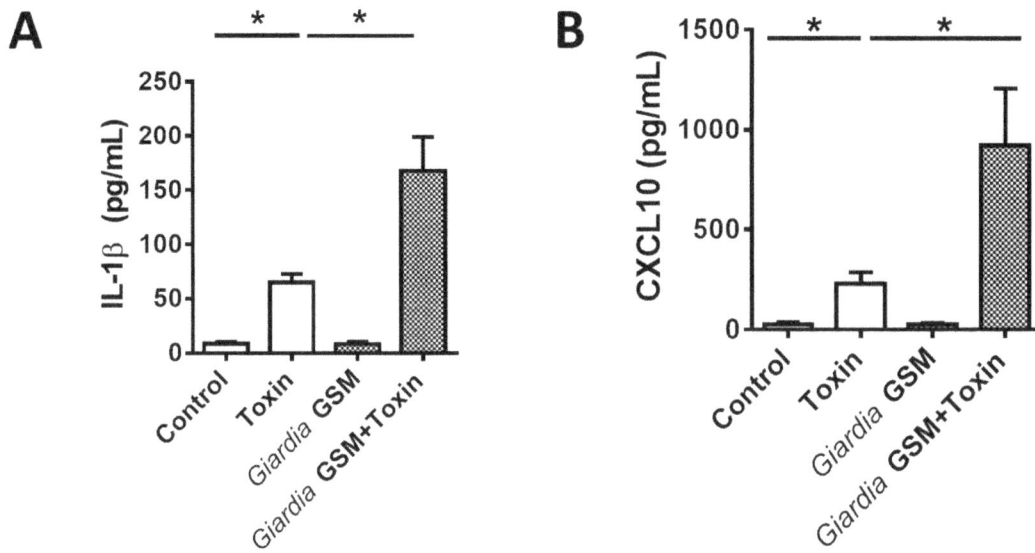

Figure 5. *Giardia* **GS/M infections increase colonic expression levels of IL-1β and CXCL10 following administration of** *C. difficile* **TcdAB.** Male C57BL/6 mice, aged 4 to 6 weeks, were infected with *Giardia* GS/M trophozoites. 7 days post-infection, animals were given 100 μg of TcdAB or PBS i.r. and, after 3 hours, animals were euthanized. Levels of (A) IL-1β and (B) CXCL10 were compared between *Giardia* GS/M-infected animals and uninfected animals given i.r. 100 μg of TcdAB or PBS. Values were determined via bead-based cytokine assay. All data are representative of two independent experiments (n = 5–11/group) and represented as mean ± SEM. * p<0.05.

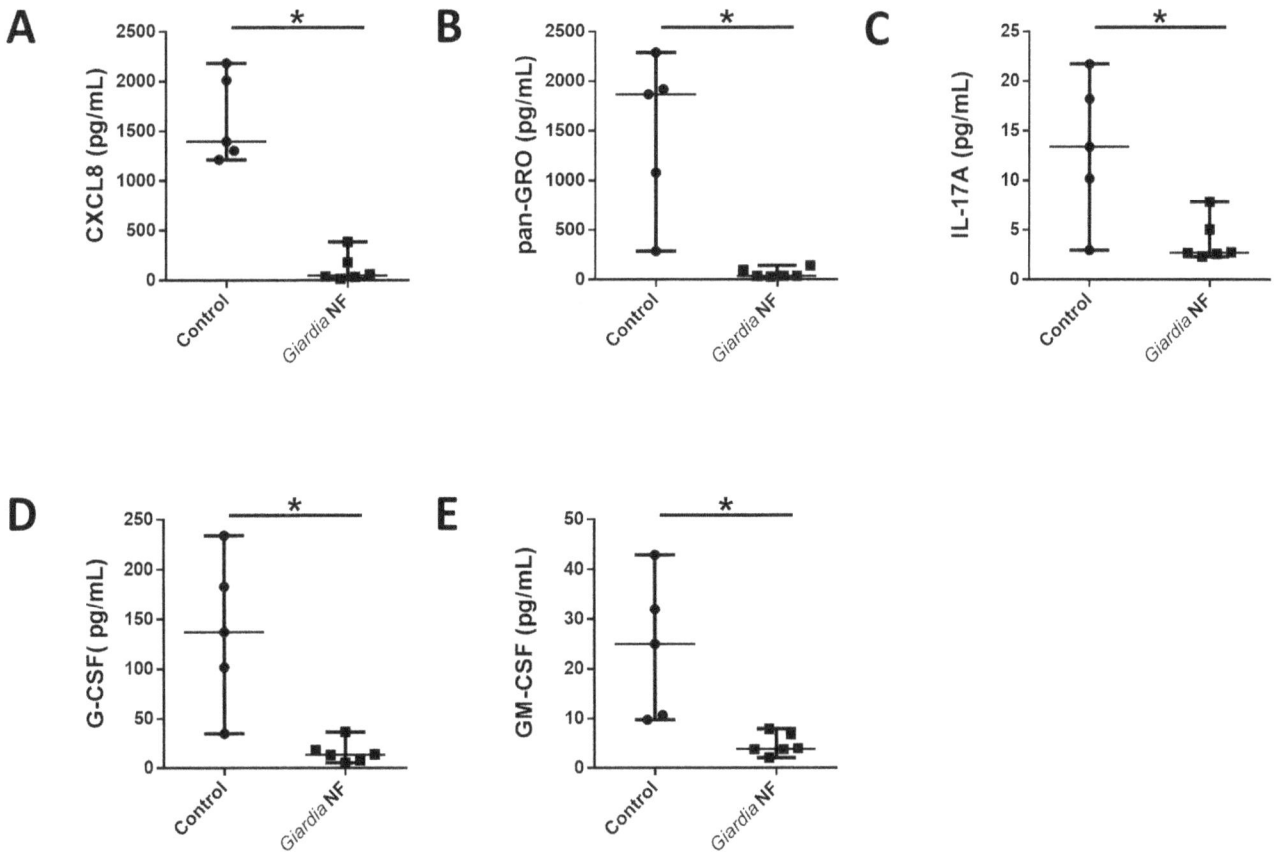

Figure 6. *Giardia* **NF trophozoites attenuate several PMN-associated mediators released from human colonic mucosal biopsies.** Human descending colon mucosal biopsy tissues obtained from areas of active inflammation from patients with active Crohn's disease (CD) were incubated with 5.0×10⁶ *Giardia* NF trophozoites for 6 hours. Supernatant levels of (A) IL-8/CXCL8, (B) GRO, (C) CCL3, (D) IL-17A, (E) G-CSF, and (F) GM-CSF were determined via a bead-based cytokine assay. All data are representative of three independent experiments (n = 4–6/group) and represented as median with the range. * p<0.05.

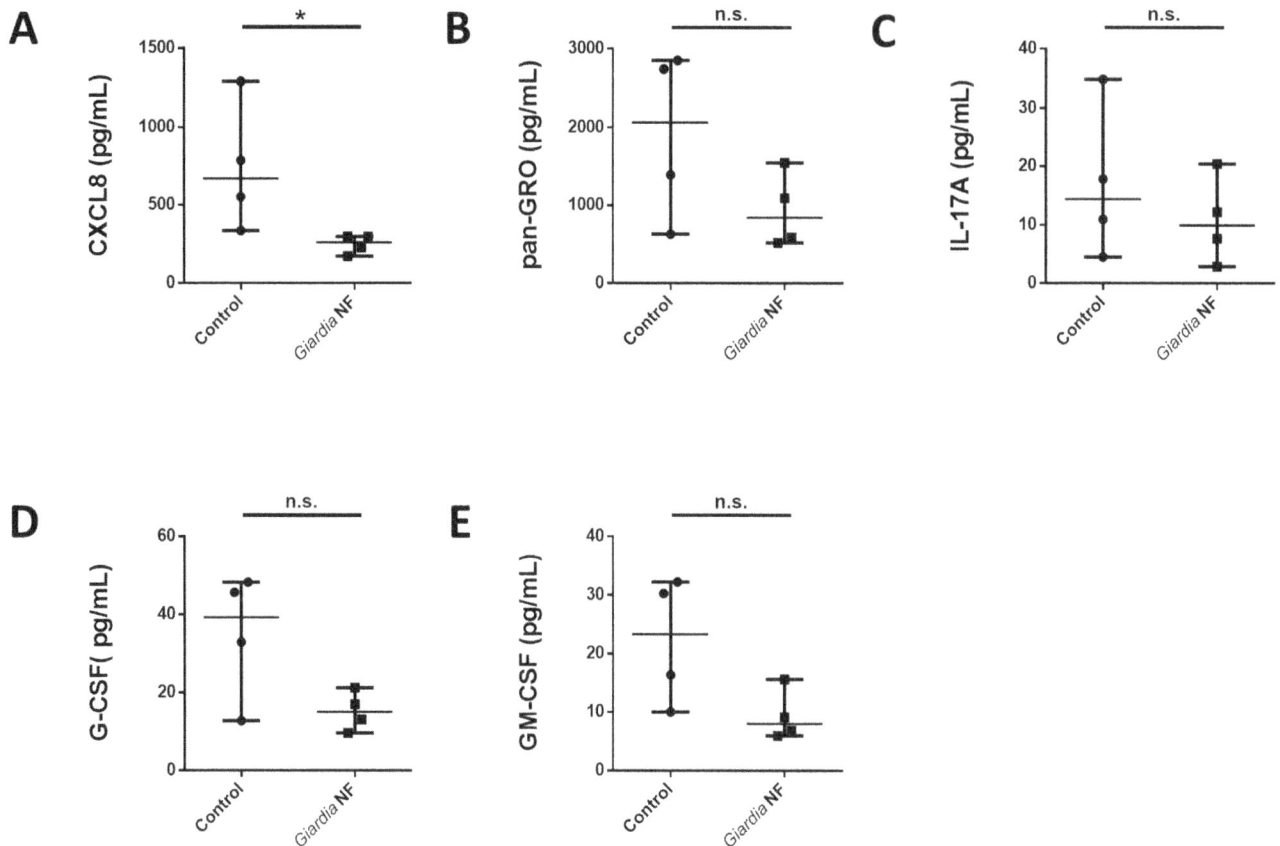

Figure 7. *Giardia* NF trophozoites attenuate the expression of IL-8/CXCL8 assessed in colonic mucosal biopsy homogenates. Human descending colon mucosal biopsy tissues obtained from areas of active inflammation from patients with active Crohn's disease (CD) were incubated with 5.0×10^6 *Giardia* NF trophozoites for 6 hours. Tissue homogenate levels of (A) CXCL8, (B) GRO, (C) CCL3, (D) IL-17A, (E) G-CSF, and (F) GM-CSF were determined via a bead-based cytokine assay. All data are representative of three independent experiments (n = 4–6/group) and represented as median with the range. * $p < 0.05$.

retrieval solution (Tris/EDTA, pH 9.0 solution for 30 minutes at 95 degrees C. Slides were then blocked in tissue blocking buffer (1% BSA, 0.1% Triton X-100 in PBS) for 1 hour (3×20 minute washes) and incubated with an anti-MPO antibody (abcam ab9535) (1 in 200 dilution) overnight at $4°C$. At room temperature, slides were washed for 1 hour (3×20 minutes), incubated for 1 hour with secondary antibody (1 in 1000), and then washed for 1 hour (3×20 minutes). After this, slides were mounted with Fluoroshield™ containing DAPI (Sigma-Aldrich) and visualized using a Leica DMR fluorescence microscope at $40 \times$ magnification.

Statistics

Parametric data are expressed as means ± standard error of measurement, while non-parametric data is expressed as a dotplots with median and range. All statistical analyses were performed using GraphPad Prism 6 software, and normality of the data was assessed prior to analysis. Parametric comparisons were made using one-way ANOVA with Tukey's post hoc analysis. Non-parametric comparisons were made using a Mann Whitney test. Statistical significance was established at $p < 0.05$ (*).

Results

Giardia infections attenuate granulocyte infiltration during TcdAB-induced colitis in an isolate-dependent manner

Experiments were performed to determine whether *in vivo* assemblage A *Giardia* NF or assemblage B *Giardia* GS/M infections were capable of attenuating acute intestinal inflammatory responses to intracolonic administration of *C. difficile* toxin A (TcdA) and toxin B (TcdB). This model of intestinal inflammation was selected because it induces a rapid, acute intestinal inflammatory response resulting from the administration of bacterial toxins, and *Giardia* infections have been reported to occur concurrently with *C. difficile* [16]. Small intestinal trophozoites numbers in *Giardia* GS/M-infected animals were higher than numbers from NF-infected animals (Figure 1A). These results are consistent with previous observations that assemblage B *Giardia* isolates, such as GS/M, have higher parasite burdens compared to assemblage A isolates, such as NF trophozoites [50,54]. Administration of *C. difficile* TcdAB for 3 hours did not change trophozoite numbers (Figure 1A).

As previously demonstrated [52], intracolonic administration of 100 μg of *C. difficile* TcdAB significantly increased colonic tissue MPO activity (Figure 1B and C). As determined via immunofluorescence, numbers of MPO-positive cells increased in colonic tissues of TdcAB-trated animals compared to control animals

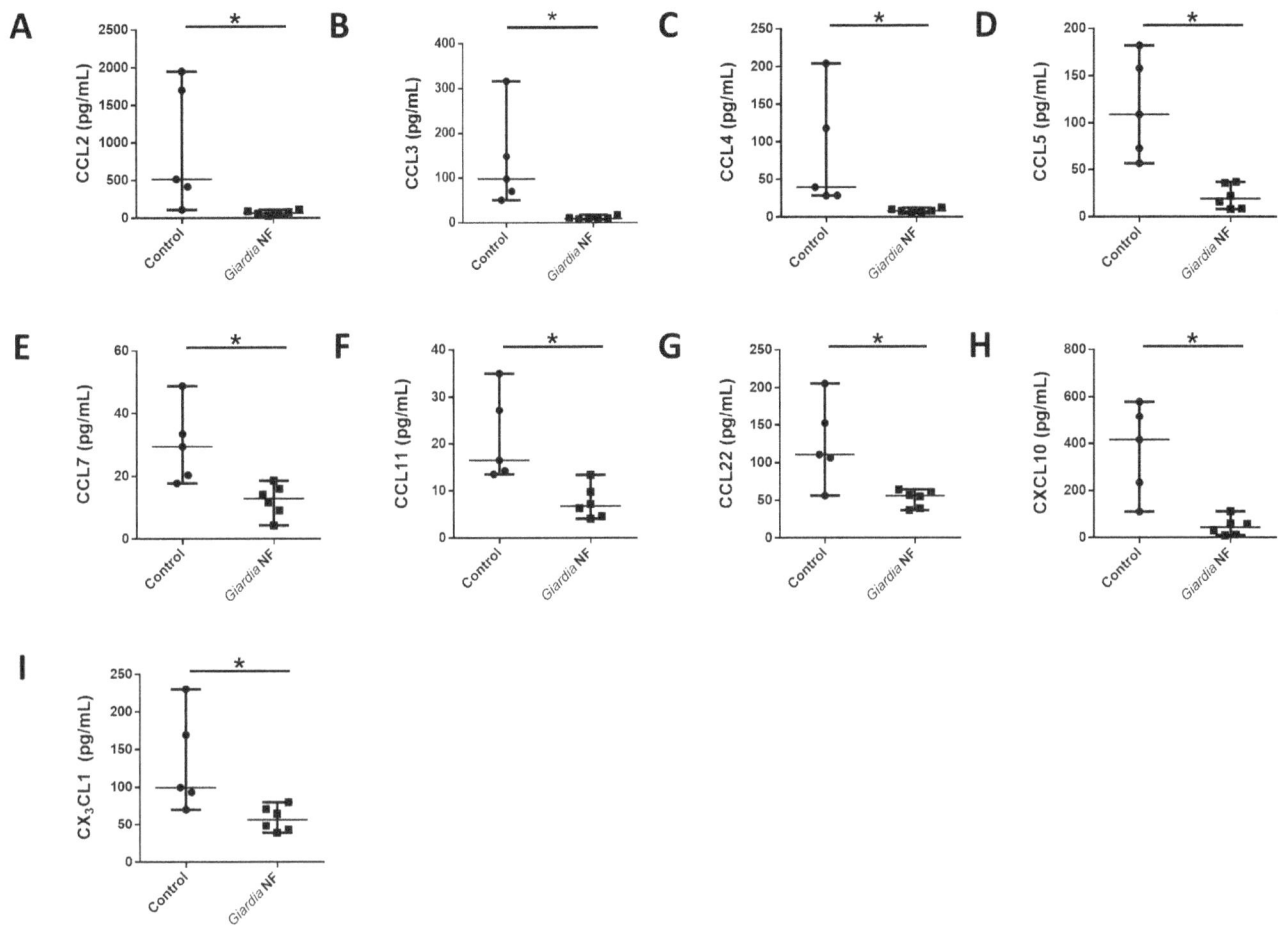

Figure 8. *Giardia* NF trophozoites attenuate the release of several chemokines from colonic mucosal biopsies. Human descending colon mucosal biopsy tissues obtained from areas of active inflammation from patients with active Crohn's disease (CD) were incubated with 5.0×10^6 *Giardia* NF trophozoites for 6 hours. Supernatant levels of (A) CCL2, (B) CCL4, (C) CCL5, (D) CCL7, (E) CCL11, (F) CCL22, (G) CXCL10, and (H) CX$_3$CL1 were determined via a bead-based cytokine assay. All data are representative of three independent experiments (n = 4–6/group) and represented as median with the range. * p<0.05.

(Figure 2). In *Giardia* NF-infected animals administered intracolonically with 100 µg of TcdAB, colonic MPO activity (Figure 1B) and numbers of MPO-positive cells were lower when compared against control animals (Figure 2). Contrastingly, MPO activity levels (Figure 1C) and numbers of MPO-positive cells (Figure 2) in *Giardia* GS/M-infected animals intracolonically instilled with 100 µg of TcdAB were not significantly different from TcdAB-administered controls.

Giardia NF infections attenuate neutrophil-associated cytokines during TcdAB-induced colitis

A bead-based cytokine assay on colonic tissue samples was performed to determine whether *Giardia* NF infections attenuated protein levels of pro-inflammatory cytokines and chemokines associated with granulocyte tissue recruitment. Compared to control animals, intracolonic instillation of 100 µg of TcdAB significantly increased tissue levels of several PMN-associated mediators, including CXCL1, CXCL2, IL-17 and G-CSF (Figure 3). Colonic tissues collected from *Giardia* NF infected-animals administered TcdAB. demonstrated significantly lower protein levels of CXCL1, CXCL2, and IL-17 (Figure 3A to C), while G-CSF protein levels were similar from uninfected TcdAB-treated animals (Figure 3D). Notably, colonic protein levels of

CXCL1, CXCL2, IL-17, and G-CSF in *Giardia* GS/M-infected animals were similar from uninfected TcdAB-treated animals (Figure 3E to H). These results support above observations that granulocyte infiltration is not attenuated in *Giardia* GS/M infected animals instilled with TcdAB, and suggest attenuation of granulocyte infiltration in *Giardia* NF-infected animals given TcdAB may result from decreased expression of PMN-associated mediators. These results also indicate that *Giardia*-mediated attenuation of PMN-associated mediators induced via TcdAB occurs in an isolate-dependent manner.

Giardia GSM reduce colonic expression of several pro-inflammatory cytokines during TcdAB-induced colitis

After focusing on neutrophil-released cytokines, we investigated whether *Giardia* infections attenuated colonic expression of other pro-inflammatory cytokines. Compared to uninfected animals given PBS, intracolonic instillation of TcdAB to uninfected animals also resulted in heightened protein levels of CCL2, IL-6, and leukocyte inhibitory factor (LIF) (Figure 4) and IL-1β, IL-5, CXCL10, and CCL11 (Figure S1). Colonic protein levels of CCL2, IL-6, and LIF were significantly reduced in *Giardia* NF-infected animals given intracolonically with TcdAB (Figure 4A to C). In addition, colonic IL-12p70 levels were significantly greater

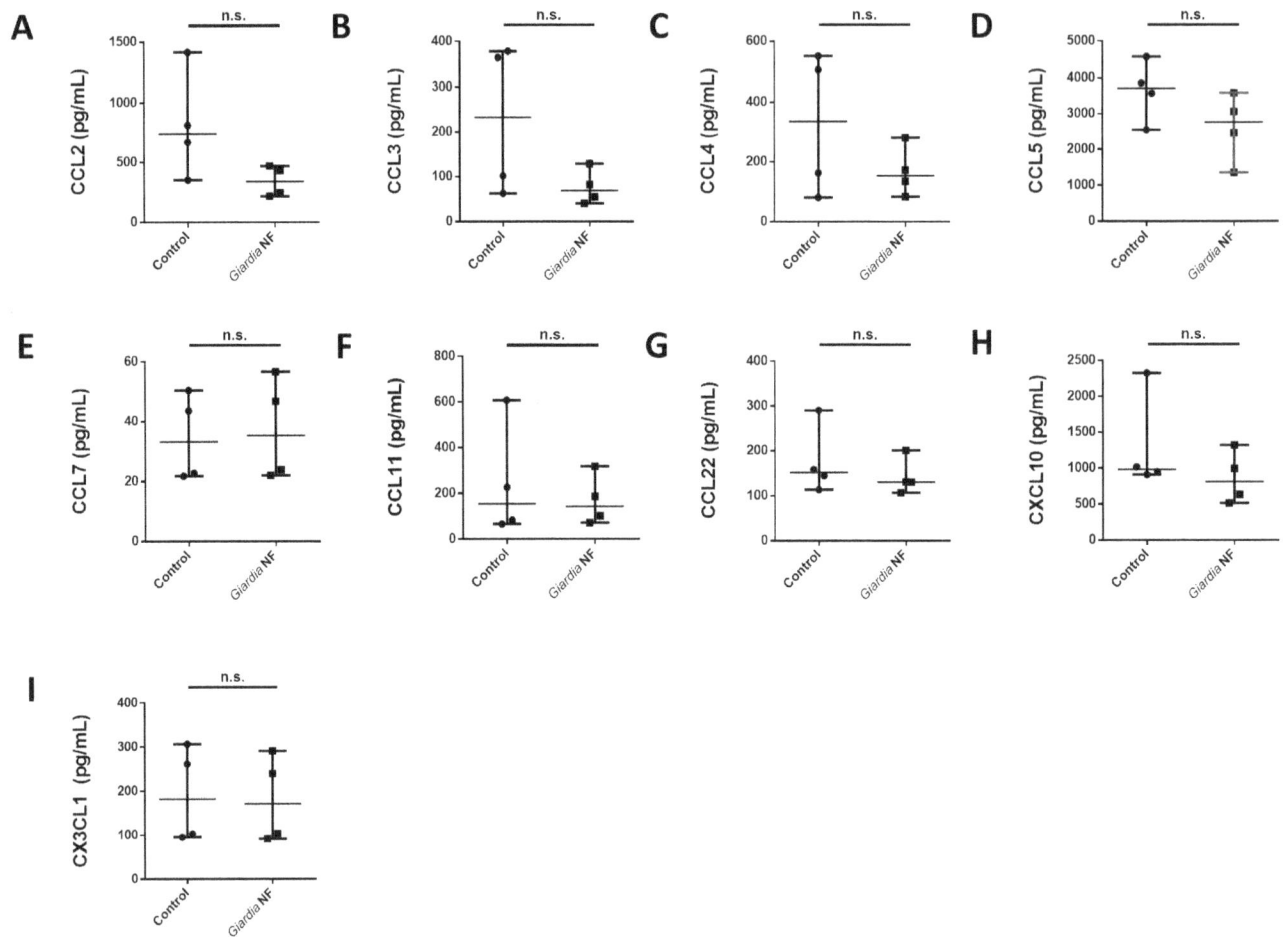

Figure 9. *Giardia* **NF trophozoites do not attenuate the expression of several chemokines assessed in colonic mucosal biopsy tissue homogenates.** Human descending colon mucosal biopsy tissues obtained from areas of active inflammation from patients with active Crohn's disease (CD) were incubated with 5.0×10^6 *Giardia* NF trophozoites for 6 hours. Tissue homogenate levels of (A) CCL2, (B) CCL4, (C) CCL5, (D) CCL7, (E) CCL11, (F) CCL22, (G) CXCL10, and (H) CX$_3$CL1 were determined via a bead-based cytokine assay. All data are representative of three independent experiments (n = 4–6/group) and represented as median with the range. * $p < 0.05$.

in uninfected TcdAB-treated controls when compared to *Giardia* NF infected animals treated with TcdAB (Figure 4D). Conversely, CCL2, IL-6, LIF, and IL-12p70 levels in *Giardia* GS/M-infected animals treated with TcdAB were not significantly different from animals given TcdAB without giardiasis (Figure 4E to H). These results demonstrate that *G. duodenalis* NF attenuate colonic levels of CCL2, IL-6, LIF, and, potentially, IL-12p70 induced by 100 μg intracolonically of TcdAB in an isolate-dependent manner. *Giardia* NF-infections was unable to attenuate colonic expression levels of IL-1β, IL-5, CXCL10, and CCL11 (Fig. S1) Protein levels of IL-1α, IL-7, IL-9, IL-10, IL-12p40, IL-15, IFN-γ, CCL3, CCL4, CXCL9, VEGF remained unchanged between all animal groups (Figure S2). Therefore, following TcdAB instillation *Giardia* NF specifically targeted and decreased colonic expression of pro-inflammatory cytokines that recruit PMNs (CXCL1, CXCL2, and IL-17) and other cytokines (CCL2, IL-6, and LIF).

Bead-based cytokine analysis of colonic tissues indicated *Giardia* GS/M infections upregulated colonic expression levels of a subset of pro-inflammatory mediators, following TcdAB. In uninfected animals, colonic administration of TcdAB significantly increased colonic expression levels of IL-1β and CXCL10 (Figure 5), and protein levels of these mediators were significantly increased in *Giardia* GS/M-infected animals compared to uninfected TcdAB-

treated animals (Figure 5). *Giardia* GS/M-infected animals given TcdAB also demonstrated increased colonic protein levels of several inflammatory mediators not initially increased via colonic instillation of TcdAB (Figure S3). This did not appear to be a global increase in expression of inflammatory mediators, as CCL3 and CCL11 concentrations were similar between *Giardia* GS/M-infected animals given TcdAB and uninfected TcdAB controls (Figure S4). Several inflammatory mediators also remained unchanged between all experimental groups (Figure S5). These results suggest *Giardia* GS/M infections enhance expression of pro-inflammatory cytokines following colonic instillation of TcdAB. These data indicate that *Giardia*-infections can enhance or decrease the expression of inflammatory mediators following administration of a bacterial toxin, events which are isolate-specific.

Giardia NF trophozoites attenuate various inflammatory mediators from inflamed human mucosal biopsy tissues *ex vivo*

Our above results demonstrated *in vivo* that *Giardia* NF attenuated PMN accumulation and expression of pro-inflammatory cytokines following colonic administration of TcdA/TcdB. Therefore, we next assessed whether this same isolate was capable

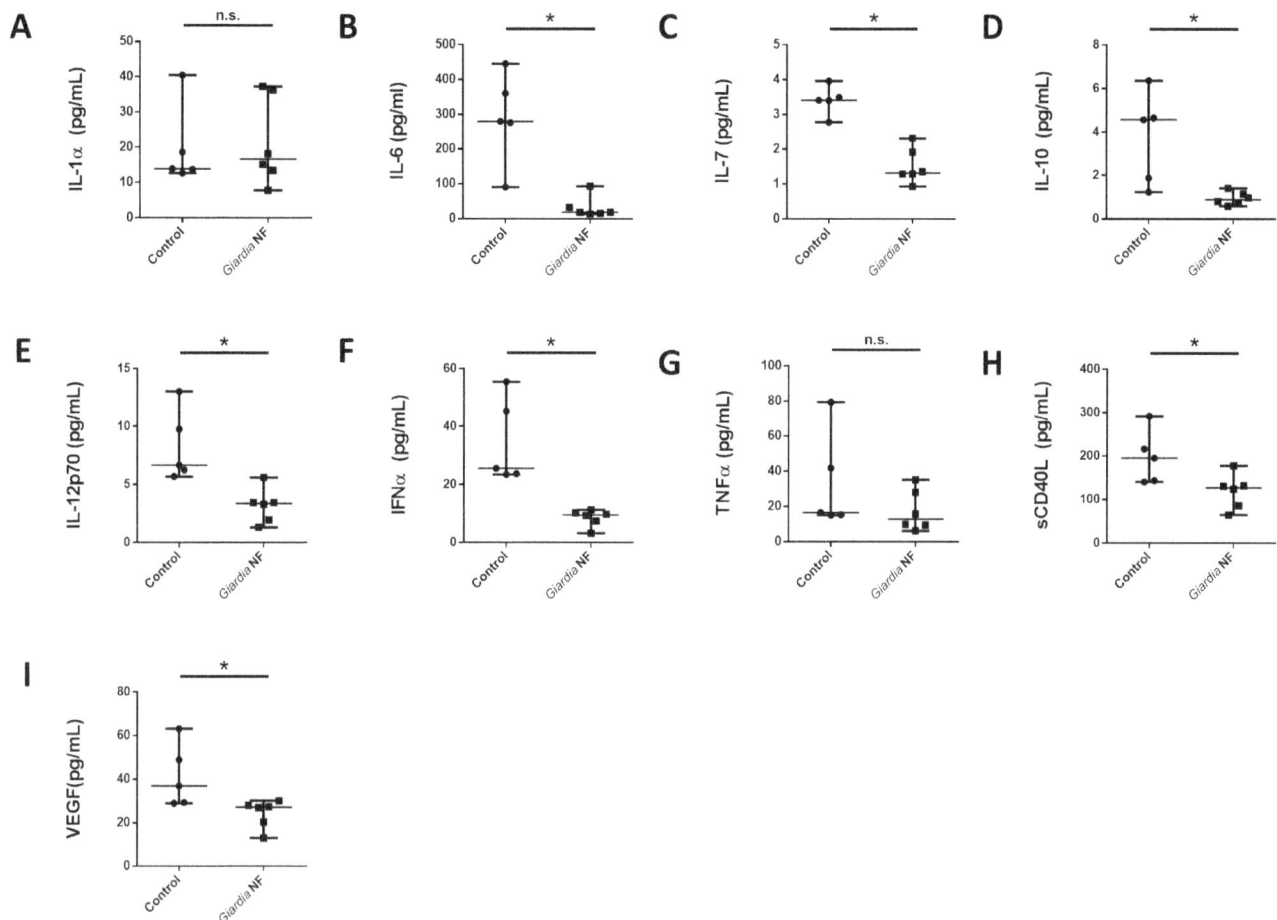

Figure 10. *Giardia* **NF trophozoites attenuate the release of several pro-inflammatory mediators from colonic mucosal biopsies.** Human descending colon mucosal biopsy tissues obtained from areas of active inflammation from patients with active Crohn's disease (CD) were incubated with 5.0×10^6 *Giardia* NF trophozoites for 6 hours. Supernatant levels of (A) IL-1α, (B) IL-6, (C) IL-7, (D) IL-10, (E) IL-12p70, (F) IFNα, (G) TNFα, (H) sCD40L, and (I) VEGF were determined via a bead-based cytokine assay. All data are representative of three independent experiments (n = 4–6/group) and represented as median with the range. * p<0.05.

of attenuating similar inflammatory mediators from inflamed human colonic tissue. Using a previously published model [41], *Giardia* NF trophozoites were co-incubated *ex vivo* with colonic mucosal biopsy tissues collected the descending colon of patients with active CD, where supernatants and biopsy tissue homogenates were analyzed via a bead-based cytokine assay. Supernatants collected from co-incubation of *Giardia* NF trophozoites with inflamed biopsy tissues displayed significantly reduced supernatant levels of cytokines associated with granulocyte recruitment, including IL-8/CXCL8, growth-related oncogene (GRO) family proteins (CXCL1-3), IL-17A, G-CSF, and GM-CSF (Figure 6). Co-incubation of *Giardia* NF trophozoites with biopsy tissues also resulted in decreased IL-8/CXCL8 levels detected within biopsy tissue homogenates (Figure 7A), but not G-CSF and GM-CSF.

As *in vivo Giardia* NF attenuated colonic levels of CCL2, IL-6, LIF, and, potentially, IL-12p70 (Figure 4), we investigated whether live trophozoites were capable of attenuating protein expression of other chemokines. Supernatant levels of CCL2-5, CCL7, CCL11, CCL22, CXCL10, and CX$_3$CL1 were significantly reduced in biopsy tissues co-incubated with *Giardia* NF trophozoites compared to control biopsy tissues (Figure 8). Tissue homogenate levels of these mediators were similar when biopsy tissues were incubated in the presence or absence of *Giardia* NF

trophozoites (Figure 9). Therefore, attenuation of these mediators by *G. duodenalis* NF trophozoites appears to occur following their release into supernatants. These results also suggest that *Giardia* NF trophozoites are capable of attenuating CCL2 in different experimental models of inflammation. Compared to control biopsy supernatant, co-incubation of *Giardia* NF trophozoites with inflamed biopsy tissues *ex vivo* resulted in significant attenuation of IL-6, IL-7, IL-10, IL-12p70, IFN-α, soluble CD40 ligand (sCD40L) and vascular endothelial growth factor (VEGF) (Figure 10). Tissue homogenate levels of IL-1α, IL-6, and TNF-α were significantly reduced in biopsy tissue homogenates co-incubated with *Giardia* NF trophozoites, compared to biopsy tissue homogenates incubated in the absence of *Giardia* NF trophozoites (Figure 11). These results suggest that *Giardia* NF trophozoites attenuate secreted IL-7, IL-10, IL-12p70, IFN-α, sCD40L, and VEGF from inflamed mucosal biopsy tissues *ex vivo*, while also attenuating tissue levels of IL-1α, IL-6, and TNF-α. These results also suggest that *Giardia* infections are capable of attenuating IL-6 and IL-12p70 in during an *in vivo* infection and following incubation with human intestinal biopsy tissues *ex vivo*. Our data also indicated that *Giardia* NF trophozoites did not attenuate the expression of pro-inflammatory cytokines (Figure S6

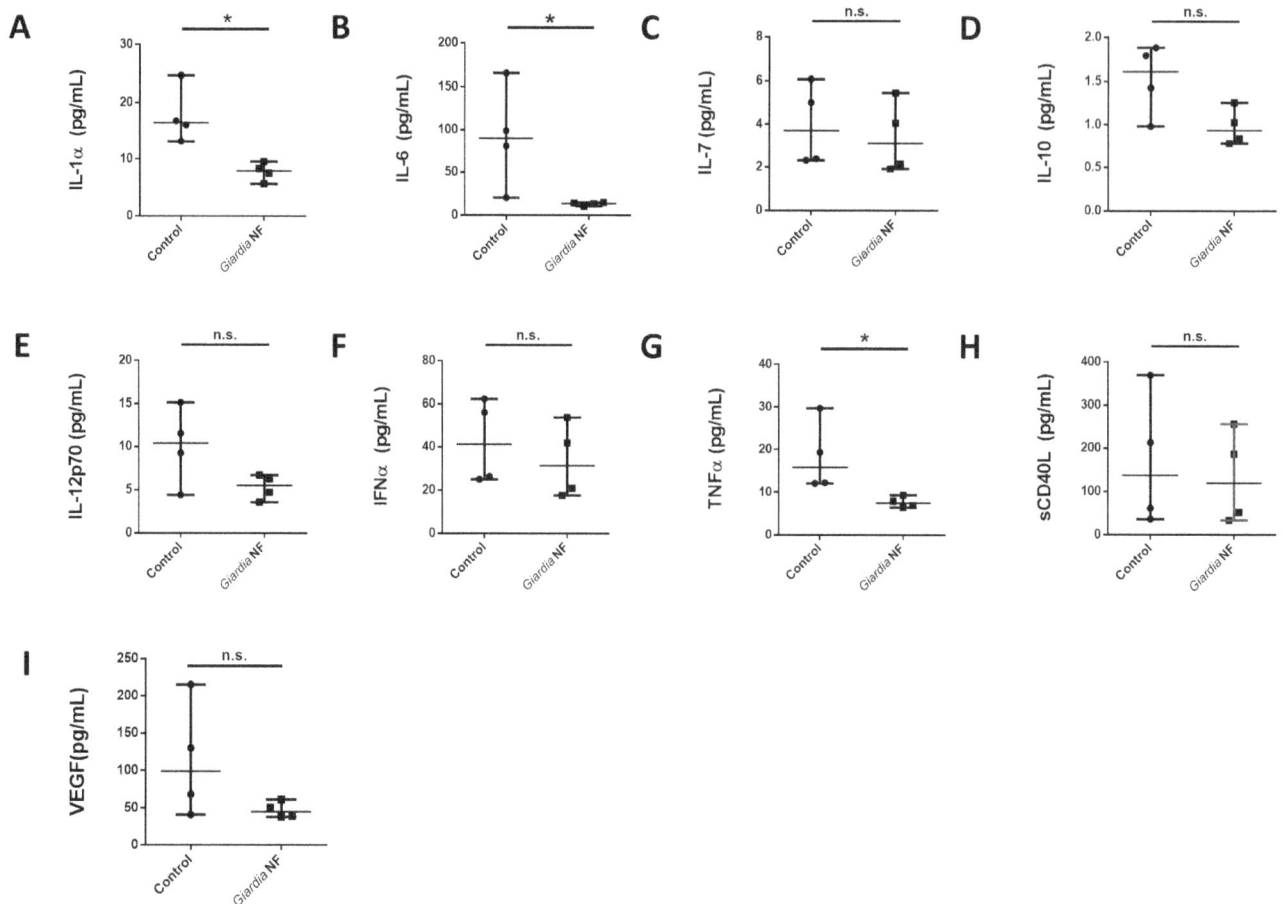

Figure 11. *Giardia* **NF trophozoites attenuate the expression of certain pro-inflammatory mediators assessed in colonic mucosal biopsy tissue homogenates.** Human descending colon mucosal biopsy tissues obtained from areas of active inflammation from patients with active Crohn's disease (CD) were incubated with 5.0×10^6 *Giardia* NF trophozoites for 6 hours. Tissue homogenate levels of (A) IL-1α, (B) IL-6, (C) IL-7, (D) IL-10, (E) IL-12p70, (F) IFNα, (G) TNFα, (H) sCD40L, and (I) VEGF were determined via a bead-based cytokine assay. All data are representative of three independent experiments (n = 4–6/group) and represented as median with the range. * $p < 0.05$.

and S7) or growth factors (Figure S8 and S9) within supernatants or biopsy tissue homogenates.

Discussion

Results from the present study demonstrate novel mechanisms through which certain *Giardia* infections were able to attenuate pro-inflammatory responses elicited by an inflammatory bacterial toxin *in vivo,* or from inflamed human intestinal tissues *ex vivo.* Attenuation of granulocyte infiltration in *Giardia* NF infected animals given TcdAB occurred concomitantly with reduced colonic expression of several inflammatory mediators, including those associated with granulocyte tissue recruitment. Several of these factors were also decreased following co-incubation of *Giardia* NF trophozoites and *ex vivo* inflamed human colonic mucosal biopsy tissues, and, therefore, support our *in vivo* observations that *Giardia* NF trophozoites are capable of attenuating factors associated with PMN recruitment. These results also reinforce previous observations that *Giardia* trophozoites are capable of attenuating IL-8/CXCL8 production from intestinal tissues *in vitro* and *ex vivo* [41]. Furthermore, our data indicate that *Giardia* trophozoites can attenuate the production of additional inflammatory mediators from intestinal tissues of differing origin, strengthening the notion that infection with

specific *Giardia* isolates may modulate host immune responses in the gut. Interestingly *Giardia* GS/M-infected animals did not display reduced granulocyte infiltration nor decreased expression of inflammatory mediators. On the contrary, *Giardia* GS/M infections enhanced the expression of several factors initially upregulated following TcdAB administration and increased the expression of several factors not initially induced by TcdAB exposure. Therefore, our data suggest that *Giardia* infections attenuate intestinal pro-inflammatory responses in an isolate-dependent manner, and attenuation of such responses is associated with reduced tissue granulocyte infiltration. The results show that *Giardia* NF attenuate a variety of pro-inflammatory mediators when co-incubated with inflamed colonic mucosal biopsy tissues *ex vivo,* but the inability to decrease expression of all mediators examined suggests this process may be targeted towards certain inflammatory mediators. Moreover, observations that some inflammatory mediators are preferentially degraded within supernatants while others within tissues suggests that *Giardia* NF trophozoites possess multiple mechanisms capable of attenuating expression of inflammatory mediators in inflamed intestinal mucosal biopsy tissues.

Excessive PMN infiltration and activation contribute to several gastrointestinal inflammatory disorders, including the inflammatory bowel diseases (IBD) and *C. difficile* colitis [55–57]. The present findings indicate that *Giardia* NF infections *in vivo* attenuate PMN tissue recruitment induced by a pro-inflammatory bacterial toxin by reducing tissue expression of mediators that can contribute to the tissue recruitment of PMNs. Release of ELR+ chemokines, such as CXCL1, CXCL2, and IL-8/CXCL8, from a variety of intestinal tissue-resident cells promotes PMN accumulation and chemotaxis through intestinal tissues. In mice, these chemokines include CXCL1 and CXCL2, while in humans this also includes IL-8/CXCL8 [58–61]. Separately, enhanced expression of G-CSF promotes the release of PMNs into the bloodstream [31,62], in a process that can be initiated via IL-17 production [32,33]. Our results suggest that *Giardia* NF infections attenuate granulocyte infiltration into colonic tissues following colonic. TcdA/B by reducing colonic expression levels of CXCL1, CXCL2, and IL-17; these results are supported by observations that *Giardia* GS/M infections failed to attenuate colonic granulocyte infiltration and expression of PMN-associated mediators. In addition, *Giardia* NF parasites attenuated supernatant levels of IL-8/CXCL8, GRO-family proteins (CXCL1-3), IL-17, G-CSF, and GM-CSF when co-incubated with *ex vivo* descending colon mucosal biopsy tissues. Together, these data show for the first time that *Giardia* infections attenuate the release of a variety of mediators associated with PMN recruitment in different experimental models of inflammation, in an isolate-dependent manner.

Further research is needed to demonstrate whether the results from this study explain how *Giardia* infections can, in a strain-dependent manner, modulate the incidence of diarrheal disease [20,22], and create an environment that is favourable for colonization by other gastrointestinal pathogens [18]. Indeed, the data suggest that *Giardia* may attenuate pro-inflammatory responses elicited by other co-infecting pathogens that can cause diarrheal disease, while other strains may enhance its development. Following their recruitment to tissues, PMNs are known to induce multiple pathophysiological events within intestinal tissues capable of causing diarrhea [28,42–44]. Therefore, *Giardia* NF infections may protect against the development of diarrheal disease by attenuating PMN recruitment. *Giardia* GS/M infections did not enhance granulocyte infiltration following colonic of TcdAB; however, these infections resulted in enhanced expression of several pro-inflammatory mediators that could, potentially, enhance the development of diarrheal disease via other pathways. For example, IL-1β and IFNγ have been shown to modulate intestinal barrier dysfunction and ion secretion, and, resultantly, cause diarrhea [63–68]. As PMNs are essential to host defence against a variety of pathogens, more research is needed to determine whether attenuation of PMN recruitment by *Giardia* may also facilitate colonization by other pathogens. To date, research *in vitro* and *in vivo* has demonstrated differences in the pathogenesis of *Giardia* isolates, but has failed to clearly establish whether these differences are assemblage-specific. [10,50,54,69,70].

Data from this study also demonstrate that *Giardia* infections reduce the expression of pro-inflammatory mediators not associated with PMN recruitment, in an isolate-dependent manner. *Giardia* NF infections attenuated colonic expression of IL-6 and the related cytokine LIF induced following i.r. TcdAB, whereas *Giardia* GS/M did not. Moreover, co-incubation of *Giardia* NF trophozoites with inflamed human intestinal mucosal biopsy tissues *ex vivo* also resulted in attenuation of supernatant and tissue IL-6. The ability of *Giardia sp.* to modulate IL-6 during *in vivo* infection remains incompletely understood. These observations are consistent with previous studies that have shown that animals infected with *Giardia muris* display decreased IL-6 levels

in jejunal tissues 5 days post-infection [71], while mice infected with *Giardia* GS/M display elevated IL-6 intestinal tissue levels [72]. The release of IL-6 and the related cytokine LIF can induce the expression of acute phase response proteins, including CRP [73,74]. As Tanzanian children infected with *Giardia* were found to have lower serum CRP than their non-infected counterparts [22], future studies should examine whether *Giardia* infections can modulate the acute phase response in a strain-dependent manner, and establish whether the parasite-mediated attenuation of IL-6 expression is involved in this process.

The present findings illustrate that *Giardia* NF infections attenuate colonic expression of CCL2 and IL-12p70 *in vivo*, as well as in inflamed human intestinal mucosal biopsy tissues *ex vivo*. CCL2 is an important chemokine for monocytes and macrophages (reviewed in [75]). Conversely, co-incubation of *Giardia* trophozoites with *in vitro* Caco-2 monolayers results in increased mRNA expression of CCL2 [76]. Additional research is required in order to understand how *Giardia* infections may modulate monocyte/macrophage recruitment. IL-12p70 is composed of IL-12p35 and IL-12p40, and is the bioactive form responsible for inducing helper T1 (T$_H$1) adaptive immune responses ([77]). Recent research has demonstrated that *Giardia* trophozoites products are capable of attenuating IL-12p70 expression from dendritic cells (DCs) stimulated with bacterial lipopolysaccharide [78]. In contrast other studies found that *Giardia* products may increase DC IL-12p70 expression [79,80]. More research is required in order to determine how *Giardia* infections may modulate CCL2 and IL-12p70 expression in the context of intestinal inflammatory responses. We have recently shown that *Giardia* cathepsin B (catB) cysteine proteases degrade intestinal epithelial IL-8/CXCL8 and can attenuate PMN chemotaxis [41]. It remains to be determined whether *Giardia* catB proteases are involved in attenuating PMN accumulation *in vivo* and during human giardiasis. Genetic manipulation of the *Giardia* genome with Cre/loxP system has recently been demonstrated [81]. This would be extremely beneficial in determining the anti-inflammatory potential of *Giardia* catB proteases *in vivo*. Whether *Giardia* may exert immune-modulatory effects using other secretory-excretory products apart from *Giardia* catB proteases remains largely unknown. *Giardia* arginine deiminase has been shown to attenuate nitric oxide production from intestinal epithelial cells [82–84], to inhibit intestinal epithelial proliferation [85], to induce expression of IL-12p70 from DCs [80], and to inhibit T-cell proliferation [84], however these effects have yet to be characterized *in vivo*.

In conclusion, results from this study demonstrate that *Giardia* infections can attenuate granulocyte infiltration and expression of PMN chemokines in an isolate-dependent manner in an *in vivo* model of infectious colitis. *Giardia* trophozoites were also able to reduce the expression of a variety of inflammatory mediators released from inflamed colonic mucosal biopsy tissues from CD patients. These findings establish the strain-dependent anti-inflammatory properties of *Giardia* in the intestine.

Supporting Information

Figure S1 *Giardia* NF infections *in vivo* do not attenuate colonic expression levels of IL-1β, IL-5, CXCL10, or CCL11 during *C. difficile* TcdA/TcdB-induced colitis. Male C57BL/6 mice, aged 4 to 6 weeks, were infected with *Giardia* NF trophozoites. 7 days post-infection, animals were given 100 μg of TcdA/TcdB or PBS i.r. and, after 3 hours, animals were euthanized. Levels of (A) IL-1β, (B) IL-5, (C) CXCL10, and (D) CCL11 were compared between *Giardia* NF-infected animals and uninfected animals given i.r. 100 μg of TcdA/TcdB or PBS.

Values were determined via bead-based cytokine assay. All data are representative of two independent experiments (n = 5–11/group) and represented as mean ± SEM. n.s. = not significant * p<0.05.

Figure S2 *Giardia* NF infections *in vivo* do not attenuate colonic expression levels of inflammatory mediators not increased during *C. difficile* TcdA/TcdB-induced colitis. Male C57BL/6 mice, aged 4 to 6 weeks, were infected with *Giardia* NF trophozoites. 7 days post-infection, animals were given 100 μg of TcdA/TcdB or PBS i.r. and, after 3 hours, animals were euthanized. Levels of IL-1α (A), IL-2 (B), IL-7 (C), (D) IL-9, (E) IL-10, (F) IL-12p40, (G) IL-15, (H) IFNγ, (I) CCL3, (J) CCL4, (K) CXCL9 and (L) VEGF were compared between *Giardia* NF-infected animals and uninfected animals given i.r. 100 μg of TcdA/TcdB or PBS. Values were determined via bead-based cytokine assay. All data are representative of two independent experiments (n = 5–11/group) and represented as mean ± SEM.

Figure S3 *Giardia* GS/M infections *in vivo* upregulate colonic expression levels of several inflammatory mediators not increased during *C. difficile* TcdA/TcdB-induced colitis. Male C57BL/6 mice, aged 4 to 6 weeks, were infected with *Giardia* GS/M trophozoites. 7 days post-infection, animals were given 100 μg of TcdA/TcdB or PBS i.r. and, after 3 hours, animals were euthanized. Levels of (A) IL-2, (B) IL-5, (C) IL-15, (D) IL-17, (E) IFNγ, (F) CXCL9, (G) CCL4, and (H) VEGF were compared between *Giardia* GS/M-infected animals and uninfected animals given i.r. 100 μg of TcdA/TcdB or PBS. Values were determined via bead-based cytokine assay. All data are representative of two independent experiments (n = 5–11/group) and represented as mean ± SEM. * p<0.05.

Figure S4 *Giardia* GS/M infections *in vivo* do not modulate colonic expression levels of CCL3 and CCL11 during *C. difficile* TcdA/TcdB-induced colitis. Male C57BL/6 mice, aged 4 to 6 weeks, were infected with *Giardia* GS/M trophozoites. 7 days post-infection, animals were given 100 μg of TcdA/TcdB or PBS i.r. and, after 3 hours, animals were euthanized. Levels of (A) CCL3 and (B) CCL11 were compared between *Giardia* GS/M-infected animals and uninfected animals given i.r. 100 μg of TcdA/TcdB or PBS. Values were determined via bead-based cytokine assay. All data are representative of two independent experiments (n = 5–11/group) and represented as mean ± SEM. * p<0.05.

Figure S5 *Giardia* GS/M infections *in vivo* do not attenuate colonic expression levels of inflammatory mediators not increased during *C. difficile* TcdA/TcdB-induced colitis. Male C57BL/6 mice, aged 4 to 6 weeks, were infected with *Giardia* GS/M trophozoites. 7 days post-infection, animals were given 100 μg of TcdA/TcdB or PBS i.r. and, after 3 hours, animals were euthanized. Levels of (A) IL-1α, (B) IL-7, (C) IL-9, (D) IL-10, (E) IL-12p40, and (F) TNFα were compared between *Giardia* GS/M-infected animals and uninfected animals given i.r. 100 μg of TcdA/TcdB or PBS. Values were determined via bead-based

cytokine assay. All data are representative of two independent experiments (n = 5–11/group) and represented as mean ± SEM.

Figure S6 *Giardia* NF trophozoites do attenuate supernatant levels of several inflammatory mediators released from descending colon mucosal biopsy tissues. Human descending colon mucosal biopsy tissues obtained in areas of active inflammation from patients with active Crohn's disease (CD) were incubated with 5.0×10^6 *Giardia* NF trophozoites for 6 hours. Supernatant levels of (A) IL-1ra, (B) IL-1β, (C) IL-2, (D) IL-12p40, (E) IL-13, (F) IL-15, (G) IFNγ, and (H) TNFβ were determined via a bead-based cytokine assay. All data are representative of three independent experiments (n = 4–6/group) and represented as median with the range. n.s. = not significant.

Figure S7 *Giardia* NF trophozoites do attenuate levels of several inflammatory mediators released from descending colon mucosal biopsy homogenates. Human descending colon mucosal biopsy tissues obtained in areas of active inflammation from patients with active Crohn's disease (CD) were incubated with 5.0×10^6 *Giardia* NF trophozoites for 6 hours. Tissue homogenate levels of (A) IL-1ra, (B) IL-1β, (C) IL-2, (D) IL-12p40, (E) IL-13, (F) IL-15, (G) IFNγ, and (H) TNFβ were determined via a bead-based cytokine assay. All data are representative of three independent experiments (n = 4–6/group) and represented as median with the range. n.s. = not significant.

Figure S8 *Giardia* NF trophozoites do attenuate supernatant levels of several growth factors released from descending colon mucosal biopsy tissues. Human descending colon mucosal biopsy tissues obtained in areas of active inflammation from patients with active Crohn's disease (CD) were incubated with 5.0×10^6 *Giardia* NF trophozoites for 6 hours. Supernatant levels of (A) FGF, (B) Flt-3L, (C) PDGF-AA, (D) PDGF-BB, and (E) TGFβ were determined via a bead-based cytokine assay. All data are representative of three independent experiments (n = 4–6/group) and represented as median with the range. n.s. = not significant.

Figure S9 *Giarida* NF trophozoites do attenuate levels of several growth factors released from descending colon mucosal biopsy tissue homogenates. Human descending colon mucosal biopsy tissues obtained in areas of active inflammation from patients with active Crohn's disease (CD) were incubated with 5.0×10^6 *G. duodenalis* NF trophozoites for 6 hours. Tissue homogenate levels of (A) FGF, (B) Flt-3L, (C) PDGF-AA, (D) PDGF-BB, and (E) TGFβ were determined via a bead-based cytokine assay. All data are representative of three independent experiments (n = 4–6/group) and represented as median with the range. n.s. = not significant.

Author Contributions

Conceived and designed the experiments: JAC SAH AGB. Performed the experiments: JAC JPM. Analyzed the data: JAC AGB. Contributed reagents/materials/analysis tools: PLB AGB LPS. Contributed to the writing of the manuscript: JAC AGB.

References

1. Monis PT, Andrews RH, Mayrhofer G, Ey PL (1999) Molecular systematics of the parasitic protozoan Giardia intestinalis. Mol Biol Evol 16: 1135–1144.
2. Lasek-Nesselquist E, Welch DM, Sogin ML (2010) The identification of a new Giardia duodenalis assemblage in marine vertebrates and a preliminary analysis

of G. duodenalis population biology in marine systems. Int J Parasitol 40: 1063–1074.
3. Franzen O, Jerlstrom-Hultqvist J, Castro E, Sherwood E, Ankarklev J, et al. (2009) Draft genome sequencing of giardia intestinalis assemblage B isolate GS: is human giardiasis caused by two different species? PLoS Pathog 5: e1000560.

4. Jerlstrom-Hultqvist J, Ankarklev J, Svard SG (2010) Is human giardiasis caused by two different Giardia species? Gut Microbes 1: 379–382.

5. Cotton JA, Beatty JK, Buret AG (2011) Host parasite interactions and pathophysiology in Giardia infections. Int J Parasitol 41: 925–933.

6. Wensaas KA, Langeland N, Hanevik K, Morch K, Eide GE, et al. (2012) Irritable bowel syndrome and chronic fatigue 3 years after acute giardiasis: historic cohort study. Gut 61: 214–219.

7. Halliez MC, Buret AG (2013) Extra-intestinal and long term consequences of Giardia duodenalis infections. World J Gastroenterol 19: 8974–8985.

8. Oberhuber G, Kastner N, Stolte M (1997) Giardiasis: a histologic analysis of 567 cases. Scand J Gastroenterol 32: 48–51.

9. Scott KG, Meddings JB, Kirk DR, Lees-Miller SP, Buret AG (2002) Intestinal infection with Giardia spp. reduces epithelial barrier function in a myosin light chain kinase-dependent fashion. Gastroenterology 123: 1179–1190.

10. Chin AC, Teoh DA, Scott KG, Meddings JB, Macnaughton WK, et al. (2002) Strain-dependent induction of enterocyte apoptosis by Giardia lamblia disrupts epithelial barrier function in a caspase-3-dependent manner. Infect Immun 70: 3673–3680.

11. Troeger H, Epple HJ, Schneider T, Wahnschaffe U, Ullrich R, et al. (2007) Effect of chronic Giardia lamblia infection on epithelial transport and barrier function in human duodenum. Gut 56: 328–335.

12. Hanevik K, Hausken T, Morken MH, Strand EA, Morch K, et al. (2007) Persisting symptoms and duodenal inflammation related to Giardia duodenalis infection. J Infect 55: 524–530.

13. Chen TL, Chen S, Wu HW, Lee TC, Lu YZ, et al. (2013) Persistent gut barrier damage and commensal bacterial influx following eradication of Giardia infection in mice. Gut Pathog 5: 26.

14. Ankarklev J, Jerlstrom-Hultqvist J, Ringqvist E, Troell K, Svard SG (2010) Behind the smile: cell biology and disease mechanisms of Giardia species. Nat Rev Microbiol 8: 413–422.

15. Hagel I, Cabrera M, Puccio F, Santaella C, Buvat E, et al. (2011) Co-infection with Ascaris lumbricoides modulates protective immune responses against Giardia duodenalis in school Venezuelan rural children. Acta Trop 117: 189–195.

16. Wang L, Xiao L, Duan L, Ye J, Guo Y, et al. (2013) Concurrent infections of Giardia duodenalis, Enterocytozoon bieneusi, and Clostridium difficile in children during a cryptosporidiosis outbreak in a pediatric hospital in China. PLoS Negl Trop Dis 7: e2437.

17. Ankarklev J, Hestvik E, Lebbad M, Lindh J, Kaddu-Mulindwa DH, et al. (2012) Common coinfections of Giardia intestinalis and Helicobacter pylori in non-symptomatic Ugandan children. PLoS Negl Trop Dis 6: e1780.

18. Mukherjee AK, Chowdhury P, Rajendran K, Nozaki T, Ganguly S (2014) Association between Giardia duodenalis and Coinfection with Other Diarrhea-Causing Pathogens in India. Biomed Res Int 2014: 786480.

19. Oberhelman RA, Flores-Abuxapqui J, Suarez-Hoil G, Puc-Franco M, Heredia-Navarrete M, et al. (2001) Asymptomatic salmonellosis among children in day-care centers in Merida, Yucatan, Mexico. Pediatr Infect Dis J 20: 792–797.

20. Bilenko N, Levy A, Dagan R, Deckelbaum RJ, El-On Y, et al. (2004) Does co-infection with Giardia lamblia modulate the clinical characteristics of enteric infections in young children? Eur J Epidemiol 19: 877–883.

21. Bhavnani D, Goldstick JE, Cevallos W, Trueba G, Eisenberg JN (2012) Synergistic effects between rotavirus and coinfecting pathogens on diarrheal disease: evidence from a community-based study in northwestern Ecuador. Am J Epidemiol 176: 387–395.

22. Veenemans J, Mank T, Ottenhof M, Baidjoe A, Mbugi EV, et al. (2011) Protection against diarrhea associated with Giardia intestinalis Is lost with multi-nutrient supplementation: a study in Tanzanian children. PLoS Negl Trop Dis 5: e1158.

23. Kostman R (1975) Infantile genetic agranulocytosis. A review with presentation of ten new cases. Acta Paediatr Scand 64: 362–368.

24. Zeidler C, Germeshausen M, Klein C, Welte K (2009) Clinical implications of ELA2-, HAX1-, and G-CSF-receptor (CSF3R) mutations in severe congenital neutropenia. Br J Haematol 144: 459–467.

25. Kuijpers T, Lutter R (2012) Inflammation and repeated infections in CGD: two sides of a coin. Cell Mol Life Sci 69: 7–15.

26. Szabady RL, McCormick BA (2013) Control of neutrophil inflammation at mucosal surfaces by secreted epithelial products. Front Immunol 4: 220.

27. Fournier BM, Parkos CA (2012) The role of neutrophils during intestinal inflammation. Mucosal Immunol 5: 354–366.

28. Chin AC, Parkos CA (2007) Pathobiology of neutrophil transepithelial migration: implications in mediating epithelial injury. Annu Rev Pathol 2: 111–143.

29. Summers C, Rankin SM, Condliffe AM, Singh N, Peters AM, et al. (2010) Neutrophil kinetics in health and disease. Trends Immunol 31: 318–324.

30. Kolaczkowska E, Kubes P (2013) Neutrophil recruitment and function in health and inflammation. Nat Rev Immunol 13: 159–175.

31. Wengner AM, Pitchford SC, Furze RC, Rankin SM (2008) The coordinated action of G-CSF and ELR+CXC chemokines in neutrophil mobilization during acute inflammation. Blood 111: 42–49.

32. Mei J, Liu Y, Dai N, Hoffmann C, Hudock KM, et al. (2012) Cxcr2 and Cxcl5 regulate the IL-17/G-CSF axis and neutrophil homeostasis in mice. J Clin Invest 122: 974–986.

33. Schwarzenberger P, Huang W, Ye P, Oliver P, Manuel M, et al. (2000) Requirement of endogenous stem cell factor and granulocyte-colony-stimulating factor for IL-17-mediated granulopoiesis. J Immunol 164: 4783–4789.

34. Semerad CL, Liu F, Gregory AD, Stumpf K, Link DC (2002) G-CSF is an essential regulator of neutrophil trafficking from the bone marrow to the blood. Immunity 17: 413–423.

35. Massena S, Christoffersson G, Hjertstrom E, Zcharia E, Vlodavsky I, et al. (2010) A chemotactic gradient sequestered on endothelial heparan sulfate induces directional intraluminal crawling of neutrophils. Blood 116: 1924–1931.

36. Ley K, Laudanna C, Cybulsky MI, Nourshargh S (2007) Getting to the site of inflammation: the leukocyte adhesion cascade updated. Nat Rev Immunol 7: 678–689.

37. Chou RC, Kim ND, Sadik CD, Seung E, Lan Y, et al. (2010) Lipid-cytokine-chemokine cascade drives neutrophil recruitment in a murine model of inflammatory arthritis. Immunity 33: 266–278.

38. McDonald B, Pittman K, Menezes GB, Hirota SA, Slaba I, et al. (2010) Intravascular danger signals guide neutrophils to sites of sterile inflammation. Science 330: 362–366.

39. Lee HY, Hyung S, Lee NY, Yong TS, Han SH, et al. (2012) Excretory-secretory products of Giardia lamblia induce interleukin-8 production in human colonic cells via activation of p38, ERK1/2, NF-kappaB and AP-1. Parasite Immunol 34: 183–198.

40. Fisher BS, Estrano CE, Cole JA (2013) Modeling long-term host cell-Giardia lamblia interactions in an in vitro co-culture system. PLoS One 8: e81104.

41. Cotton JA, Bhargava A, Ferraz JG, Yates RM, Beck PL, et al. (2014) Giardia duodenalis cathepsin B proteases degrade intestinal epithelial interleukin-8 and attenuate interleukin-8-induced neutrophil chemotaxis. Infect Immun.

42. Madara JL, Patapoff TW, Gillece-Castro B, Colgan SP, Parkos CA, et al. (1993) 5′-adenosine monophosphate is the neutrophil-derived paracrine factor that elicits chloride secretion from T84 intestinal epithelial cell monolayers. J Clin Invest 91: 2320–2325.

43. Strohmeier GR, Lencer WI, Patapoff TW, Thompson LF, Carlson SL, et al. (1997) Surface expression, polarization, and functional significance of CD73 in human intestinal epithelia. J Clin Invest 99: 2588–2601.

44. Weissmuller T, Campbell EL, Rosenberger P, Scully M, Beck PL, et al. (2008) PMNs facilitate translocation of platelets across human and mouse endothelium and together alter fluid homeostasis via epithelial cell-expressed ecto-NTPDases. J Clin Invest 118: 3682–3692.

45. Hirota SA, Fines K, Ng J, Traboulsi D, Lee J, et al. (2010) Hypoxia-inducible factor signaling provides protection in Clostridium difficile-induced intestinal injury. Gastroenterology 139: 259–269 e253.

46. Teoh DA, Kamienecki D, Pang G, Buret AG (2000) Giardia lamblia rearranges F-actin and alpha-actinin in human colonic and duodenal monolayers and reduces transepithelial electrical resistance. J Parasitol 86: 800–806.

47. Aggarwal A, Merritt JW Jr, Nash TE (1989) Cysteine-rich variant surface proteins of Giardia lamblia. Mol Biochem Parasitol 32: 39–47.

48. Diamond LS, Harlow DR, Cunnick CC (1978) A new medium for the axenic cultivation of Entamoeba histolytica and other Entamoeba. Trans R Soc Trop Med Hyg 72: 431–432.

49. Keister DB (1983) Axenic culture of Giardia lamblia in TYI-S-33 medium supplemented with bile. Trans R Soc Trop Med Hyg 77: 487–488.

50. Solaymani-Mohammadi S, Singer SM (2011) Host immunity and pathogen strain contribute to intestinal disaccharidase impairment following gut infection. J Immunol 187: 3769–3775.

51. Singer SM, Nash TE (2000) The role of normal flora in Giardia lamblia infections in mice. J Infect Dis 181: 1510–1512.

52. Hirota SA, Iablokov V, Tulk SE, Schenck LP, Becker H, et al. (2012) Intrarectal instillation of Clostridium difficile toxin A triggers colonic inflammation and tissue damage: development of a novel and efficient mouse model of Clostridium difficile toxin exposure. Infect Immun 80: 4474–4484.

53. Arndt H, Kubes P, Grisham MB, Gonzalez E, Granger DN (1992) Granulocyte turnover in the feline intestine. Inflammation 16: 549–559.

54. Bartelt LA, Roche J, Kolling G, Bolick D, Noronha F, et al. (2013) Persistent G. lamblia impairs growth in a murine malnutrition model. J Clin Invest 123: 2672–2684.

55. Kumar NB, Nostrant TT, Appelman HD (1982) The histopathologic spectrum of acute self-limited colitis (acute infectious-type colitis). Am J Surg Pathol 6: 523–529.

56. Chin AC, Parkos CA (2006) Neutrophil transepithelial migration and epithelial barrier function in IBD: potential targets for inhibiting neutrophil trafficking. Ann N Y Acad Sci 1072: 276–287.

57. Kelly CP, Becker S, Linevsky JK, Joshi MA, O'Keane JC, et al. (1994) Neutrophil recruitment in Clostridium difficile toxin A enteritis in the rabbit. J Clin Invest 93: 1257–1265.

58. Spehlmann ME, Dann SM, Hruz P, Hanson E, McCole DF, et al. (2009) CXCR2-dependent mucosal neutrophil influx protects against colitis-associated diarrhea caused by an attaching/effacing lesion-forming bacterial pathogen. J Immunol 183: 3332–3343.

59. Ranganathan P, Jayakumar C, Manicassamy S, Ramesh G (2013) CXCR2 knockout mice are protected against DSS-colitis-induced acute kidney injury and inflammation. Am J Physiol Renal Physiol 305: F1422–1427.

60. Jamieson T, Clarke M, Steele CW, Samuel MS, Neumann J, et al. (2012) Inhibition of CXCR2 profoundly suppresses inflammation-driven and spontaneous tumorigenesis. J Clin Invest 122: 3127–3144.

61. Buanne P, Di Carlo E, Caputi L, Brandolini L, Mosca M, et al. (2007) Crucial pathophysiological role of CXCR2 in experimental ulcerative colitis in mice. J Leukoc Biol 82: 1239–1246.

62. Eash KJ, Greenbaum AM, Gopalan PK, Link DC (2010) CXCR2 and CXCR4 antagonistically regulate neutrophil trafficking from murine bone marrow. J Clin Invest 120: 2423–2431.

63. Utech M, Ivanov AI, Samarin SN, Bruewer M, Turner JR, et al. (2005) Mechanism of IFN-gamma-induced endocytosis of tight junction proteins: myosin II-dependent vacuolarization of the apical plasma membrane. Mol Biol Cell 16: 5040–5052.

64. Wang F, Schwarz BT, Graham WV, Wang Y, Su L, et al. (2006) IFN-gamma-induced TNFR2 expression is required for TNF-dependent intestinal epithelial barrier dysfunction. Gastroenterology 131: 1153–1163.

65. Al-Sadi R, Guo S, Ye D, Dokladny K, Alhmoud T, et al. (2013) Mechanism of IL-1beta modulation of intestinal epithelial barrier involves p38 kinase and activating transcription factor-2 activation. J Immunol 190: 6596–6606.

66. Paul G, Marchelletta RR, McCole DF, Barrett KE (2012) Interferon-gamma alters downstream signaling originating from epidermal growth factor receptor in intestinal epithelial cells: functional consequences for ion transport. J Biol Chem 287: 2144–2155.

67. Sugi K, Musch MW, Field M, Chang EB (2001) Inhibition of Na+, K+-ATPase by interferon gamma down-regulates intestinal epithelial transport and barrier function. Gastroenterology 120: 1393–1403.

68. Bertelsen LS, Eckmann L, Barrett KE (2004) Prolonged interferon-gamma exposure decreases ion transport, NKCC1, and Na+-K+-ATPase expression in human intestinal xenografts in vivo. Am J Physiol Gastrointest Liver Physiol 286: G157–165.

69. Panaro MA, Cianciulli A, Mitolo V, Mitolo CI, Acquafredda A, et al. (2007) Caspase-dependent apoptosis of the HCT-8 epithelial cell line induced by the parasite Giardia intestinalis. FEMS Immunol Med Microbiol 51: 302–309.

70. Cevallos A, Carnaby S, James M, Farthing JG (1995) Small intestinal injury in a neonatal rat model of giardiasis is strain dependent. Gastroenterology 109: 766–773.

71. Scott KG, Logan MR, Klammer GM, Teoh DA, Buret AG (2000) Jejunal brush border microvillous alterations in Giardia muris-infected mice: role of T lymphocytes and interleukin-6. Infect Immun 68: 3412–3418.

72. Zhou P, Li E, Zhu N, Robertson J, Nash T, et al. (2003) Role of interleukin-6 in the control of acute and chronic Giardia lamblia infections in mice. Infect Immun 71: 1566–1568.

73. Pepys MB, Hirschfield GM (2003) C-reactive protein: a critical update. J Clin Invest 111: 1805–1812.

74. Mayer P, Geissler K, Ward M, Metcalf D (1993) Recombinant human leukemia inhibitory factor induces acute phase proteins and raises the blood platelet counts in nonhuman primates. Blood 81: 3226–3233.

75. Deshmane SL, Kremlev S, Amini S, Sawaya BE (2009) Monocyte chemoattractant protein-1 (MCP-1): an overview. J Interferon Cytokine Res 29: 313–326.

76. Roxstrom-Lindquist K, Ringqvist E, Palm D, Svard S (2005) Giardia lamblia-induced changes in gene expression in differentiated Caco-2 human intestinal epithelial cells. Infect Immun 73: 8204–8208.

77. Gee K, Guzzo C, Che Mat NF, Ma W, Kumar A (2009) The IL-12 family of cytokines in infection, inflammation and autoimmune disorders. Inflamm Allergy Drug Targets 8: 40–52.

78. Kamda JD, Singer SM (2009) Phosphoinositide 3-kinase-dependent inhibition of dendritic cell interleukin-12 production by Giardia lamblia. Infect Immun 77: 685–693.

79. Banik S, Renner Viveros P, Seeber F, Klotz C, Ignatius R, et al. (2013) Giardia duodenalis arginine deiminase modulates the phenotype and cytokine secretion of human dendritic cells by depletion of arginine and formation of ammonia. Infect Immun 81: 2309–2317.

80. Obendorf J, Renner Viveros P, Fehlings M, Klotz C, Aebischer T, et al. (2013) Increased expression of CD25, CD83, and CD86, and secretion of IL-12, IL-23, and IL-10 by human dendritic cells incubated in the presence of Toll-like receptor 2 ligands and Giardia duodenalis. Parasit Vectors 6: 317.

81. Wampfler PB, Faso C, Hehl AB (2014) The Cre/loxP system in Giardia lamblia: genetic manipulations in a binucleate tetraploid protozoan. Int J Parasitol 44: 497–506.

82. Ringqvist E, Palm JE, Skarin H, Hehl AB, Weiland M, et al. (2008) Release of metabolic enzymes by Giardia in response to interaction with intestinal epithelial cells. Mol Biochem Parasitol 159: 85–91.

83. Eckmann L, Laurent F, Langford TD, Hetsko ML, Smith JR, et al. (2000) Nitric oxide production by human intestinal epithelial cells and competition for arginine as potential determinants of host defense against the lumen-dwelling pathogen Giardia lamblia. J Immunol 164: 1478–1487.

84. Stadelmann B, Hanevik K, Andersson MK, Bruserud O, Svard SG (2013) The role of arginine and arginine-metabolizing enzymes during Giardia - host cell interactions in vitro. BMC Microbiol 13: 256.

85. Stadelmann B, Merino MC, Persson L, Svard SG (2012) Arginine consumption by the intestinal parasite Giardia intestinalis reduces proliferation of intestinal epithelial cells. PLoS One 7: e45325.

19

Plasmodium falciparum Malaria in Children Aged 0-2 Years: The Role of Foetal Haemoglobin and Maternal Antibodies to Two Asexual Malaria Vaccine Candidates (MSP3 and GLURP)

David Tiga Kangoye[1], Issa Nebie[1], Jean-Baptiste Yaro[1], Siaka Debe[1], Safiatou Traore[1], Oumarou Ouedraogo[1], Guillaume Sanou[1], Issiaka Soulama[1], Amidou Diarra[1], Alfred Tiono[1], Kevin Marsh[2,3], Sodiomon Bienvenu Sirima[1*⑨], Philip Bejon[2,3⑨]

1 Centre National de Recherche et de Formation sur le Paludisme, Ouagadougou, Burkina Faso, 2 Kenyan Medical Research Institute, Centre for Geographic Medicine Research (Coast), Kilifi, Kenya, 3 Nuffield Department of Medicine, Centre for Clinical Vaccinology and Tropical Medicine, University of Oxford, Churchill Hospital, Oxford, United Kingdom

Abstract

Background: Children below six months are reported to be less susceptible to clinical malaria. Maternally derived antibodies and foetal haemoglobin are important putative protective factors. We examined antibodies to *Plasmodium falciparum* merozoite surface protein 3 (MSP3) and glutamate-rich protein (GLURP), in children in their first two years of life in Burkina Faso and their risk of malaria.

Methods: A cohort of 140 infants aged between four and six weeks was recruited in a stable transmission area of southwestern Burkina Faso and monitored for 24 months by active and passive surveillance. Malaria infections were detected by examining blood smears using light microscopy. Enzyme-linked immunosorbent assay was used to quantify total Immunoglobulin G to *Plasmodium falciparum* antigens MSP3 and two regions of GLURP (R0 and R2) on blood samples collected at baseline, three, six, nine, 12, 18 and 24 months. Foetal haemoglobin and variant haemoglobin fractions were measured at the baseline visit using high pressure liquid chromatography.

Results: A total of 79.6% of children experienced one or more episodes of febrile malaria during monitoring. Antibody titres to MSP3 were prospectively associated with an increased risk of malaria while antibody responses to GLURP (R0 and R2) did not alter the risk. Antibody titres to MSP3 were higher among children in areas of high malaria risk. Foetal haemoglobin was associated with delayed first episode of febrile malaria and haemoglobin CC type was associated with reduced incidence of febrile malaria.

Conclusions: We did not find any evidence of association between titres of antibodies to MSP3, GLURP-R0 or GLURP-R2 as measured by enzyme-linked immunosorbent assay and early protection against malaria, although anti-MSP3 antibody titres may reflect increased exposure to malaria and therefore greater risk. Foetal haemoglobin was associated with protection against febrile malaria despite the study limitations and its role is therefore worthy further investigation.

Editor: Érika Martins Braga, Universidade Federal de Minas Gerais, Brazil

Funding: This work was funded by the European and Developing Countries Clinical Trials Partnership, grant number IP.2008.31100.001 to the Malaria Vectored Vaccines Consortium (MVVC), and coordinated by the European Vaccine Initiative. PB is jointly funded by the UK Medical Research Council (MRC) and the UK Department for International Development (DFID) under the MRC/DFID Concordat agreement. The funders had no role in study design, data collection and analysis, decision to publish, or preparation of the manuscript.

* Email: s.sirima.cnlp@fasonet.bf

⑨ These authors contributed equally to this work.

Introduction

Children under five years of age bear the majority of the malaria burden [1], however infants have been shown to be less likely to develop clinical malaria within their first six months of life [2–5].

This apparent protection has been linked to passive transfer of anti-malaria antibodies by earlier experimental studies [6–8]. A study in older children confirmed the protective effect of passive transfer of antibodies [9]. Other possible protective factors include foetal haemoglobin, as suggested by in vitro studies [10,11] and mouse models [12]. Rodent malaria experiments have provided evidence of the protective effect of a para-aminobenzoic acid-poor diet [13–16], which can be expected in exclusive breast-feeding. Breast milk constituents might also contribute to the protection of infants against malaria as suggested by in vitro studies [17]. Finally, studies also suggested that infants are less exposed to mosquito bites [18], thus reducing their risk of contracting malaria.

Correlates of protection of specific antibodies against malaria have been extensively studied in adults and children [19–32], however fewer studies have specifically addressed the risk of malaria in infancy. Anti-malaria antibodies present during infancy are initially passively received from the mother in utero (mainly IgG) [33], and then endogenously produced by the infant (IgM and IgG) as a response to repeated exposure to infective mosquito bites [34]. Nonetheless, it has been evidenced that earlier in life, some foetuses are able to mount an immune response to prenatal exposure [35] through transplacental transfer of soluble malaria antigens [35,36]. Most sero-epidemiological studies investigating these potential immune correlates of protection in infants did not find the hypothesised protective effect of antibodies to the individual antigens tested (CSP, LSA-1, crude schizont extract, MSP1, MSP2, Pf155/RESA and the vaccine candidate SPf66) [37–41]. However, two studies conducted independently in Liberia and Kenya [42,43] identified a protective effect of antibodies to MSP1–19.

We performed the present longitudinal prospective infant cohort study in the south-western region of Burkina Faso, where the transmission of malaria is high and stable. Our study aimed at investigating the role of transplacentally acquired anti-malaria antibodies, while adjusting for foetal haemoglobin and other factors, in the susceptibility of children to *P. falciparum* malaria in their first years of life. We specifically investigated three merozoite surface proteins, MSP3-LSP and two regions of GLURP (R0 and R2), as they have been used in a malaria vaccine that is currently undergoing testing [44,45].

Methods

Ethical consideration

This study was approved by the Institutional Review Board of Centre National de Recherche et de Formation sur le Paludisme (CNRFP) in Burkina Faso. The study was conducted according to the principles of the Declaration of Helsinki. Individual written informed consent was obtained from the parents of each child before any study procedure was performed. The parents who could not read and write in the language used in the informed consent form, signed the form with their thumb print after it was completed by an independent witness on their behalf. The IRB approved this consent procedure.

Study site

The study area is described elsewhere [46]. Briefly, the study area encompasses four health catchment areas served by four dispensaries reporting to Banfora health district. Two (Flantama and Korona) are located within Banfora town with water and power supply and mainly cement brick, iron sheet roofing houses. The other two (Nafona and Bounouna) are Banfora sub-urban villages with mainly adobe-walled, thatched roofing houses. The annual rainfall is above 900 mm with the rains lasting from May to October. The malaria incidence in children under five is 1.18 episodes/child/year using an active case detection [46] with *Plasmodium falciparum* being responsible for more than 90% malaria cases.

Study population

A cohort of 140 infants aged between four and six weeks was recruited into the study. The parents were informed of the study aims and procedures during the early post-natal visits at the dispensaries, prior to the children reaching one month of age. Recruitment was carried out simultaneously at the four health catchment areas of the study site from November 2010 to February 2011. Since measuring time to the first infection was one of the objectives of the study, infants who had either a documented previous episode of malaria or a positive blood smear at the baseline visit were excluded from the study; nevertheless, some infants might have had malaria infections that were unobserved. After the recruitment of the study participants, the geodetic coordinates (longitude, latitude) of their homesteads were recorded using Global Positioning System (GPS) devices. The approximate centre of each family compound was the reference point to record these coordinates.

Surveillance of malaria morbidity

To detect malaria infections, the children were followed up actively with weekly home visits and passively with dispensary monitoring for two years. The active follow-up consisted of weekly home visits performed by fieldworkers whose main tasks were to check the children's health, collect blood samples and start anti-malarial treatment when indicated according to the National Malaria Control Program guidelines. The time window for the home visits was ±2 days. In the passive follow-up, the caregivers were encouraged to bring their children to the nearby dispensary or the clinical research unit at any time should the child appear unwell. In both surveillance methods a blood smear was collected in case of fever (i.e. a reported history of fever in the past 24 hours and/or axillary temperature ≥37.5°C). In addition, to monitor the occurrence of asymptomatic parasitaemia, bloods smears were systematically collected on a monthly basis until the detection of the first malaria infection regardless of the axillary temperature. The children who became ill over the course of the study received health care free of charge either at the dispensaries, the clinical research unit or the paediatric unit of the regional referral hospital when necessary.

Parasitological examination

Light microscopy was used to examine the blood smears, which were collected, air dried and GIEMSA-stained as described elsewhere [46]. The parasite density of each blood smear was assessed by two independent microscopists and their results were compared for consistency. When their results were concordant, the average was recorded as the final result; otherwise a third microscopist was involved. The final result was the average of the two most concordant parasite densities.

Haemoglobin typing

High Pressure Liquid Chromatography (HPLC) was used to quantify the fractions of foetal haemoglobin and haemoglobin variants in the children.

Antibody quantification

Enzyme-linked immunosorbent assay (ELISA) was used to quantify anti-malaria total IgG in capillary blood samples collected by finger or heel prick at baseline, three, six, nine, 12, 18 and 24 months of follow up. Two *P. falciparum* blood-stage antigens MSP3-LSP [47], GLURP (R0 and R2) [48] were used in these assays. The ELISA was performed as described elsewhere [24].

Statistical analysis

Fever was defined as an axillary temperature $\geq 37.5°C$ and/or a reported history of fever in the past 24 hours. Malaria infection was defined as any positive parasitaemia regardless of the axillary temperature. Two definitions were set for febrile malaria; definition 1 included all febrile episodes with any level of asexual *P. falciparum* parasitaemia, and definition 2 included only febrile episodes associated with asexual *P. falciparum* parasite density $\geq 10000/\mu L$. This latter definition was derived after examining the distributions of the log-transformed parasite densities among children with and without fever in cross-sectional surveys (Figure S1). Febrile malaria episodes occurring within 21 days were considered a single episode.

The antibody titres expressed in arbitrary units were log-transformed to approximate a normal distribution. We fitted a multiple fractional polynomial regression of antibody titres on age to estimate the non-linear relationship between anti-malaria antibody titres and age. We fitted a linear regression model to estimate the relationship between a set of potential predictors and anti-malaria antibody titres. In this linear model, antibody titre was included as a time-changing outcome, i.e. the antibody titre measured at the end of each time interval within which malaria infections were recorded, and age as fractional polynomial covariate.

We adapted a previously published method for calculating exposure indexes (EI) to our cohort with time-to-event data [49]. An individual malaria exposure index (EI) was computed as the median time to the first malaria infection of the surrounding neighbours of each index child within a circle of a given radius around him/her at the middle of the circle. We found that a 1.5 km radius best predicted risk in our dataset (Figure S2). EIs were transformed to negative values so that the most exposed has the highest exposure index.

We fitted a Cox regression model to estimate the relationship between the time to first febrile malaria episode and a set of covariates of interest. We also fitted a negative binomial regression model to estimate the association between multiple malaria episodes and a set of potential explanatory covariates. Antibodies were fit in two ways; a) applying the baseline antibody titre throughout the period of monitoring and b) applying time-varying antibody titre, i.e. the antibody titre measured at the most recent time point, which therefore changed throughout the period of monitoring. The log likelihood ratio test was used to test the significance of variables with multiple levels. The assumption of proportional hazards for Cox regression was tested based on the Kaplan Meier method and the Schoenfeld residuals. The assumption of a linear effect of antibody titres on malaria episodes was investigated using fractional polynomial modelling. We used the Huber-White Sandwich estimator to adjust for clustering by individual in linear and negative binomial regression models. The

data were analysed using Stata 13 (StataCorp, College Station, Texas).

Results

Study population characteristics

A total of 216 mothers of new-borns were invited to attend study screening visits during early post-natal visits at the four dispensaries of the study area. Of these, 148 (68.5%) attended the study screening visits with their infants. A total of 140 infants were recruited during a three- month period from mid-November 2010 to mid-February 2011. The baseline characteristics of the infants and their mothers are summarised in Table 1.

Follow-up of participants and malaria morbidity

Twenty-three children (16.4%) were lost to follow-up before completing 24 months with a median [Inter Quartile Range-IQR] follow up time of 9.8 [2, 14.75] months. Among them, six migrated out of the study area, 10 withdrew their consent, four died and three dropped out of the study and were no longer reachable.

One or more episodes of febrile malaria were experienced by 79.6% of all children during follow-up. Sixty-three children (45.98%) had at least one malaria infection in their first year of life and 46 (76.66%) of the remaining 60 children in their second year.

Anti-malaria antibody kinetics

Antibody titres from one time point were weakly to moderately correlated with antibodies at the next time point (rho ranging from –0.28 to 0.55; 0.31 to 0.58 and 0.15 to 0.60, respectively for anti-MSP3, anti-GLURP R0 and anti-GLURP R2 antibody titres) with the strongest correlations observed from baseline to month 3, and months 9 to 12 (Table S1). The time-course of individual antibody kinetics is shown in Figure S3. Anti-GLURP R0 and anti-GLURP R2 antibodies were more closely correlated to each other ($r = 0.52$, $p < 0.001$) than either GLURP sub-unit antibody was correlated to anti-MSP3 antibody titres ($r = 0.35$, $p < 0.001$ and $r = 0.4$, $p < 0.001$ respectively).

Figure 1 shows individual antibody titres and a best-fit line using multiple fractional polynomials. There is an overall decline of anti-malaria total IgG titres to the three antigens from one to four months of age, presumably indicating the waning of maternally-derived anti-malaria antibodies. Thereafter, both the anti-GLURP R2 and anti-GLURP R0 total IgG titres rise slightly with increasing age while anti-MSP3 total IgG titres remain constant.

Predictive factors for the changing antibody titres

The univariate analysis of all potential covariates is displayed in Table S2. The final multi-variable model showed strong associations with age and season of measurement for each antigen tested (Table 2). In addition, associations were observed between the number of malaria episodes recorded immediately before the blood sample collection and the titres of antibodies to GLURP.

Antibody titres and risk of febrile malaria

Anti-MSP3 antibody titres were significantly associated with an increased hazard of the first febrile malaria episode on univariate (Table S4) and multivariable analysis (Table 3). There was no evidence of association between anti-GLURP antibody titres and varying malaria risk. The baseline anti-malaria antibody titres for both antigens did not show any significant association with the risk of febrile malaria. The wet season and individual malaria exposure index were associated with higher risk of febrile malaria. The

Table 1. Baseline characteristics of the study population.

Characteristic	Statistic
Male_freq(%)	67 (47.9%)
Age infants(days)_median [min, max]	33 [27,42]
Weight (kg)_ median [min, max][1]	4.1 [2.8, 5.9]
Length (cm)_ median [min, max][1]	54 [48, 61]
MUAC (cm)_ median [min, max]	12 [8.5, 16]
Hb rate (g/dL)_ median [min, max]	12.6 [8.7, 17.6]
Foetal Hb (%)_median [min, max][3]	59.6 [20.1, 89.6]
Hb phenotype_freq (%)[3]	
AA	113 (80.71)
AC	18 (13.14)
AS	1 (0.73)
CC	5 (3.65)
Delivery way[2]	
Natural	133 (95)
Ceasarian section	5 (3.6)
Neonatal rescucitation_freq (%)	12 (8.6)
Neonatal infection_freq (%)	2 (1.43)
EPI (up to date at 1 month)_freq (%)	124 (88.6)
Age groups mothers (years)_freq (%)	
≤19	10 (7.1)
20–29	89 (63.6)
≥30	37 (26.4)
ITN use (during pregnancy)_freq (%)[1]	123 (87.9)
IPTp courses_freq (%)	
0	10 (7.14)
1	28 (20)
2	100 (71.43)
3	2 (1.43)
Gravidity status[2] _freq (%)	
primigravidae	32 (22.86)
multigravidae	106 (75.71)
Education level (mothers)[1]_freq (%)	
No formal education	78 (55.71)
Primary school	39 (27.86)
Secondary school or above	22 (15.71)
Distribution of study population	
Bounouna	34 (22.97)
Nafona	41 (27.7)
Korona	16 (10.81)
Flantama	49 (33.11)

[1, 2, 3]Number of missing data.

baseline foetal haemoglobin fraction showed a protective effect in the multivariable model (HR = 0.97, $p = 0.003$, 95% CI [0.96, 0.99]) which was not significant in univariate analysis (Table S3). This was dependent on adjusting for exposure index in the multivariable model, and we noted a non-significant association between foetal haemoglobin and exposure index ($r = 0.17$, $p = 0.064$). Belonging to the haemoglobin CC type group was also significantly associated with decreased malaria incidence (IRR = 0.44, $p = 0.046$, 95% CI [0.19, 0.99]) in the multivariable but not in the univariate analysis.

We did not detect significant variation in the proportionality of hazards over time (Figure 2, Figure S4), but there was borderline variation in varying hazards for the exposure index ($p = 0.082$, in the direction of decreasing hazard over time) and for foetal haemoglobin ($p = 0.098$, in the direction of increasing hazard over time).

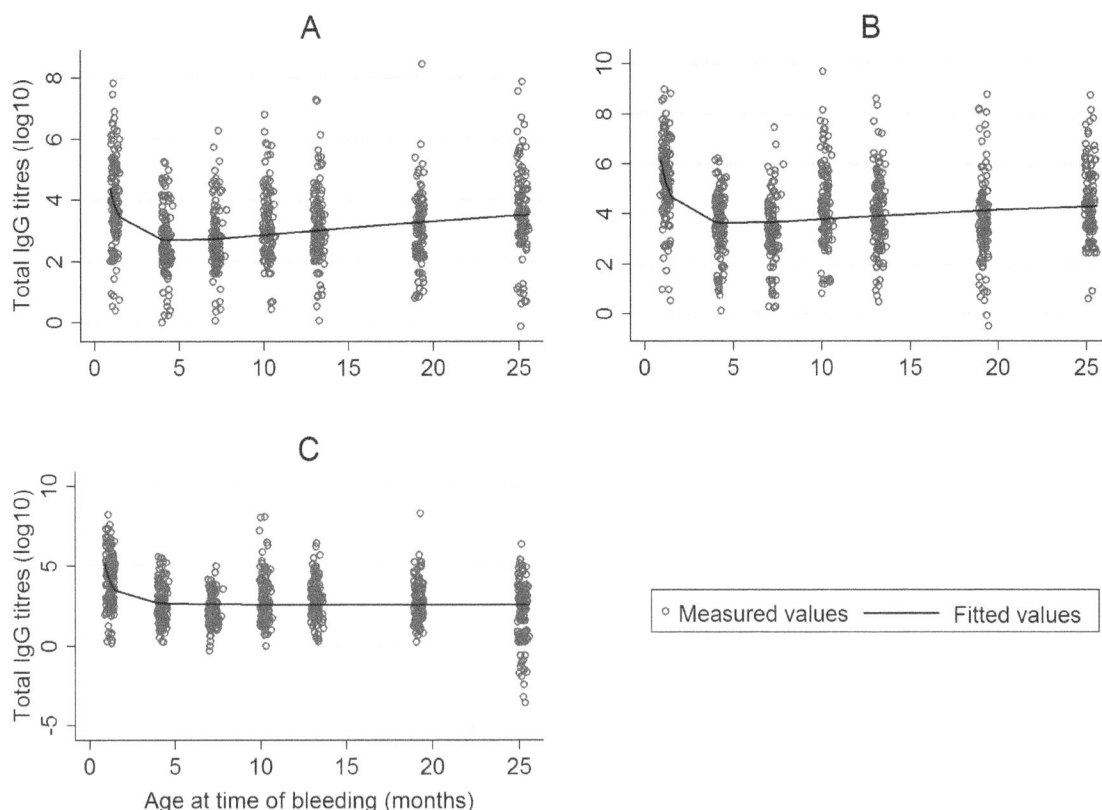

Figure 1. Antibody kinetics trends over the 24 months of observation. (A) Antibodies to GLURP R0, (B) Antibodies to GLURP R2, (C) Antibodies to MSP3.

The results of negative binomial regression were consistent with the Cox regression analysis (Table 3). The best fit lines from the multivariable fractional polynomial regression are consistent with a linear effect of antibody titres on febrile malaria episodes (Figure S6). Changing anti-MSP3 antibody titres were significantly associated with increased incidence of febrile malaria episodes (IRR = 1.17, p = 0.007, 95% CI [1.04, 1.30]), but baseline titres were not. Changing Anti-GLURP R0 antibody titres were significantly associated with increased incidence of febrile malaria on univariate (Table S4) but not multivariable analysis (Table 3). Age, season and individual malaria exposure index were significantly associated with increased incidence of febrile malaria episodes. We examined interaction terms: we found that the effects of total IgG to the malaria antigens we tested were not additive, and noted an interaction between malaria exposure index and total IgG to MSP3 of marginal significance (Table S7).

Restricting the observation period to six weeks post bleeding for plasma, to take into account the fact that anti-malaria antibodies are more likely to be short-lived in young children [50], yielded stronger associations in univariate analysis but not in the multivariable analysis and not in the direction of protection (Table S5).

Discussion

In this study, transplacentally acquired anti-malaria antibodies and foetal haemoglobin were investigated in relation to suscepti-bility to malaria in a cohort of 140 infants. Antibody titres to GLURP and MSP3 were found to decline in the first four months of life, presumably due to the loss of maternal antibodies. Endogenous production was responsible for the subsequent

increase in the case of GLURP and stabilisation of the loss of antibodies for MSP3. Contrary to our expectations, there was no association between antibody titres to GLURP (R0 and R2) and the risk of febrile malaria in the first two years of life; antibody titres to MSP3 even appeared as a marker of exposure since it was statistically significantly positively associated with the incidence of febrile malaria and inversely associated with time to first febrile malaria episode. Confounding between malaria risk and antibody titre by variation in exposure has previously been reported [51,52].

In the investigation of the role of antibodies to MSP3 and GLURP (R0 and R2) against *P. falciparum* malaria, most of the previous studies did not specifically target infants. A number of prospective sero-epidemiological studies have investigated the role of antibodies to these merozoite surface antigens in older children and adults in West Africa, East Africa as well as South-East Asia.

The lack of association between antibodies to GLURP (R0 and R2) and protection against clinical malaria in our study is partially concordant with the findings of a previous study in older children in Burkina Faso. Nebie and colleagues investigated total IgG to GLURP (R0 and R2) among others antigens in children aged 6 months to 10 years and found a protective effect for antibodies to GLURP R0 but not to GLURP R2 when the antibody titres were analysed individually. When antibody titres to all the four antigens studied (NANP, GLURP R0 and R2, MSP3) were included in a multivariable model, antibodies to GLURP (R0 and R2) were not associated with protection against malaria [24]. In Ghana, Dodoo and colleagues reported a protective effect of total IgG and IgG subclasses to GLURP (R0 and R2) in the univariate analysis. However neither total IgG nor IgG subclasses to GLURP (R0 and R2) were significantly protective when all the serological covariates

Table 2. Multivariable predictive model for changing anti-malaria antibody titres.

Predictor	IgG anti-MSP3			IgG anti-GLURP R0			IgG anti-GLURP R2		
	Coef.	95% CI	p	Coef.	95% CI	p	Coef.	95% CI	p
Age power (−2/−5/−5)[a]	0.02	[0.02, 0.03]	<0.001	2	[1.49, 2.50]	<0.001	−2.04	[−2.98, −1.10]	<0.001
Age power (NA/0/−5)	-	-	-	1.19	[0.79, 1.58]	<0.001	−0.92	[−1.21, −0.62]	<0.001
ITN use (pregnancy)									
Yes	0	-	-	0	-	-	0	-	-
No	−0.11	[−0.49, 0.26]	0.548	0.16	[−0.22, 0.53]	0.413	0.02	[−0.44, 0.48]	0.918
Season									
Dry season	0	-	-	0	-	-	0	-	-
Rains	0.35	[0.13, 0.57]	0.002	0.33	[0.19, 0.47]	<0.001	0.28	[0.07, 0.50]	0.011
Malaria Exposure index	0.02	[−0.003, 0.04]	0.098	−0.01	[−0.03, 0.006]	0.202	0.01	[−0.01, 0.04]	0.393
Number previous infections[b]	0.12	[−0.002, 0.24]	0.053	0.28	[0.17, 0.39]	<0.001	0.31	[0.17, 0.45]	<0.001

[a]Age is transformed in multiple fractional polynomials with the corresponding powers for antibodies to MSP3, GLURP R0 and GLURP R2 indicated in brackets.
[b]Number of malaria infections recorded between two consecutive time points for antibody titres measurement.

Table 3. Multivariable models of risk of malaria using changing anti-malaria antibody titres.

Predictor[f]	Cox regression			Negative binomial regression		
	HR[d]	95%CI	P	IRR[e]	95%CI	P
Age	NA	-	-	1.11	[1.09, 1.13]	<0.001
MUAC[a] (baseline)	-	-	-	1.26	[1.09, 1.45]	0.002
Foetal Hb[b] rate (baseline)	0.97	[0.96, 0.99]	0.003	0.98	[0.97, 0.99]	0.013
Haemoglobin type						
AA	1	-	-	1	-	-
AS[c]	NA[c]	-	-	NA[c]	-	-
AC	1.29	[0.68, 2.47]	0.430	1.14	[0.75, 1.72]	0.540
CC	0.52	[0.12, 2.19]	0.370	0.44	[0.19, 0.99]	0.046
Anti-MSP3 (changing)	1.34	[1.08, 1.66]	0.007	1.17	[1.04, 1.30]	0.007
Anti-GLURP R0 (changing)	1.15	[0.91, 1.44]	0.233	1.003	[0.88, 1.14]	0.968
Anti-GLURP R2 (changing)	0.98	[0.83, 1.16]	0.859	0.92	[0.82, 1.03]	0.149
ITN use (pregnancy)						
Yes	1	-	-	1	-	-
No	0.86	[0.41, 1.79]	0.687	1.23	[0.80, 1.88]	0.348
Season						
Dry season	1	-	-	1	-	-
Rains	10.85	[2.80, 42.15]	0.001	1.4	[1.02, 1.92]	0.037
Malaria Exposure index	1.08	[1.04, 1.13]	<0.001	1.06	[1.03, 109]	<0.001

[a]MUAC: mid upper arm circumference.
[b]Hb: haemoglobin.
[c]NA not applicable; only one child had haemoglobin AS type.
[d]HR: hazard ratio.
[e]IRR: incidence rate ratio.
[f]There was no collinearity between the predictors (Figure S5, Table S6).

were taken into account in the final multivariable model [19,25]. In a study conducted in Tanzania by Lusingu and colleagues, total IgG to GLURP R0 was not associated with protection against febrile malaria; among IgG subclasses only IgG1 was associated with protection. In contrast, other studies have demonstrated a protective association for antibodies to GLURP (R0 and R2) in older children and adults [20,26,27,30]. In a study in Myanmar that investigated antibodies to MSP1, MSP3, GLURP (R0, R1, R2), only antibodies to GLURP R0 showed a protective effect when all the antibodies were considered together in the analysis [21].

Antibodies to MSP3 have been associated with protection in previous sero-epidemiological studies [21,23,27,53] and a vaccine trial [44] even if the assessment of efficacy was not the primary objective in the latter. However other studies did not find a protective effect of antibodies to MSP3 [19,25,26] but none of these studies concluded on antibodies to MSP3 appearing as a marker of exposure.

Although antibodies to MSP3 and GLURP have not been previously studied in newborn cohorts, antibodies to other *P. falciparum* malaria antigens have been investigated. In sero-epidemiological newborn cohort studies, antibodies to crude *P. falciparum* schizont extract and MSP2 were found to be associated with higher risk of malaria infection in infants [38,40] indicating higher exposure. Only antibodies to MSP1-19 were associated with protection against clinical malaria [42,43].

Antibody titres to MSP3 and GLURP (R0 and R2) were not associated with protection in our study, and we suggest that confounding due to exposure led to an apparent association with increased susceptibility for antibody response to MSP3. Limitations of our study include the fact that the high malaria transmission season began 5–7 months after recruitment. Therefore the majority of the maternal antibodies were likely gone by the time that febrile malaria episodes began, and children were exposed during a period of lower antibody titres. We did not use an external control to quantify malaria antibodies as performed elsewhere [54], however we speculate that antibody titres at 5 months and beyond were lower than those previously reported to be protective [42,43].

Interestingly, foetal haemoglobin was significantly inversely associated to febrile malaria incidence although the effect size was relatively small. The effect was only statistically significant on multivariable analysis, and appeared to depend on adjusting by exposure index in the multivariable model. Furthermore the effect seems to be evident after 6 months of age, when we would expect foetal haemoglobin to have been lost from the circulation. We speculate that an interaction between malaria exposure and foetal haemoglobin may be responsible for a delayed protective effect, perhaps due to an early but controlled infection in the presence of high levels of foetal haemoglobin leading to more rapid acquisition of immunity [55].

Children who carried the haemoglobin CC type appeared to have a significantly lower risk of malaria as compared to haemoglobin AA type children, as has been previously reported [56,57].

In conclusion, the present study did not find any evidence of association between antibody titres to MSP3 and GLURP (R0 and R2) and protection against *P. falciparum* febrile malaria in

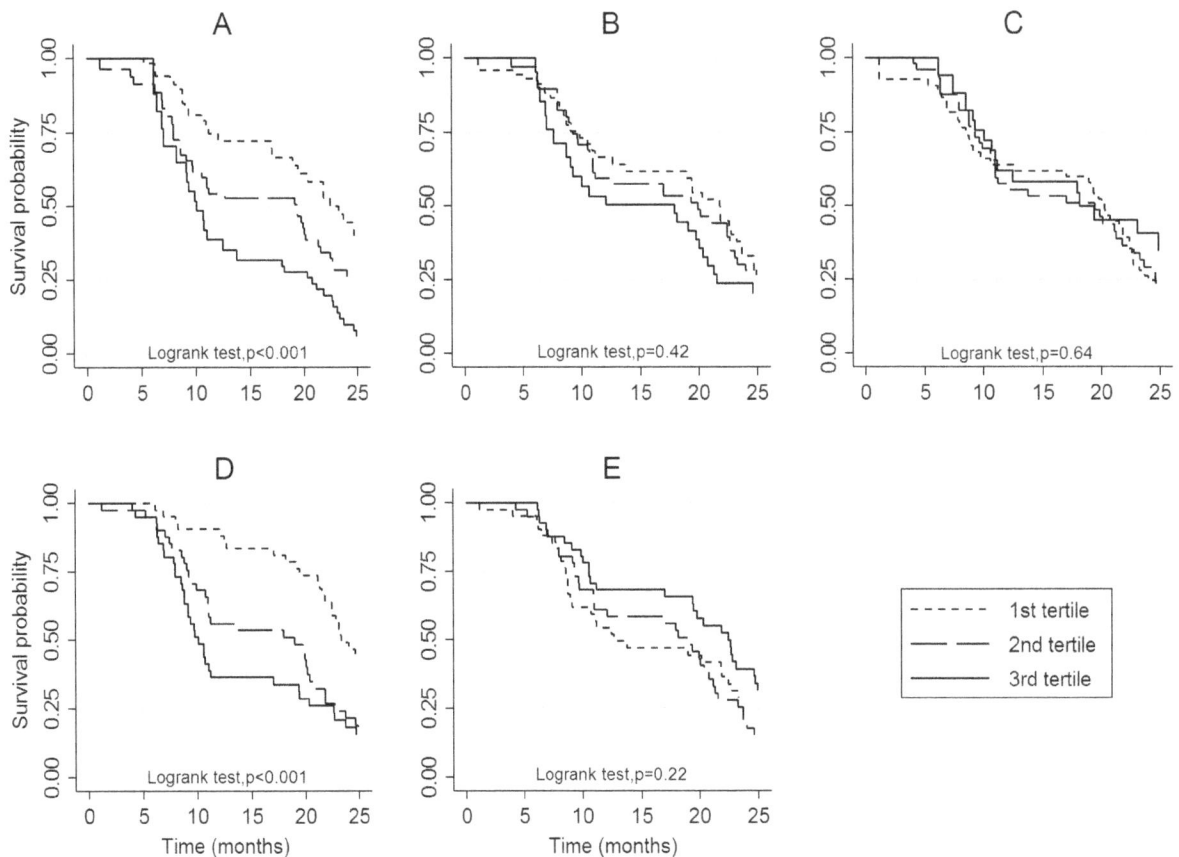

Figure 2. Effect of antibody titres, individual malaria exposure index and baseline foetal haemoglobin rate on survival to malaria: Kaplan Meier estimates. (A) Antibodies to MSP3, (B) Antibodies to GLURP R0, (C) Antibodies to GLURP R2, (D) Individual malaria exposure index, (E) Foetal haemoglobin fraction.

children in their first few months of life. Despite the above mentioned limitations of the study, the baseline fraction of foetal was associated with protection against febrile malaria. Its role in the protection of children against malaria in their first few months of life is therefore worthy of further investigation. Finally, our study also underlines that the role of haemoglobinopathies should be taken into account in the exploration of protective factors in the low susceptibility of infants to malaria.

Acknowledgments

We are grateful to the parents/guardians of the children in Banfora whose commitment and patience made this study possible. We thank the fieldworkers' team and nurses in the dispensaries who monitored these children for two years as well as all the lab staff. We would also like to especially thank Dr Michael Theisen who kindly provided the malaria antigens used in this study, MSP3-LSP, GLURP R0 and GLURP R2.

Supporting Information

Figure S1 Distribution of parasitemia in febrile and afebrile children for the whole study period. The box-and-whisker plot represents the median and the inter-quartile range of the parasite density in a log 10 scale in the febrile and afebrile children groups.

Figure S2 Selection of the best radius for individual malaria exposure index calculation. The lowest log likelihood in the univariate Cox regression analysis was the selection criteria for the radius to be used.

Figure S3 Dynamics of total IgG to MSP3 over the whole study period using a 25% random sample representing 23 children. Each line represents a child. The antibody titres are in log 10 scale.

Figure S4 Test of proportional hazards assumption: Schoenfeld residuals plots. (A) Baseline total IgG to MSP3, (B) Baseline total IgG to GLURP R0, (C) Baseline total IgG to GLURP R2, (D) Changing total IgG to MSP3, (E) Changing total IgG to GLURP R0, (F) Changing total IgG to GLURP R2, (G) Individual malaria exposure index, (H) Baseline Foetal haemoglobin rate.

Figure S5 Scatter plot matrix of the continuous independent variables used used in multivariable regression models.

Figure S6 Multivariable fractional polynomial plots for antibodies to MSP3 (A), GLURP R0 (B) and GLURP R2 (C).

Table S1 Variability of antibody titres dynamics. The correlations (r) are examined between every two consecutive time points for antibody measurement.

Table S2 Predictive model for changing anti-malaria antibody titers using linear regression. Univariate analysis.

Table S3 Cox regression analysis using changing antibody titres.

Table S4 Predictive model for the occurrence of febrile malaria episodes using changing antibody titres.

Table S5 Predictive model for the occurrence of febrile malaria episodes using changing antibody titres with a

restricted observation period (6 weeks) post-bleeding for plasma.

Table S6 Multicollinearity diagnostics for continuous independent variables used in multivariable regression models.

Table S7 Testing for interactions between predictors in multivariable negative binomial regression model.

Author Contributions

Conceived and designed the experiments: IN AT SBS. Performed the experiments: DK JY SD ST OO GS. Analyzed the data: DK PB. Contributed to the writing of the manuscript: DK IN JY SD ST OO GS IS AD AT KM SBS PB.

References

1. WHO (2012) World Malaria Report 2012. Available: http://www.who.int/malaria/publications/world_malaria_report_2012/report/en/. Accessed 2013 December.
2. Macdonald G (1950) The analysis of malaria parasite rates in infants. Trop Dis Bull 47: 915–938.
3. Bruce-Chwatt LJ (1952) Malaria in African infants and children in Southern Nigeria. Ann Trop Med Parasitol 46: 173–200.
4. Brabin B (1990) An analysis of malaria parasite rates in infants: 40 years after MacDonald. Trop Dis Bull 87: 20.
5. Snow RW, Nahlen B, Palmer A, Donnelly CA, Gupta S, et al. (1998) Risk of severe malaria among African infants: direct evidence of clinical protection during early infancy. J Infect Dis 177: 819–822.
6. Cohen S, Mc GI, Carrington S (1961) Gamma-globulin and acquired immunity to human malaria. Nature 192: 733–737.
7. Edozien JC, Gilles HM, Udeozo IOK (1962) Adult And Cord-Blood Gamma-Globulin and Immunity to Malaria in Nigerians. The Lancet 280: 951–955.
8. McGregor IA, Carrington SP, Cohen S (1963) Treatment of East African P. falciparum malaria with West African human γ-globulin. Transactions of The Royal Society of Tropical Medicine and Hygiene 57: 170–175.
9. Sabcharoen A, Burnouf T, Ouattara D, Attanath P, Bouharoun-Tayoun H, et al. (1991) Parasitologic and clinical human response to immunoglobulin administration in falciparum malaria. Am J Trop Med Hyg 45: 297–308.
10. Pasvol G, Weatherall DJ, Wilson RJ, Smith DH, Gilles HM (1976) Fetal haemoglobin and malaria. Lancet 1: 1269–1272.
11. Amaratunga C, Lopera-Mesa TM, Brittain NJ, Cholera R, Arie T, et al. (2011) A role for fetal hemoglobin and maternal immune IgG in infant resistance to Plasmodium falciparum malaria. PLoS One 6: e14798.
12. Shear HL, Grinberg L, Gilman J, Fabry ME, Stamatoyannopoulos G, et al. (1998) Transgenic mice expressing human fetal globin are protected from malaria by a novel mechanism. Blood 92: 2520–2526.
13. Kicska GA, Ting LM, Schramm VL, Kim K (2003) Effect of dietary p-aminobenzoic acid on murine Plasmodium yoelii infection. J Infect Dis 188: 1776–1781.
14. Maegraith BG, Deegan T, Jones ES (1952) Suppression of malaria (P. berghei) by milk. Br Med J 2: 1382–1384.
15. Hawking F (1954) Milk, p-aminobenzoate, and malaria of rats and monkeys. Br Med J 1: 425–429.
16. Jacobs RL (1964) Role of P-Aminobenzoic Acid in Plasmodium Berghei Infection in the Mouse. Exp Parasitol 15: 213–225.
17. Kassim O, Ako-Anai K, Torimiro S, Hollowell G, Okoye V, et al. (2000) Inhibitory factors in breastmilk, maternal and infant sera against in vitro growth of Plasmodium falciparum malaria parasite. Journal of Tropical Pediatrics 46: 92–96.
18. Muirhead-Thomson RC (1951) The distribution of anopheline mosquito bites among different age groups; a new factor in malaria epidemiology. Br Med J 1: 1114–1117.
19. Dodoo D, Atuguba F, Bosomprah S, Ansah NA, Ansah P, et al. (2011) Antibody levels to multiple malaria vaccine candidate antigens in relation to clinical malaria episodes in children in the Kasena-Nankana district of Northern Ghana. Malar J 10: 108.
20. Oeuvray C, Theisen M, Rogier C, Trape JF, Jepsen S, et al. (2000) Cytophilic immunoglobulin responses to Plasmodium falciparum glutamate-rich protein are correlated with protection against clinical malaria in Dielmo, Senegal. Infect Immun 68: 2617–2620.
21. Soe S, Theisen M, Roussilhon C, Aye KS, Druilhe P (2004) Association between protection against clinical malaria and antibodies to merozoite surface antigens in an area of hyperendemicity in Myanmar: complementarity between responses

to merozoite surface protein 3 and the 220-kilodalton glutamate-rich protein. Infect Immun 72: 247–252.
22. Lusingu JP, Vestergaard LS, Alifrangis M, Mmbando BP, Theisen M, et al. (2005) Cytophilic antibodies to Plasmodium falciparum glutamate rich protein are associated with malaria protection in an area of holoendemic transmission. Malar J 4: 48.
23. Roussilhon C, Oeuvray C, Muller-Graf C, Tall A, Rogier C, et al. (2007) Long-term clinical protection from falciparum malaria is strongly associated with IgG3 antibodies to merozoite surface protein 3. PLoS Med 4: e320.
24. Nebie I, Tiono AB, Diallo DA, Samandoulougou S, Diarra A, et al. (2008) Do antibody responses to malaria vaccine candidates influenced by the level of malaria transmission protect from malaria? Trop Med Int Health 13: 229–237.
25. Dodoo D, Aikins A, Kusi KA, Lamptey H, Remarque E, et al. (2008) Cohort study of the association of antibody levels to AMA1, MSP119, MSP3 and GLURP with protection from clinical malaria in Ghanaian children. Malar J 7: 142.
26. Courtin D, Oesterholt M, Huismans H, Kusi K, Milet J, et al. (2009) The quantity and quality of African children's IgG responses to merozoite surface antigens reflect protection against Plasmodium falciparum malaria. PLoS One 4: e7590.
27. Meraldi V, Nebie I, Tiono AB, Diallo D, Sanogo E, et al. (2004) Natural antibody response to Plasmodium falciparum Exp-1, MSP-3 and GLURP long synthetic peptides and association with protection. Parasite Immunol 26: 265–272.
28. Mamo H, Esen M, Ajua A, Theisen M, Mordmuller B, et al. (2013) Humoral immune response to Plasmodium falciparum vaccine candidate GMZ2 and its components in populations naturally exposed to seasonal malaria in Ethiopia. Malar J 12: 51.
29. Osier FHA, Fegan G, Polley SD, Murungi L, Verra F, et al. (2008) Breadth and Magnitude of Antibody Responses to Multiple Plasmodium falciparum Merozoite Antigens Are Associated with Protection from Clinical Malaria. Infection and Immunity 76: 2240–2248.
30. Dodoo D, Theisen M, Kurtzhals JAL, Akanmori BD, Koram KA, et al. (2000) Naturally Acquired Antibodies to the Glutamate-Rich Protein Are Associated with Protection against Plasmodium falciparum Malaria. Journal of Infectious Diseases 181: 1202–1205.
31. Dziegiel M, Rowe P, Bennett S, Allen SJ, Olerup O, et al. (1993) Immunoglobulin M and G antibody responses to Plasmodium falciparum glutamate-rich protein: correlation with clinical immunity in Gambian children. Infection and Immunity 61: 103–108.
32. Hogh B, Petersen E, Dziegiel M, David K, Hanson A, et al. (1992) Antibodies to a Recombinant Glutamate-Rich Plasmodium Falciparum Protein: Evidence for Protection of Individuals Living in a Holoendemic Area of Liberia. The American Journal of Tropical Medicine and Hygiene 46: 307–313.
33. Williams AI, McFarlane H (1969) Distribution of malarial antibody in maternal and cord sera. Arch Dis Child. 511–514.
34. Nhabomba A, Guinovart C, Jimenez A, Manaca M, Quinto L, et al. (2014) Impact of age of first exposure to Plasmodium falciparum on antibody responses to malaria in children: a randomized, controlled trial in Mozambique. Malaria Journal 13: 121.
35. Metenou S, Suguitan AL, Long C, Leke RGF, Taylor DW (2007) Fetal Immune Responses to Plasmodium falciparum Antigens in a Malaria-Endemic Region of Cameroon. The Journal of Immunology 178: 2770–2777.
36. May K, Grube M, Malhotra I, Long CA, Singh S, et al. (2009) Antibody-Dependent Transplacental Transfer of Malaria Blood-Stage Antigen Using a Human Ex Vivo Placental Perfusion Model. PLoS ONE 4: e7986.

37. Achidi EA, Salimonu LS, Perlmann H, Perlmann P, Berzins K, et al. (1996) Lack of association between levels of transplacentally acquired Plasmodium falciparum-specific antibodies and age of onset of clinical malaria in infants in a malaria endemic area of Nigeria. Acta Trop 61: 315–326.

38. Wagner G, Koram K, McGuinness D, Bennett S, Nkrumah F, et al. (1998) High incidence of asymptomatic malara infections in a birth cohort of children less than one year of age in Ghana, detected by multicopy gene polymerase chain reaction. The American Journal of Tropical Medicine and Hygiene 59: 115–123.

39. Kitua AY, Urassa H, Wechsler M, Smith T, Vounatsou P, et al. (1999) Antibodies against Plasmodium falciparum vaccine candidates in infants in an area of intense and perennial transmission: relationships with clinical malaria and with entomological inoculation rates. Parasite Immunol 21: 307–317.

40. Riley EM, Wagner GE, Ofori MF, Wheeler JG, Akanmori BD, et al. (2000) Lack of association between maternal antibody and protection of African infants from malaria infection. Infect Immun 68: 5856–5863.

41. Zhou Z, Xiao L, Branch OH, Kariuki S, Nahlen BL, et al. (2002) Antibody responses to repetitive epitopes of the circumsporozoite protein, liver stage antigen-1, and merozoite surface protein-2 in infants residing in a Plasmodium falciparum-hyperendemic area of western Kenya. XIII. Asembo Bay Cohort Project. Am J Trop Med Hyg 66: 7–12.

42. Hogh B, Marbiah NT, Burghaus PA, Andersen PK (1995) Relationship between maternally derived anti-Plasmodium falciparum antibodies and risk of infection and disease in infants living in an area of Liberia, west Africa, in which malaria is highly endemic. Infect Immun 63: 4034–4038.

43. Branch OH, Udhayakumar V, Hightower AW, Oloo AJ, Hawley WA, et al. (1998) A longitudinal investigation of IgG and IgM antibody responses to the merozoite surface protein-1 19-kiloDalton domain of Plasmodium falciparum in pregnant women and infants: associations with febrile illness, parasitemia, and anemia. Am J Trop Med Hyg 58: 211–219.

44. Sirima SB, Cousens S, Druilhe P (2011) Protection against malaria by MSP3 candidate vaccine. N Engl J Med 365: 1062–1064.

45. Belard S, Issifou S, Hounkpatin AB, Schaumburg F, Ngoa UA, et al. (2011) A randomized controlled phase Ib trial of the malaria vaccine candidate GMZ2 in African children. PLoS One 6: e22525.

46. Tiono AB, Kangoye DT, Rehman AM, Kargougou DG, Kabore Y, et al. (2014) Malaria incidence in children in South-west burkina faso: comparison of active and passive case detection methods. PLoS One 9: e86936.

47. Druilhe P, Spertini F, Soesoe D, Corradin G, Mejia P, et al. (2005) A malaria vaccine that elicits in humans antibodies able to kill Plasmodium falciparum. PLoS Med 2: e344.

48. Theisen M, Vuust J, Gottschau A, Jepsen S, Hogh B (1995) Antigenicity and immunogenicity of recombinant glutamate-rich protein of Plasmodium falciparum expressed in Escherichia coli. Clin Diagn Lab Immunol 2: 30–34.

49. Olotu A, Fegan G, Wambua J, Nyangweso G, Ogada E, et al. (2012) Estimating individual exposure to malaria using local prevalence of malaria infection in the field. PLoS One 7: e32929.

50. Kinyanjui S, Bejon P, Osier F, Bull P, Marsh K (2009) What you see is not what you get: implications of the brevity of antibody responses to malaria antigens and transmission heterogeneity in longitudinal studies of malaria immunity. Malaria Journal 8: 242.

51. Bejon P, Cook J, Bergmann-Leitner E, Olotu A, Lusingu J, et al. (2011) Effect of the pre-erythrocytic candidate malaria vaccine RTS,S/AS01E on blood stage immunity in young children. J Infect Dis 204: 9–18.

52. Greenhouse B, Ho B, Hubbard A, Njama-Meya D, Narum DL, et al. (2011) Antibodies to Plasmodium falciparum antigens predict a higher risk of malaria but protection from symptoms once parasitemic. J Infect Dis 204: 19–26.

53. Osier FH, Fegan G, Polley SD, Murungi L, Verra F, et al. (2008) Breadth and magnitude of antibody responses to multiple Plasmodium falciparum merozoite antigens are associated with protection from clinical malaria. Infect Immun 76: 2240–2248.

54. Murungi LM, Kamuyu G, Lowe B, Bejon P, Theisen M, et al. (2013) A threshold concentration of anti-merozoite antibodies is required for protection from clinical episodes of malaria. Vaccine 31: 3936–3942.

55. Pombo DJ, Lawrence G, Hirunpetcharat C, Rzepczyk C, Bryden M, et al. (2002) Immunity to malaria after administration of ultra-low doses of red cells infected with Plasmodium falciparum. Lancet 360: 610–617.

56. Modiano D, Luoni G, Sirima BS, Simpore J, Verra F, et al. (2001) Haemoglobin C protects against clinical Plasmodium falciparum malaria. Nature 414: 305–308.

57. Bougouma EC, Tiono AB, Ouedraogo A, Soulama I, Diarra A, et al. (2012) Haemoglobin variants and Plasmodium falciparum malaria in children under five years of age living in a high and seasonal malaria transmission area of Burkina Faso. Malar J 11: 154.

20

Hematopoietic Stem/Progenitor Cell Sources to Generate Reticulocytes for *Plasmodium vivax* Culture

Florian Noulin[1]*, Javed Karim Manesia[2], Anna Rosanas-Urgell[1], Annette Erhart[1], Céline Borlon[1], Jan Van Den Abbeele[3], Umberto d'Alessandro[4], Catherine M. Verfaillie[2]

1 Unit of Malariology, Institute of Tropical Medicine, Antwerp, Belgium, **2** Department of development and regeneration, Stem Cell Institute, Leuven, Belgium, **3** Unit of Veterinary Protozoology, Institute of Tropical Medicine, Antwerp, Belgium, **4** Medical Research Council Unit, Fajara, The Gambia

Abstract

The predilection of *Plasmodium vivax* (*P. vivax*) for reticulocytes is a major obstacle for its establishment in a long-term culture system, as this requires a continuous supply of large quantities of reticulocytes, representing only 1–2% of circulating red blood cells. We here compared the production of reticulocytes using an established *in vitro* culture system from three different sources of hematopoietic stem/progenitor cells (HSPC), i.e. umbilical cord blood (UCB), bone marrow (BM) and adult peripheral blood (PB). Compared to CD34$^+$-enriched populations of PB and BM, CD34$^+$-enriched populations of UCB produced the highest amount of reticulocytes that could be invaded by *P. vivax*. In addition, when CD34$^+$-enriched cells were first expanded, a further extensive increase in reticulocytes was seen for UCB, to a lesser degree BM but not PB. As invasion by *P. vivax* was significantly better in reticulocytes generated *in vitro*, we also suggest that *P. vivax* may have a preference for invading immature reticulocytes, which should be confirmed in future studies.

Editor: Ana Paula Arez, Instituto de Higiene e Medicina Tropical, Portugal

Funding: The funding was provided by ITM Secondary Research Funding (SOFI-B), http://www.itg.be/itgtool_v2/Projecten/Project.asp?PNr = 755023. The funders had no role in study design, data collection and analysis, decision to publish, or preparation of the manuscript.

Competing Interests: The authors have declared that no competing interests exist.

* Email: flo_noulin@hotmail.com

Introduction

Plasmodium vivax (*P. vivax*) is the most widespread malaria parasite outside sub-Saharan Africa, and accounts for 80 to 300 million of malaria cases per year [1]. The predilection of *P. vivax* for reticulocytes is a major obstacle for the establishment of a long-term *P. vivax* culture system [2,3]. As reticulocytes represent only 1–2% of the circulating red blood cells (RBCs) with a half-life of 2 days (including 1 day in the peripheral blood), collecting sufficient reticulocytes to maintain a *P. vivax* culture is a challenge [3].

It has been previously shown that reticulocytes can be obtained by concentrating adult peripheral blood (PB) or umbilical cord blood (UCB) using either a 70% percoll solution [4] or a plasma autologous ultra-centrifugation [5]. One study suggested that *P. vivax* could be maintained in culture for up to 85 days with reticulocytes concentrated from umbilical blood cord (UCB) [6]; however, parasites did not develop beyond one schizogony cycle and parasite densities were very low [7]. In addition, it is possible to culture hematopoietic stem/progenitor cells (HSPC)/CD34$^+$ cells to induce erythroid differentiation and consequently produce reticulocytes *in vitro* [8]. Reticulocytes generated from CD34$^+$ cells from both bone marrow (BM) and peripheral blood mononuclear cells (PBMC) have been previously used for culturing *P. vivax*; but the authors did not provide data regarding the efficiency of invasion and the development of the parasites *in vitro* [9].

In this report, we compared different sources of hematopoietic stem/progenitor cells (HPSC), namely UCB, BM and peripheral blood, for their capacity to produce reticulocytes that allow invasion by *P. vivax*.

Materials and Methods

HSPCs expansion and reticulocytes differentiation

CD34$^+$ cell isolation. The differentiation of HSPC into reticulocytes was done according to a previously described protocol [10]. UCB was obtained from the Belgian Cord Blood Bank at the Gasthuisberg Hospital Leuven, BM samples were obtained from volunteer donors; and human peripheral blood from the Antwerp Red Cross. Mononuclear cells from peripheral blood (PBMC), UCB and BM were isolated on a Ficoll gradient (GE Healthcare), by 30 minutes centrifugation at 400 g. The mononuclear cells were collected and washed twice with PBS. CD34$^+$ enriched HSPCs were isolated using Magnetic Assorting Cell Sorting (MACS, Biotenyl Biotech). HSPC cell purity after MACS selection was assessed by FACS analysis, using CD34 and CD45 antibodies (eBioscience).

Reticulocytes differentiation

CD34$^+$-enriched cells were dispensed in a 6-well plate with IMDM medium (Gibco) supplemented with L-glutamine (4 M, Sigma), penicillin-streptomycin (1%, Invitrogen), folic acid (10 µg/mL, Sigma), inositol (40 µg/mL, Sigma), transferrin (120 µg/mL, Sigma), monothioglycerol (1.6 10^{-4} M, Sigma), insulin (10 mg/mL, Sigma) and 10% human plasma. During the first 8 days, the medium was supplemented with the following factors: hydrocor-

Table 1. Hematopoietic stem progenitor cells (HSPC) expansion and reticulocyte differentiation for three different sources HSPCs (6 independent experiments were carried out for each HSPC source).

HSPC sources	Mean proportion (%) of reticulocytes at D14 (± SD) (n = 6)	Cell number mean fold increase (± SD) (n = 3)	
		After 5 days of expansion	After 7 days of differentiation
UCB	18.3±1.3	11.5±2.3	33.5±2.4
BM	20.5±1.5	3.1±0.3	8.6±0.5
PBMC	32±6	1.3±0.2	3.4±0.2

Results show the mean proportions of reticulocytes observed at the peak of enucleation after 14 days of culture, as well as the increase in total cell number after 5 days of expansion and 7 days of differentiation (3 independent experiments). After 5 days of expansion, the number of cells was counted and divided by the initial number of plated MACS/CD34+ cells, while after 7 days of HSPC differentiation, the cell count was compared to the number of cells not previously expanded, results are expressed in mean fold increase (± SD).

tisone (HDS, 10^{-6} M, Sigma), interleukin-3 (IL-3, 5 ng/mL, R&D system), stem cell factor (SCF, 100 ng/mL, Bioke) and erythropoietin (EPO, 3 IU/mL, R&D system) and placed at 37°C in a 5% CO_2 incubator. The initial volume of medium was 4 mL and after 4 days, an extra 3 mL was added. After 8 days, the cells were centrifuged for 5 minutes at 300 g, fresh IMDM medium supplemented with EPO (3 IU/mL) was added, and the cells were transferred in a 25 cm² flask. On day 11, the medium was changed and complete medium was added without EPO. Afterwards, medium was refreshed every 3 days and 10% heat inactivated human serum was added to protect the viability of the cells.

For CD34+ cell expansion, CD34+ enriched cells were dispensed in a 6-well plate with 4 mL Serum-free expansion medium (SFEM, Sigma) with SCF (50 ng/mL), thrombopoietin (TPO, 50 ng/mL, R&D system), FMS-like tyrosine kinase 3 (FLT3, 50 ng/mL, R&D system) and IL-6 (50 ng/mL, R&D system) for 5 days at 37°C, and 5% CO_2. On day 5, the cells were counted, and transferred into a new 6-well plate to induce the reticulocyte differentiation (using the protocol described above).

Reticulocyte concentration

Reticulocytes were enriched from UCB or PB by loading on a 70% isotonic percoll cushion which was spun for 15 minutes at 400 g. After two washes with PBS, reticulocytes were counted as described below.

Reticulocyte count

Cells were spun at 300 g for 5 minutes and re-suspended in 50 µL of PBS; 50 µL of Cresyl blue (Roche) diluted 1:1000 was added and cells were incubated at room temperature for 30 minutes. After a cytospin centrifugation (3 minutes at 700 rpm), the cells were fixed with methanol, and stained for 10 minutes with Giemsa (Sigma). The slides were then examined by microscopy (immersion objective, 630× magnification), and reticulocytes were counted against a minimum of 500 RBCs and the density per 100 RBCs was computed. A reticulocyte was morphologically defined as an enucleated cell with at least 3 dots of cresyl blue RNA.

Plasmodium vivax invasion assays

Cryopreserved *P. vivax* isolates [11] from infected patients were provided by the Shoklo Malaria Research Unit (SMRU, Mae Sot, Thailand). The samples were thawed with NaCl solutions and cultured for 36 to 40 hours with McCoy's medium (Gibco) supplemented with glucose (2%) and 20% heat inactivated human serum. *P. vivax* mature forms were concentrated on a 45% percoll

after a 5 minutes treatment with 0.05% trypsin. After 15 minutes of centrifugation at 1600 g, cells above the 45% percoll were collected and washed twice before checking the quality of the concentration. If more than 90% of the cells contained parasites, they were mixed with our previously differentiated and cryopreserved reticulocytes (chosen to contain the same percentage of reticulocytes for all the conditions tested) in a 96-well plate and the initial parasite density was adjusted on a 1:6 ratio (final volume 100 µL, hematocrit 2–5%). Cells were checked at 24 hours post-invasion by doing a cytospin slide stained with Giemsa. The parasite densities were computed after examining a minimum of 500 RBCs.

Data analysis

Data were entered and analyzed with STATA12 (StataCorp, Texas). Reticulocytes were counted after 14 days of differentiation and the mean±SD calculated for each source of HSPC. The Kruskall-Wallis test was used to compare population means. Means and standard deviations were calculated to summarize HSPC expansion rates.

Ethics statement

P. vivax samples collection was approved by the ethics committees of the faculty of tropical medicine, Mahidol University, Bangkok, Thailand (number MUTM-2008-15) and the University of Oxford, Centre for Clinical Vaccinology and Tropical Medicine, United Kingdom (Ethics approval number: OXTREC 027-025). UCB were collected from the cord blood bank at the Gasthuisberg Hospital, Leuven, Belgium (Ethics approval number ML6620). Bone marrow samples were taken from voluntary donors at the Gasthuisberg hospital, Leuven, Belgium (Ethics approval number B322201112107). Adult peripheral blood samples were bought from the Antwerp Red Cross blood bank.

A written inform consent was signed by each donor.

Results

Reticulocyte production from BM, PB and UCB CD34+-enriched cell populations

Reticulocyte differentiation was successfully induced from magnetically sorted CD34+-enriched populations from UCB, PBMC and BM in three independent experiments (n = 3). The enrichment for CD34+ cells in the sorted populations was 55% (SD±6) for UCB, 35% (SD±8) for BM and 16% (SD±6), for PBMC (3 independent experiments) as determined by FACS. The peak of enucleation occurred after 14 days of differentiation,

D0 of expansion

D5 of expansion

UCB:

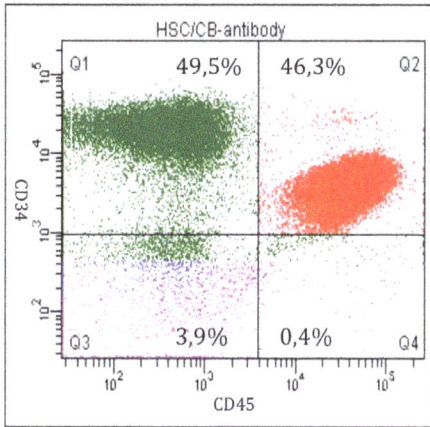

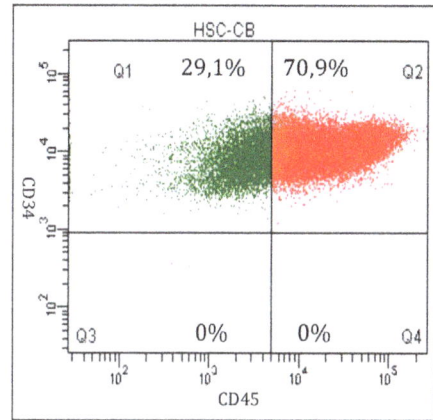

BM

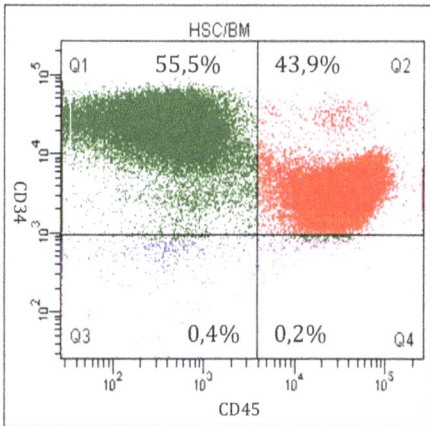

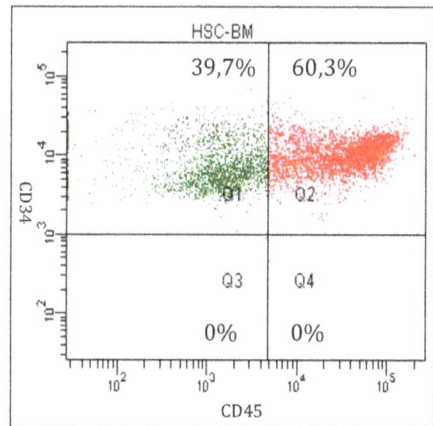

PBMC

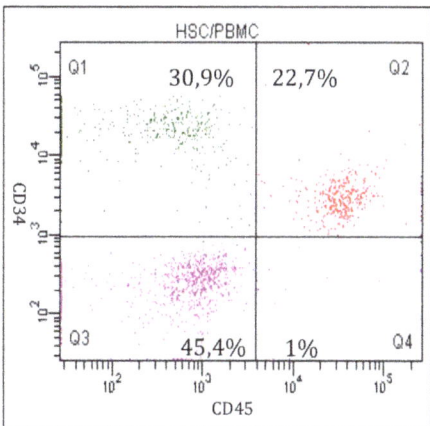

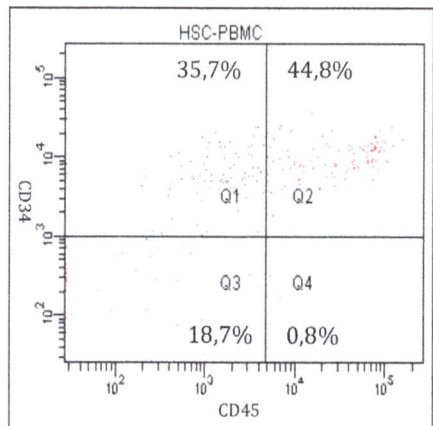

Figure 1. FACS analyses of the CD34$^+$/CD45$^+$ cells from UCB, PBMNC and BM, after isolation (Day 0) and following 5 days of expansion. The Q2 gate represents the population double positive for CD34 (APC) and CD45 (PE).

regardless of the source. The enucleation of erythroid cells from PBMC (mean = 32, SD±6) was significantly higher (p = 0.002) than that of UCB (18%, SD±1.3 and BM (21%, SD±1.5) (Table 1, 6 independent experiments).

Reticulocyte production from *ex vivo* expanded BM, PB and UCB CD34-enriched cell populations

We next tested if larger numbers of reticulocytes could be obtained from *ex vivo* expanded CD34$^+$-enriched cell populations. After 5 days of expansion in serum-free medium with TPO and SCF, the total cell populations increased >10-fold in cultures initiated with UCB/CD34$^+$-enriched cells, 3-fold for BM/CD34$^+$-enriched cells while for PBMC no expansion was observed (Table 1, 3 independent experiments). FACS analysis demonstrated an increase in the CD34$^+$/CD45$^+$ population between Day 0 and Day 5 for all three cell sources: from 55% to 70% (SD±2) for UCB, 35% to 55% (SD±5) for BM, and 16% to 29% (SD±16) for PBMC (n = 3 for CB and BM, n = 2 for PBMC)(Figure 1).

Following expansion, a similar number of cells (irrespective of the CD34$^+$ content or expansion) were cultured under reticulocyte differentiation conditions. After 7 days of erythroid differentiation, the total number of cells, previously subject to an expansion step, was 3 times higher compared to CD34$^+$ cells that were immediately induced to differentiate. After 14 days of differentiation, expanded cells expressed high levels of CD235a and CD71 receptors, regardless of cell source (respectively 87.4% for UCB, 81.7% for BM and 70.6% for PB; Figure 2). Compared to unexpanded cells, the proportion of reticulocytes obtained at day 14 from *in vitro* expanded CD34$^+$ cells was 5 to 10-fold higher.

P. vivax invasion

P. vivax parasites invaded reticulocytes derived from CD34$^+$ cells or directly obtained from enriched blood (PB or UCB;

Figure 3). The mean invasion rates for the different sources of HSPC sources were 3.05%, 3.05% and 3.15% respectively for UCB, BM and PB (n = 4). The means for UCB-concentrated reticulocytes (n = 4) and PB-concentrated reticulocytes (n = 3) were respectively 1.4% and 0.2%.

When using the same *P. vivax* isolate, the invasion rate between different HSPC sources did not differ for all of the 4 *P. vivax* isolates tested (Figure 4a); however, the invasion rate observed varied for each of the *P. vivax* isolates used. When we compared two different *P. vivax* isolates using the same HSPC-derived reticulocytes, the invasion rate varied significantly by isolate (Figure 4b; PV1 and PV2, p<0.001). After 3 days of culture, only few rings (parasite density <0.05%) could be observed and none survived longer than 72 hours, regardless of the HSPC source. Interestingly, the invasion rate of *P. vivax* in HSPC-derived reticulocytes appeared to be higher when compared with reticulocytes isolated directly from PB. For the same *P. vivax* isolate, the parasite density 24 hours post-invasion in UCB/HSPC-derived reticulocytes was up to 9-fold higher than in UCB-concentrated reticulocytes (1.8% *versus* 0.2%, respectively), and 18-fold higher than adult PB-concentrated reticulocytes (2.1% *versus* 0.1% respectively). Parasite densities were not significantly different between HSPC-derived reticulocytes (5%), UCB-concentrated reticulocytes (4.6% p = 0.056) and PB-concentrated reticulocytes (3.5% p = 0.06) when the reticulocyte percentage was 20% for HSPC-derived reticulocytes and respectively 60% and 70% for reticulocytes concentrated from PB and UCB.

Discussion

In this study, we compare for the first time different source of HSPC to generate reticulocytes suitable for *P. vivax* studies. We could demonstrate that compared with CD34$^+$-enriched populations from PB and BM, CD34$^+$-enriched populations from UCB

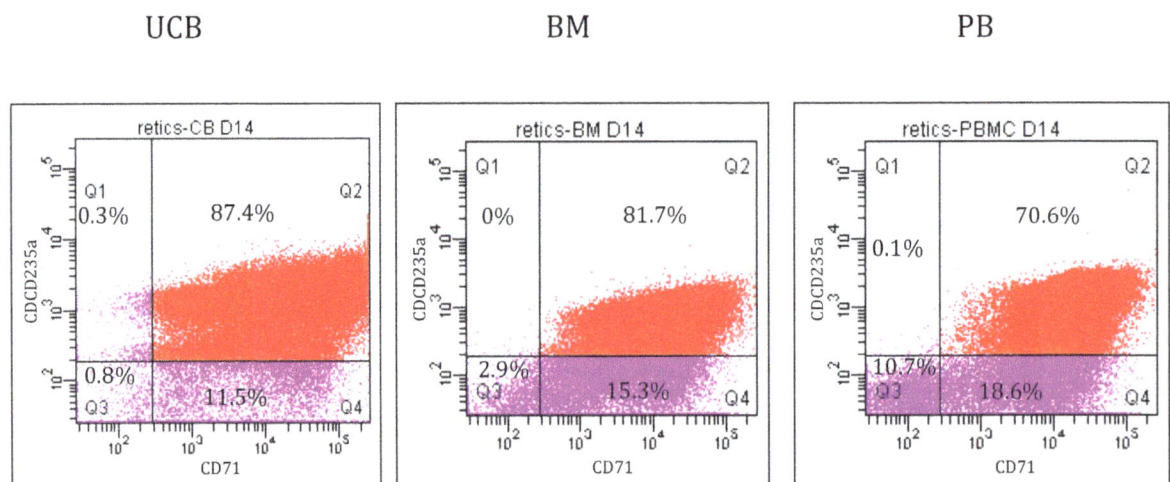

UCB   **BM**   **PB**

Figure 2. FACS analyses of the CD235a$^+$/CD71$^+$ cells from UCB, PBMNC and BM, after 5 days of expansion and 14 days of differentiation. The Q2 gate represents the population positive for CD235a (Per-CP-Cy5-5) and CD71 (PE).

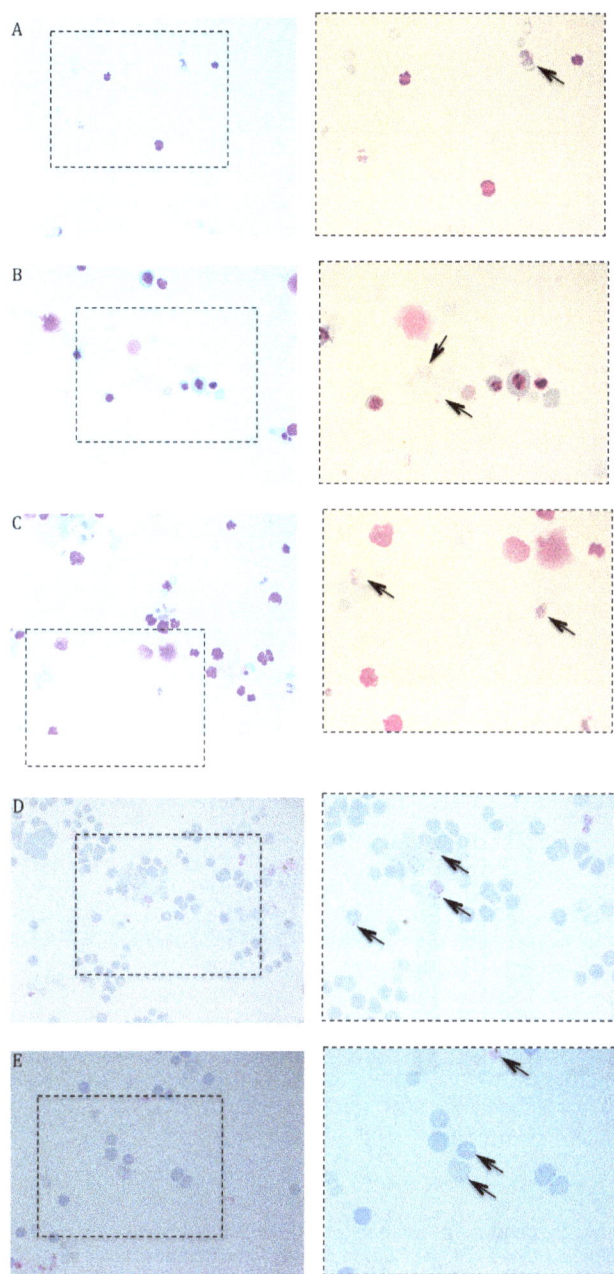

Figure 3. Cytopsin of the *P. vivax* culture 24 h post-invasion for different sources of reticulocytes. A) UCB/HSPC, B) BM/HSPC, C) PB/HSPC, D) reticulocytes-enriched UCB and E) reticulocytes-enriched PB. The left panels represent pictures with a 63× magnification and the corresponding right panels represent a 100× of the left picture square. *P. vivax* infected cells are under arrow.

produced the highest number of reticulocytes that can be invaded by *P. vivax*. Second, when CD34$^+$-enriched cells were first expanded, a further extensive increase in reticulocytes was generated from UCB, to a lesser degree BM but not PB.

The number of reticulocytes derived from UCB CD34$^+$ enriched cell populations could be substantially increased when the CD34$^+$ cells were first expanded for 5 days. Noteworthy, in our experiments PB/HSPC cells showed a limited increase of the population after 5 days of expansion compared to UCB/HSPC and BM/HSPC. This is likely due to the low frequency of CD34$^+$

cell in non-mobilized PBMC [12] and the more mature fate of HSPC in PB or BM compared to cells from UCB [13].

Reticulocytes derived from magnetically sorted CD34$^+$ cells from either PBMC, BM or UCB could be invaded by *P. vivax* with similar efficiencies, while invasion was significantly influenced by the type of *P. vivax* isolate. Despite successful invasion, none of the produced reticulocyte population could support the full development and long-term culture of *P. vivax*. Reticulocytes generated in this study were more permissive for *P. vivax* invasion compared with the study published by Panichakul *et al* [6], who observed only 0.0015% of invasion, mainly due to a very low percentage of reticulocytes (0.5%); or Furuya *et al* [14] who used frozen erythroblast derived from UCB/HSPC (0.8% parasitemia). The 3.5% parasitemia rate observed in our study are in line with the invasion rate observed by Borlon *et al* (2.1%) [11] and Russell *et al* (3.7%) [4], both using reticulocyte-enriched UCB (percentage of reticulocytes greater than 50%).

Our observations might also suggest a preference of *P. vivax* for more immature reticulocytes as we observed a higher invasion rate of reticulocytes derived from any HSPC source compared to those concentrated from either UCB or PB. The higher invasion rate in HSPC derived reticulocytes concentrated from UCB compared to those from PB could be explained by the distribution of their reticulocytes populations. Indeed, Paterakis *et al* [15] classified reticulocytes according to their RNA content by FACS analysis and divided them into 3 categories, i.e. immature reticulocytes (high amount of RNA), median, and old reticulocytes (medium and low amount of RNA, respectively). They found that among the reticulocyte population, UCB contains more immature reticulocytes (13.6%) than adult peripheral blood (1%). Furthermore, HSPC-derived reticulocytes were collected at the peak of enucleation, i.e. when they were at an immature stage of development. This provides additional evidence for preference of *P. vivax* for immature reticulocytes and should be further investigated.

When reticulocyte-enriched blood from UCB and PB was used at 3-fold higher reticulocyte concentrations (70% and 60%, respectively), the invasion rates became similar to that obtained with HSPC-derived reticulocytes (20% for HSPC). This is in line with a recently published report by Martín-Jaular *et al* [16] wherein the authors observed a predominant invasion of CD71^{high}-expression cells (CD71 being a marker of reticulocyte maturation, as their expression decrease while the reticulocyte maturate to the RBC stage) by *P. yoelii* (a mouse *Plasmodium* species also invading preferentially reticulocytes).

Our preliminary observations need further in-depth investigation to provide new insights into the invasion mechanisms of *P. vivax*, and more specifically on the critical stage-specific receptors. If confirmed, this hypothesis would justify the use of reticulocytes derived from HSPC instead of reticulocyte-enriched blood as target cells for the establishment of continuous *P. vivax* cultures.

In conclusion, our results demonstrate that it is possible to produce large amounts of immature reticulocytes that can be efficiently invaded by *P. vivax*. The ability to derive reticulocytes from UCB/HSPCs in in larger quantities than from PB or BM/HSPCs without loosing the permissiveness to *P. vivax* make this source of HSPC as more suitable and likely to develop into a standardized and continuous source of reticulocytes for the long-term culture of *P. vivax*.

The possibility to efficiently expand the CD34$^+$ population to generate more reticulocytes coupled with the possibility to cryopreserve those HSPC-derived reticulocytes [7] opens also new perspectives to create stocks of reticulocytes to use as target cells for the establishment of an *in vitro* culture of *P. vivax*.

a) Parasite density for different sources of HSPC-derived reticulocytes with 1 *P. vivax*

isolate.

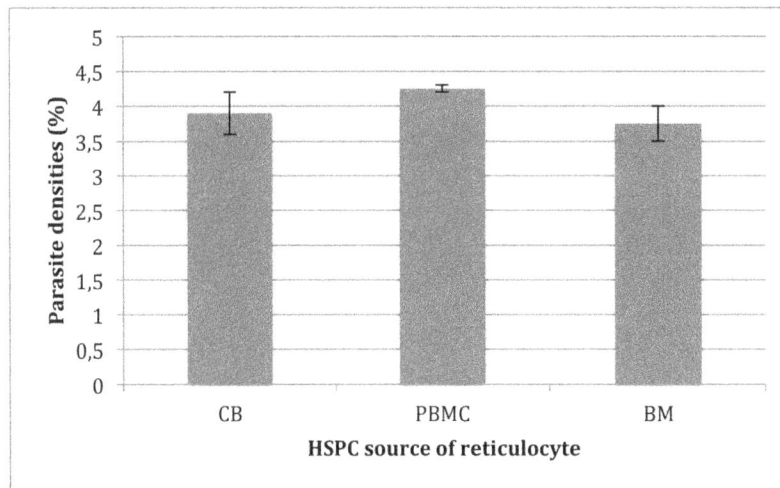

b) Parasite density for each source of reticulocytes with different *P. vivax* isolates.

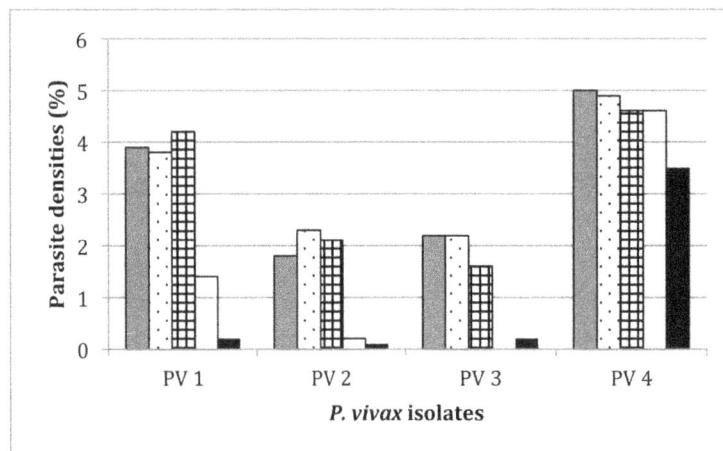

Figure 4. Parasite densities 24 hours post-invasion with *P. vivax*. The parasite density was counted for at least 500 red blood cells, dividing the number of infected enucleated cells by the total number of cells and multiplied by 100 (%). a) Parasite density for different sources of HSPC-derived reticulocytes with 1 *P. vivax* isolate. The mean and SD of 2 different batches of differentiated reticulocytes was calculated for each source of HSPC and tested for invasion with the same *P. vivax* isolate. b) Parasite density for each source of reticulocytes with different *P. vivax* isolates. Parasite densities (%) were counted by dividing the number of *P. vivax* ring-infected cells by the total number of counted RBCs and multiplying the result by 100. Different reticulocyte sources were tested: grey = UCB/HSPC-derived reticulocytes; dotted = BM/HSPC-derived reticulocytes; squared = PBMC/HSPC-derived reticulocytes; white = reticulocytes concentrated from UCB, black = reticulocytes concentrated from adult peripheral blood. PV1 and PV2 were tested with the same batches of HSPC-derived reticulocytes for the 3 different sources (UCB, BM and PBMC). For PV4, the proportion of reticulocytes was 20% for HSCP-derived reticulocytes and respectively 70% and 60% for reticulocytes concentrated from UCB adult peripheral blood.

Acknowledgments

We would like to thanks Prof. Michel Delforge for kindly providing us with the bone marrow samples and Thomas Vanwelden for isolation of bone marrow mononuclear cells. We also thank Prof François Nosten and his team (SMRU) for providing the *plasmodium vivax* isolates.

Author Contributions

Conceived and designed the experiments: FN JKM. Performed the experiments: FN. Analyzed the data: FN CMV AE. Contributed reagents/materials/analysis tools: ARU AE. Wrote the paper: FN JKM JVDA UDA AE CB ARU CMV.

References

1. Price RN, Tjitra E, Guerra CA, Yeung S, White NJ, et al. (2007) Vivax malaria: neglected and not benign. The American journal of tropical medicine and hygiene 77: 79–87.

2. Noulin F, Borlon C, Van Den Abbeele J, D'Alessandro U, Erhart A (2013) 1912–2012: a century of research on Plasmodium vivax in vitro culture. Trends in parasitology 29: 286–294.

3. Moreno-Perez DA, Ruiz JA, Patarroyo MA (2013) Reticulocytes: Plasmodium vivax target cells. Biology of the cell/under the auspices of the European Cell Biology Organization 105: 251–260.

4. Russell B, Suwanarusk R, Borlon C, Costa FT, Chu CS, et al. (2011) A reliable ex vivo invasion assay of human reticulocytes by Plasmodium vivax. Blood 118: e74–81.

5. Golenda CF, Li J, Rosenberg R (1997) Continuous in vitro propagation of the malaria parasite Plasmodium vivax. Proceedings of the National Academy of Sciences of the United States of America 94: 6786–6791.

6. Panichakul T, Sattabongkot J, Chotivanich K, Sirichaisinthop J, Cui L, et al. (2007) Production of erythropoietic cells in vitro for continuous culture of Plasmodium vivax. International journal for parasitology 37: 1551–1557.

7. Noulin F, Borlon C, van den Eede P, Boel L, Verfaillie CM, et al. (2012) Cryopreserved Reticulocytes Derived from Hematopoietic Stem Cells Can Be Invaded by Cryopreserved Plasmodium vivax Isolates. PloS one 7: e40798.

8. Douay L, Andreu G (2007) Ex vivo production of human red blood cells from hematopoietic stem cells: what is the future in transfusion? Transfusion medicine reviews 21: 91–100.

9. Fernandez-Becerra C, Lelievre J, Ferrer M, Anton N, Thomson R, et al. (2013) Red blood cells derived from peripheral blood and bone marrow CD34(+) human haematopoietic stem cells are permissive to Plasmodium parasites infection. Memorias do Instituto Oswaldo Cruz 108: 801–803.

10. Giarratana MC, Kobari L, Lapillonne H, Chalmers D, Kiger L, et al. (2005) Ex vivo generation of fully mature human red blood cells from hematopoietic stem cells. Nature biotechnology 23: 69–74.

11. Borlon C, Russell B, Sriprawat K, Suwanarusk R, Erhart A, et al. (2012) Cryopreserved Plasmodium vivax and cord blood reticulocytes can be used for invasion and short term culture. International journal for parasitology 42: 155–160.

12. Bender JG, Unverzagt K, Walker DE, Lee W, Smith S, et al. (1994) Phenotypic analysis and characterization of CD34+ cells from normal human bone marrow, cord blood, peripheral blood, and mobilized peripheral blood from patients undergoing autologous stem cell transplantation. Clinical immunology and immunopathology 70: 10–18.

13. Steidl U, Kronenwett R, Rohr UP, Fenk R, Kliszewski S, et al. (2002) Gene expression profiling identifies significant differences between the molecular phenotypes of bone marrow-derived and circulating human CD34+ hematopoietic stem cells. Blood 99: 2037–2044.

14. Furuya T, Sa JM, Chitnis CE, Wellems TE, Stedman TT (2014) Reticulocytes from cryopreserved erythroblasts support Plasmodium vivax infection in vitro. Parasitology international 63: 278–284.

15. Paterakis GS, Lykopoulou L, Papassotiriou J, Stamulakatou A, Kattamis C, et al. (1993) Flow-cytometric analysis of reticulocytes in normal cord blood. Acta haematologica 90: 182–185.

16. Martin-Jaular L, Elizalde-Torrent A, Thomson-Luque R, Ferrer M, Segovia JC, et al. (2013) Reticulocyte-prone malaria parasites predominantly invade CD71hi immature cells: implications for the development of an in vitro culture for Plasmodium vivax. Malaria journal 12: 434.

PERMISSIONS

LIST OF CONTRIBUTORS

Jan B. W. J. Cornelissen and Henk J. Wisselink
Central Veterinary Institute of Wageningen UR, Department of Infection Biology, Lelystad, The Netherlands

Joke W. B. van der Giessen
Central Veterinary Institute of Wageningen UR, Department of Bacteriology and TSEs, Lelystad, The Netherlands
National Institute of Public Health and the Environment (RIVM), Centre for Zoonoses and Environmental Microbiology, Bilthoven, The Netherlands

Katsuhisa Takumi
National Institute of Public Health and the Environment (RIVM), Centre for Zoonoses and Environmental Microbiology, Bilthoven, The Netherlands

Peter F. M. Teunis
National Institute for Public Health and the Enviroment (RIVM), Centre for Epidemiology, Bilthoven, The Netherlands

Natalie G. Sanders, David J. Sullivan, Godfree Mlambo, George Dimopoulos, Abhai K. Tripathi and W. Harry Feinstone
Department of Molecular Microbiology and Immunology, Johns Hopkins Bloomberg School of Public Health, Johns Hopkins University, Baltimore, Maryland, United States of America

Gisely C. Melo, André M. Siqueira, Belisa M. L. Magalhães and Aline C. C. Alencar
Universidade do Estado do Amazonas, Manaus, Amazonas, Brazil

Wuelton M. Monteiro and Marcus V. G. Lacerda
Universidade do Estado do Amazonas, Manaus, Amazonas, Brazil
Fundação de Medicina Tropical Dr. Heitor Vieira Dourado, Manaus, Amazonas, Brazil

Siuhelem R. Silva
Universidade Paulista UNIP, Manaus, Amazonas, Brazil

Andrea Kuehn
Fundação de Medicina Tropical Dr. Heitor Vieira Dourado, Manaus, Amazonas, Brazil
Barcelona Centre for International Health Research (CRESIB, Hospital Clínic-Universitat de Barcelona), Barcelona, Spain

Hernando A. del Portillo
Barcelona Centre for International Health Research (CRESIB, Hospital Clínic-Universitat de Barcelona), Barcelona, Spain
Instituciá Catalana de Recerca i Estudis Avanc‚ats (ICREA), Barcelona, Spain

Carmen Fernandez-Becerra
Barcelona Centre for International Health Research (CRESIB, Hospital Clínic-Universitat de Barcelona), Barcelona, Spain

Selina Kern
Research Center for Infectious Diseases, University of Würzburg, Würzburg, Germany
Institute of Molecular Biotechnology, RWTH Aachen University, Aachen, Germany

Shruti Agarwal and Thomas Brügl
Research Center for Infectious Diseases, University of Würzburg, Würzburg, Germany,

Kilian Huber, Franz Bracher, André P. Gehring and Benjamin Strödke
Department of Pharmacy – Center for Drug Research, Ludwig-Maximillians University, Munich, Germany

Christine C. Wirth, Gabriele Pradel, Rainer Fischer and Liliane Onambele Abodo
Institute of Molecular Biotechnology, RWTH Aachen University, Aachen, Germany

Thomas Dandekar
Bioinformatics, Biocenter, University of Würzburg, Würzburg, Germany

Christian Doerig
INSERM U609, Global Health Institute, Ecole Polytechnique Fédérale de Lausanne (EPFL), Lausanne, Switzerland
Department of Microbiology, Monash University, Clayton, Victoria, Australia

Andrew B. Tobin and Mahmood M. Alam
Department of Cell Physiology and Pharmacology, MRC Toxicology Unit, University of Leicester, Leicester, United Kingdom

Raul J. Cano
Center for Applications in Biotechnology, California Polytechnic State University, San Luis Obispo, California, United States of America

Jessica Rivera-Perez, Gary A. Toranzos, Erileen García-Roldán and Steven E. Massey
Department of Biology, University of Puerto Rico, San Juan, Puerto Rico

Tasha M. Santiago-Rodriguez
Department of Pathology, University of California San Diego, San Diego, California, United States of America

Yvonne M. Narganes-Storde and Luis Chanlatte-Baik
Center for Archaeological Research, University of Puerto Rico, Rio Piedras Campus, San Juan, Puerto Rico

Lucy Bunkley-Williams
Department of Biology, University of Puerto Rico, Mayaguez Campus, San Juan, Puerto Rico

Szymon M. Drobniak, Joanna Sudyka and Mariusz Cichoń
Institute of Environmental Sciences, Jagiellonian University, Kraków, Poland

Andrzej Dyrcz
Department of Behavioural Ecology, Wroclaw University, Wroclaw, Poland

Sándor Hornok, Balázs Tanczoś and Róbert Farkas
Department of Parasitology and Zoology, Faculty of Veterinary Science, Szent István University, Budapest, Hungary

Getachew Abichu
National Research Center, Department of Parasitology, Arachnoentomology Unit, Sebeta, Ethiopia

Marina L. Meli, Regina Hofmann-Lehmann and Enikő Gönczi
Clinical Laboratory and Center for Clinical Studies, Vetsuisse Faculty, University of Zurich, Zurich, Switzerland

Kinga M. Sulyok and Miklós Gyuranecz
Institute for Veterinary Medical Research, Centre for Agricultural Research, Hungarian Academy of Sciences, Budapest, Hungary

Michael Coleman, Marlize Coleman, Janet Hemingway, Abdiasiis Omar, Michelle C. Stanton, Eddie K. Thomsen and Phillip J. McCall
Vector Biology Department, Liverpool School of Tropical Medicine, Liverpool, United Kingdom

Mohammed H. Al-Zahrani, Adel A. Alsheikh and Raafat F. Alhakeem
Public Health Directorate, Ministry of Health, Riyadh, Kingdom of Saudi Arabia

Abdullah A. Al Rabeeah
Vector Biology Department, Liverpool School of Tropical Medicine, Liverpool, United Kingdom

Ziad A. Memish
Vector Biology Department, Liverpool School of Tropical Medicine, Liverpool, United Kingdom
Public Health Directorate, Ministry of Health, Riyadh, Kingdom of Saudi Arabia
College of Medicine, Alfaisal University, Riyadh, Kingdom of Saudi Arabia

Emily P. Lane
Department of Research and Scientific Services, National Zoological Gardens of South Africa, Pretoria, South Africa

Mornéde Wet and Peter Thompson
Epidemiology Section, Department of Production Animal Studies, Faculty of Veterinary Science, University of Pretoria, Pretoria, South Africa

Ursula Siebert
Institute for Terrestrial and Aquatic Wildlife Research, University of Veterinary Medicine, Hannover, Foundation, Germany

Peter Wohlsein
Department of Pathology, University of Veterinary Medicine, Hannover, Foundation, Germany

Stephanie Plö
South African Institute for Aquatic Biodiversity, c/o Port Elizabeth Museum/Bayworld, Port Elizabeth, South Africa

Neil Portman
Sir William Dunn School of Pathology, University of Oxford, Oxford, United Kingdom
Faculty of Veterinary Science, University of Sydney, Sydney, Australia

Keith Gull
Sir William Dunn School of Pathology, University of Oxford, Oxford, United Kingdom

Thomas Mouveaux, Gabrielle Oria, Elisabeth Werkmeister and Stanislas Tomavo
Center for Infection and Immunity of Lille, CNRS UMR 8204, INSERM U 1019, Institut Pasteur de Lille, Université Lille Nord de France, Lille, France

Christian Slomianny
Laboratory of Cell Physiology, INSERM U 1003, UniversitéLille Nord de France, Villeneuve d'Ascq, France

Barbara A. Fox and David J. Bzik
Department of Microbiology and Immunology, The Geisel School of Medicine at Dartmouth, Lebanon, New Hampshire, United States of America

Amy R. Noe, Ramses Ayala, Brian Roberts, Scott B. Winram, Steve Giardina and Gabriel M. Gutierrez
Leidos Inc., Frederick, Maryland, United States of America

Diego Espinosa and Fidel Zavala
Johns Hopkins Malaria Research Institute and Department of Molecular Microbiology and Immunology, Johns Hopkins Bloomberg School of Public Health, Johns Hopkins University, Baltimore, Maryland, United States of America

Xiangming Li, Jordana G. A. Coelho-dos-Reis, Ryota Funakoshi and Moriya Tsuji
HIV and Malaria Vaccine Program, Aaron Diamond AIDS Research Center, Affiliate of The Rockefeller University, New York, New York, United States of America

Hongfan Jin, Diane M. Retallack, Ryan Haverstock and Jeffrey R. Allen
Pfenex Inc., San Diego, California, United States of America

Thomas S. Vedvick, Christopher B. Fox and Steven G. Reed
Infectious Disease Research Institute, Seattle, Washington, United States of America

John Sacci
Department of Microbiology and Immunology, University of Maryland School of Medicine, Baltimore, Maryland, United States of America

Manuel Soler and Tomás Pérez-Contreras
Departamento de Zoología, Facultad de Ciencias, Universidad de Granada, Granada, Spain
Grupo Coevolución, Unidad Asociada al Consejo Superior de Investigaciones Científicas (CSIC), Universidad de Granada, Granada, Spain

Juan Diego Ibáñez-Álamo, Gianluca Roncalli and Elena Macías-Sádnchez
Departamento de Zoología, Facultad de Ciencias, Universidad de Granada, Granada, Spain

Liesbeth de Neve
Departamento de Zoología, Facultad de Ciencias, Universidad de Granada, Granada, Spain
Department of Biology, Terrestrial Ecology Unit, Ghent University, Gent, Belgium

Nora Adriana Hernández-Cuevas, Christian Weber, Chung-Chau Hon and Nancy Guillen
Institut Pasteur, Unité Biologie Cellulaire du Parasitisme, Paris, France
INSERM U786, Paris, France

Marilis Rodriguez, Andy Alhassan, Rosalynn L. Ord, Jeny R. Cursino-Santos, Manpreet Singh and Cheryl A. Lobo
Department of Blood-Borne Parasites, New York Blood Center, New York, New York, United States of America

Jeremy Gray
University College Dublin School of Biology and Environmental Science, Dublin, Republic of Ireland

Souheila Guerbouj and Ikram Guizani
Laboratory of Molecular Epidemiology and Experimental Pathology, Pasteur Institute of Tunis, Université de Tunis el Manar, Tunis, Tunisia
Laboratory of Epidemiology and Ecology of Parasitic Diseases, Pasteur Institute of Tunis, Tunis, Tunisia

Fattouma Djilani, Mohamed Fethi Diouani and Riadh Ben Ismail
Laboratory of Epidemiology and Ecology of Parasitic Diseases, Pasteur Institute of Tunis, Tunis, Tunisia

Jihene Bettaieb and Afif Ben Salah
Laboratory of Medical Epidemiology, Pasteur Institute of Tunis, Tunis, Tunisia

Bronwen Lambson
Molteno Institute for Parasitology, Department of Pathology, University of Cambridge, Cambridge, United Kingdom

Agnieszka A. Religa and Andrew P. Waters
Wellcome Trust Centre for Molecular Parasitology, Institute of Infection, Immunity and Inflammation, University of Glasgow, Glasgow, United Kingdom

Jai Ramesar and Chris J. Janse
Leiden Malaria Research Group, Parasitology, Leiden University Medical Centre, Leiden, the Netherlands

Artur Scherf
Biology of Host-Parasite Interactions Unit, Institut Pasteur, Paris, France

James A. Cotton, Jean-Paul Motta and Andre G. Buret
Department of Biological Sciences, University of Calgary, Calgary, Alberta, Canada
Inflammation Research Network, University of Calgary, Calgary, Alberta, Canada
Host-Parasite Interactions, University of Calgary, Calgary, Alberta, Canada

L. Patrick Schenck
Department of Biochemistry and Molecular Biology, University of Calgary, Calgary, Alberta, Canada
Department of Medicine, University of Calgary, Calgary, Alberta, Canada

Simon A. Hirota
Department of Physiology and Pharmacology, University of Calgary, Calgary, Alberta, Canada
Department of Immunology, Microbiology and Infectious Diseases, University of Calgary, Calgary, Alberta, Canada

Paul L. Beck
Department of Medicine, University of Calgary, Calgary, Alberta, Canada

David Tiga Kangoye, Issa Nebie, Jean-Baptiste Yaro, Siaka Debe, Safiatou Traore, Oumarou Ouedraogo, Guillaume Sanou, Issiaka Soulama, Amidou Diarra, Alfred Tiono and Sodiomon Bienvenu Sirima
Centre National de Recherche et de Formation sur le Paludisme, Ouagadougou, Burkina Faso

Kevin Marsh and Philip Bejon
Kenyan Medical Research Institute, Centre for Geographic Medicine Research (Coast), Kilifi, Kenya
Nuffield Department of Medicine, Centre for Clinical Vaccinology and Tropical Medicine, University of Oxford, Churchill Hospital, Oxford, United Kingdom

Florian Noulin, Anna Rosanas-Urgell, Annette Erhart and Céline Borlon
Unit of Malariology, Institute of Tropical Medicine, Antwerp, Belgium

Javed Karim Manesia and Catherine M. Verfaillie
Department of development and regeneration, Stem Cell Institute, Leuven, Belgium

Jan Van Den Abbeele
Unit of Veterinary Protozoology, Institute of Tropical Medicine, Antwerp, Belgium

Umberto d'Alessandro
Medical Research Council Unit, Fajara, The Gambia

Index